Analysis of Computer and Communication Networks

Fayez Gebali

Analysis of Computer and Communication Networks

Fayez Gebali
Department of Electrical & Computer Engineering
University of Victoria
P.O. Box 3055 STN CSC
Victoria, B.C.
V8W 3P6 CANADA
e-mail: fayez@uvic.ca

ISBN: 978-1-4419-4502-0 e-ISBN: 978-0-387-74437-7

DOI: 10.1007/978-0-387-74437-7

Printed on acid-free paper

9 8 7 6 5 4 3 2 1

springer.com

Trademarks

The following list includes commercial and intellectual trademarks belonging to holders whose products are mentioned in this book. Omissions from this list are inadvertent.

CANVAS is a registered trademark of Deneba Systems, Inc.
MAPLE is a registered trademark of Waterloo Maple, Inc.
MATLAB is a registered trademark of The Mathworks, Inc.
Promina is a registered trademark of Network Equipment Technologies, Inc.

All other trademarks, registered trademarks, products, applications, service names mentioned herein are the property of their respective holders/owners.

Disclaimer

Product information contained in this book is primarily based on technical reports and documentation and publicly available information received from sources believed to be reliable. However, neither the author nor the publisher guarantees the accuracy and completeness of information published herein. Neither the publisher nor the author shall be responsible for any errors, omissions, or damages arising out of use of this information. No information provided in this book is intended to be or shall be construed to be an endorsement, certification, approval, recommendation, or rejection of any particular supplier, product, application, or service.

A Note to Instructors

A password protected, solutions manual is available online for qualified instructors utilizing this book in their courses. Please see www.Springer.com for more details or contact Dr. Fayez Gebali directly.

Preface

The purpose of this book is to give the reader two things, to paraphrase Mark Twain: Roots to know the basics of modeling networks and Wings to fly away and attempt modeling other proposed systems of interest.

The Internet phenomenon is affecting us all in the way we communicate, conduct business, and access information and entertainment. More unforeseen applications are still to come. All of this is due to the existence of an efficient global high-performance network that connects millions of users and moves information at a high rate with small delay.

High-Performance Networks

A high-performance network is characterized by two performance measures *bandwidth* and *delay*. Traditional network design focused mainly on bandwidth planning; the solution to network problems was to add more bandwidth. Nowadays, we have to consider message delay particularly for delay-sensitive applications such as voice and real-time video. Both bandwidth and delay contribute to the performance of the network. Bandwidth can be easily increased by compressing the data, by using links with higher speed, or by transmitting several bits in parallel using sophisticated modulation techniques. Delay, however, is not so easily improved. It can only be reduced by the use of good scheduling protocols, very fast hardware and switching equipment throughout the network. The increasing use of optical fibers means that the transmission channel is close to ideal with extremely high bandwidth and low delay (speed of light). The areas that need optimization are the interfaces and devices that connect the different links together such as hubs, switches, routers, and bridges. The goal of this book is to explore the design and analysis techniques of these devices. There are indications, however, that the optical fiber channel is becoming less than ideal due to the increasing bit rates. Furthermore, the use of wireless mobile networking is becoming very popular. Thus new and improved techniques for transmitting across the noisy, and band-limited, channel become very essential. The work to be done to optimize the physical level of communication is devising algorithms and hardware for adaptive data coding and compression. Thus digital signal processing is finding an increasing and pivotal role in the area of networking.

Scope

The three main building blocks of high-performance networks are the links, the switching equipment connecting the links together, and the software employed at the nodes and switches. The purpose of this book is to provide the basic techniques for modeling and analyzing the last two components: the software and the switching equipment. The book also reviews the design options used to build efficient switching equipment. For this purpose, different topics are covered in the book such as Markov chains and queuing analysis, traffic modeling, interconnection networks, and switch architectures and buffering strategies.

There are many books and articles dealing with continuous-time Markov chains and queuing analysis. This is because continuous-time systems are thought to be easily modeled and analyzed. However, digital communications are discrete in nature. Luckily, discrete-time Markov chains are simple, if not even easier, to analyze. The approach we chose to present Markov chains and queuing analysis is to start with explaining the basic concepts, then explain the analytic and numerical techniques that could be used to study the system. We introduce many worked examples throughout to get a feel as to how to apply discrete-time Markov chains to many communication systems.

We employ MATLAB ® throughout this book due to its popularity among engineers and engineering students. There are many equally useful mathematical packages available nowadays on many workstations and personal computers such as Maple ® and Mathematica ®.

Organization

This book covers the mathematical theory and techniques necessary for analyzing telecommunication systems. Queuing and Markov chain analyses are provided for many protocols that are used in networking. The book then discusses in detail applications of Markov chains and queuing analysis to model over 15 communications protocols and hardware components. Several appendices are also provided that round up the discussion and provide a handy reference for the necessary background material.

Chapter 1 discusses probability theory and random variables. There is discussion of sample spaces and how to count the number of outcomes of a random experiment. Also discussed is probability density function and expectations. Important distributions are discussed since they will be used for describing traffic in our analysis. The Pareto distribution is discussed in this chapter, which is usually not discussed in standard engineering texts on probability. Perhaps what is new in this chapter is the review of techniques for generating random numbers that obey a desired probability distribution. Inclusion of this material rounds up the chapter and helps the designer or researcher to generate the network traffic data needed to simulate a switch under specified conditions.

Chapter 2 discusses random processes and in particular Poisson and exponential processes. The chapter also discusses concepts associated with random processes such as ensemble average, time average, autocorrelation function, and cross-correlation function.

Chapter 3 discusses discrete-time Markov chains. Techniques for constructing the state transition matrix are explored in detail as well as how the time step is determined since all discrete-time Markov chains require awareness of the time step value. The chapter also discusses transient behavior of Markov chains and explains the various techniques for studying it such as diagonalization, expansion of the initial distribution vector, Jordan canonic form, and using the z-transform.

Chapter 4 discusses Markov chains at equilibrium, or steady state. Analytic techniques for finding the equilibrium distribution vector are explained such as finding the eigenvalues and eigenvectors of the state transition matrix, solving difference equations, and the z-transform technique. Several numerical techniques for finding the steady-state distribution are discussed such as use of forward- and backward-substitution, and iterative equations. The concepts of balance equations and flow balance are also explained.

Chapter 5 discusses reducible Markov chains and explains the concept of closed and transient states. The transition matrix for a reducible Markov chain is partitioned into blocks and the closed and transient states are related to each partitioning block. An expression is derived for the state of a Markov chain at any time instant n and also at equilibrium. The chapter also discusses how a reducible Markov chain could be identified by studying its eigenvalues and eigenvectors. It is shown that the eigenvectors enable us to identify all sets of closed and transient states.

Chapter 6 discusses periodic Markov chains. Two types of periodic Markov chains are identified and discussed separately. The eigenvalues of periodic Markov chains are discussed and related to the periodicity of the system. Transient analysis of a periodic Markov chain is discussed in detail and asymptotic behavior is analyzed.

Chapter 7 discusses discrete-time queues and queuing analysis. Kendall's notation is explained and several discrete-time queues are analyzed such as the infinite-sized $M/M/1$ queue and the finite-sized $M/M/1/B$ queue. Equally important queues encountered in this book are also considered such as $M^m/M/1/B$ and $M/M^m/1/B$ queues. The important performance parameters considered for each queue are the throughput, delay, average queue size, loss probability, and efficiency. The chapter also discusses how to analyze networks of queues using two techniques: the flow balance approach and the merged approach.

Chapter 8 discusses the modeling of several flow control protocols using Markov chains and queuing analysis. Three traffic management protocols are considered: leaky bucket, token bucket, and the virtual scheduling (VS) algorithm.

Chapter 9 discusses the modeling of several error control protocols using Markov chains and queuing analysis. Three error control using automatic repeat request algorithms are considered: stop-and-wait (SW ARQ), go-back-N (GBN ARQ), and selective repeat protocol (SRP ARQ).

Chapter 10 discusses the modeling of several medium access control protocols using Markov chains and queuing analysis. Several media access protocols are discussed: IEEE Standard 802.1p (static priority), pure and slotted ALOHA, IEEE Standard 802.3 (CSMA/CD, Ethernet), Carrier sense multiple access with collision avoidance (CSMA/CA), IEEE Standard 802.4 (token bus) & 802.5 (Token ring), IEEE Standard 802.6 (DQDB), IEEE Standard 802.11 distributed coordination function for ad hoc networks, and IEEE Standard 802.11 point coordination function for infrastructure networks (1-persistent and p-persistent cases are considered).

Chapter 11 discusses the different models used to describe telecommunication traffic. The topics discussed deal with describing the data arrival rates, data destinations, and packet length variation. The interarrival time for Poisson traffic is discussed in detail and a realistic model for Poisson traffic is proposed. Extracting the parameters of the Poisson traffic model is explained given a source average and burst rates. The interarrival time for Bernoulli sources is similarly treated and a realistic model is proposed together with a discussion on how to determine the Bernoulli model parameters. Self-similar traffic is discussed and the Pareto model is discussed. Extracting the parameters of the Pareto traffic model is explained given a source average and burst rates. Modulated Poisson traffic models are also discussed such as the on–off model and the Markov modulated Poisson process. In addition to modeling data arrival processes, the chapter also discusses the traffic destination statistics for uniform, broadcast, and hot-spot traffic types. The chapter finishes by discussing packet length statistics and how to model them.

Chapter 12 discusses scheduling algorithms. The differences and similarities between scheduling algorithms and media access protocols are discussed. Scheduler performance measures are explained and scheduler types or classifications are explained. The concept of max–min fairness is explained since it is essential for the discussion of scheduling algorithms. Twelve scheduling algorithms are explained and analyzed: first-in/first-out (FIFO), static priority, round robin (RR), weighted round robin (WRR), processor sharing (PS), generalized processor sharing (GPS), fair queuing (FQ), packet-by-packet GPS (PGPS), weighted fair queuing (WFQ), frame-based fair queuing (FFQ), core-stateless fair queuing (CSFQ), and finally random early detection (RED).

Chapter 13 discusses network switches and their design options. Media access techniques are first discussed since networking is about sharing limited resources using a variety of multiplexing techniques. Circuit and packet-switching are discussed and packet switching hardware is reviewed. The basic switch components are explained and the main types of switches are discussed: input queuing, output queuing, shared buffer, multiple input queue, multiple output queue, multiple input and output queue, and virtual routing/virtual queuing (VRQ). A qualitative discussion of the advantages and disadvantages of each switch type is provided. Detailed quantitative analyses of the switches is discussed in Chapter 15.

Chapter 14 discusses interconnection networks. Time division networks are discussed and random assignment time division multiple access (TDMA) is analyzed. Several space division networks are studied: crossbar network, generalized cube network (GCN), banyan network, augmented data manipulator network (ADMN), and improved logical neighborhood network (ILN). For each network, a detailed explanation is provided for how a path is established and, equally important, the packet acceptance probability is derived. This last performance measure will prove essential to analyze the performance of switches.

Chapter 15 discusses modeling techniques for input buffer, output buffer, and shared buffer switches. Equations for the performance of each switch are obtained to describe packet loss probability, average delay within the switch, the throughput, and average queue size.

Chapter 16 discusses the design of two next-generation high-performance network switches. The first Promina 4000 switch developed by N.E.T. Inc. The second is the VRQ switch which was developed at the University of Victoria and is being continually improved. The two designs are superficially similar and a comparative study is reported to show how high-performance impacted the design decisions in each switch.

Appendix A provides a handy reference for many formulas that are useful while modeling the different queues considered here. The reader should find this information handy since it was difficult to find all the formulas in a single source.

Appendix B discusses techniques for solving difference equations or recurrence relations. These recurrence relations crop up in the analysis of queues and Markov chains.

Appendix C discusses how the z-transform technique could be used to find a closed-form expression for the distribution vector $\mathbf{s}(n)$ at any time value through finding the z-transform of the transition matrix $\mathbf{P}$.

Appendix D discusses vectors and matrices. Several concepts are discussed such as matrix inverse, matrix nullspace, rank of a matrix, matrix diagonalization, and eigenvalues and eigenvectors of a matrix. Techniques for solving systems of linear equations are discussed since these systems are encountered in several places in the book. Many special matrices are discussed such as circulant matrix, diagonal matrix, echelon matrix, Hessenberg matrix, identity matrix, nonnegative matrix, orthogonal matrix, plane rotation, stochastic (Markov) matrix, substochastic matrix, and tridiagonal matrix.

Appendix E discusses the use of MATLAB in engineering applications. A brief introduction to MATLAB is provided since it is one of the more common mathematical packages used.

Appendix F discusses design of databases. A database is required in a switch to act as the lookup table for important properties of transmitted packets. Hashing and B-trees are two of the main techniques used to construct the fast routing or lookup tables used in switches and routers. The performance of the hashing function and average lookup delay are analyzed. The B-tree data structure is discussed and the advantages of B-trees over regular binary trees and multiway trees are explained.

Advanced Topics

I invested special effort in making this book useful to practicing engineers and students. There are many interesting examples and models throughout the book. However, I list here some interesting topics:

- Chapter 1 discusses heavy-tailed distribution in Section 1.20 and generation of random numbers in Section 1.35.
- Chapter 3 discusses techniques for finding higher powers for Markov chain state transition matrix in Sections 3.13 and 3.14.
- Chapter 5 discusses reducible Markov chains at steady state in Section 5.7 and transient analysis of reducible Markov chains in Section 5.6. Also, there is a discussion on how to identify a reducible Markov chain by examining its eigenvalues and eigenvectors.
- Chapter 6 discusses transient analysis of periodic Markov chains in Section 6.15 and asymptotic behavior of periodic Markov chains in Section 6.15. Also, there is a discussion on how to identify a periodic Markov chain and how to determine its period by examining its eigenvalues.
- Chapter 7 discusses developing performance metrics for the major queue types.
- Chapter 8 discusses how to model three flow control protocols dealing with traffic management.
- Chapter 9 discusses how to model three flow control protocols dealing with error control.
- Chapter 10 discusses how to model three flow control protocols dealing with medium access control.
- Chapter 11 discusses developing realistic models for source traffic using Poisson description (Section 11.3.2), Bernoulli (Section 11.4.3), and Pareto traffic (Section 11.8). There is also discussion on packet destination and length modeling.
- Chapter 12 discusses 12 scheduling algorithms and provides Markov chain analysis for many of them.
- Chapter 13 discusses seven types of switches based on their buffering strategies and the advantages and disadvantages of each choice.
- Chapter 14 discusses many types of interconnection networks and also provides, for the first time, analysis of the performance of each network.

Web Resource

The website *http://www.ece.uvic.ca/~fayez/Book*, www.springer.com/978-0-387-74437-7 contains information about the textbook and any related web resources.

Errors

This book covers a wide range of topics related to communication networks and provides an extensive set of analyses and worked examples. It is "highly probable" that it contains errors and omissions. Other researchers and/or practicing engineers might have other ideas about the content and organization of this book. We welcome receiving any constructive comments and suggestions for inclusion in the next edition. If you find any errors, we would appreciate hearing from you. We also welcome ideas for examples and problems (along with their solutions if possible) to include in the next edition with proper citation.

You can send your comments and bug reports electronically to fayez@uvic.ca, or you can fax or mail the information to

Dr. Fayez Gebali
Elec. and Comp. Eng. Dept.
University of Victoria Victoria, B.C., Canada V8W 3P6
Tel: (250) 721-6509
Fax: (250)721-6052.

Acknowledgment

I was motivated to write this book as a result of reading two very good works. One book is of course the famous *Queueing Systems – Volume 1: Theory* by Leonard Kleinrock. The other book is *Communication and Computer Networks: Modeling with Discrete-Time Queues* by Michael E. Woodward. The title of the last book sums it all. Discrete-time queuing systems proved to be very useful in dealing with communications systems.

I felt that there were more practical topics and important definitions that either were not discussed or were a bit vague. This book provides a minor contribution to this area by summarizing in book form the techniques I advise my students to use while doing their graduate degrees under my supervision. I would like to thank Dr. Mohamed Watheq El-Kharashi for suggesting that I put everything in a book format. I also value his excellent suggestions and help. His vision, suggestions, and his time were absolutely helpful and appreciated.

I would like to thank Dr. W.-S. Lu of the Electrical and Computer Engineering Department at the University of Victoria for his extremely valuable help and insight in all the matrix topics covered in this book. I would also like to thank Dr. G.A. Shoja and Dr. D. Olesky of the Computer Science Department at the University of Victoria for their suggestions, the references they provided, and their encouragement during preparing the material for this book. The constant help and feedback of some of my past and current graduate students was particularly helpful. In particular, I note Mr. Maher Fahmi, Dr. A. Sabba, and Dr. Esam Abdel-Raheem.

Gratitude must go to PMC-Sierra, Inc. of Vancouver for the meetings and consultations I regularly had with their directors, managers, and engineering staff. The financial support of PMC-Sierra throughout the years have contributed enormously to my research program. Without that support I would not have been able to build my research group or acquire the necessary equipment.

The MathWorks, Inc. has provided me with their MATLAB, Simulink, and Stateflow software packages. This is a great help in developing the material and presentations for the examples and problems of this book. For that I am obliged.

Thanks must go to my children Michael Monir, Tarek Joseph, Aleya Lee, and Manel Alia for being such responsible and outstanding young men and women.

Victoria, B.C.
Canada

Dr. Fayez Gebali, Ph.D., P.Eng.

About the Author

Dr. Fayez Gebali has been the Associate Dean (undergraduate programs) from July 2002 – June 2007 at the Faculty of Engineering, University of Victoria. He is also a professor at the Electrical and Computer Engineering Department. He joined the department after its inception in 1983 where he was an assistant professor from 1984 to 1986, associate professor from 1986 to 1991, and professor from 1991 to the present. Before that he was a research associate with Dr. M.I. Elmasry in the Department of Electrical Engineering at the University of Waterloo from 1983 to 1984. He was also a research associate with Dr. K. Colbow in the Physics Department of Simon Fraser University from 1979 to 1983.

Dr. Gebali's research interests include computer networks, communication protocol modeling, and computer arithmetic. He has published over 170 articles in refereed journals and conferences in his areas of expertise.

Dr. Gebali received his B.Sc. in Electrical Engineering (first-class honors) from Cairo University, his B.Sc. (first-class honors) in Applied Mathematics from Ain Shams University, and his Ph.D. degree in Electrical Engineering from the University of British Columbia. Dr. Gebali is a senior member of the IEEE since 1983 and a member of the Society of Professional Engineers and Geoscientists of British Columbia since 1985.

Contents

Chapter 1
Probability

1.1 Introduction

The goal of this chapter is to provide a review of the principles of probability, random variables, and distributions. Probability is associated with conducting a *random experiment* or *trial* and checking the resulting *outcome*.

Definition 1 *Outcome:* An *outcome* is any possible observation of a random experiment.

For example, the random experiment might be flipping a coin and the outcome would be heads or tails depending on whether the coin lands face up or down. The possible outcomes of a random experiment could be discrete or continuous. An example of a random experiment with discrete outcomes is rolling a die since the possible outcomes would be the numbers 1, 2, ... , 6. An example of a random experiment with continuous outcomes is spinning a pointer and measuring its angle relative to some reference direction. The angle we measure could be anything between 0° and 360°.

Definition 2 *Sample space:* The collection of all possible, mutually exclusive outcomes is called the *sample space* S.

For the random experiment of rolling a die, the sample space will be the set

$$S = \{1, 2, 3, 4, 5, 6\}.$$

This sample space is discrete since the possible outcomes were discrete. For the example of spinning a pointer, the sample space is specified by the equation

$$S = \{\theta \mid 0^\circ \leq \theta < 360^\circ\}.$$

Definition 3 *Event:* An *event* is a set of outcomes sharing a common characteristic.

In that sense, an event could correspond to one or several outcomes of an experiment. An event could be thought of as a *subset* of the sample space S. On the other hand, an outcome is an *element* of the subset space.

F. Gebali, *Analysis of Computer and Communication Networks*,
DOI: 10.1007/978-0-387-74437-7_1,

Definition 4 *Event space:* The *event space* is the collection of all possible, mutually exclusive events of a random experiment.

In that sense, the event space is the partitioning of S into subsets that do not share their elements.

For the case of die rolling experiment, we might be interested in the event E that the outcome is an even number. Thus we can specify E: the number is even or $\overline{E}$: the number is odd. In that case, there are six outcomes and two events. The event space will be composed of two events

$$S = \{E, \overline{E}\}$$

Another event could be that the outcome is less than or equal to 2, say. In that case, there are two events again for the experiment—E: the number $=$ 1 or 2 and $\overline{E}$: the number $>$ 2. The event space will also consist of two events.

1.2 Applying Set Theory to Probability

We saw above that sets are used to describe aspects of random experiments such as sample space, outcomes, and events. Table 1.1 shows the correspondence between set theory terminology and probability definitions.

Assume A and B are two events in a random experiment. These two events might be graphically shown using the Venn diagram of Fig. 1.1. The event defined as any outcome that belongs to either A or B is called the *union* of A and B and is represented by the expression

$$A \cup B$$

Figure 1.1(a) shows the union operation as the shaded area.

The event defined as any outcome that belongs to both A and B is called the *intersection* of A and B and is represented by the expression

$$A \cap B$$

Figure 1.1(b) shows the intersection operation as the shaded area.

Table 1.1 Correspondence between set theory and probability definitions

Probability	Set theory
Outcome	Element of a set
Sample space S	Universal set U
Event	Set
Impossible event	Null set Ø

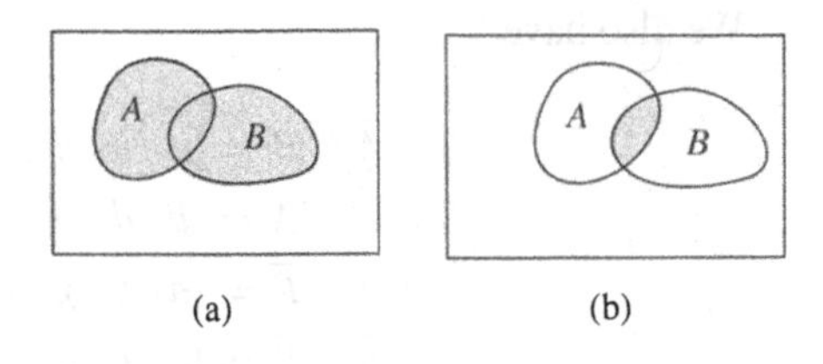

Fig. 1.1 Venn diagram for two events A and B in a sample space S. (**a**) The union operation $A \cup B$. (**b**) The intersection operation $A \cap B$

Definition 5 *Complementary event:* Given an event A, the *complementary event* $\overline{A}$ is a set of all outcomes that do not belong to the set A.

Figure 1.2 shows that the universal set U is partitioned into two sets A and $\overline{A}$. These two sets are not overlapping in the sense that there is not a single outcome that belongs to both A and $\overline{A}$ simultaneously.

We can write the following equations describing the operations on events.

$$\overline{A} = U - A$$
$$A \cup \overline{A} = U$$
$$A \cap \overline{A} = \emptyset$$

De Morgan's law for sets applies also for events and we can write

$$\overline{A \cap B} = \overline{A} \cup \overline{B}$$

Example 1 Let $U = \{a,\ b,\ c,\ d,\ e,\ f,\ g,\ h,\ i,\ j,\ k\}$, $A = \{a,\ c,\ e,\ h,\ j\}$, and $B = \{c,\ d,\ e,\ f,\ k\}$. Find the events

1. $A \cup B$
2. $A \cap B$

Prove also that $\overline{A \cap B} = \overline{A} \cup \overline{B}$.

$$A \cup B = \{a,\ c,\ d,\ e,\ f,\ h,\ j,\ k\}$$
$$A \cap B = \{c,\ e\}$$

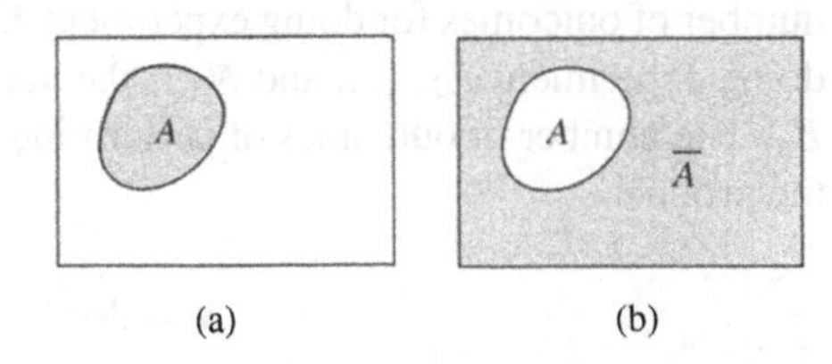

Fig. 1.2 Event A and its complementary event $\overline{A}$

We also have

$$\overline{A \cap B} = \{a,\ b,\ d,\ f,\ g,\ h,\ i,\ j,\ k\}$$
$$\overline{A} = \{b,\ d,\ f,\ g,\ i,\ k\}$$
$$\overline{B} = \{a,\ b,\ g,\ h,\ i,\ j\}$$
$$\overline{A} \cup \overline{B} = \{a,\ b,\ d,\ f,\ g,\ h,\ i,\ j,\ k\}$$

Hence De Morgan's law is proved by direct counting. ■

Definition 6 *Mutually exclusive events:* Events A and B are said to be *mutually exclusive* or *disjoint events* if they have no elements in common.

From that definition, we can write

$$A \cap B = \phi \tag{1.1}$$

1.3 Counting Sample Space Points

In many discussions of probability, we need to determine the number of possible outcomes of a given experiment or trial. The multiplication principle, permutations, and combinations, to be discussed in the following sections, will prove useful.

1.4 The Multiplication Principle

The fundamental principle of counting is useful in finding the number of points in the sample space. Suppose that the number of outcomes for doing experiment E_1 is x and the number of outcomes for doing experiment E_2 is y. Then the number of outcomes for doing experiment E_1 followed by experiment E_2 is the product $x\ y$.

Example 2 Assume there are 3 different ways for a data packet to travel through network A and 15 different ways to travel through network B. Use the multiplication principle to find the number of ways the packet could travel through network A followed by network B.

The multiplication principle states that there are 3×15 or 45 ways for the packet to travel through network A followed by network B. ■

We can generalize the multiplication principle as follows. Suppose that N_1 is the number of outcomes for doing experiment E_1, and N_2 is the number of outcomes for doing experiment E_2, …, and N_n is the number of outcomes for doing experiment E_n. The number of outcomes of performing experiments E_1, E_2, …, E_n is given by the product

$$N = N_1\ N_2\ \cdots\ N_n$$

Example 3 A die is thrown three times and the sequence of numbers is recorded. Determine the number of 3-digit sequences that could result.

We perform three experiments where each one is the act of throwing the die. One possible outcome would be the number sequence 312, which corresponds to obtaining 3 on the first throw, 1 on the second, and 2 on the third. The number of outcomes of each experiment is 6. Therefore, the total number of outcomes is

$$N = 6 \times 6 \times 6 = 216$$

■

Example 4 In time division multiplexing (also known as synchronous transmission mode), each slot in a frame is reserved to a certain user. That slot could be occupied or empty when the user is busy or idle, respectively. Assuming the frame consists of 10 slots, how many slot occupancy patterns can be received in a single frame?

Each time slot can be treated as an experiment with only two outcomes, busy or idle. The experiment is repeated 10 times to form the frame. The total number of possible outcomes is

$$N = 2^{10} = 1024$$

■

1.5 Permutations

Permutations arise when we are given n objects and we want to arrange them according to a certain order. In other words, we arrange the objects by randomly picking one, then the next, etc. We have n choices for picking the first item, $n - 1$ choices for picking the second item, etc.

1.5.1 n Distinct Objects Taken n at a Time

The number of permutations of n distinct objects taken n at a time is denoted by $P(n, n)$ and is given by

$$P(n, n) = n! = n \times (n - 1) \times (n - 2) \times \cdots \times 3 \times 2 \times 1$$

The function $n!$ is called *Factorial-n* and can be obtained using the MATLAB function `factorial(n)`.

Example 5 Packets are sent through the Internet in sequence; however, they are received out of sequence since each packet could be sent through a different route. How many ways can a 5-packet sequence be received?

If we number our packets as 1, 2, 3, 4, and 5, then one possible sequence of received packets could be 12543 when packet 1 arrives first followed by packet 2, then packet 5, etc. The number of possible received packet sequences is given by

$$P(5, 5) = 5! = 120$$

■

1.5.2 n Distinct Objects Taken k at a Time

A different situation arises when we have n distinct objects but we only pick k objects to arrange. In that case, the number of permutations of n distinct objects taken k at a time is given by

$$\begin{aligned} P(n, k) &= n \times (n-1) \times \cdots \times (n-k+1) \\ &= \frac{n!}{(n-k)!} \end{aligned} \tag{1.2}$$

Example 6 Assume that 10 packets are sent in sequence, but they are out of sequence when they are received. If we observe only a 3-packet sequence, how many 3-packet sequences could we find?

A possible observed packet arrival sequence could be 295. Another might be 024, etc. We have $n = 10$ and $k = 3$.

$$P(10, 3) = \frac{10!}{7!} = 720$$

■

1.6 Permutations of Objects in Groups

Now, assume we have n objects in which n_1 objects are alike, n_2 objects are alike, and so on, and n_k objects are alike such that

$$n = \sum_{i=1}^{k} n_i \tag{1.3}$$

Here we classify the objects not by their individual labels but by the group in which they belong. An output sequence will be distinguishable from another if the objects picked happen to belong to different groups. As a simple example, suppose we have 20 balls that could be colored red, green, or blue. We are now interested not in picking a particular ball, but in picking *a* ball of a certain color.

In that case, the number of permutations of these n objects taken n at a time is given by

$$x = \frac{n!}{n_1!\, n_2!\, \cdots\, n_k!} \tag{1.4}$$

This number is smaller than $P(n, n)$ since several of the objects are alike and this reduces the number of *distinguishable* combinations.

Example 7 Packets arriving at a terminal could be one of three possible service classes: class A, class B, or class C. Assume that we received 10 packets and we found out that there were 2 packets in class A, 5 in class B, and 3 in class C. How many possible service class arrival order could we have received?

We are not interested here in the sequence of received packets. Instead, we are interested only in the arrival order of the service classes.

We have $n_1 = 2$, $n_2 = 5$, and $n_3 = 3$ such that $n = 10$. The number of service class patterns is

$$x = \frac{10!}{2!\, 5!\, 3!} = 2,520$$

In other words, there are 10 possibilities for receiving 10 packets such that exactly 2 of them belonged to class A, 5 belonged to class B, and 3 belonged to class C. ■

Example 8 In time division multiplexing, each time slot in a frame is reserved to a certain user. That time slot could be occupied or empty when the user is busy or idle, respectively. If we know that each frame contains 10 time slots and 4 users are active and 6 are idle, how many possible active slot patterns could have been received?

We are interested here in finding the different ways we could have received 4 active slots out of 10 possible slots. Thus we "color" our slots as active or idle without regard to their location in the frame.

We have $n_1 = 4$ and $n_2 = 6$ such that $n = 10$.

$$x = \frac{10!}{4!\, 6!} = 210$$

■

Example 9 A bucket contains 10 marbles. There are 5 red marbles, 2 green marbles, and 3 blue marbles. How many different color permutations could result if we arranged the marbles in one straight line?

We have $n_1 = 5$, $n_2 = 2$, and $n_3 = 3$ such that $n = 10$. The number of different color permutations is

$$x = \frac{10!}{5!\, 2!\, 3!} = 2,520$$

■

1.7 Combinations

The above permutations took the *order* of choosing the objects into consideration. If the order of choosing the objects is not taken into consideration then *combinations* are obtained.

The number of combinations of n objects taken k at a time is called the *binomial coefficient* and is given by

$$C(n,k) = \binom{n}{k} = \frac{n!}{k!\,(n-k)!} \tag{1.5}$$

MATLAB has the function *nchoosek (n,k)* for evaluating the above equation where $0 \le k \le n$.

Example 10 Assume 10 packets are received with 2 packets in error. How many combinations are there for this situation?

We have $n = 10$ and $k = 2$.

$$C(10,2) = \binom{10}{2} = \frac{10!}{2!\,8!} = 45$$

■

Example 11 Assume 50 packets are received but 4 packets are received out of sequence. How many combinations are there for this situation?

We have $n = 50$ and $k = 4$.

$$C(50,4) = \binom{50}{4} = \frac{50!}{4!\,46!} = 230,300$$

■

1.8 Probability

We define probability using the *relative-frequency approach*. Suppose we perform an experiment like the tossing of a coin for N times. We define event A is when the coin lands head up and define N_A as the number of times that event A occurs when the coin tossing experiment is repeated N times. Then the probability that we will get a head when the coin is tossed is given by

$$p(A) = \lim_{N\to\infty} \frac{N_A}{N} \tag{1.6}$$

This equation defines the relative frequency that event A happens.

1.9 Axioms of Probability

We defined our sample space S as the set of all possible outcomes of an experiment. An impossible outcome defines the empty set or null event $\emptyset$. Based on this we can state four basic axioms for the probability.

1. The probability $p(A)$ of an event A is a nonnegative fraction in the range $0 \leq p(A) \leq 1$. This can be deduced from the basic definition of probability in (1.6).
2. The probability of the null event $\emptyset$ is zero, $p(\emptyset) = 0$.
3. The probability of all possible events S is unity, $p(S) = 1$.
4. If A and B are *mutually exclusive* events (cannot happen at the same time), then the probability that event A *or* event B occurs is

$$p(A \cup B) = p(A) + p(B) \tag{1.7}$$

Event E and its complement E^c are mutually exclusive. By applying the above axioms of probability, we can write

$$p(E) + p(E^c) = 1 \tag{1.8}$$

$$p(E \cap E^c) = 0 \tag{1.9}$$

1.10 Other Probability Relationships

If A and B are two events (they need not be mutually exclusive), then the probability that event A **or** event B occurring is

$$p(A \cup B) = p(A) + p(B) - p(A \cap B) \tag{1.10}$$

The probability that event A occurs *given that* event B occurred is denoted by $P(A|B)$ and is sometimes referred to as the probability of A conditioned by B. This is given by

$$p(A|B) = \frac{p(A \cap B)}{p(B)} \tag{1.11}$$

Now, if A and B are two *independent* events, then we can write $p(A|B) = p(A)$ because the probability of event A taking place will not change whether event B occurs or not. From the above equation we can now write the probability that event A *and* event B occurs is

$$p(A \cap B) = p(A) \times p(B) \tag{1.12}$$

provided that the two events are *independent*.

The probability of the complement of an event is given by

$$p(A^c) = 1 - p(A) \tag{1.13}$$

1.11 Random Variables

Many systems based on random phenomena are best studied using the concept of random variables. A random variable allows us to employ mathematical and numerical techniques to study the phenomenon of interest. For example, measuring the length of packets arriving at random at the input of a switch produces as outcome a number that corresponds to the length of that packet.

According to references [1–4], a *random variable* is simply a numerical description of the outcome of a random experiment. We are free to choose the *function* that maps or assigns a numerical value to each outcome depending on the situation at hand. Later, we shall see that the choice of this function is rather obvious in most situations. Figure 1.3 graphically shows the steps leading to assigning a numerical value to the outcome of a random experiment. First we run the experiment, then we observe the resulting outcome. Each outcome is assigned a numerical value.

Assigning a numerical value to the outcome of a random experiment allows us to develop uniform analysis for many types of experiments independent of the nature of their specific outcomes [1].

We denote a random variable by a capital letter (the name of the function) and any particular value of the random variable is denoted by a lower case letter (the value of the function).

The following are the examples of random variables and their numerical values.

1. Number of arriving packets at a given time instance is an example of a *discrete random variable* N with possible values $n = 0, 1, 2, \cdots$.
2. Tossing a coin and assigning 0 when a tail is obtained and 1 when a head is obtained is an example of a discrete random variable X with values $x \in \{0, 1\}$.
3. The weight of a car in kilograms is an example of a *continuous random variable* W with values in the range $1000 \leq w \leq 2000$ kg typically.
4. The temperature of a day at noon is an example of random variable T. This random variable could be discrete of continuous depending on the type of thermometer (analog or digital).

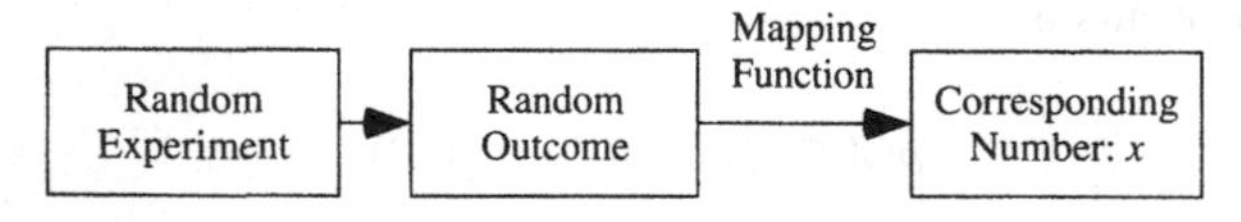

Fig. 1.3 The steps leading to assigning a numerical value to the outcome of a random experiment

5. The atmospheric pressure at a given location is an example of a random variable P. This random variable could be discrete of continuous depending on the accuracy of the barometer (analog or digital).

1.12 Cumulative Distribution Function (cdf)

The cumulative distribution function (cdf) for a random variable X is denoted by $F_X(x)$ and is defined as the *probability* that the random variable is less than or equal to x. Thus the event of interest is $X \leq x$ and we can write

$$F_X(x) = p(X \leq x) \tag{1.14}$$

The subscript X denotes the random variable associated with the function while the argument x denotes a numerical value. For simplicity we shall drop the subscript and write $F(x)$ when we are dealing with a single random variable and there is no chance of confusion. Because $F(x)$ is a probability, it must have the same properties of probabilities. In addition, $F(x)$ has other properties as shown below.

$$F(-\infty) = 0 \tag{1.15}$$

$$F(\infty) = 1 \tag{1.16}$$

$$0 \leq F(x) \leq 1 \tag{1.17}$$

$$F(x_1) \leq F(x_2) \qquad \text{when} x_1 \leq x_2 \tag{1.18}$$

$$p(x_1 < X \leq x_2) = F(x_2) - F(x_1) \tag{1.19}$$

The cdf is a monotonically increasing function of x. From the last equation, the probability that x lies in the region $x_0 < x \leq x_0 + \epsilon$ (where ϵ is arbitrarily small) is given by

$$p(x_0 < X \leq x_0 + \epsilon) = F(x_0 + \epsilon) - F(x_0) \tag{1.20}$$

Thus the amount of jump in cdf at $x = x_0$ is the probability that $x = x_0$.

Example 12 Consider the random experiment of spinning a pointer around a circle and measuring the angle it makes when it stops. Plot the cdf $F_\Theta(\theta)$.

Obviously, the random variable Θ is continuous since the pointer could point at any angle. The range of values for θ is between 0° and 360°. Thus the function $F_\Theta(\theta)$ has the following extreme values

$$F_\Theta(-0^\circ) = p(\theta \leq -0^\circ) = 0$$

$$F_\Theta(360^\circ) = p(\theta \leq 360^\circ) = 1$$

There is no preference for the pointer to settle at any angle in particular and the cdf will have the distribution as shown in Fig. 1.4. ■

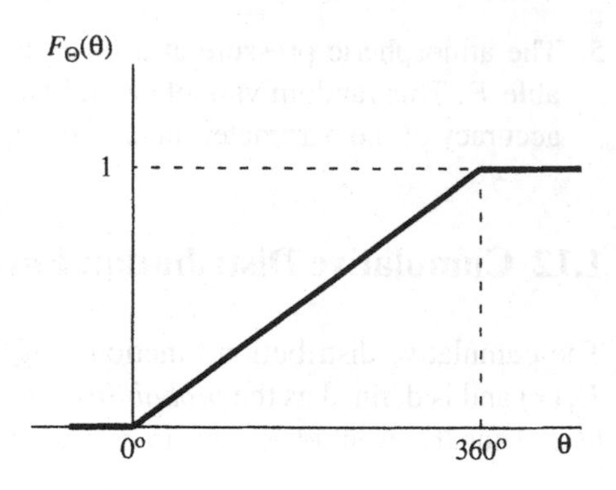

Fig. 1.4 Cumulative distribution function for a continuous random variable

1.12.1 cdf in the Discrete Case

For the case of a discrete random variable, we make use of the cdf property in (1.20). The cdf for a discrete random variable will be a staircase as illustrated in the following example.

Example 13 Consider again the case of the spinning pointer experiment but define the discrete random variable Q which identifies the quadrant in which the pointer rests in. The quadrants are assigned the numerical values 1, 2, 3, and 4. Thus the random variable Q will have the values $q = 1, 2, 3$, or 4.

Since the pointer has equal probability of stopping in any quadrant, we can write

$$p(q = 1) = \frac{1}{4}$$
$$p(q = 2) = \frac{1}{4}$$
$$p(q = 3) = \frac{1}{4}$$
$$p(q = 4) = \frac{1}{4}$$

The cdf for this experiment is shown in Fig. 1.5. ■

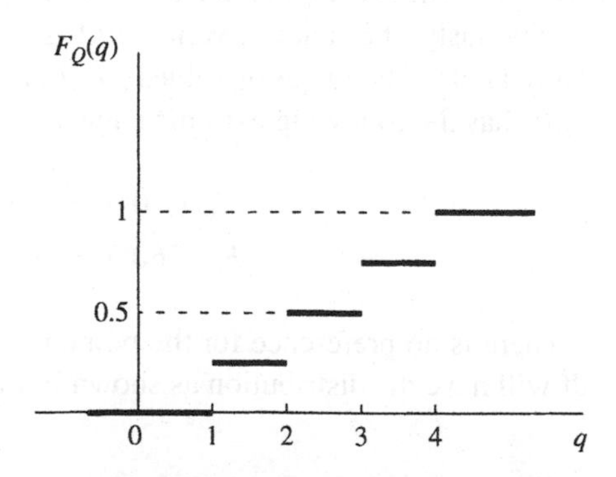

Fig. 1.5 Cumulative distribution function for a discrete random variable

1.13 Probability Density Function (pdf)

The probability density function (pdf) for a *continuous random variable* X is denoted by $f_X(x)$ and is defined as the derivative of $F_X(x)$

$$f_X(x) = \frac{dF_X(x)}{dx} \tag{1.21}$$

Because $F_X(x)$ is a monotonically increasing function of x, we conclude that $f_X(x)$ will never be negative. It can, however, be zero or even greater than 1.

We will follow our simplifying convention of dropping the subscript when there is no chance of confusion and write the pdf as $f(x)$ instead of $f_X(x)$. By integrating the above equation, we obtain

$$\int_{x_1}^{x_2} f(x)\, dx = F(x_2) - F(x_1) \tag{1.22}$$

Thus we can write

$$p(x_1 < X \leq x_2) = \int_{x_1}^{x_2} f(x)\, dx \tag{1.23}$$

The area under the pdf curve is the probability $p(x_1 < X \leq x_2)$

$f(x)$ has the following properties:

$$f(x) \geq 0 \qquad \text{for all } x \tag{1.24}$$

$$\int_{-\infty}^{\infty} f(x)\, dx = 1 \tag{1.25}$$

$$\int_{-\infty}^{x} f(y)\, dy = F(x) \tag{1.26}$$

$$\int_{x_1}^{x_2} f(x)\, dx = p(x_1 < X \leq x_2) \tag{1.27}$$

$$f(x)\, dx = p(x < X \leq x + dx) \tag{1.28}$$

1.14 Probability Mass Function

For the case of a *discrete random variable*, the cdf is discontinuous in the shape of a staircase. Therefore, its slope will be zero everywhere except at the discontinuities where it will be infinite.

The pdf in the discrete case is called the *probability mass function* (pmf) [5]. The pmf is defined as the probability that the random variable X has the value x and is denoted by $p_X(x)$. We can write

$$p_X(x) \equiv p(X = x) \tag{1.29}$$

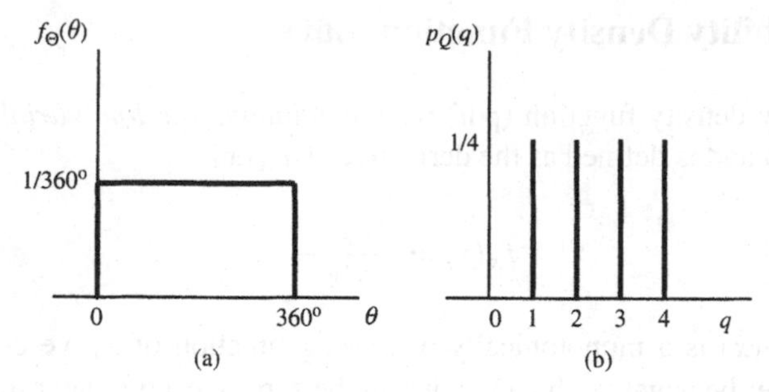

Fig. 1.6 Continuous and discrete random variables. (**a**) pdf for the continuous case. (**b**) pmf for the discrete case

where the expression on the right-hand side (RHS) indicates the probability that the random variable X has the value x.

We will follow our simplifying convention of dropping the subscript when there is no chance of confusion and write the pmf as $p(x)$ instead of $p_X(x)$. $p_X(x)$ has the following properties:

$$p_X(x) \geq 0 \qquad \text{for all } x \tag{1.30}$$

$$\sum_x p_X(x) = 1 \tag{1.31}$$

Example 14 The pointer spinning experiment was considered for the continuous case (Example 12) and the discrete case (Example 13). Plot the pdf for the continuous random variable Θ and the corresponding pmf for the discrete random variable Q.

Figure 1.6(a) shows the pdf for the continuous case where the random variable Θ measures the angle of the pointer. Figure 1.6(b) shows the pmf for the discrete case where the random variable Q measures the quadrant where the pointer is located. ■

1.15 Expected Value and Variance

The pdf and pmf we obtained above help us find the expected value $E[X]$ of a random variable X. For the continuous case, the expected value is given by

$$E[X] = \int_{-\infty}^{\infty} x\, f(x)\, dx \tag{1.32}$$

For the discrete case, the expected value is given by the weighted sum

$$E[X] = \sum_i x_i\, p(x_i) \tag{1.33}$$

The expectation is sometimes referred to as the first moment of the random variable. Sometimes μ is used as another symbol for the expected value.

$$\mu = E[X]$$

The *mean* (m) of a set of random variable samples is defined as

$$m = \frac{1}{n} \sum_{i=1}^{n} x_i \tag{1.34}$$

The mean is not exactly equal to the expected value μ since m changes its value depending on how many samples we take. However, as $n \to \infty$, the two quantities become equal [5]. Higher moments are also useful and we define the *variance*, or second central moment, of the random variable as

$$\sigma^2 = E\left[(X - \mu)^2\right] \tag{1.35}$$

The variance describes how much of the mass of the distribution is close to the expected value. A small value for σ^2 indicates that most of the random variable values lie close to the expected value μ. In other words, small variance means that the pdf is large only in regions close to the expected value μ. For an archery target practice experiment, this might mean that most of the arrows were clustered together and landed very close at some spot on the target (not necessarily dead center).

Conversely, a large variance means that the pdf is large for values of X far away from μ. Again for the archery experiment, this means that most of the arrows were not clustered together and landed at different spots on the target. The *standard deviation* σ is simply the square root of the variance.

Example 15 Assume a random variable A from a binary random experiment in which only two events result. A has two values a and 0. The probability that the value a is obtained is p and the probability that the value 0 is obtained is $q = 1 - p$. Find the expected value of A.

This is a discrete random variable and the pmf for A is

$$p(a) = \begin{cases} q & \text{when } A = 0 \\ p & \text{when } A = a \end{cases} \tag{1.36}$$

The expected value is obtained from (1.33) as

$$E[A] = q \times 0 + p \times a = p\, a \tag{1.37}$$

Notice that the expected value will be between 0 and a since p is a nonnegative fraction. ■

1.16 Common Continuous RVs

We discuss in the following sections some continuous random variables that are useful for network simulations. Discussion of common discrete random variables is found in later sections.

1.17 Continuous Uniform (Flat) RV

The uniform random variable, or uniform distribution, usually arises in physical situations where there is no preferred value for the random variable. For example, the value of a signal during analog-to-digital conversion could lie anywhere within each quantization level. This distribution is also useful in our studies because it is often used to obtain random numbers that obey the more sophisticated distributions to be discussed below. These random numbers are then considered to be the "traffic" generated at the inputs of our communication networks.

A uniform distribution is characterized by a random variable that spans in the range a–b such that $a < b$,

$$f(x) = \begin{cases} 1/(b-a) & a \leq x < b \\ 0 & \text{otherwise} \end{cases} \tag{1.38}$$

and the corresponding cdf is given by

$$F(x) = \begin{cases} 0 & x < a \\ (x-a)/(b-a) & a \leq x < b \\ 1 & x \geq b \end{cases} \tag{1.39}$$

Typically $a = 0$ and $b = 1$. Figure 1.7(a) shows the pdf for the uniform distribution and Fig. 1.7(b) shows the corresponding cdf. The mean and variance for X are given by

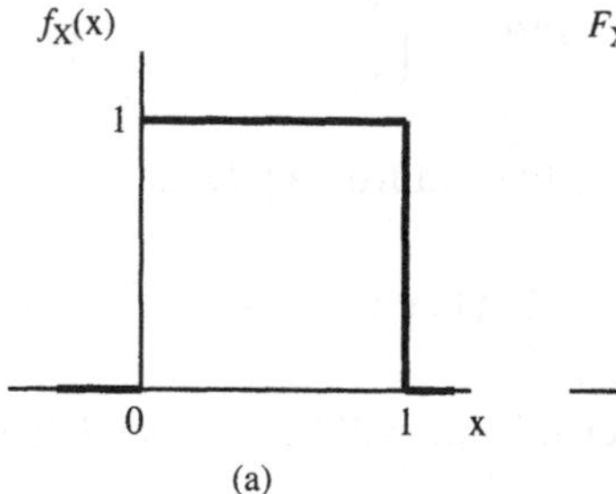

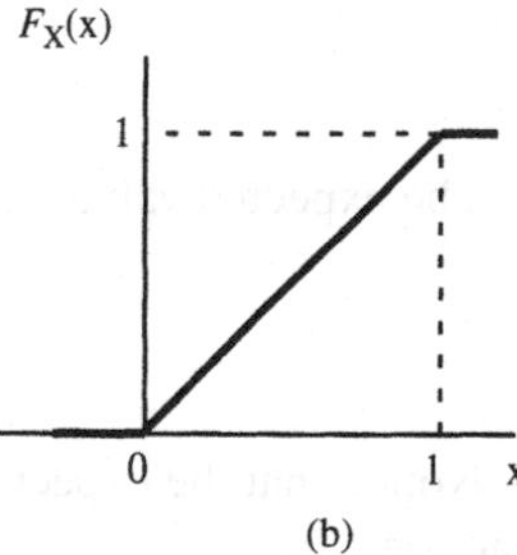

Fig. 1.7 The uniform distribution for a continuous random variable. (**a**) The pdf and (**b**) is the corresponding cdf

$$\mu = \frac{b+a}{2} \tag{1.40}$$

$$\sigma^2 = \frac{(b-a)^2}{12} \tag{1.41}$$

The following MATLAB code generates and plots a random variable having a uniform distribution in the range $0 \le x < 1$.

```
%uniform.m
n=1000 % number of samples is 1000
x=rand(1,n)
subplot(1,2,1)
plot(x,'k')
box off, axis square
xlabel('Sample index'), ylabel('Random number value')
subplot(1,2,2)
hist(y)
title('pdf plot')
box off, axis square
xlabel('Bins'),ylabel('Number of samples)
```

Figure 1.8 shows the result of running the code. Figure 1.8(a) shows the samples and Figure 1.8(b) shows the histogram of the random variable. Notice that the distribution of the samples in the different bins is almost equal. If we chose the number of samples to be larger than 1000, the histogram would show more equal distribution among the bins.

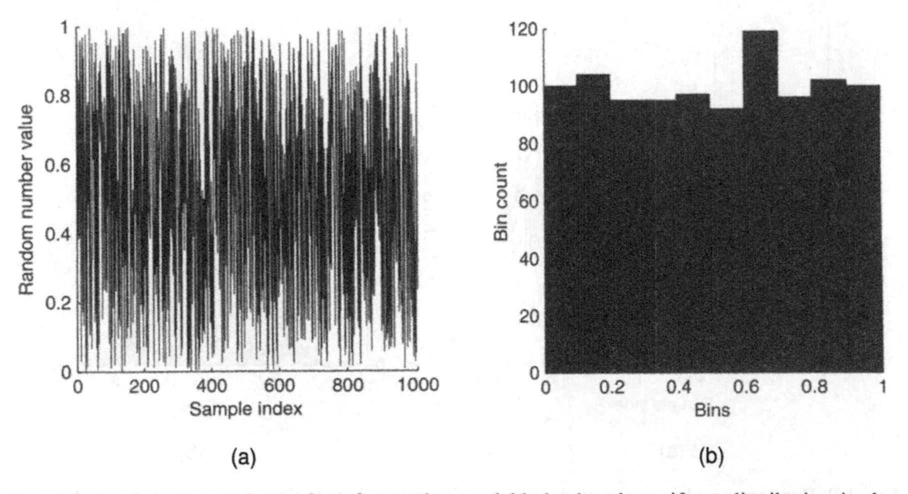

Fig. 1.8 **(a)** One thousand samples of a random variable having the uniform distribution in the range 0–1. **(b)** Histogram for the samples showing a uniform distribution

1.18 Gaussian RV

This distribution arises in many random variables used in electrical engineering such as the noise in a wireless channel. The Gaussian distribution applies for the case of a continuous random variable X that is allowed to have the values ranging from $-\infty$ to $+\infty$. The pdf for this distribution is given by

$$f(x) = \frac{1}{\sigma\sqrt{2\pi}}\, e^{-(x-\mu)^2/(2\sigma^2)} \qquad (1.42)$$

where μ is the mean and σ is the standard deviation of the distribution. The cdf for this distribution is given by

$$F(x) = \int_{-\infty}^{x} f(y)\; dy \qquad (1.43)$$

There is no closed-form formula for the cdf associated with the Gaussian distribution but that function is tabulated in many textbooks on statistics. The *standard* or *normal* random variable is a Gaussian RV with $\mu = 0$ and $\sigma = 1$ [5]. Figure 1.9 shows the output of a Gaussian random variable with zero mean and unity variance using the `randn` function of MATLAB. One thousand samples were generated in this experiment.

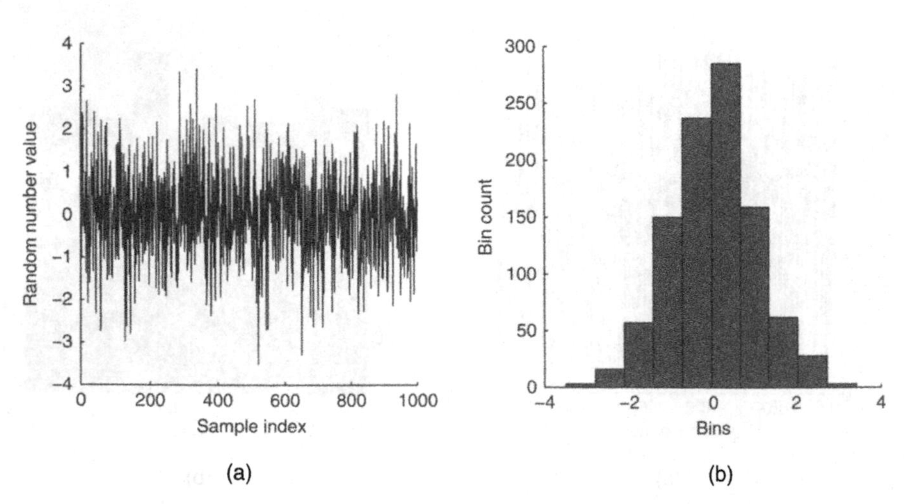

Fig. 1.9 Random numbers generated using the Gaussian distribution with zero mean and unity variance. (**a**) shows the random samples and (**b**) shows their histogram

1.19 Exponential RV

The exponential distribution applies for the case of a continuous random variable X that is allowed to have the values ranging from 0 to $+\infty$. The pdf for this distribution is given by

$$f(x) = be^{-b\,x} \qquad x \geq 0 \tag{1.44}$$

where $b > 0$. The corresponding cdf is given by

$$F(x) = 1 - e^{-b\,x} \tag{1.45}$$

The mean and variance for X are

$$\mu = \frac{1}{b} \tag{1.46}$$

$$\sigma^2 = \frac{1}{b^2} \tag{1.47}$$

A famous example of exponential RV is the radioactive decay where we have

$$f(t) = \lambda e^{-\lambda t} \qquad t \geq 0 \tag{1.48}$$

Here λ is the rate of decay of an element. In that case $1/\lambda$ is called the *lifetime* when the radioactive material decreases by the ratio $1/e$.

1.20 Pareto RV

The Poisson and binomial distributions have been traditionally employed to model traffic arrival at networks. Recent work has shown that such models may be inadequate because the traffic may exhibit periods of high data rates even when the average data rate is low. This type of traffic is described as being self-similar (fractal) [6–8]. Self-similar traffic has distributions with very high variance. Sometimes such distributions are described as being *heavy-tailed* since the pdf has large values for x far away from the mean μ. This type of distribution might then prove more accurate in describing the pdf for the rate of data produced by a bursty source.

The Pareto distribution could be made to be a heavy-tailed distribution by proper choice of its parameters. The Pareto distribution is described by the pdf

$$f(x) = \frac{b\,a^b}{x^{b+1}} \qquad \text{with } a \leq x < \infty \tag{1.49}$$

where a is the position parameter and $b > 0$ is the shape parameter. Figure 1.10 shows the pdf distribution for the case when $a = 2$ and $b = 3$ (solid line) and $b = 5$

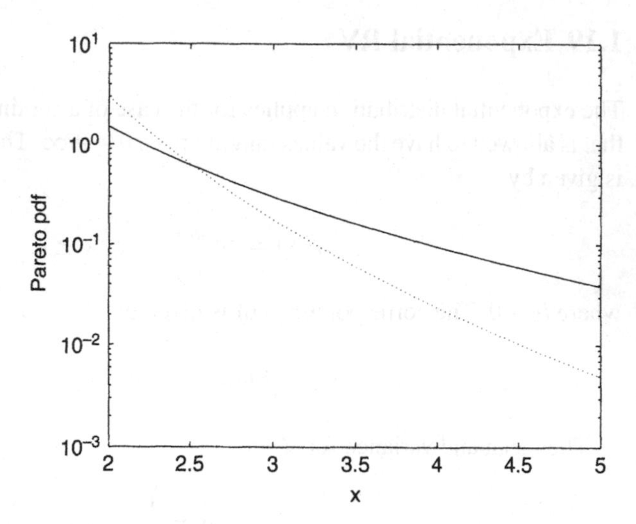

Fig. 1.10 Pareto pdf distribution for the case when $a = 2$ and $b = 3$ (*solid line*) and $b = 5$ (*dashed line*)

(dashed line). For the smaller value of shape parameter b, the pdf becomes flatter and has higher values at larger values of x. This results in larger variance for X. The corresponding cdf is

$$F(x) = 1 - \left(\frac{a}{x}\right)^b \tag{1.50}$$

The mean and variance for X are

$$\mu = \frac{ba}{b-1} \tag{1.51}$$

$$\sigma^2 = \frac{b\,a^2}{(b-1)^2\,(b-2)} \tag{1.52}$$

The mean is always positive as long as $b > 1$. The variance is meaningful only when $b > 2$. From 1.50, we can write

$$\begin{aligned} p(X > x) &= 1 - p(X \leq x) \\ &= \left(\frac{a}{x}\right)^b \end{aligned} \tag{1.53}$$

which means that the probability that the random variable has a value greater than x decreases geometrically [9].

Pareto distribution is typically used to generate network traffic that shows bursty behavior. This means that when a traffic burst is encountered, it is very probable that the burst will continue. For such traffic, the shape parameter b is typically chosen in the range 1.4–1.6.

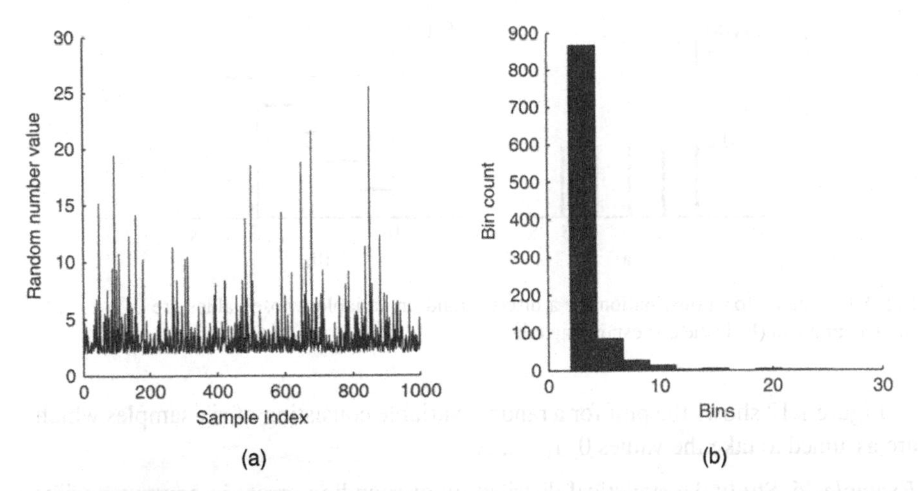

Fig. 1.11 Random numbers generated using the Pareto distribution with $a = 2$ and $b = 2.5$. (**a**) shows the random samples and (**b**) shows their histogram

Figure 1.11 shows the output of a Pareto random variable with position parameter $a = 2$ and shape parameter $b = 2.5$. One thousand samples were generated in this experiment using the inversion method as described later in Section 1.35.2.

1.21 Common Discrete RVs

We discuss in the following sections some discrete random variables that are useful for network simulations.

1.22 Discrete Uniform RV

Assume N is a random variable such that there are k distinct sample points. The pmf for the discrete uniform RV is defined by

$$p(n) = \begin{cases} 1/k & 1 \le n \le k \\ 0 & \text{otherwise} \end{cases}$$

where n is the sample index. Alternatively, the pmf can be expressed in the form

$$p(n) = \frac{1}{k} \sum_{i=1}^{k} \delta(n - i) \tag{1.54}$$

where $\delta(j)$ is the Dirac delta function which is one when $j = 0$ and zero for all other values of $j \neq 0$.

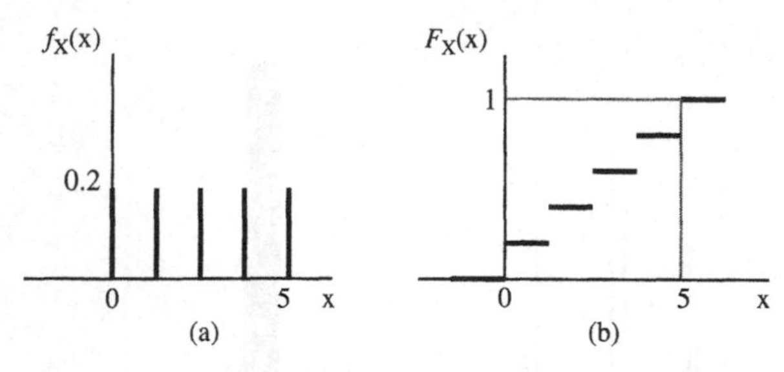

Fig. 1.12 The uniform distribution for a discrete random variable whose values are 0, 1, ..., 5. (**a**) The pmf and (**b**) is the corresponding cdf

Figure 1.12 shows the pmf for a random variable consisting of six samples which are assumed to take the values 0, 1, ..., 5.

Example 16 Study the statistical distribution of rounding errors in computer arithmetic.

The truncation or rounding operation is required after fixed and floating point number multiplication and also after floating point number addition. Rounding or truncation is employed to reduce the number of bits back to their original size n, say. Without loss of generality, we assume the inputs to be fractions with the binary point at the left such that the weight of the most significant bit is 2^{-1} and the weight of the least significant bit (LSB) is 2^{-n}. Figure 1.13 shows the location of the binary point for the fixed-point number and also shows the bits that will be truncated or rounded.

Truncating the extra bits to the right of LSB results in an error e whose magnitude varies approximately in the range

$$-2^{-n} \leq e \leq 2^{-n} \tag{1.55}$$

Assume the number of bits to be truncated is m. In that case, our truncated data has m bits and the number of possible error samples is $k = 2 \times 2^m - 1$. The factor 2 comes from the fact that the error could be positive or negative.

Define the discrete random variable E that corresponds to the rounding or truncation error. Since the probability of any truncation error is equally likely, we have

$$p(e) = \frac{1}{k} = \frac{1}{2^{(m+1)} - 1}$$

■

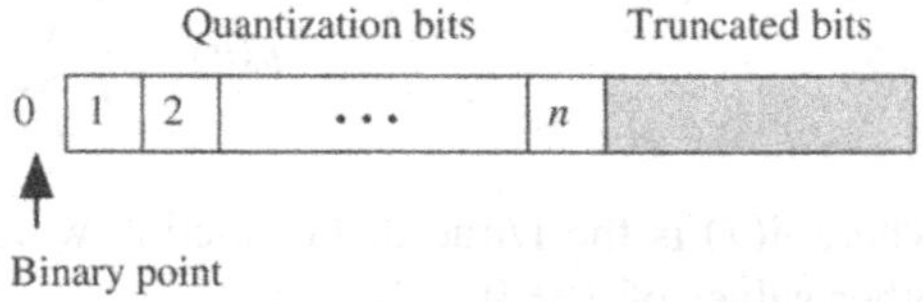

Quantization bits Truncated bits

0 | 1 | 2 | ... | n |

Binary point

Fig. 1.13 Truncation of a fractional number from $n + m$ bits to n bits

1.23 Bernoulli (Binary) RV

Many systems in communications have two outcomes. For example, a received packet might be error-free or it might contain an error. For a router or a switch, a packet might arrive at a given time step or no packet arrives. Consider a chance experiment that has two mutually exclusive outcomes A and $\overline{A}$ that occur with probabilities p and q, respectively. We define the discrete random variable X where $X = 1$ when A occurs and $X = 0$ when $\overline{A}$ occurs. We can write

$$p(1) = p \tag{1.56}$$

$$p(0) = q \tag{1.57}$$

where $q = 1 - p$. Alternatively, $p(x)$ can be expressed by a single equation in the form

$$p(x) = q\ \delta(x) + p\ \delta(x-1) \tag{1.58}$$

The mean and variance for X are

$$\mu = p \tag{1.59}$$

$$\sigma^2 = p\,(1-p) \tag{1.60}$$

Figure 1.14(a) shows the pmf for the binary distribution and Fig. 1.14(b) shows the cdf.

Example 17 The 50/50 draw is one of the traditions of a typical minor lacrosse or baseball sports events. Spectators purchase numbered tickets. One of the tickets is picked at random and half the proceeds goes to the winner and the rest goes to support the team (or the executive council might just use the money for their own purposes). Assume you purchased one ticket and there was a total of 100 entries at the start of the draw. What are your chances of winning or losing? How much would your winnings be?

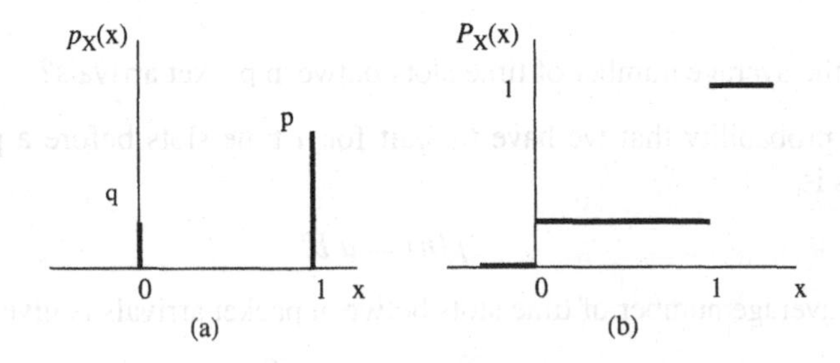

Fig. 1.14 The binary distribution. (**a**) The pmf and (**b**) is the corresponding cdf

The total number of entries is 100, and the probabilities of winning or losing are

$$p = \frac{1}{100} = 0.01$$
$$q = 1 - p = 0.99$$

Assuming the purchase price of the ticket is \$1, the winner takes \$50. The money won would be \$49. ■

1.24 Geometric RV

This distribution is encountered when success, event A, occurs after n failures. This is the case when several devices are attached to a bus and compete for access. The probability of success is a and the probability of failure is $b = 1 - a$.

The pmf is the probability of success after n repeated failures and is given by

$$p(N = n) = a\, b^n \qquad \text{for } n \geq 0 \tag{1.61}$$

Alternatively, $p(n)$ can be expressed by a single equation in the form

$$p(n) = \sum_{i \geq 0} a\, b^i\, \delta(n - i) \qquad \text{for } n \geq 0 \tag{1.62}$$

The mean and variance for N are

$$\mu = \frac{b}{a} \tag{1.63}$$

$$\sigma^2 = \frac{b}{a^2} \tag{1.64}$$

Example 18 Assume packets arrive at a certain device with probability a at a given time slot. The probability that a packet does not arrive at a time slot is $b = 1 - a$.

(a) What is the probability that we have to wait for n time slots before a packet arrives?
(b) What is the average number of time slots between packet arrivals?

(a) The probability that we have to wait for n time slots before a packet arrives is

$$p(n) = a\, b^n$$

(b) The average number of time slots between packet arrivals is given by

$$E[n] = \sum_{i=0}^{\infty} n\, p(n) = \sum_{i=0}^{\infty} n\, a\, b^n$$

From Appendix A, the above equation becomes

$$E[n] = \frac{a\,b}{(1-b)^2} = \frac{b}{a}$$

■

1.25 Binomial RV

Consider a chance experiment that has two mutually exclusive outcomes A and $\overline{A}$ that occur with probabilities a and b, respectively. We define the discrete random variable K which equals the number of times that A occurs in any order in N trials or repetitions of the random experiment. We can write the pmf for this distribution as

$$p(K = k) = \binom{N}{k} a^k\, b^{N-k} \qquad \text{for } 0 \le k \le N \qquad (1.65)$$

where $b = 1 - a$. Basically, the probability of event A occurring k times during N trials and not occurring for $N - k$ times is $a^k\, b^{N-k}$, and the number of ways of this event taking place is the binomial coefficient $C(N, k)$.

Alternatively, $p(k)$ can be expressed by a single equation in the form

$$p(k) = \sum_{i=0}^{N} \binom{N}{i} a^i\, b^{N-i}\, \delta\,(k-i) \qquad \text{for } 0 \le k \le N \qquad (1.66)$$

The mean and variance for K are

$$\mu = N\,a \qquad (1.67)$$

$$\sigma^2 = N\,a\,b \qquad (1.68)$$

The binomial distribution is sometimes referred to as sampling with replacement as, for example, when we have N objects and each one has a certain property (color, for example). We pick an object, inspect its property, then place it back in the population. If the selected object is removed from the population after inspection, then we would have the case of the *hypergeometric distribution*, which is also known as sampling without replacement, but this is outside our present interest.

Example 19 Assume a classroom has n students. What are the chances that m students have a birthday today?

The probability that a student has a birthday on a given day is $a = 1/365$ and the probability that a student does not have a birthday on a given day is $b = 1 - a = 364/365$. The probability that m students have a birthday today is

$$p(m) = \binom{n}{m} a^m\, b^{n-m}$$

For example, when $n = 20$ and $m = 2$, we get

$$p(2) = \binom{20}{2} a^2\, b^{18} = 1.35 \times 10^{-3} \tag{1.69}$$

■

1.25.1 Approximating the Binomial Distribution

We saw in the binomial distribution that $p(k)$ is given by

$$p(k) = \binom{N}{k} a^k\, b^{N-k} \tag{1.70}$$

where $b = 1 - a$. When N is large, it becomes cumbersome to evaluate the above equation. Several approximations exist for evaluating the binomial distribution using simpler expressions. We discuss two techniques in the following two subsections.

DeMoivre-Laplace Theorem

We can approximate $p(k)$ by the Gaussian distribution provided that the following condition is satisfied [2].

$$n\, a\, b \gg 1 \tag{1.71}$$

When this condition is true we can write

$$p(k) \approx \frac{1}{\sigma\sqrt{\pi}} e^{-(k-\mu)^2/\sigma^2} \tag{1.72}$$

where we have to choose the two parameters μ and σ according to the following two equations.

$$\mu = N\, a \tag{1.73}$$

$$\sigma = \sqrt{2N\, a\, b} \tag{1.74}$$

The above approximation is known as the DeMoivre-Laplace Theorem and is satisfactory even for moderate values of n such as when $n \approx 5$.

Poisson Theorem

If Na is of the order of one (i.e., $Na \approx 1$), then DeMoivre-Laplace approximation is no longer valid. We can still obtain a relatively simple expressions for the binomial distribution if the following condition applies [2].

$$n\, a \approx 1 \tag{1.75}$$

When this condition is true we can write

$$p(k) \approx \frac{(n\,a)^k}{k!} e^{-n\,a} \tag{1.76}$$

Thus we are able to replace the binomial distribution with the Poisson distribution which is discussed in Section 1.26.

Example 20 A file is being downloaded from a remote site and 500 packets are required to transmit the entire file. It has been estimated that on the average 5 percent of received packets through the channel are in error. Determine the probability that 10 received packets are in error using the binomial distribution and its approximations using DeMoivre-Laplace and Poisson approximations.

The parameters for our binomial distribution are

$$\begin{aligned} N &= 500 \\ a &= 0.05 \\ b &= 0.95 \end{aligned}$$

The probability that 5 packets are in error is

$$p(10) = \binom{500}{10} (0.05)^{10}\,(0.95)^{490} = 2.9165 \times 10^{-4}$$

Use the DeMoivre-Laplace Theorem with the parameters

$$\begin{aligned} \mu &= N\,a = 25 \\ \sigma &= \sqrt{N\,a\,b} = 6.892 \end{aligned}$$

In that case, the required probability is

$$p(10) = 7.1762 \times 10^{-4}$$

Use the Poisson Theorem which results in the probability

$$p(10) = 3.6498 \times 10^{-4}$$

Under the above parameters, the Poisson approximation gives better results than the DeMoivre-Laplace approximation. ■

1.26 Poisson RV

The Poisson distribution defines the probability $p(k)$ that an event A occurs k times during a certain interval. The probability is given by

$$p(K = k) = \frac{a^k \, e^{-a}}{k!} \tag{1.77}$$

where $a > 0$ and $k = 0, 1, \ldots$.

The parameter a in the above formula is usually expressed as

$$a = \lambda \, t$$

where λ is the rate of event A and t is usually thought of as time. Because we talk about rates, we usually associate Poisson distribution with time or with average number of occurrences of an event. So let us derive the expression for Poisson distribution based on this method of thinking.

Consider a chance experiment where an event A occurs at a rate λ events/second. In a small time interval (Δt), the probability that the event A occurs is $p = \lambda \Delta t$. We chose Δt so small so that event A occurs at most only once. For example, λ might indicate the average number of packets arriving at a link per unit time. In that case, the variable t will indicate time. λ might also refer to the average number of bits in error for every 1000 bits, say. In that case, the variable t would indicate the number of bits under consideration. In these situations, we express the parameter a in the form $a = \lambda t$, where λ expresses the rate of some event and t expresses the size or the period under consideration.

Typically, the Poisson distribution is concerned with the probability that a specified *number* of occurrences of event A take place in a specified interval t. By assuming a discrete random variable K that takes the values $0, 1, 2, \ldots$, then the probability that k events occur in a time t is given by

$$p(k) = \frac{(\lambda t)^k \, e^{-\lambda t}}{k!} \tag{1.78}$$

Alternatively, $p(k)$ can be expressed by a single equation in the form

$$p(k) = \sum_{i \geq 0} \frac{(\lambda \, t)^i e^{-\lambda \, t}}{i!} \delta(k - i) \qquad \text{for } k \geq 0 \tag{1.79}$$

The mean and variance for K are

$$\mu = \lambda \, t \tag{1.80}$$

$$\sigma^2 = \lambda \, t \tag{1.81}$$

Deriving Poisson Distribution from Binomial Distribution

We saw in the binomial distribution that $p(k)$ is given by

$$p(k) = \binom{N}{k} a^k \, b^{N-k} \tag{1.82}$$

where $b = 1 - a$. When N is large, it becomes cumbersome to evaluate the above equation. We can easily evaluate $p(k)$ in the special case when the number of trials N becomes very large and a becomes very small such that

$$N\, a = \mu \tag{1.83}$$

where μ is non-zero and finite. This gives rise to Poisson distribution

$$p(k) = \lim_{N\to\infty, a\to 0} \binom{N}{k} a^k \, b^{N-k} \approx \frac{(\mu)^k \, e^{-\mu}}{k!} \tag{1.84}$$

We can prove the above equation using the following two simplifying expressions. We start by using Stirling's formula

$$N! \approx \sqrt{2\pi}\, N^N e^{-N} \tag{1.85}$$

and the following limit from calculus

$$\lim_{N\to\infty} (1 - \frac{a}{N})^N = e^{-a} \tag{1.86}$$

Using the above two expressions, we can write [10]

$$p(k) = \binom{N}{k} a^k \, b^{N-k} \tag{1.87}$$

$$= \frac{N!}{k!(N-k)!} a^k \, (1-b)^{N-k} \tag{1.88}$$

$$= \frac{N^N e^{-N}}{k!\,(N-k)^{N-k}\, e^{-(N-k)}} \left(\frac{\mu}{N}\right)^k \left(1 - \frac{\mu}{N}\right)^{N-k} \tag{1.89}$$

$$= \frac{\mu^k}{k!} \left(\frac{N}{N-k}\right)^{N-k} \left(1 - \frac{\mu}{N}\right)^{N-k} \tag{1.90}$$

$$\approx \frac{\mu^k}{k!} e^{-\mu} \tag{1.91}$$

where $\mu = N\, a$. This gives the expression for the Poisson distribution. This is especially true when $N > 50$ and $a < 0.1$ in the binomial distribution.

Example 21 Packets arrive at a device at an average rate of 500 packets per second. Determine the probability that four packets arrive during 3 ms.

We have $\lambda = 500$, $t = 3 \times 10^{-3}$ and $k = 4$

$$p(4) = \frac{(1.5)^4 \, e^{-1.5}}{4!} = 4.7 \times 10^{-2}$$

■

1.27 Systems with Many Random Variables

We reviewed in Section 1.11 the concept of a random variable where the outcome of a random experiment is mapped to a single number. Here, we consider random experiments whose output is mapped to two or more numbers. Many systems based on random phenomena are best studied using the concept of multiple random variables. For example, signals coming from a Quadrature Amplitude Modulation (QAM) system are described by the equation

$$v(t) = a \cos(\omega t + \phi) \tag{1.92}$$

Incoming digital signals are modulated by assigning different values to a and ϕ. In that sense, QAM modulation combines amplitude and phase modulation techniques. The above signal contains two pieces of information: viz, the amplitude a and the phase ϕ that correspond to two random numbers A and Φ. So every time we sample the signal $v(t)$, we have to find two values for the associated random variables A and Φ.

Figure 1.15 graphically shows the sequence of events leading to assigning multiple numerical values to the outcome of a random experiment. First we run the experiment then we observe the resulting outcome. Each outcome is assigned multiple numerical values.

As an example, we could monitor the random events of packet arrival at the input of a switch. Several outcomes of this random event could be observed such as (1) packet length; (2) packet type (voice, video, data, etc.); (3) interarrival time—i.e. the time interval between arriving packets; and (4) destination address. In these situations, we might want to study the relationships between these random variables in order to understand the underlying characteristics of the random experiment we are studying.

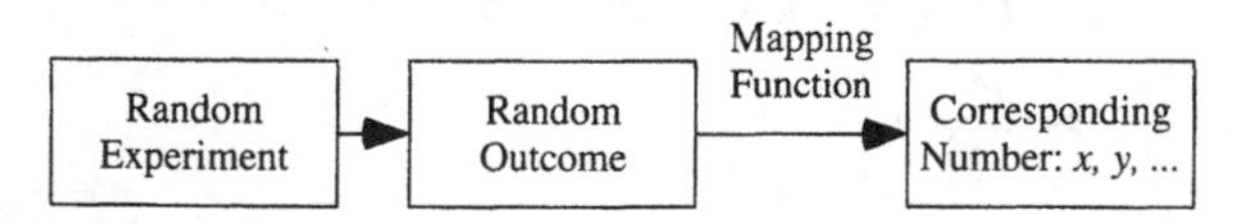

Fig. 1.15 The sequence of events leading to assigning multiple numerical values to the outcome of a random experiment

1.28 Joint cdf and pdf

Assume our random experiment gives rise to two discrete random variables X and Y. We define the joint cdf of X and Y as

$$F_{XY}(x, y) = p(X \leq x, Y \leq y) \quad (1.93)$$

When the two random variables are independent, we can write

$$p(X \leq x, Y \leq y) = p(X \leq x)\; p(Y \leq y) \quad (1.94)$$

Thus for independent random variables the joint cdf is simply the product of the individual cdf's.

$$F_{XY}\,(x, y) = F_X\,(x)\;F_Y\,(y) \quad (1.95)$$

For continuous RVs, the joint pdf is defined as

$$f_{XY}(x, y) = \frac{\partial^2 F_{XY}}{\partial x\; \partial y} \quad (1.96)$$

The joint pdf satisfies the following relation.

$$\int_{x,y} f_{XY}(x, y)\; dx\; dy = 1 \quad (1.97)$$

The two random variables are *independent* when the joint pdf can be expressed as the product of the individual pdf's.

$$f_{XY}(x, y) = f_X(x)\; f_Y(y) \quad (1.98)$$

For discrete RVs, the joint pmf is defined as the probability that $X = x$ and $Y = y$

$$p_{XY}(x, y) = p(X = x,\;\; Y = y) \quad (1.99)$$

The joint pmf satisfies the following relation.

$$\sum_x \sum_y p_{XY}(x, y) = 1 \quad (1.100)$$

Two random variables are *independent* when the joint pmf can be expressed as the product of the individual pmf's.

$$p_{XY}(x, y) = p(X = x)\; p(Y = y) \quad (1.101)$$

Example 22 Assume arriving packets are classified according to two properties: packet length (short or long) and packet type (voice or data). Random variable $X = 0, 1$ is used to describe packet length and random variable $Y = 0, 1$ is used to describe packet type. The probability that the received packet is short is 0.9 and probability that it is long is 0.1. When a packet is short, the probability that it is a voice packet is 0.4 and probability that it is a data packet is 0.6. When a packet is long, the probability that it is a voice packet is 0.05 and probability that it is a data packet is 0.95. Find the joint pmf of X and Y.

We have the mapping:

$$X = \begin{cases} 0 & \text{short packet} \\ 1 & \text{long packet} \end{cases}$$

$$Y = \begin{cases} 0 & \text{voice packet} \\ 1 & \text{data packet} \end{cases}$$

Based on the above mapping, and assuming independent RVs, we get

$$\begin{aligned} p_{XY}(0,0) &= 0.9 \times 0.4 &= 0.36 \\ p_{XY}(0,1) &= 0.9 \times 0.6 &= 0.54 \\ p_{XY}(1,0) &= 0.1 \times 0.05 &= 0.005 \\ p_{XY}(1,1) &= 0.1 \times 0.95 &= 0.095 \end{aligned}$$

Note that the sum of all the probabilities must add up to 1. ■

1.29 Individual pmf From a Given Joint pmf

Sometimes we want to study an individual random variable even though our random experiment generates multiple RVs. An individual random variable is described by its pmf which is obtained as

$$p_X(x) = p(X = x) \tag{1.102}$$

If our random experiment generates two RVs X and Y, then we have two individual pmf's given by

$$p_X(x) = \sum_y p_{XY}(x, y) \tag{1.103}$$

$$p_Y(y) = \sum_x p_{XY}(x, y) \tag{1.104}$$

From the definition of pmf we must have

$$\sum_x p_X(x) = 1 \tag{1.105}$$

$$\sum_y p_Y(y) = 1 \tag{1.106}$$

Example 23 Assume a random experiment that generates two random variables X and Y with the given joint pmf.

$p_{XY}(x, y)$	$x = 1$	$x = 2$	$x = 3$
$y = 0$	0.1	0.05	0.1
$y = 3$	0.1	0.05	0
$y = 7$	0	0.1	0.1
$y = 9$	0.2	0	0.2

Find the individual pmf for X and Y. Are X and Y independent RVs?
We have

$p_X(x)$	$x = 1$	$x = 2$	$x = 3$
	0.4	0.2	0.4

and

$p_Y(y)$	$y = 0$	$y = 3$	$y = 7$	$y = 9$
	0.25	0.15	0.2	0.4

■

1.30 Expected Value

The joint pmf helps us find the expected value of one of the random variables.

$$\mu_X = E(X) = \sum_x \sum_y x \; p_{XY}(x, y) \tag{1.107}$$

Example 24 Using values in Example 23, find the expected values for the random variables X and Y.

We can write

$$\mu_X = 1 \times 0.4 + 2 \times 0.2 + 3 \times 0.4 \qquad = 2$$
$$\mu_y = 0 \times 0.25 + 3 \times 0.15 + 7 \times 0.2 + 9 \times 0.4 \qquad = 5.45$$

■

1.31 Correlation

The correlation between two random variables is defined as

$$\begin{aligned} r_{XY} &= E[XY] \\ &= \sum_x \sum_y x\, y\, p_{XY}(x, y) \end{aligned} \tag{1.108}$$

We say that the two random variables X and Y are *orthogonal* when $r_{XY} = 0$.

1.32 Variance

The variance of a random variable is defined as

$$\begin{aligned} \sigma_X^2 &= E[(X - \mu_X)^2] \\ &= \sum_x \sum_y (x - \mu_X)^2\, p_{XY}(x, y) \end{aligned} \tag{1.109}$$

The following equation can be easily proven.

$$\sigma_X^2 = E[X^2] - \mu_X^2 \tag{1.110}$$

1.33 Covariance

The covariance between two random variables is defined as

$$\begin{aligned} c_{XY} &= E[(X - \mu_X)(Y - \mu_Y)] \\ &= \sum_x \sum_y (x - \mu_X)(y - \mu_Y)\, p_{XY}(x, y) \end{aligned} \tag{1.111}$$

The following equation can be easily proven.

$$c_{XY} = r_{XY} - \mu_X\, \mu_Y \tag{1.112}$$

We say that the two random variables X and Y are *uncorrelated* when $c_{XY} = 0$. The *correlation coefficient* ρ_{XY} is defined as

$$\rho_{XY} = c_{XY} / \sqrt{\sigma_X^2\, \sigma_Y^2} \tag{1.113}$$

When we are dealing with two random variables obtained from the same random process, the correlation coefficient would be written as

$$\rho_{X(n)} = c_X(n)/\sigma_X^2 \tag{1.114}$$

We expect that the correlation coefficient would decrease as the value of n becomes large to indicate that the random process "forgets" its past values.

1.34 Transforming Random Variables

Mathematical packages usually have functions for generating random numbers following the uniform and normal distributions only. However, when we are simulating communication systems, we need to generate network traffic that follows other types of distributions. How can we do that? Well, we can do that through transforming random variables, which is the subject of this section. Section 1.35 will show how to actually generate the random numbers using the techniques of this section.

1.34.1 Continuous Case

Suppose we have a random variable X with known pdf and cdf and we have another random variable Y that is a function of X

$$Y = g(X) \tag{1.115}$$

X is named the *source* random variable and Y is named the *target* random variable. We are interested in finding the pdf and cdf of Y when the pdf and cdf of X are known. The probability that X lies in the range x and $x + dx$ is given from (1.28) by

$$p(x \leq X \leq x + dx) = f_X(x)\, dx \tag{1.116}$$

But this probability must equal the probability that Y lies in the range y and $y + dy$. Thus we can write

$$f_Y(y)\, dy = f_X(x)\, dx \tag{1.117}$$

where $f_Y(y)$ is the pdf for the random variable Y and it was assumed that the function g was monotonically increasing with x. If g was monotonically decreasing with x, then we would have

$$f_Y(y)\, dy = -f_X(x)\, dx \tag{1.118}$$

The above two equations define the *fundamental law of probabilities*, which is given by

$$|f_Y(y)\,dy| = |f_X(x)\,dx| \tag{1.119}$$

or

$$f_Y(y) = f_X(x)\left|\frac{dx}{dy}\right| \tag{1.120}$$

since $f_Y(y)$ and $f_X(x)$ are always positive.
In the discrete case the fundamental law of probability gives

$$p_Y(y) = p_X(x) \tag{1.121}$$

where $p_X(x)$ is the given pmf of the source random variable and $p_Y(y)$ is the pmf of the target random variable.

Example 25 Given the random variable X whose pdf has the form

$$f_X(x) = e^{-x^2} \qquad x \geq 0 \tag{1.122}$$

Find the pdf for the random variable Y that is related to X by the relation

$$Y = X^2 \tag{1.123}$$

We have

$$x = \pm\sqrt{y} \tag{1.124}$$

$$\frac{dx}{dy} = \pm\frac{1}{2\sqrt{y}} \tag{1.125}$$

From (1.120) we can write

$$f_Y(y) = \frac{1}{2\sqrt{y}} \times e^{-x^2} \tag{1.126}$$

By substituting the value of x in terms of y, we get

$$f_Y(y) = \frac{1}{2\sqrt{y}}e^{-y} \qquad y \geq 0 \tag{1.127}$$

■

Example 26 Assume the sinusoidal signal

$$x = \cos\omega t$$

where the signal has a random frequency ω that varies uniformly in the range $\omega_1 \le \omega \le \omega_2$. The frequency is represented by the random variable Ω. What are the expected values for the random variables Ω and X?

We can write

$$f_\Omega = 1/(\omega_2 - \omega_1)$$

and $E[\Omega]$ is given by the expression

$$E[\Omega] = \frac{1}{\omega_2 - \omega_1} \int_{\omega_1}^{\omega_2} \omega \, d\omega \tag{1.128}$$

$$= \frac{\omega_2 + \omega_1}{2} \tag{1.129}$$

We need to find f_X and $E[X]$. For that we use the fundamental law of probability. First, we know that $-1 \le x \le 1$ from the definition of the sine function, so $f_X(x) = 0$ for $|x| > 1$. Now we can write

$$f_X(x) = f_\Omega(\omega) \left| \frac{d\omega}{dx} \right| \qquad |x| \le 1 \tag{1.130}$$

Now

$$\omega = \frac{1}{t} \cos^{-1} x \tag{1.131}$$

and

$$\left| \frac{d\omega}{dx} \right| = \frac{1}{t\sqrt{1 - x^2}} \qquad |x| \le 1 \tag{1.132}$$

Thus we get

$$f_X(x) = \frac{1}{\omega_2 - \omega_1} \times \frac{1}{t\sqrt{1 - x^2}} \qquad |x| \le 1 \tag{1.133}$$

The expected value for X is given by

$$E[X] = \frac{1}{(\omega_2 - \omega_1)t} \int_{-1}^{1} \frac{x}{\sqrt{1 - x^2}} \, dx = 0 \tag{1.134}$$

due to the odd symmetry of the function being integrated. ■

Example 27 Suppose our source random variable is uniformly distributed and our target random variable Y is to have a pdf given by

$$f_Y(y) = \lambda e^{-\lambda y} \qquad \lambda > 0 \tag{1.135}$$

Derive Y as a function of X.

This example is fundamentally important since it shows how we can find the *transformation* that allows us to obtain random numbers following a desired pdf given the random numbers for the uniform distribution. This point will be further discussed in the next section.

We use (1.120) to write

$$\lambda e^{-\lambda y} = f_X(x) \left| \frac{dx}{dy} \right| \tag{1.136}$$

Assume the source random variable X is confined to the range 0–1. From Section 1.17 we have $f_X(x) = 1$ and we can write

$$\lambda e^{-\lambda y} = \frac{dx}{dy} \tag{1.137}$$

By integrating (1.137), we get

$$e^{-\lambda y} = x \tag{1.138}$$

and we obtain the dependence of Y on X as

$$Y = g(X) = -\frac{\ln X}{\lambda} \qquad 0 \le x \le 1 \tag{1.139}$$

■

1.34.2 Discrete Case

In the discrete case the fundamental law of probability gives

$$p_Y(y) = p_X(x) \tag{1.140}$$

where $p_X(x)$ is the given pmf of the source random variable and $p_Y(y)$ is the pmf of the target random variable. The values of Y are related to the values of X by the equation

$$Y = g(X) \tag{1.141}$$

The procedure for deriving the pmf for Y given the pmf for X is summarized as follows:

1. For each value of x obtain the corresponding value $p(X = x)$ through observation or modeling.
2. Calculate $y = g(x)$.
3. Associate y with the probability $p(X = x)$.

1.35 Generating Random Numbers

We review briefly in this section how to generate sequences of random numbers obeying one of the distributions discussed in the previous section. This background is useful to know even if packages exist that fulfill our objective.

Before we start, we are reminded of Von Neumann's remark on the topic:

> Anyone who considers arithmetical methods of producing random digits is, of course, in a state of sin. (John Von Neumann 1951)

1.35.1 Uniform Distribution

To generate a sequence of integer random numbers obeying the uniform distribution using C programming, one invokes the function `srand(seed)`. This function returns an integer in the range 0 to `RAND_MAX` [11]. When a continuous random number is desired, `drand` is used. The function `rand` in MATLAB is used to generate random numbers having uniform distribution.

1.35.2 Inversion Method

To generate a number obeying a given distribution, we rely on the fundamental law of probabilities summarized by (1.120). When X has a uniform distribution over the range $0 \leq x \leq 1$, we can use the following procedure for obtaining y. We have

$$\frac{dx}{dy} = f(y) \tag{1.142}$$

where $f(x) = 1$ and $f(y)$ is the function describing the pdf of the target distribution. By integrating (1.142), we get

$$\int_0^y f(z)\, dz = x \tag{1.143}$$

Thus we have

$$F(y) = x \tag{1.144}$$

$$y = F^{-1}(x) \tag{1.145}$$

The procedure then to obtain the random numbers corresponding to y is to generate x according to any standard random number generator producing a uniform distribution. Then use the above equation to provide us with y. Thus the target random number value y is obtained according to the following steps:

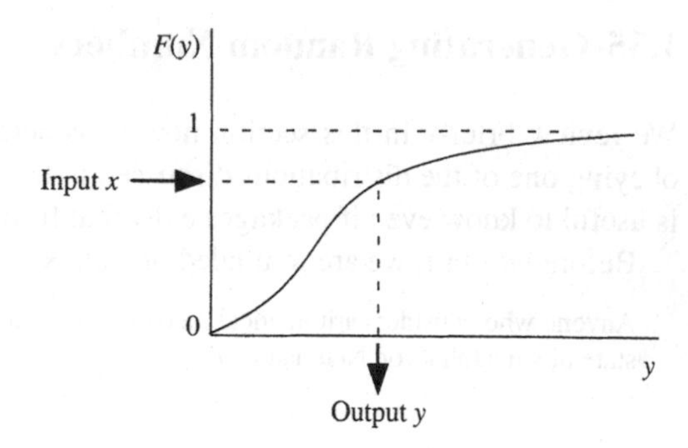

Fig. 1.16 Transformation method for finding y given x

1. Obtain the source random number x having a uniform distribution.
2. Lookup the value of $F(y)$ that is numerically equal to x according to (1.144). $F(y)$ is either computed or stored in a lookup table.
3. Find the corresponding value of y according to (1.145). If the inverse function is not available, then a lookup table is used and y is chosen according to the criterion $F(y) \leq x \leq F(y + \epsilon)$, where $y + \epsilon$ denotes the next entry in the table.

Graphically, the procedure is summarized in Fig. 1.16.

The inversion technique was used to generate random numbers that obey the Pareto distribution using MATLAB. In that case, when x is the trial source input, the output y is given from the F^{-1} by the equation

$$y = -\frac{a}{\exp\left[(1/b)\ln(1-x)\right]} \tag{1.146}$$

Figure 1.11 shows the details of the method for generating random numbers following the Pareto distribution. The Pareto parameters chosen were $a = 2$, $b = 2.5$, minimum value for data was $x_{\min} = a$. One thousand samples were chosen to generate the data. Note that the minimum value of the data equals a according to the restrictions of the Pareto distribution.

1.35.3 Rejection Method

The previous method requires that the target cdf be known and computable such that the inverse of the function can be determined either analytically or through using a lookup table. The rejection method is more general since it only requires that the target pdf is known and computable. We present here a simplified version of this method.

Assume we want to generate the random numbers y that lie in the range $a \leq y < b$ and follow the target pdf distribution specified by $f(y)$. We choose the uniform distribution $g(y)$ that covers the same range a–b such that the following condition is satisfied for y in the range $a \leq y < b$

$$g(y) = \frac{1}{b-a} > f(y) \tag{1.147}$$

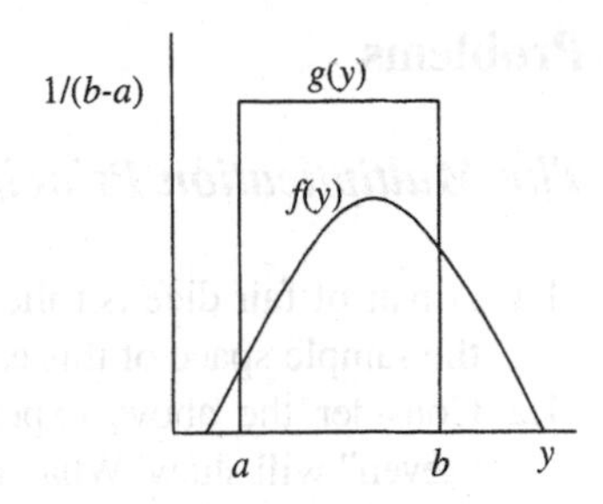

Fig. 1.17 The rejection method

If this condition cannot be satisfied, then a different $g(y)$ must be chosen that might not follow the uniform distribution. We proceed as follows.

1. Obtain the source random number x using any random number generator having the uniform distribution.
2. Accept the candidate value x as the target value $y = x$ with probability

$$p\,(\text{accept}x) = \frac{f(y)}{g(y)} \tag{1.148}$$

$$= (b-a)\, f(y) \tag{1.149}$$

We are assured by (1.147) that the above expression for the probability is always valid. This technique is not efficient when $f(y)$ is mostly small with few large peaks since most of the candidate points will be rejected. Graphical representation of the rejection method is given in Fig. 1.17.

1.35.4 Importance Sampling Method

Importance sampling is a generalization of the rejection method. A trial pdf, $g\,(y)$, is chosen as in the rejection method. It is *not* necessary here to choose $g\,(y)$ such that it is larger than $f(y)$. Each point y has a weight associated with it given by

$$w(y) = \frac{f(y)}{g\,(y)} \tag{1.150}$$

Based on this array of weights, we choose a weight W that is slightly larger than the maximum value of $w(y)$

$$W = \max\,[w(y)] + \epsilon \qquad \epsilon > 0 \tag{1.151}$$

The method is summarized in the following steps.

1. Obtain the source random number x that has a uniform distribution.
2. Apply the inversion method using $g\,(y)$ to obtain a candidate value for y.
3. Accept the candidate y value with probability

$$p\,(\text{accept } y) = \frac{w(y)}{W} \tag{1.152}$$

Problems

The Multiplication Principle

1.1 A pair of fair dice is rolled once. Identify one possible outcome and identify the sample space of this experiment.

1.2 Consider the above experiment where we are interested in the event that "seven" will show. What are the outcomes that constitute the event?

1.3 A student has to take three courses from three different fields. Field A offers 5 courses, field B offers 10 courses, and field C offers 3 courses. In how many ways can the student select the course load?

1.4 A car dealer specializes in selling three types of vehicles: sedans, trucks, and vans. Each vehicle could be rated as excellent, okay, lemon, or "bring your own jumper cables". In how many different ways can a customer buy a vehicle?

1.5 Car license plates in British Columbia consist of three letters followed by three numbers. How many different license plate numbers could be formed?

Permutations

1.6 Internet packets have different lengths. Assume 10 packets have been received such that two are over-length, four have medium length, and four are short. How many different packet patterns are possible?

1.7 In Time Division Multiple Access (TDMA) communication, a time frame is divided into ten time slots such that each time slot can be used by any user. Suppose that a frame is received that contains three packets due to user 1, two due to user 2, and the rest of the time slots were empty. How many frame patterns are possible?

1.8 A router receives 15 packets from 15 different sources. How many ways could these packets be received?

1.9 In Problem 1.8, of the 15 packets received, five were due to one user and the rest were due to 10 different users. How many ways could the packets types be received?

Combinations

1.10 A packet buffer has 50 packets. If it is known that 15 of those packets belong to a certain user and the rest belong to different users then in how many possibilities can these packets be stored in the buffer?

1.11 A router receives 10 packets such that four of them are in error. How many different packet patterns are possible?

1.12 In a cellular phone environment, a cell has 100 users that want access and there are only 16 available channels. How many different possibilities exist for choosing among these users?

Probability

1.13 A signal source randomly moves from being active to being idle. The active period lasts 5 seconds and the idle period lasts 10 seconds. What is the probability that the source will be active when it is sampled?

1.14 In a wireless channel a certain user found that for each 1000 packets transmitted 10 were lost, 100 were in error, and 50 were delayed. Find the following.

(a) Probability that a packet is lost.
(b) Probability that a packet is received without delay.
(c) Probability that a packet is received without delay and without errors.
(d) Probability that a packet is received without delay or without errors.

1.15 Assume a gambler plays double or nothing game using a fair coin and starts with one dollar. What is the probability that he/she will wind up winning $1,024?

1.16 Four friends decide to play the following game. A bottle is spun and the person that the bottle points to is unceremoniously thrown out of the game. What is the probability that you are still in the game after n spins? Is there an upper limit on the value of n?

1.17 A bird breeder finds that the probability that a chick is male is 0.2 and a female is 0.8. If the nest has three eggs, what is the probability that two male chicks will be produced?

Random Variables

1.18 Indicate the range of values that the random variable X may assume and classify the random variable as finite/infinite and continuous/discrete in the following experiments.

(a) The number of times a coin is thrown until a head appears.
(b) The wait time in minutes until a bus arrives at the bus stop.
(c) The duration of a telephone conversation.
(d) The number of students in a classroom.
(e) The number of retransmissions until a packet is received error free.

1.19 A packet is received either correctly or in error on a certain channel. A random variable X is assigned a value equal to the number of error free packets in a group of three packets. Assume that the error in a packet occurs independent of the other packets.

(a) List all the possible outcomes for the reception of three packets.
(b) List all the possible values of X.
(c) Express the probability for the event $x = 2$ in terms of $f_X(x)$ and $F_X(x)$.

1.20 In some communication scheme, when a packet is received in error, a request for retransmission is issued until the packet is received error free. Let the random variable Y denote the number of retransmission requests. What are the values of Y?

1.21 Packets arrive at a certain input randomly at each time step (a time step is defined here as the time required to transmit or receive one complete packet). Let the random variable W denote the wait time (in units of time steps) until a packet is received. What values may W assume?

The Cumulative Distribution Function (cdf)

1.22 Assume a random variable X whose cdf is $F(x)$. Express the probability $p(X > x)$ as a function of $F(x)$.

1.23 A system monitors the times between packet arrivals, starting at time $t = 0$. This time is called the *interarrival time* of packets. The interarrival time is a random variable T with cdf $F_T(t)$. The probability that the system receives a packet in the time interval $(t, t + \delta t)$ is given by $p\delta t$.

(a) Find the probability that the system receives a packet in a time less than or equal to t.
(b) Find the probability that the system receives a packet in a time greater than t.

1.24 Plot the cdf for the random variable in Problem 1.19.

1.25 Explain the meaning of equations (1.15) to (1.19) for the cdf function. Note that (1.19) is really a restatement of (1.7) since the events $X \leq x_1$ and $x_1 \leq X \leq x_2$ are mutually exclusive.

1.26 A buffer contains ten packets. Four packets contain an error in their payload and six are error free. Three packets are picked at random for processing. Let the random variable X denote the number of error-free packets selected.

(a) List all possible outcomes of the experiment.
(b) Find the value of X for each outcome.
(c) Find the probability associated with each value of X.
(d) Plot the cdf for this random variable.

Note that this problem deals with sampling without replacement: i.e. we pick a packet but do not put it back in the buffer. Hence the probability of picking an error-free packet will vary depending on whether it was picked first, second, or third.

1.27 Sketch the pdf associated with the random variable in Problem 1.26.

1.28 A router has ten input ports and at a given time instance each input could receive a packet with probability a or it could be idle (with probability $b = 1 - a$). Let the random variable X denote the number of active input ports.

(a) List all possible outcomes of the experiment.
(b) Find the value of X for each outcome.
(c) Find the probability associated with each value of X.
(d) Sketch the cdf for this random variable.

1.29 Sketch the pdf associated with the random variable in Problem 1.28.

1.30 A traffic source generates a packet at a certain time step with probability a. Let the random variable X denote the number of packets produced in a period $T = 5$ time steps.

(a) List all possible outcomes of the experiment.
(b) Find the value of X for each outcome.
(c) Find the probability associated with each value of X.
(d) Sketch the cdf for this random variable assuming $a = 0.2$.

The Probability Density Function (pdf)

1.31 Sketch the pdf and cdf for the binomial distribution. Assume $a = 0.3$ and $N = 5$.

1.32 Sketch the pdf associated with the random variable in Problem 1.19.

1.33 Sketch the pdf associated with the random variable in Problem 1.23.

1.34 Sketch the pdf associated with the random variable in Problem 1.24.

1.35 Sketch the pdf associated with the random variable in Problem 1.26.

Expected Value

1.36 Prove that (1.34) on page 15 converges to (1.33) as $n \to \infty$. Start your analysis by (a) assuming that n samples are grouped such that n_j samples produce the same outcome. (b) Use the definition of probability in (1.6) on page 8 to complete your proof.

1.37 What are the expected values for the random variables Θ and Q in Example 14?

1.38 Assume a Poisson process, rate parameter is λ, that gives rise to two random variables X_1 and X_2 which correspond to k packets received at times t_1 and t_2, respectively where $t_2 \geq t_1$. (a) Find the mean and variance for these two random variables. (b) Now define a new random variable $Y = X_2 - X_1$ and find its mean and variance.

1.39 Find the expected value for the random variable in Problem 1.19.

1.40 Find the expected value for the random variable in Problem 1.23.

1.41 Find the expected value for the random variable in Problem 1.24.

1.42 Find the expected value for the random variable in Problem 1.26.

1.43 Repeat Example 16 when the random variable E is treated as a discrete random variable.

The Uniform Distribution

1.44 Write down the pdf and cdf for the uniform distribution of a continuous random variable X that spans the range $a \leq x < b$.

1.45 Repeat the above problem when the random variable is discrete and has n discrete values in the same range given above.

1.46 Find the average value and variance for the random variable in Problem 1.44.

1.47 Find the average value and variance for the random variable in Problem 1.45.

The Binomial Distribution

1.48 Prove that the mean and standard deviation of the binomial distribution are Na and $\sqrt{a(1-a)}$, respectively.

1.49 Sketch the pdf for the binomial distribution. Assume values for a and N.

1.50 The probability of error in a single packet is 10^{-5}. What is the probability that three errors occur in 1,000 packets assuming binomial distribution.

1.51 Assume q as the probability that people making reservations on a certain flight will not show up. The airline then sells t tickets for a flight that takes only s passengers ($t > s$). Write an expression for the probability that there will be a seat available for every passenger that shows up. What is that probability for the special case when only one seat is oversold (i.e., $t = s + 1$)?

The Poisson Distribution

1.52 Verify that the mean and standard deviation for the binomial and Poisson distributions become almost identical for large N and small p as was discussed in Section 1.26.

1.53 The probability of a defective electronic component is 0.1. Find the following.

(a) The mean and standard deviation for the distribution of defective components in a batch of 500 components using the Poisson and binomial distributions.

(b) The probability that 2 components are defective in a batch of 500 components.

1.54 Sketch the pdf for the Poisson distribution for different values of λt and k. Comment on your results.

1.55 Sketch on one graph the binomial and Poisson distributions for the case $N = 5$ and $p = 0.1$. Assume $\lambda t = 0.5$.

1.56 Repeat question 1.50 assuming Poisson distribution with $\lambda t = 0.1$, where we assumed the "rate" λ of occurrence of error per packet is 10^{-4} and the "duration" of interest is $t = 1000$ packets.

1.57 Five percent of the rented videos are worth watching. Find the probability that in a sample of 10 videos chosen at random, exactly two will be worth watching using

(a) binomial distribution and
(b) Poisson distribution.

The Exponential Distribution

1.58 Derive the cdf for the exponential distribution.

1.59 Prove that the Pareto distribution formula given by (1.49) is a valid pdf (i.e., the area under the curve is 1).

Joint pmf

1.60 Consider the random experiment of throwing a dart at a target. The point of impact is characterized by two random variables X and Y to indicate the location of the dart assuming the center is the point of origin. We can justifiably assume that X and Y are statistically independent.

(a) Write down the joint cdf $F_{XY}(x, y)$ as a function of the individual cdfs $F_X(x)$ and $F_Y(y)$.
(b) Write down the joint pdf $f_{XY}(x, y)$ as a function of the individual pdfs $f_X(x)$ and $f_Y(y)$.
(c) Write down an expression for $f_X(x)$ and $f_Y(y)$ assuming that each follows the normal (Gaussian) distribution.

1.61 Find the correlation between the two random variables X and Y in Example 22 on page 32.

1.62 Find the variance of random variable X in Example 22 on page 32.

1.63 Find the covariance between the two random variables X and Y in Example 22 on page 32.

1.64 Prove (1.110) on page 34.

1.65 Find the correlation coefficient for the two random variables X and Y in Example 22 on page 32.

Random Numbers

1.66 Use the inversion method to generate y in the range 1–5 such that the target pdf is $f(y) \propto 1/\sqrt{y}$.

1.67 Generate the random number y that has pdf $f(y) = (1 + y)/\sqrt{y}$ using the importance sampling method.

References

1. P.Z. Peebles, *Probability, Random Variables, and Random Signal Principles*, McGraw–Hill, New York, 1993.
2. A. Papoulis, *Probability, Random Variables, and Stochastic Processes*, McGraw-Hill, New York, 1984.
3. G.R. Cooper and C.D. McGillem, *Probability Methods of Signal and System Analysis*, Oxford University Press, New York, 1999.
4. "statistics" Encyclopedia Britannica Online. http://search.eb.com/bol/topic?eu=115242&sctn=8
5. R.D. Yates and D.J. Goodman, *Probability and Stochastic Processes*, John Wiley, New York, 1999.
6. A. Erramilli, G. Gordon, and W. Willinger, "Applications of fractals in engineering for realistic traffic processes", *International Teletraffic Conference*, Vol. 14, pp. 35–44, 1994.
7. W. Leland, M. Taqqu, W. Willinger, and D. Wilson, "On the self-similar nature of Ethernet traffic", *IEEE/ACM Transactions on Networking*, vol. 2, pp. 1-15, 1994.
8. W. Willinger, M. Taqqu, R. Sherman, and D. Wilson, "Self-similarity through high variability statistical analysis of Ethernet LAN traffic at the source level", *IEEE/ACM Transactions on Networking*, vol. 2, 1997.
9. T. Hettmansperger and M. Keenan,"Tailweight, statistical interference and families of distributions – A brief survey" in *Statistical Distributions in Scientific Work*, vol. 1, G.P. Patil et al. ed., Kluwer, Boston, 1980.
10. R. Syski, *Random Processes: A First Look*, Marcel Dekker, New York, 1979.
11. W.H. Press, S.T. Teukolsky, W.T. Vetterling, and B.P. Fhannery, *Numerical Recipes in C: The Art of Scientific Computing*, Cambridge University Press, Cambridge, 1992.

Chapter 2
Random Processes

2.1 Introduction

We saw in Section 1.11 on page 10 that many systems are best studied using the concept of random variables where the outcome of a random experiment was associated with some numerical value. Next, we saw in Section 1.27 on page 30 that many more systems are best studied using the the concept of multiple random variables where the outcome of a random experiment was associated with multiple numerical values. Here we study *random processes* where the outcome of a random experiment is associated with a *function of time* [1]. Random processes are also called *stochastic processes*. For example, we might study the output of a digital filter being fed by some random signal. In that case, the filter output is described by observing the output waveform at random times.

Thus a *random process* assigns a random *function of time* as the outcome of a random experiment. Figure 2.1 graphically shows the sequence of events leading to assigning a function of time to the outcome of a random experiment. First we run the experiment, then we observe the resulting outcome. Each outcome is associated with a time function $x(t)$.

A random process $X(t)$ is described by

- the *sample space* S which includes all possible outcomes s of a random experiment
- the *sample function* $x(t)$ which is the time function associated with an outcome s. The values of the sample function could be discrete or continuous
- the *ensemble* which is the set of all possible time functions produced by the random experiment
- the time parameter t which could be continuous or discrete
- the statistical dependencies among the random processes $X(t)$ when t is changed.

Based on the above descriptions, we could have four different types of random processes:

1. Discrete time, discrete value: We measure time at discrete values $t = nT$ with $n = 0, 1, 2, \ldots$. As an example, at each value of n we could observe the number of cars on the road $x(n)$. In that case, $x(n)$ is an integer between 0 and 10, say.

F. Gebali, *Analysis of Computer and Communication Networks*,
DOI: 10.1007/978-0-387-74437-7_2, © Springer Science+Business Media, LLC 2008

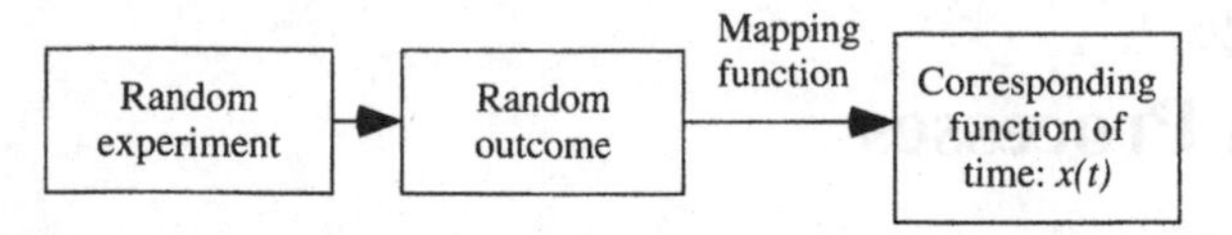

Fig. 2.1 The sequence of events leading to assigning a time function $x(t)$ to the outcome of a random experiment

Each time we perform this experiment, we would get a totally different sequence for $x(n)$.

2. Discrete time, continuous value: We measure time at discrete values $t = nT$ with $n = 0, 1, 2, \ldots$. As an example, at each value of n we measure the outside temperature $x(n)$. In that case, $x(n)$ is a real number between $-30°$ and $+45°$, say. Each time we perform this experiment, we would get a totally different sequence for $x(n)$.
3. Continuous time, discrete value: We measure time as a continuous variable t. As an example, at each value of t we store an 8-bit digitized version of a recorded voice waveform $x(t)$. In that case, $x(t)$ is a binary number between 0 and 255, say. Each time we perform this experiment, we would get a totally different sequence for $x(t)$.
4. Continuous time, continuous value: We measure time as a continuous variable t. As an example, at each value of t we record a voice waveform $x(t)$. In that case, $x(t)$ is a real number between 0 V and 5 V, say. Each time we perform this experiment, we would get a totally different sequence for $x(t)$.

Figure 2.2 shows a discrete time, discrete value random process for an observation of 10 samples where only three random functions are generated. We find that for $n = 2$, the values of the functions correspond to the random variable $X(2)$.

Therefore, random processes give rise to random variables when the time value t or n is fixed. This is equivalent to sampling all the random functions at the specified time value, which is equivalent to taking a vertical slice from all the functions shown in Fig. 2.2.

Example 1 A time function is generated by throwing a die in three consecutive throws and observing the number on the top face after each throw. Classify this random process and estimate how many sample functions are possible.

This is a discrete time, discrete value process. Each sample function will be have three samples and each sample value will be from the set of integers 1 to 6. For example, one sample function might be 4, 2, 5. Using the multiplication principle for probability, the total number of possible outputs is $6^3 = 216$. ■

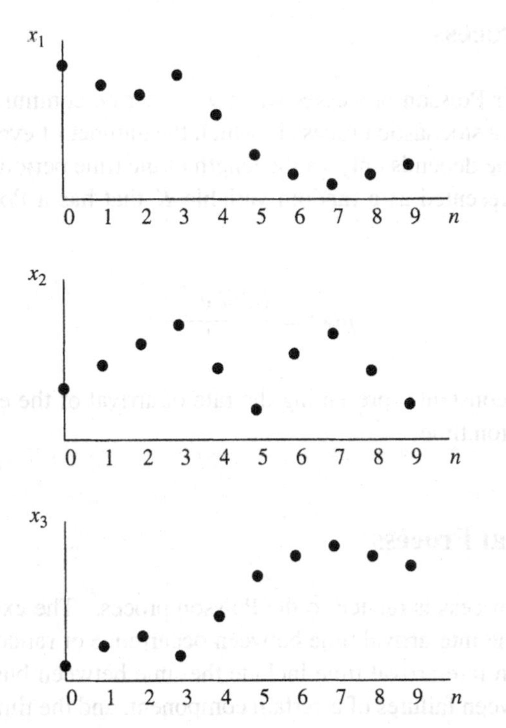

Fig. 2.2 An example of a discrete time, discrete value random process for an observation of 10 samples where only three random functions are possible.

2.2 Notation

We use the notation $X(t)$ to denote a continuous-time random process and also to denote the random variable measured at time t. When $X(t)$ is continuous, it will have a pdf $f_X(x)$ such that the probability that $x \leq X \leq x + \varepsilon$ is given by

$$p(X = x) = f_X(x)\,dx \tag{2.1}$$

When $X(t)$ is discrete, it will have a pmf $p_X(x)$ such that the probability that $X = x$ is given by

$$p(X = x) = p_X(x) \tag{2.2}$$

Likewise, we use the notation $X(n)$ to denote a discrete-time random process and also to denote the random variable measured at time n. That random variable is statistically described by a pdf $f_X(x)$ when it is continuous, or it is described by a pmf $p_X(x)$ when it is discrete.

2.3 Poisson Process

We shall encounter Poisson processes when we describe communication traffic. A Poisson process is a stochastic process in which the number of events occurring in a given period of time depends only on the length of the time period [2]. This number of events k is represented as a random variable K that has a Poisson distribution given by

$$p(k) = \frac{(\lambda\, t)^k e^{-\lambda\, t}}{k!} \tag{2.3}$$

where $\lambda > 0$ is a constant representing the rate of arrival of the events and t is the length of observation time.

2.4 Exponential Process

The exponential process is related to the Poisson process. The exponential process is used to model the interarrival time between occurrence of random events. Examples that lead to an interarrival time include the time between bus arrivals at a bus stop, the time between failures of a certain component, and the time between packet arrival at the input of a router.

The random variable T could be used to describe the interarrival time. The probability that the interarrival time lies in the range $t \leq T \leq t + dt$ is given by

$$\lambda e^{-\lambda\, t}\, dt \tag{2.4}$$

where λ is the average rate of the event under consideration.

2.5 Deterministic and Nondeterministic Processes

A *deterministic* process is one where future values of the sample function are known if the present value is known. An example of a deterministic process is the modulation technique known as quadrature amplitude modulation (QAM) for transmitting groups of binary data. The transmitted analog waveform is given by

$$v(t) = a \cos(\omega t + \phi) \tag{2.5}$$

where the signal amplitude a and phase angle ϕ change their value depending on the bit pattern that has been received. The analog signal is transmitted for the time period $0 \leq t < T_0$. Since the arriving bit pattern is random, the values of the corresponding two parameters a and ϕ are random. However, once a and ϕ are determined, we would be able to predict the shape of the resulting waveform.

A *nondeterministic* random process is one where future values of the sample function cannot be known if the present value is known. An example of a nondeterministic random process is counting the number of packets that arrive at the input of a switch every one second and this observation is repeated for a certain time. We would not be able to predict the pattern even if we know the present number of arriving packets.

2.6 Ensemble Average

The random variable $X(n_1)$ represents all the possible values x obtained when time is frozen at the value n_1. In a sense, we are sampling the ensemble of random functions at this time value.

The expected value of $X(n_1)$ is called the *ensemble average* or statistical average $\mu(n_1)$ of the random process at n_1. The ensemble average is expressed as

$$\mu_X(t) = E[X(t)] \qquad \text{continuous-time process} \tag{2.6}$$

$$\mu_X(n) = E[X(n)] \qquad \text{discrete-time process} \tag{2.7}$$

The ensemble average could itself be another random variable since its value could change at random with our choice of the time value t or n.

Example 2 The modulation scheme known as frequency-shift keying (FSK) can be modeled as a random process described by

$$X(t) = a \cos \omega t$$

where a is a constant and ω corresponds to the random variable Ω that can have one of two possible values ω_1 and ω_2 that correspond to the input bit being 0 or 1, respectively. Assuming that the two frequencies are equally likely, find the expected value $\mu(t)$ of this process.

Our random variable Ω is discrete with probability 0.5 when $\Omega = \omega_1$ or $\Omega = \omega_2$. The expected value for $X(t)$ is given by

$$\begin{aligned} E[X(t)] &= 0.5\, a\, \cos \omega_1 t + 0.5\, a\, \cos \omega_2 t \\ &= a\, \cos\left[\frac{(\omega_1 + \omega_2)\,t}{2}\right] \times \cos\left[\frac{(\omega_1 - \omega_2)\,t}{2}\right] \end{aligned}$$

■

Example 3 The modulation scheme known as pulse amplitude modulation (PAM) can be modeled as a random process described by

$$X(n) = \sum_{i=0}^{\infty} g(n)\, \delta(n - i)$$

where $g(n)$ is the amplitude of the input signal at time n. $g(n)$ corresponds to the random variable G that is uniformly distributed in the range 0–A. Find the expected value $\mu(t)$ of this process.

This is a discrete time, continuous value random process. Our random variable G is continuous and the expected value for $X(n)$ is given by

$$E[X(n)] = \frac{1}{A} \int_0^A g\, dg$$
$$= \frac{A}{2}$$

■

2.7 Time Average

Figure 2.2 helps us find the *time average* of the random process. The time average is obtained by finding the average value for *one* sample function such as $X_1(n)$ in the figure. The time average is expressed as

$$\overline{X} = \tfrac{1}{T} \int_0^T X(t)\, dt \qquad \text{continuous-time process} \qquad (2.8)$$

$$\overline{X} = \tfrac{1}{N} \sum_0^{N-1} X(n)] \qquad \text{discrete-time process} \qquad (2.9)$$

In either case we assumed we sampled the function for a period T or we observed N samples. The time average $\overline{X}$ could itself be a random variable since its value could change with our choice of the random function under consideration.

2.8 Autocorrelation Function

Assume a discrete-time random process $X(n)$ which produces two random variables $X_1 = X(n_1)$ and $X_2 = X(n_2)$ at times n_1 and n_2 respectively. The *autocorrelation function* for these two random variables is defined by the following equation:

$$r_{XX}(n_1, n_2) = E[X_1\, X_2] \qquad (2.10)$$

In other words, we consider the two random variables X_1 and X_2 obtained from the same random process at the two different time instances n_1 and n_2.

Example 4 Find the autocorrelation function for a second-order finite-impulse response (FIR) digital filter, sometimes called moving average (MA) filter, whose output is given by the equation

$$y(n) = a_0 x(n) + a_1 x(n-1) + a_2 x(n-2) \tag{2.11}$$

where the input samples $x(n)$ are assumed to be zero mean independent and identically distributed (iid) random variables.

We assign the random variable Y_n to correspond to output sample $y(n)$ and X_n to correspond to input sample $x(n)$. Thus we can have the following autocorrelation function

$$r_{YY}(0) = E\left[Y_n Y_n\right] = a_0^2 E\left[X_0^2\right] + a_1^2 E\left[X_1^2\right] + a_2^2 E\left[X_2^2\right] E\left[X_0^2\right] \tag{2.12}$$

$$= \left(a_0^2 + a_1^2 + a_2^2\right) \sigma^2 \tag{2.13}$$

Similarly, we can write

$$r_{YY}(1) = E\left(Y_n Y_{n+1}\right) = 2a_0 a_1\, \sigma^2 \tag{2.14}$$

$$r_{YY}\,(2) = E\left(Y_n Y_{n+2}\right) = a_0 a_2\, \sigma^2 \tag{2.15}$$

$$r_{YY}\,(k) = 0; \qquad k > 2 \tag{2.16}$$

where σ^2 is the input sample variance. Figure 2.3 shows a plot of the autocorrelation assuming all the filter coefficients are equal. ■

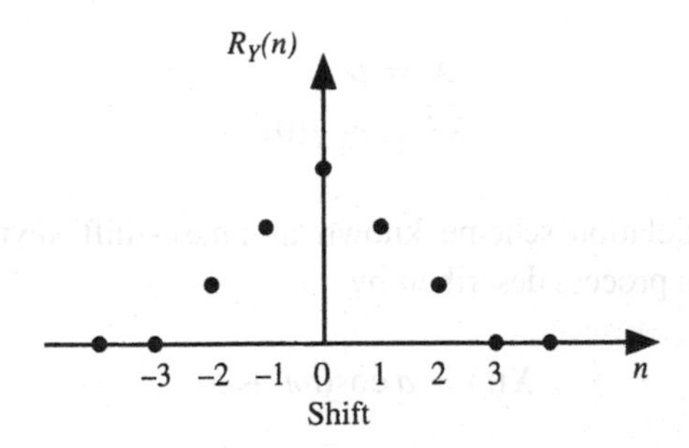

Fig. 2.3 Autocorrelation function of a second-order digital filter whose input is uncorrelated samples

2.9 Stationary Processes

A **wide-sense stationary** random process has the following two properties [3]

$$E\left[X(t)\right] = \mu = \text{ constant} \tag{2.17}$$

$$E\left[X(t)\, X(t+\tau)\right] = r_{XX}\left(t, t+\tau\right) = r_{XX}(\tau) \tag{2.18}$$

Such a process has a constant expected value and the autocorrelation function depends on the time difference between the two random variables.

The above equations apply to a continuous time random process. For a discrete-time random process, the equations for a wide-sense stationary random process become

$$E\left[X(n)\right] = \mu = \text{ constant} \tag{2.19}$$

$$E\left[X\left(n_1\right)\; X\left(n_1 + n\right)\right] = r_{XX}\left(n_1, n_1 + n\right) = r_{XX}(n) \tag{2.20}$$

The autocorrelation function for a wide-sense stationary random process exhibits the following properties [1].

$$r_{XX}(0) = E\left[X^2(n)\right] \geq 0 \tag{2.21}$$

$$|r_{XX}(n)| \leq r_{XX}(0) \tag{2.22}$$

$$r_{XX}(-n) = r_{XX}(n) \qquad \text{even symmetry} \tag{2.23}$$

A stationary random process is *ergodic* if all time averages equal their corresponding statistical averages [3]. Thus if $X(n)$ is an ergodic random process, then we could write

$$\overline{X} = \mu \tag{2.24}$$

$$\overline{X^2} = r_{XX}(0) \tag{2.25}$$

Example 5 The modulation scheme known as phase-shift keying (PSK) can be modeled as a random process described by

$$X(t) = a\cos(\omega t + \phi)$$

where a and ω are constant and ϕ corresponds to the random variable Φ with two values 0 and π which are equally likely. Find the autocorrelation function $r_{XX}(t)$ of this process.

The phase pmf is given by

$$p(0) = 0.5$$
$$p(\pi) = 0.5$$

The autocorrelation is found as

$$\begin{aligned} r_{XX}(\tau) &= E\left[a\cos(\omega t + \Phi)\, a\cos(\omega t + \omega\tau + \Phi)\right] \\ &= 0.5\, a^2 \, \cos(\omega\tau)\, E\left[\cos(2\omega t + \omega\tau + 2\Phi)\right] \\ &= 0.5\, a^2 \, \cos(\omega\tau)\, \cos(2\omega t + \omega\tau) \end{aligned}$$

We notice that this process is not wide-sense stationary since the autocorrelation function depends on t. ∎

2.10 Cross-Correlation Function

Assume two discrete-time random processes $X(n)$ and $Y(n)$ which produce two random variables $X_1 = X(n_1)$ and $Y_2 = Y(n_2)$ at times n_1 and n_2, respectively. The *cross-correlation function* is defined by the following equation.

$$r_{XY}(n_1, n_2) = E\left[X_1\, Y_2\right] \tag{2.26}$$

If the cross-correlation function is zero, i.e. $r_{XY} = 0$, then we say that the two processes are *orthogonal*. If the two processes are *statistically independent*, then we have

$$r_{XY}(n_1, n_2) = E\left[X(n_1)\right] \times E\left[Y(n_2)\right] \tag{2.27}$$

Example 6 Find the cross-correlation function for the two random processes

$$X(t) = a\cos\omega t$$
$$Y(t) = b\sin\omega t$$

where a and b are two independent and identically distributed random variables with mean μ and variance σ^2.

The cross-correlation function is given by

$$\begin{aligned} r_{XY}(t, t+\tau) &= E\left[a\cos\omega t\; b\sin(\omega t + \omega\tau)\right] \\ &= 0.5[\sin\omega\tau + \sin(2\omega t + \omega\tau)]\; E[a]\; E[b] \\ &= 0.5\, \mu^2\, [\sin\omega\tau + \sin(2\omega t + \omega\tau)] \end{aligned}$$

∎

2.11 Covariance Function

Assume a discrete-time random process $X(n)$ which produces two random variables $X_1 = X(n_1)$ and $X_2 = X(n_2)$ at times n_1 and n_2, respectively. The *autocovariance function* is defined by the following equation:

$$c_{XX}(n_1, n_2) = E[(X_1 - \mu_1)(X_2 - \mu_2)] \tag{2.28}$$

The autocovariance function is related to the autocorrelation function by the following equation:

$$c_{XX}(n_1, n_2) = r_X(n_1, n_2) - \mu_1 \mu_2 \tag{2.29}$$

For a wide-sense stationary process, the autocovariance function depends on the difference between the time indices $n = n_2 - n_1$.

$$c_{XX}(n) = E[(X_1 - \mu)(X_2 - \mu)] = r_{XX}(n) - \mu^2 \tag{2.30}$$

Example 7 Find the autocovariance function for the random process $X(t)$ given by

$$X(t) = a + b \cos \omega t$$

where ω is a constant and a and b are iid random variables with zero mean and variance σ^2.

We have

$$\begin{aligned} c_{XX} &= E\{(A + B\cos \omega t)[A + B\cos \omega(t+\tau)]\} \\ &= E\left[a^2\right] + E[ab]\,[\cos \omega t + \cos \omega(t+\tau)] + E\left[b^2\right]\cos^2 \omega(t+\tau) \\ &= \sigma^2 + E[a]\,E[b]\,[\cos \omega t + \cos \omega(t+\tau)] + \sigma^2 \cos^2 \omega(t+\tau) \\ &= \sigma^2\left[1 + \cos^2 \omega(t+\tau)\right] \end{aligned}$$

■

The *cross-covariance function* for two random processes $X(n)$ and $Y(n)$ is defined by

$$\begin{aligned} c_{XY}(n) &= E[(X(n_1) - \mu_X)(Y(n_1 + n) - \mu_Y)] \\ &= r_{XY}(n) - \mu_X \mu_Y \end{aligned} \tag{2.31}$$

Two random processes are called *uncorrelated* when their cross-covariance function vanishes.

$$c_{XY}(n) = 0 \tag{2.32}$$

Example 8 Find the cross-covariance function for the two random processes $X(t)$ and $Y(t)$ given by

$$X(t) = a + b \cos \omega t$$
$$Y(t) = a + b \sin \omega t$$

where ω is a constant and a and b are iid random variables with zero mean and variance σ^2.

We have

$$\begin{aligned} c_{XY}(n) &= E\left\{(A + B\cos\omega t)[A + B\sin\omega(t+\tau)]\right\} \\ &= E\left[A^2\right] + E[AB]\ [\cos\omega t + \sin\omega(t+\tau)] + E\left[B^2\right]\cos\omega t\ \sin\omega(t+\tau) \\ &= \sigma^2 + E[A]\ E[B]\ [\cos\omega t + \sin\omega(t+\tau)] + \sigma^2\cos\omega t\ \sin\omega(t+\tau) \\ &= \sigma^2\left[1 + \cos\omega t\ \sin\omega(t+\tau)\right] \end{aligned}$$

■

2.12 Correlation Matrix

Assume we have a discrete-time random process $X(n)$. At each time step i we define the random variable $X_i = X(i)$. If each sample function contains n components, it is convenient to construct a vector representing all these random variables in the form

$$\mathbf{x} = \left[X_1\ X_2\ \cdots\ X_n\right]^t \tag{2.33}$$

Now we would like to study the correlation between each random variable X_i and all the other random variables. This would give us a comprehensive understanding of the random process. The best way to do that is to construct a *correlation matrix*.

We define the $n \times n$ correlation matrix $\mathbf{R}_X$, which gives the correlation between all possible pairs of the random variables as

$$\mathbf{R}_X = E\left[\mathbf{x}\,\mathbf{x}^t\right] = E\begin{bmatrix} X_1X_1 & X_1X_2 & \cdots & X_1X_n \\ X_2X_1 & X_2X_2 & \cdots & X_2X_n \\ \vdots & \vdots & \ddots & \vdots \\ X_nX_1 & X_nX_2 & \cdots & X_nX_n \end{bmatrix} \tag{2.34}$$

We can express $\mathbf{R}_X$ in terms of the individual correlation functions

$$\mathbf{R}_X = \begin{bmatrix} r_{XX}(1,1) & r_{XX}(1,2) & \cdots & r_{XX}(1,n) \\ r_{XX}(1,2) & r_{XX}(2,2) & \cdots & r_{XX}(2,n) \\ \vdots & \vdots & \ddots & \vdots \\ r_{XX}(1,n) & r_{XX}(2,n) & \cdots & r_{XX}(n,n) \end{bmatrix} \tag{2.35}$$

Thus we see that the correlation matrix is symmetric. For a wide-sense stationary process, the correlation functions depend only on the difference in times and we get an even simpler matrix structure:

$$\mathbf{R}_X = \begin{bmatrix} r_{XX}(0) & r_{XX}(1) & \cdots & r_{XX}(n-1) \\ r_{XX}(1) & r_{XX}(0) & \cdots & r_{XX}(n-2) \\ \vdots & \vdots & \ddots & \vdots \\ r_{XX}(n-1) & r_{XX}(n-2) & \cdots & r_{XX}(0) \end{bmatrix} \tag{2.36}$$

Each diagonal in this matrix has identical elements and our correlation matrix becomes a *Toeplitz matrix.*

Example 9 Assume the autocorrelation function for a stationary random process is given by

$$r_{XX}(\tau) = 5 + 3e^{-|\tau|}$$

Find the autocorrelation matrix for $\tau = 0$, 1, and 2.
The autocorrelation matrix is given by

$$\mathbf{R}_{XX} = \begin{bmatrix} 8 & 6.1036 & 5.4060 \\ 6.1036 & 8 & 6.1036 \\ 5.4060 & 6.1036 & 6 \end{bmatrix}$$

■

2.13 Covariance Matrix

In a similar fashion, we can define the *covariance matrix* for many random variables obtained from the same random process as

$$\mathbf{C}_{XX} = E\left[(\mathbf{x} - \overline{\mu})(\mathbf{x} - \overline{\mu})^t\right] \tag{2.37}$$

where $\overline{\mu} = \left[\, \mu_1 \; \mu_2 \; \cdots \; \mu_n \,\right]^t$ is the vector whose components are the expected values of our random variables. Expanding the above equation we can write

$$\mathbf{C}_{XX} = E\left[\mathbf{X}\mathbf{X}^{\mathrm{t}}\right] - \overline{\mu}\,\overline{\mu}^{t} \tag{2.38}$$
$$= \mathbf{R}_{X} - \overline{\mu}\,\overline{\mu}^{t} \tag{2.39}$$

When the process has zero mean, the covariance matrix equals the correlation matrix:

$$\mathbf{C}_{XX} = \mathbf{R}_{XX} \tag{2.40}$$

The covariance matrix can be written explicitly in the form

$$\mathbf{C}_{XX} = \begin{bmatrix} C_{XX}(1,1) & C_{XX}(1,2) & \cdots & C_{XX}(1,n) \\ C_{XX}(1,2) & C_{XX}(2,2) & \cdots & C_{XX}(2,n) \\ \vdots & \vdots & \ddots & \vdots \\ C_{XX}(1,n) & C_{XX}(2,n) & \cdots & C_{XX}(n,n) \end{bmatrix} \tag{2.41}$$

Thus we see that the covariance matrix is symmetric. For a wide-sense stationary process, the covariance functions depend only on the difference in times and we get an even simpler matrix structure:

$$\mathbf{C}_{XX} = \begin{bmatrix} C_{XX}(0) & C_{XX}(1) & \cdots & C_{XX}(n-1) \\ C_{XX}(1) & C_{XX}(0) & \cdots & C_{XX}(n-2) \\ \vdots & \vdots & \ddots & \vdots \\ C_{XX}(n-1) & C_{XX}(n-2) & \cdots & C_{XX}(0) \end{bmatrix} \tag{2.42}$$

Using the definition for covariance in (1.114) on page 35, we can write the above equation as

$$\mathbf{C}_{XX} = \sigma_X^2 \begin{bmatrix} 1 & \rho(1) & \rho(2) & \cdots & \rho(n-1) \\ \rho(1) & 1 & \rho(1) & \cdots & \rho(n-2) \\ \rho(2) & \rho(1) & 1 & \cdots & \rho(n-3) \\ \vdots & & & & \vdots \\ \rho(n-1) & \rho(n-2) & \rho(n-3) & \cdots & 1 \end{bmatrix} \tag{2.43}$$

Example 10 Assume the autocovariance function for a wide-sense stationary random process is given by

$$c_{XX}(\tau) = 5 + 3e^{-|\tau|}$$

Find the autocovariance matrix for $\tau = 0$, 1, and 2.

Since the process is wide-sense stationary, the variance is given by

$$\sigma^2 = c_{XX}(0) = 8$$

The autocovariance matrix is given by

$$\mathbf{C}_{XX} = 8 \begin{bmatrix} 1 & 0.7630 & 0.6758 \\ 0.7630 & 1 & 0.7630 \\ 0.6758 & 0.7630 & 1 \end{bmatrix}$$

■

Problems

2.1 Define deterministic and nondeterministic processes. Give an example for each type.

2.2 Let X be the random process corresponding to observing the noon temperature throughout the year. The number of sample functions are 365 corresponding to each day of the year. Classify this process.

2.3 Let X be the random process corresponding to reporting the number of defective lights reported in a building over a period of one month. Each month we would get a different pattern. Classify this process.

2.4 Let X be the random process corresponding to measuring the total tonnage (weight) of ships going through the Suez canal in one day. The data is plotted for a period of one year. Each year will produce a different pattern. Classify this process.

2.5 Let X be the random process corresponding to observing the number of cars crossing a busy intersection in one hour. The number of sample functions are 24 corresponding to each hour of the day. Classify this process.

2.6 Let X be the random process corresponding to observing the bit pattern in an Internet packet. Classify this process.

2.7 Amplitude-shift keying (ASK) can be modeled as a random process described by

$$X(t) = a \cos \omega t$$

where ω is constant and a corresponds to the random variable A with two values a_0 and a_1 which occur with equal probability. Find the expected value $\mu(t)$ of this process.

2.8 A modified ASK uses two bits of the incoming data to generate a sinusoidal waveform and the corresponding random process is described by

$$X(t) = a \cos \omega t$$

where ω is a constant and a is a random variable with four values a_0, a_1, a_2, and a_3. Assuming that the four possible bit patterns are equally likely find the expected value $\mu(t)$ of this process.

2.9 Phase-shift keying (PSK) can be modeled as a random process described by

$$X(t) = a\cos(\omega t + \phi)$$

where a and ω are constant and ϕ corresponds to the random variable Φ with two values 0 and π which occur with equal probability. Find the expected value $\mu(t)$ of this process.

2.10 A modified PSK uses two bits of the incoming data to generate a sinusoidal waveform and the corresponding random process is described by

$$X(t) = a\cos(\omega t + \phi)$$

where a and ω are constants and ϕ is a random variable Φ with four values $\pi/4$, $3\pi/4$, $5\pi/4$, and $7\pi/4$ [4]. Assuming that the four possible bit patterns occur with equal probability, find the expected value $\mu(t)$ of this process.

2.11 A modified frequency-shift keying (FSK) uses three bits of the incoming data to generate a sinusoidal waveform and the random process is described by

$$X(t) = a\cos\omega t$$

where a is a constant and ω corresponds to the random variable Ω with eight values $\omega_0, \omega_1, \ldots, \omega_7$. Assuming that the eight frequencies are equally likely, find the expected value $\mu(t)$ of this process.

2.12 A discrete-time random process $X(n)$ produces the random variable $X(n)$ given by

$$X(n) = a^n$$

where a is a uniformly distributed random variable in the range 0–1. Find the expected value for this random variable at any time instant n.

2.13 Define a wide-sense stationary random process.

2.14 Prove (2.23) on page 56.

2.15 Define an ergodic random process.

2.16 Explain which of the following functions represent a valid autocorrelation function.

$$r_{XX}(n) = a^n \quad 0 \le a < 1 \qquad r_{XX}(n) = |a|^n \quad 0 \le a < 1$$
$$r_{XX}(n) = a^{n^2} \quad 0 \le a < 1 \qquad r_{XX}(n) = |a|^{n^2} \quad 0 \le a < 1$$
$$r_{XX}(n) = \cos n \qquad r_{XX}(n) = \sin n$$

2.17 A random process described by

$$X(t) = a\cos(\omega t + \phi)$$

where a and ω are constant and ϕ corresponds to the random variable Φ which is uniformly distributed in the interval 0 to 2π. Find the autocorrelation function $r_{XX}(t)$ of this process.

2.18 Define what is meant by two random processes being orthogonal.

2.19 Define what is meant by two random processes being statistically independent.

2.20 Find the cross-correlation function for the following two random processes.

$$X(t) = a \cos \omega t$$
$$Y(t) = \alpha\, a \cos(\omega t + \theta)$$

where a and θ are two zero mean random variables and α is a constant.

2.21 Given two random processes X and Y, when are they uncorrelated?

References

1. R. D. Yates and D. J. Goodman, *Probability and Stochastic Processes*, John Wiley, New York, 1999.
2. Thesaurus of Mathematics. http://thesaurus.maths.org/dictionary/map/word/1656.
3. P. Z. Peebles, *Probability, Random Variables, and Random Signal Principles*, McGraw-Hill, New York, 1993.
4. S. Haykin, *An Introduction to Analog and Digital Communications*, John Wiley, New York, 1989.

Chapter 3
Markov Chains

3.1 Introduction

We explained in Chapter 1 that in order to study a stochastic system we map its random output to one or more random variables. In Chapter 2 we studied other systems where the output was mapped to random processes which are functions of time. In either case we characterized the system using the expected value, variance, correlation, and covariance functions. In this chapter we study stochastic systems that are best described using *Markov processes*. A Markov process is a random process where the value of the random variable at instant n depends *only* on its immediate past value at instant $n-1$. The way this dependence is defined gives rise to a family of sample functions just like in any other random process. In a Markov process the random variable represents the *state* of the system at a given instant n. The state of the system depends on the nature of the system under study as we shall see in that chapter. We will have a truly rich set of parameters that describe a Markov process. This will be the topic of the next few chapters. The following are the examples of Markov processes we see in many real life situations:

1. telecommunication protocols and hardware systems
2. customer arrivals and departures at banks
3. checkout counters at supermarkets
4. mutation of a virus or DNA molecule
5. random walk such as Brownian motion
6. arrival of cars at an intersection
7. bus rider population during the day, week, month, etc
8. machine breakdown and repair during use
9. the state of the daily weather

3.2 Markov Chains

If the state space of a Markov process is discrete, the Markov process is called a *Markov chain*. In that case the states are labeled by the integers 0, 1, 2, etc. We will

F. Gebali, *Analysis of Computer and Communication Networks*,
DOI: 10.1007/978-0-387-74437-7_3, © Springer Science+Business Media, LLC 2008

be concerned here with discrete-time Markov chains since they arise naturally in many communication systems.

3.3 Selection of the Time Step

A Markov chain stays in a particular state for a certain amount of time called the *hold time*. At the end of the hold time, the Markov chain moves to another state at random where the process repeats again. We have two broad classifications of Markov chains that are based on how we measure the hold time.

1. *Discrete-time Markov chain*: In a discrete-time Markov chain the hold time assumes discrete values. As a result, changes in the states occur at discrete time values. In that case time is measured at specific instances:

$$t = T_0,\ T_1,\ T_2,\ \ldots$$

   The spacing between the time steps need not be equal in the general case. Most often, however, the discrete time values are equally spaced and we can write

$$t = nT \tag{3.1}$$

$$n = 0, 1, 2, \ldots \tag{3.2}$$

   The time step value T depends on the system under study as will be explained below.
2. *Continuous-time Markov chain*: In a continuous-time Markov chain the hold time assumes continuous values. As a result, changes in the states occur at any time value. The time value t will be continuous over a finite or infinite interval.

3.3.1 Discrete-Time Markov Chains

This type of Markov chains changes state at regular intervals. The time step could be a clock cycle, start of a new day, or a year, etc.

Example 1 Consider a packet buffer where packets arrive at each time step with probability a and depart with probability c. Identify the Markov chain and specify the possible buffer states.

We choose the time step in this example to be equal to the time required to receive or transmit a packet (transmission delay). At each time step we have two independent events: packet arrival and packet departure. We model the buffer as a Markov chain where the states of the system indicate the number of packets in the buffer. Assuming the buffer size is B, then the number of states of the buffer is $B+1$ as identified in Table 3.1. ■

Table 3.1 States of buffer occupancy

State	Significance
0	buffer is empty
1	buffer has one packet
2	buffer has two packets
⋮	⋮
B	buffer has B packets (full)

Example 2 Suppose that packets arrive at random on the input port of a router at an average rate λ_a (packets/s). The maximum data rate is assumed to be σ (packets/s), where $\sigma > \lambda_a$. Study the packet arrival statistics if the port input is sampled at a rate equal to the average data rate λ_a.

The time step (seconds) is chosen as

$$T = \frac{1}{\lambda_a}$$

In one time step we could receive 0, 1, 2, ..., N packets; where N is the maximum number of packets that could arrive

$$N = \lceil \sigma \times T \rceil = \lceil \frac{\sigma}{\lambda_a} \rceil$$

where the ceiling function $f(x) = \lceil x \rceil$ gives the smallest integer larger than or equal to x.

The statistics for packet arrival follow the binomial distribution and the probability of receiving k packets in time T is

$$p(k) = \binom{N}{k} a^k \, b^{N-k}$$

where a is the probability that a packet arrives and $b = 1 - a$. Our job in almost all situations will be to find out the values of the parameters N, a, and b in terms of the given data rates.

The packet arrival probability a could be obtained using the average value of the binomial distribution. The average input traffic $N_a(in)$ is given from the binomial distribution by

$$N_a(in) = N \, a$$

But $N_a(in)$ is also determined by the average data rate as

$$N_a(in) = \lambda_a \times T = 1$$

From the above two equations we get

$$a = \frac{1}{N} \leq \frac{\lambda_a}{\sigma}$$

■

Example 3 Consider Example 2 when the input port is sampled at the rate σ.

The time step is now given by

$$T' = \frac{1}{\sigma}$$

In one time step we get either one packet or no packets. There is no chance to get more than one packet in one time step since packets cannot arrive at a rate higher than σ. Therefore, the packet arrival statistics follow the Bernoulli distribution. For a time period t, the average number of packets that arrives is

$$N_a(in) = \lambda_a \, t$$

From the Bernoulli distribution that average is given by

$$N_a(in) = a' \, \frac{t}{T'}$$

The fraction on RHS indicates the number of time steps spanning the time period t. From the above two equations we get

$$a' = \lambda_a \, T' = \frac{\lambda_a}{\sigma}$$

■

3.4 Memoryless Property of Markov Chains

In a discrete-time Markov chain, the value of the random variable $S(n)$ represents the state of the system at time n. The random variable $S(n)$ is a function of its immediate past value—i.e. $S(n)$ depends on $S(n-1)$. This is referred to as the *Markov property* or *memoryless property* of the Markov chain where the present state of the system depends only on its immediate past state [1, 2]. Alternatively, we can say that the Markov property of the Markov chain implies that the future state of the system depends only on the present sate and not on its past states [3].

The probability that the Markov chain is in state s_i at time n is a function of its past state s_j at time $n-1$ only. Mathematically, this statement is written as

$$p\left[S(n) = s_i\right] = f\left(s_j\right) \tag{3.3}$$

Fig. 3.1 The occupancy states of a buffer of size four and the possible transitions between the states

for all $i \in S$ and $j \in S$ where S is the set of all possible states of the system.

Transition from a state to the next state is determined by a *transition probability* only with no regard to how the system came to be in the present state. Many communication systems can be modeled as Markov or memoryless systems using several techniques such as introducing extra transitions, defining extra states, and adjusting the time step value. This effort is worthwhile since the memoryless property of a Markov chain will result in a *linear system* that can be easily studied.

Example 4 Consider a data buffer in a certain communication device such as a network router for example. Assume the buffer could accommodate at most four packets. We say the buffer size is $B = 4$. Identify the states of this buffer and show the possible transitions between states assuming at any time step at most one packet can arrive or leave the buffer. Finally explain why the buffer could be studied using Markov chain analysis.

Figure 3.1 shows the occupancy states of a buffer of size four and the possible transitions between the states. The buffer could be empty or it could contain 1, 2, 3, or 4 packets. Furthermore, the assumptions indicate that the size of the buffer could remain unchanged or it could increase or decrease by one.

The transition from one state to another does not depend on how the buffer happened to be in the present state. Thus the system is memoryless and could be modeled as a Markov chain. ■

3.5 Markov Chain Transition Matrix

Let us define $p_{ij}(n)$ as the probability of finding our system in state i at time step n given that the past state was state j. We equate p_{ij} to the conditional probability that the system is in state i at time n given that it it was in state j at time $n - 1$. Mathematically, we express that statement as follows.

$$p_{ij}(n) = p\left[S(n) = i \mid S(n-1) = j\right] \tag{3.4}$$

The situation is further simplified if the transition probability is independent of the time step index n. In that case we have a *homogeneous* Markov chain, and the above equation becomes

$$p_{ij} = p\left[S(n) = i \mid S(n-1) = j\right] \tag{3.5}$$

Let us define the probability of finding our system in state i at the nth step as

$$s_i(n) = p\left[X(n) = i\right] \tag{3.6}$$

where the subscript i identifies the state and n denotes the time step index. Using (3.4), we can express the above equation as

$$s_i(n) = \sum_{j=1}^{m} p_{ij} \times s_j(n-1) \tag{3.7}$$

where we assumed the number of possible states to be m and the indices i and j lie in the range $1 \leq i \leq m$ and $1 \leq j \leq m$. We can express the above equation in matrix form as

$$\mathbf{s}(n) = \mathbf{P}\,\mathbf{s}(n-1) \tag{3.8}$$

where $\mathbf{P}$ is the *state transition matrix* of dimension $m \times m$

$$\mathbf{P} = \begin{bmatrix} p_{11} & p_{12} & \cdots & p_{1,m} \\ p_{21} & p_{22} & \cdots & p_{2,m} \\ \vdots & \vdots & \ddots & \vdots \\ p_{m,1} & p_{m,2} & \cdots & p_{m,m} \end{bmatrix} \tag{3.9}$$

and $\mathbf{s}(n)$ is the *distribution vector* (or *state vector*) defined as the probability of the system being in each state at time step n:

$$\mathbf{s}(n) = \begin{bmatrix} s_1(n) & s_2(n) & \cdots & s_m(n) \end{bmatrix}^t \tag{3.10}$$

The component $s_i(n)$ of the distribution vector $\mathbf{s}(n)$ at time n indicates the *probability* of finding our system in state s_i at that time. Because it is a probability, our system could be in any of the m states. However, the probabilities only indicate the likelihood of being in a particular state. Because s describes probabilities of all possible m states, we must have

$$\sum_{i=1}^{m} s_i(n) = 1 \qquad n = 0, 1, 2, \cdots \tag{3.11}$$

We say that our vector is normalized when it satisfies (3.11). We call such a vector a *distribution vector*. This is because the vector describes the distribution of probabilities among the different states of the system.

Soon we shall find out that describing the transition probabilities in matrix form leads to great insights about the behavior of the Markov chain. To be specific, we will find that we are interested in more than finding the values of the transition probabilities or entries of the transition matrix **P**. Rather, we will pay great attention to the eignevalues and eigenvectors of the transition matrix.

Since the columns of **P** represent transitions *out of* a given state, the sum of each column must be one since this covers all the possible transition events out of the state. Therefore we have, for all values of j,

$$\sum_{i=1}^{m} p_{ij} = 1 \tag{3.12}$$

The above equation is always valid since the sum of each column in **P** is unity. For example, a 2×2 transition matrix **P** would be set up as in the following diagram:

$$\begin{array}{ccc} & & \text{Present state} \\ & & \begin{array}{cc} 1 & 2 \end{array} \\ & & \downarrow \\ \text{Next state} & \begin{array}{c} 1 \\ 2 \end{array} \leftarrow & \begin{bmatrix} p_{11} & p_{12} \\ p_{21} & p_{22} \end{bmatrix} \end{array}$$

The columns represent the present state while the rows represent the next state. Element p_{ij} represents the transition probability from state j to state i. For example, p_{12} is the probability that the system makes a transition from state s_2 to state s_1.

Example 5 An on–off source is often used in telecommunications to simulate voice traffic. Such a source has two states: The silent state s_1 where the source does not send any data packets and the active state s_2 where the source sends one packet per time step. If the source were in s_1, it has a probability s of staying in that state for one more time step. When it is in state s_2, it has a probability a of staying in that state. Obtain the transition matrix for describing that source.

The next state of the source depends only on its present state. Therefore, we can model the state of the source as a Markov chain. The state diagram for such source is shown in Fig. 3.2 and the transition matrix is given by

$$\mathbf{P} = \begin{bmatrix} s & 1-a \\ 1-s & a \end{bmatrix}$$

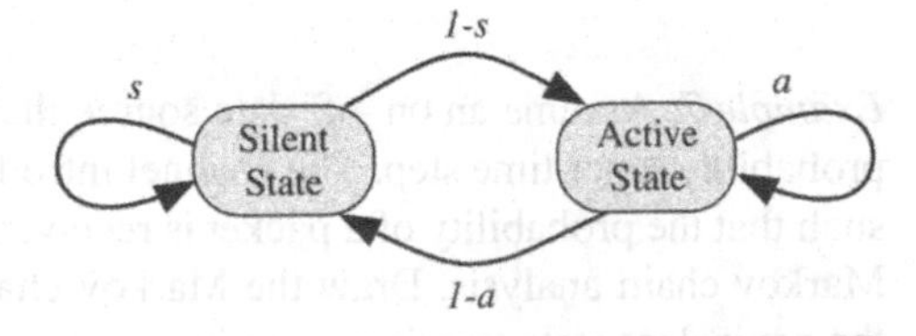

Fig. 3.2 Transition diagram for an on–off source

■

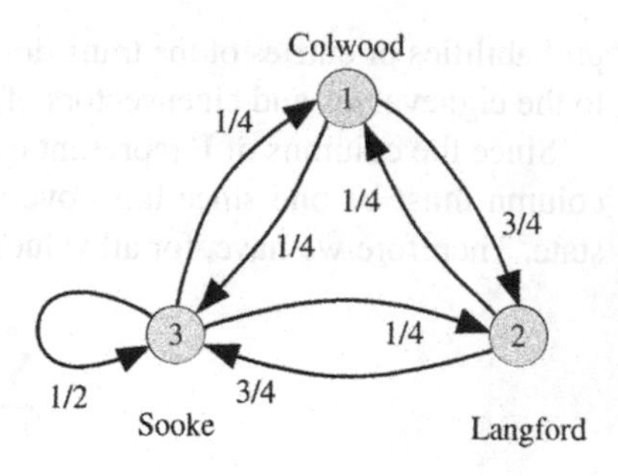

Fig. 3.3 A Markov chain involving three states

Example 6 Assume that the probability that a delivery truck moves between three cities at the start of each day is shown in Fig. 3.3. Write down the transition matrix and the initial distribution vector assuming that the truck was initially in Langford.

We assume the city at which the truck is located is the state of the truck. The next state of the truck depends only on its present state. Therefore, we can model the state of the truck as a Markov chain. We have to assign indices to replace city names. We chose the following arbitrary assignment, although any other state assignment will work as well.

City	State index
Colwood	1
Langford	2
Sooke	3

Based on the state assignment table, the transition matrix is given by

$$\mathbf{P} \begin{bmatrix} 0 & 1/4 & 1/4 \\ 3/4 & 0 & 1/4 \\ 1/4 & 3/4 & 1/2 \end{bmatrix}$$

The initial distribution vector is given by

$$\mathbf{s}(0) = \begin{bmatrix} 0 & 1 & 0 \end{bmatrix}^t$$

■

Example 7 Assume an on–off data source that generates equal length packets with probability a per time step. The channel introduces errors in the transmitted packets such that the probability of a packet is received in error is e. Model the source using Markov chain analysis. Draw the Markov chain state transition diagram and write the equivalent state transition matrix.

The Markov chain model of the source we use has four states:

State	Significance
1	Source is idle
2	Source is retransmitting a frame
3	Frame is transmitted with no errors
4	Frame is transmitted with an error

Since the next state of the source depends only on its present state, we can model the source using Markov state analysis.

Figure 3.4 shows the Markov chain state transition diagram. We make the following observations:

- The source stays idle (state s_1) with probability $1-a$.
- Transition from s_1 to s_3 occurs under two conditions: the source is active and no errors occur during transmission.
- Transition from s_1 to s_4 occurs under two conditions: the source is active and an error occurs during transmission.
- Transition from s_2 to s_3 occurs under only one condition: no errors occur.

The associated transition matrix for the system is given by

$$\mathbf{P} = \begin{bmatrix} 1-a & 0 & 1 & 0 \\ 0 & 0 & 0 & 1 \\ a(1-e) & 1-e & 0 & 0 \\ a\,e & e & 0 & 0 \end{bmatrix}$$

■

Example 8 In an ethernet network based on the carrier sense multiple access with collision detection (CSMA/CD), a user requesting access to the network starts transmission when the channel is not busy. If the channel is not busy, the user starts

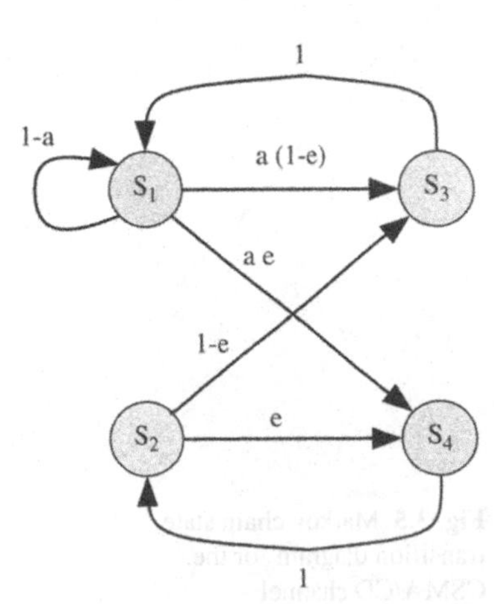

Fig. 3.4 State transition diagram for transmitting a packet over a noisy channel

transmitting. However, if one or more users sense that the channel is free, they will start transmitting and a collision will take place. If a collision from other users is detected, the user stops transmitting and returns to the idle state gain. In other words, we are not adopting any backoff strategies for simplicity in this example.

Assume the following probabilities:

u_0	Probability all users are idle
u_1	Probability only one user is transmitting
$1 - u_0 - u_1$	Probability two or more users are transmitting

(a) Justify using Markov chain analysis to describe the behavior of the channel or communication medium.
(b) Select a time step size for a discrete-time Markov chain model.
(c) Draw the Markov state transition diagram for this channel and show the state transition probabilities.

(a) A user in that system will determine its state within a time frame of twice the propagation delay on the channel. Therefore, the current state of the channel or communication medium and all users will depend only on the actions of the users in time frame of one propagation delay only. Thus our system can be described as a Markov chain.
(b) The time step T we can choose is twice the propagation delay. Assume packet transmission delay requires n time steps where all packets are assumed to have equal lengths.
(c) The channel can be in one of the following states:

(1) i: idle state
(2) t: transmitting state
(3) c: collided state

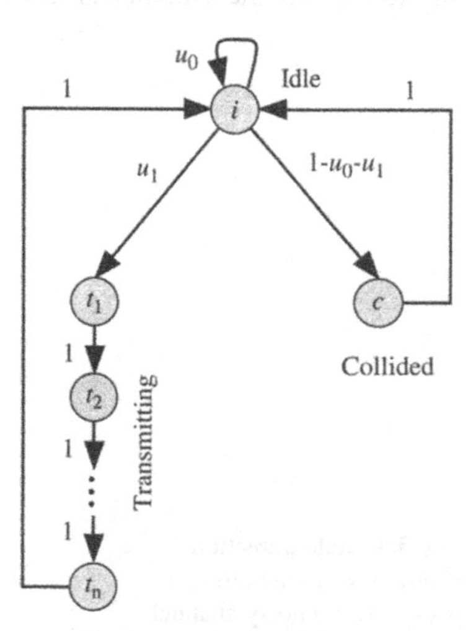

Fig. 3.5 Markov chain state transition diagram for the CSMA/CD channel

Figure 3.5 shows the state of the channel and the transition probabilities between these states.

1. The channel remains in the idle state with probability u_0.
2. The channel moves from idle to transmitting state with probability u_1.
3. The channel moves from idle to the collided state with probability $1 - u_0 - u_1$.

The other transitions are explained in the same way. We organize our state transition matrix such that first row or column corresponds to the idle state i. The second row or column correspond to the collided state. The third row or column corresponds to transmit state t_1, etc. For n transmit states, the transition matrix will have the dimension $(n+2) \times (n+2)$:

$$\mathbf{P} = \begin{bmatrix} u_0 & u_0 & 0 & 0 & \cdots & u_0 \\ 1-u_0-u_1 & 1-u_0-u_1 & 0 & 0 & \cdots & 1-u_0-u_1 \\ u_1 & u_1 & u_1 & 0 & \cdots & u_1 \\ 0 & 0 & 0 & 1 & \cdots & 0 \\ 0 & 0 & 0 & 0 & \cdots & 0 \\ \vdots & \vdots & \vdots & \vdots & \ddots & \vdots \\ 0 & 0 & 0 & 0 & \cdots & 0 \end{bmatrix}$$

■

3.6 Markov Matrices

The definition of the transition matrix **P** results in a matrix with peculiar properties:

1. The number of rows equals the number of columns. Thus **P** is a *square matrix.*
2. All the elements of **P** are real numbers. Thus **P** is a *real matrix.*
3. $0 \le p_{ij} \le 1$ for all values of i and j. Thus **P** is a *nonnegative matrix.*
4. The sum of each column is exactly 1 (i.e., $\sum_{j=1}^{m} p_{ij} = 1$).
5. The magnitude of all eigenvalues obey the condition $|\lambda_i| \le 1$. Thus the spectral radius of **P** equals 1.
6. At least one of the eigenvalues of **P** equals 1.

From all the above properties we conclude that the transition matrix is square, real, and nonnegative. Such a matrix is termed *column stochastic* matrix or *Markov* matrix. Notice that all column stochastic matrices are a subset of nonnegative matrices. Thus nonnegative matrices need not be column stochastic.

Nonnegative matrices have many interesting properties related to their eigenvalues and eigenvectors but this is beyond the scope of this book [4]. The above mentioned properties have implications on the eigenvalues of the transition matrix. The following theorem indicates one such implication.

Theorem 1 *Let* **P** *be any* $m \times m$ *column stochastic matrix. Then* **P** *has 1 as an eigenvalue* [4].

Proof We know that if λ is an eigenvalue of **P**, then the determinant of the characteristic equation must vanish, i.e.

$$\det(\mathbf{P} - \lambda \mathbf{I}) = 0 \tag{3.13}$$

where **I** is an $m \times m$ unit matrix. By Assuming that **P** has 1 as an eigenvalue, then the determinant is given by

$$\det(\mathbf{P} - 1 \times \mathbf{I}) = \begin{vmatrix} p_{11} - 1 & p_{12} & \cdots & p_{1m} \\ p_{21} & p_{22} - 1 & \cdots & p_{2m} \\ \vdots & \vdots & \ddots & \vdots \\ p_{m1} & p_{m2} & \cdots & p_{mm} - 1 \end{vmatrix} \tag{3.14}$$

where $|\mathbf{A}|$ indicates the determinant of matrix **A**. The determinant will not change if we add all the remaining rows to the first row

$$\det(\mathbf{P} - 1 \times \mathbf{I}) = \begin{vmatrix} \sum p_{i1} - 1 & \sum p_{i2} - 1 & \cdots & \sum p_{im} - 1 \\ p_{21} & p_{22} - 1 & \cdots & p_{2m} \\ \vdots & \vdots & \ddots & \vdots \\ p_{m1} & p_{m2} & \cdots & p_{mm} - 1 \end{vmatrix} \tag{3.15}$$

But the sum of the elements in each column is 1, and the first row of the above determinant will be zero. As a result, the determinant is zero and this proves that 1 is an eigenvalue of **P**.

Conversely, assume that (3.13) is satisfied for some value of λ. In that case we can write an equation similar to (3.15), namely:

$$\det(\mathbf{P} - 1 \times \mathbf{I}) = \begin{vmatrix} \sum p_{i1} - \lambda & \sum p_{i2} - \lambda & \cdots & \sum p_{im} - \lambda \\ p_{21} & p_{22} - 1 & \cdots & p_{2m} \\ \vdots & \vdots & \ddots & \vdots \\ p_{m1} & p_{m2} & \cdots & p_{mm} - 1 \end{vmatrix} \tag{3.16}$$

But we know that this determinant is zero. This is true for all values of p_{ij} which implies that the elements in the first row of the matrix must be zero. Thus we must have

$$\sum p_{i1} - \lambda = 0 \tag{3.17}$$

But the sums in the above equation is equal to 1 independent of the value of i. Thus we must have

$$1 - \lambda = 0 \tag{3.18}$$

The above equation implies that $\lambda = 1$ is a root of the equation which proves the second part of the theorem. ■

The following theorem will prove useful when we start multiplying Markov matrices to perform transient analysis on our Markov chain. The theorem essentially explains the effect of premultiplying any matrix by a column stochastic matrix.

Theorem 2 *The sum of columns of any matrix* **A** *will not change when it is premultiplied by a column stochastic matrix* **P**.

Proof When **A** is premultiplied by **P** we get matrix **B**.

$$\mathbf{B} = \mathbf{PA} \tag{3.19}$$

Element b_{ij} is given by the usual matrix product formula

$$b_{ij} = \sum_{k=1}^{m} p_{ik}\, a_{kj} \tag{3.20}$$

The sum of the jth column of matrix **B** is denoted by $\sigma_j(\mathbf{B})$ and is given by

$$\sigma_j(\mathbf{B}) = \sum_{i=1}^{m} b_{ij} = \sum_{i=1}^{m} \sum_{k=1}^{m} p_{ik}\, a_{kj} \tag{3.21}$$

Now reverse the order of summation on the right-hand side of the above equation:

$$\sigma_j(\mathbf{B}) = \sum_{k=1}^{m} a_{kj} \sum_{i=1}^{m} p_{ik} \tag{3.22}$$

Because **P** is column stochastic we have

$$\sigma_k(\mathbf{P}) = \sum_{i=1}^{m} p_{ik} = 1 \tag{3.23}$$

Therefore, (3.22) becomes

$$\sigma_j(\mathbf{B}) = \sum_{i=1}^{m} b_{ij} = \sum_{k=1}^{m} a_{kj} \quad (3.24)$$

$$= \sigma_j(\mathbf{A}) \quad (3.25)$$

Thus we proved that sum of columns of a matrix does not change when the matrix is premultiplied by a column stochastic matrix. ■

3.6.1 The Diagonals of P

We mentioned above the significance of each element p_{ij} of the transition matrix **P** as the probability of making a transition from state j to state i. Further insight can be gained for Markov chains representing queues. *Queuing systems* are a special type of Markov chains in which customers arrive and lineup to be serviced by servers. Thus a queue is characterized by the number of arriving customers at a given time step, the number of servers, the size of the waiting area for customers, and the number customers that can leave in one time step.

The state of a queuing system corresponds to the number of customers in the queue. If we take the lineup for a bank as an example, then the queue size increases when new customers arrive. The number of arriving customers could be one in some cases or many in others. This depends on the specifics of the situation. For example, if there is only one door to the bank, then we could expect at most one customer to arrive at any time. At the head of the queue, the number of servers also varies depending on how many bank tellers are ready, or disposed, to serve the customers. If there is only one teller, then we expect the size of the queue to decrease by at most one each time a customer is served. The duration of the service time also varies depending on the type of transaction being done.

The diagonals of **P** reflect the queuing system characteristics. Table 3.2 illustrates the significance of each diagonal of the matrix **P**.

Table 3.2 Significance of diagonals of **P**

Diagonal	Significance
Main	probabilities queue will retain its size
1st upper	probabilities queue size will decrease by one
2nd upper	probabilities queue size will decrease by two
3rd upper	probabilities queue size will decrease by three
⋮	⋮
1st lower	probabilities queue size will increase by one
2nd lower	probabilities queue size will increase by two
3rd lower	probabilities queue size will increase by three
⋮	⋮

3.7 Eigenvalues and Eigenvectors of P

The eigenvalues and eigenvectors of the transition matrix **P** will prove to be of utmost importance in the analyses of this book. The following theorem makes certain predictions about the eigenvector corresponding to the eigenvalue $\lambda = 1$.

Theorem 3 *Given a column stochastic matrix* **P** *and the eigenvector* **x** *corresponding to the eigenvalue* $\lambda = 1$*, the sum of the elements of* **x** *is nonzero and could be taken as unity, i.e.* $\sigma(x) = 1$.

Proof The eigenvector **x** corresponding to $\lambda = 1$ satisfies the equation

$$\mathbf{P}\,\mathbf{x} = \mathbf{x} \tag{3.26}$$

We can write the above equation as

$$(\mathbf{P} - \mathbf{I})\,\mathbf{x} = 0 \tag{3.27}$$

This is a system of linear equation with an infinite number of solutions since the matrix $(\mathbf{P} - \mathbf{I})$ is rank deficient (i.e., rank$(\mathbf{P}) < m$). To get a unique solution for **x**, we need one extra equation which we choose as

$$\sigma(x) = 1 \tag{3.28}$$

We cannot choose the sum to be zero since this is a trivial solution. Any nonzero value is acceptable. We choose to have $\sigma(x) = 1$ for reasons that will become apparent later on. This proves the theorem. ■

The following theorem makes certain predictions about the sum of elements of the other eigenvectors of **P** corresponding to the eigenvalues $\lambda < 1$.

Theorem 4 *Given a column stochastic matrix* **P** *and an eigenvector* **x** *corresponding to the eigenvalue* $\lambda \neq 1$*, the sum of the elements of* **x** *must be zero, i.e.* $\sigma(x) = 0$.

Proof The eigenvector **x** satisfies the equation

$$\mathbf{P}\,\mathbf{x} = \lambda\,\mathbf{x} \tag{3.29}$$

The sum of columns of both sides of the above equation are equal

$$\sigma(\mathbf{P}\,\mathbf{x}) = \lambda\,\sigma(\mathbf{x}) \tag{3.30}$$

From Theorem 2, on page 77, we are assured that the sum of the elements of **x** will not change after being multiplied by matrix **P**. Thus we can write

$$\sigma(\mathbf{P}\,\mathbf{x}) = \sigma(\mathbf{x}) \tag{3.31}$$

From the above two equations we have

$$\sigma(x) = \lambda\sigma(x) \tag{3.32}$$

or

$$\sigma(x)(1 - \lambda) = 0 \tag{3.33}$$

Since $\lambda \neq 1$, the only possible solution to the above equation is $\sigma(x) = 0$. This proves the theorem. ■

Example 9 Verify Theorems 3 and 4 for the Markov matrix

$$\mathbf{P} = \begin{bmatrix} 0.3 & 0.1 & 0 & 0.2 \\ 0.1 & 0.6 & 0.3 & 0.1 \\ 0.4 & 0.2 & 0.4 & 0.5 \\ 0.2 & 0.1 & 0.3 & 0.2 \end{bmatrix}$$

MATLAB gives the following eigenvectors and corresponding eigenvalues:

```
[X,D] = eig(P)

X =
 -0.1965    0.5887   -0.1002   -0.4286
 -0.6309    0.3983   -0.7720    0.1644
 -0.6516   -0.5555    0.5190   -0.4821
 -0.3724   -0.4315    0.3532    0.7463


D =

    1.0000         0         0         0
         0    0.2211         0         0
         0         0    0.3655         0
         0         0         0   -0.0866
```

We have to normalize our eigenvectors so that the sum of the components of the first column, which corresponds to $\lambda = 1$ is one.

```
X = X/sum(X(:,1))

X =
    0.1061   -0.3179    0.0541    0.2315
    0.3408   -0.2151    0.4169   -0.0888
    0.3520    0.3001   -0.2803    0.2604
    0.2011    0.2330   -0.1907   -0.4031
```

We can write, ignoring rounding errors,

$$\begin{aligned}
\sigma_1(\mathbf{X}) &= 1 \\
\sigma_2(\mathbf{X}) &= 0 \\
\sigma_3(\mathbf{X}) &= 0 \\
\sigma_4(\mathbf{X}) &= 0
\end{aligned}$$

■

3.8 Constructing the State Transition Matrix P

The state transition matrix **P** is the key to analyzing Markov chains and queues. To construct the matrix the following steps are usually followed [3].

1. Verify that the system under study displays the Markov property. In other words, ensure that transition to a new state depends only on the current state.
2. All possible states of the system are identified and labeled. The labeling of the states is arbitrary although some labeling schemes would render the transition matrix easier to visualize.
3. All possible transitions between the states are either drawn on the state transition diagram, or the corresponding elements of the state transition matrix are identified.
4. The probability of every transition in the state diagram is obtained.
5. The transition matrix is constructed.
6. Relabeling of the states is always possible. That will change the locations of the matrix elements and make the structure of the matrix more visible. This rearrangement will still produce a column stochastic matrix and will not disturb its eigenvalues or the *directions* of its eigenvectors.

Example 10 The closing price of a certain stock on a given weekday is either falling or rising compared to the previous day's price. If the price stays the same, then it is classified as rising if the previous day's trend was rising, and vice versa. The probabilities of price transitions between these two states are shown in Fig. 3.6. Construct state transition matrix.

The price of the stock has only two states, falling (s_1) or rising (s_2). The transition matrix will be

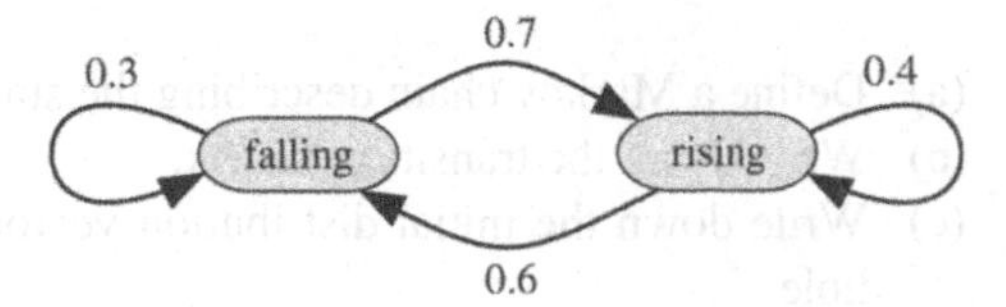

Fig. 3.6 Day-to-day closing price fluctuations of a certain stock

$$\mathbf{P} = \begin{bmatrix} 0.3 & 0.6 \\ 0.7 & 0.4 \end{bmatrix}$$

■

3.9 Transient Behavior

From (3.8) on page 70 we can write the distribution vector at time step $n = 1$ as

$$\mathbf{s}(1) = \mathbf{P}\,\mathbf{s}(0) \tag{3.34}$$

and

$$\mathbf{s}(2) = \mathbf{P}\,\mathbf{s}(1) \tag{3.35}$$
$$= \mathbf{P}\,[\mathbf{P}\,\mathbf{s}(0)] \tag{3.36}$$
$$= \mathbf{P}^2\,\mathbf{s}(0) \tag{3.37}$$

and we can generalize to express the distribution vector at step n as

$$\mathbf{s}(n) = \mathbf{P}^n\,\mathbf{s}(0) \qquad n = 0, 1, 2, ... \tag{3.38}$$

This equation allows us to calculate the distribution vector at the nth time step given the transition matrix and the initial distribution vector $\mathbf{s}(0)$.

Example 11 In Example 6, on page 72, what is the probability that our truck is in Colwood after five deliveries assuming that it was Langford initially?

We are looking for element p_{12} in matrix $\mathbf{P}^5$. Using any mathematical tool such as MAPLE or MATLAB, we get

$$\mathbf{P}^5 = \begin{bmatrix} 0.199 & 0.2 & 0.200 \\ 0.288 & 0.277 & 0.278 \\ 0.513 & 0.523 & 0.522 \end{bmatrix}$$

Hence the required probability is 0.2. ■

Example 12 Assume a ball falls on the pegs shown in Fig. 3.7. The pegs are setup such that the ball must hit a peg at each level and bounce off one of the two pegs immediately below. The location of the ball at the bottom bucket indicates the prize to be won.

(a) Define a Markov chain describing the state of the ball.
(b) Write down the transition matrix.
(c) Write down the initial distribution vector if the ball is dropped in the middle hole.

Fig. 3.7 A ball falling through a maze of pegs

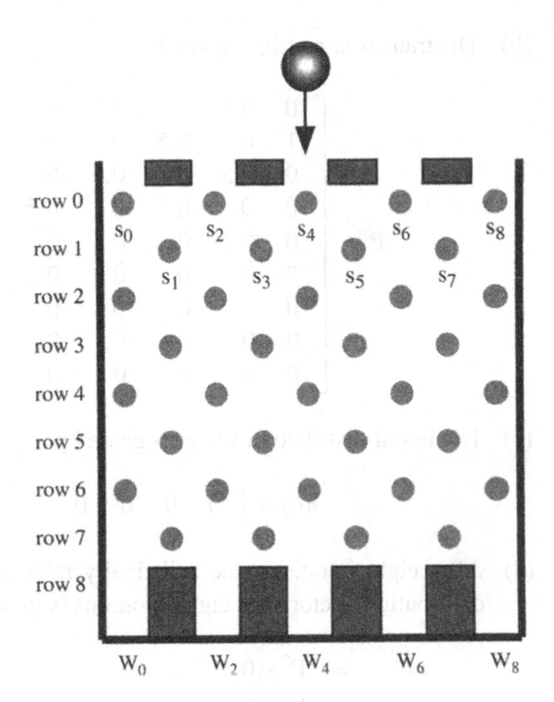

(d) Determine the probability of hitting the middle bucket.
(e) Assume the money to be won is given by the vector

$$\mathbf{w} = \begin{bmatrix} \$5 & \$1 & \$10 & \$1 & \$5 \end{bmatrix}^t$$

Determine the average amount of money that could be won.

Since the next location of the ball depends only on its present location, we can describe this system using Markov chains.

(a) We can model this system as a Markov chain with nine states s_1 to s_9, as indicated in the figure. State $s_i(n)$ indicates that the ball is at column i and row n in the game. The rows are numbered starting with zero at the top row. Thus our distribution vector could be written as

$$\mathbf{s} = \begin{bmatrix} s_0 & s_1 & s_2 & s_3 & s_4 & s_5 & s_6 & s_7 & s_8 \end{bmatrix}^t$$

where at even time steps the even states could be occupied and at odd time steps only the odd states could be occupied.

(b) The transition matrix is given by

$$\mathbf{P} = \begin{bmatrix} 0 & 0.5 & 0 & 0 & 0 & 0 & 0 & 0 & 0 \\ 1 & 0 & 0.5 & 0 & 0 & 0 & 0 & 0 & 0 \\ 0 & 0.5 & 0 & 0.5 & 0 & 0 & 0 & 0 & 0 \\ 0 & 0 & 0.5 & 0 & 0.5 & 0 & 0 & 0 & 0 \\ 0 & 0 & 0 & 0.5 & 0 & 0.5 & 0 & 0 & 0 \\ 0 & 0 & 0 & 0 & 0.5 & 0 & 0.5 & 0 & 0 \\ 0 & 0 & 0 & 0 & 0 & 0.5 & 0 & 0.5 & 0 \\ 0 & 0 & 0 & 0 & 0 & 0 & 0.5 & 0 & 1 \\ 0 & 0 & 0 & 0 & 0 & 0 & 0 & 0.5 & 0 \end{bmatrix}$$

(c) The initial distribution vector is given by

$$\mathbf{s}(0) = \begin{bmatrix} 0 & 0 & 0 & 0 & 1 & 0 & 0 & 0 & 0 \end{bmatrix}^t$$

(d) After eight iterations, the ball finally falls into one of the buckets. Thus the distribution vector after eight iterations is given by

$$\begin{aligned} \mathbf{s}(8) &= \mathbf{P}^8\, \mathbf{s}(0) \\ &= \begin{bmatrix} 0.11 & 0 & 0.25 & 0 & 0.28 & 0 & 0.25 & 0 & 0.11 \end{bmatrix}^t \end{aligned}$$

Note that the only valid states are the ones that correspond to a bucket locations at the bottom of the figure. This explains the zeros in the odd locations 1, 3, 5, and 7. The probability of settling into the middle bucket is $s_4 = 0.28$.

(e) The average winnings are given by

$$\begin{aligned} W_a &= W_0\, s_0(8) + W_2\, s_2(8) + W_4\, s_4(8) + W_6\, s_6(8) + W_8\, s_8(8) \\ &= \$4.4 \end{aligned}$$

It is interesting that if hundreds of balls are dropped at the center toward the bottom, then their distribution at the bottom barrels is bell-shaped and their number is the binomial coefficients [5] ■

Example 13 A computer memory system is composed of very fast on-chip cache, fast on-board RAM, and slow hard disk. When the computer is accessing a block from each memory system, the next block required could come from any of the three available memory systems. This is modeled as a Markov chain with the state of the system representing the memory from which the current block came from: state s_1 corresponds to the cache, state s_2 corresponds to the RAM, and state s_3 corresponds to the hard disk. The transition matrix is given by

$$\mathbf{P} = \begin{bmatrix} 0.7 & 0.1 & 0 \\ 0.2 & 0.7 & 0.1 \\ 0.1 & 0.2 & 0.9 \end{bmatrix}$$

Find the probability that after three consecutive block accesses the system will read a block from the cache.

The starting distribution vector is $\mathbf{s}(0) = \begin{bmatrix} 1 & 0 & 0 \end{bmatrix}^t$ and we have

$$\mathbf{P}^2 = \begin{bmatrix} 0.51 & 0.14 & 0.01 \\ 0.29 & 0.53 & 0.16 \\ 0.20 & 0.33 & 0.83 \end{bmatrix}$$

$$\mathbf{P}^3 = \begin{bmatrix} 0.386 & 0.151 & 0.023 \\ 0.325 & 0.432 & 0.197 \\ 0.289 & 0.417 & 0.780 \end{bmatrix}$$

The distribution vector after three iterations is

$$\mathbf{s}(3) = \mathbf{P}^3 \mathbf{s}(0) = \begin{bmatrix} 0.386 & 0.325 & 0.289 \end{bmatrix}^t$$

The probability that the system will read a block from the cache is 0.386. ■

3.9.1 Properties of $\mathbf{P}^n$

Matrix $\mathbf{P}^n$ has many interesting properties that we list below:

- $\mathbf{P}^n$ remains a column stochastic matrix according to Lemma 1.
- A nonzero element in $\mathbf{P}$ can increase or decrease in $\mathbf{P}^n$ but can never become zero.
- A zero element in $\mathbf{P}$ could remain zero or increase in $\mathbf{P}^n$ but can never become negative.

As a result of Theorem 2 on page 77, we deduce the following two lemmas.

Lemma 1 *Given a column stochastic matrix* $\mathbf{P}$, *then* $\mathbf{P}^n$, *for* $n \geq 0$, *is also column stochastic.*

The proof is easily derived from Theorem 2.

Lemma 2 *The state vector* $\mathbf{s}(n)$ *at instance n is given by*

$$\mathbf{s}(n) = \mathbf{P}^n \mathbf{s}(0) \tag{3.39}$$

This vector must be a distribution vector for all values of $n \geq 0$. The proof is easily derived from Theorem 2 after applying the theorem to the state vector $\mathbf{s}(0)$.

3.10 Finding s (n)

We can determine the state of our Markov chain at time step n if we are able to calculate $\mathbf{P}^n$. In general, performing n matrix multiplications is tedious and leads to computational noise. Besides, no insight can be gained from repeatedly multiplying a matrix. Alternative techniques for obtaining an expression for $\mathbf{s}(n)$ or $\mathbf{P}^n$ include the following:

1. Repeated multiplications to get $\mathbf{P}^n$.
2. Expanding the initial distribution vector $\mathbf{s}(0)$.
3. Diagonalizing the matrix $\mathbf{P}$.
4. Using the Jordan canonic form of $\mathbf{P}$.
5. Using the z-transform.

The first method is simple and is best done using a mathematical package such as MATLAB. We have seen examples of this technique in the previous section. We show in the following sections how the other techniques are used.

3.11 Finding s (n) by Expanding s (0)

In most cases the transition matrix $\mathbf{P}$ is simple, i.e. it has m distinct eigenvalues. In that case we can express our initial distribution vector $\mathbf{s}(0)$ as a linear combination of the m eigenvectors:

$$\mathbf{s}(0) = c_1\, \mathbf{x}_1 + c_2\, \mathbf{x}_2 + \cdots + c_m\, \mathbf{x}_m \tag{3.40}$$

where $\mathbf{x}_i$ is the ith eigenvector of $\mathbf{P}$ and c_i is the corresponding scalar expansion coefficients. We can write the above equation as a simple matrix expression:

$$\mathbf{s}(0) = \mathbf{X}\, \mathbf{c} \tag{3.41}$$

where $\mathbf{X}$ is an $m \times m$ matrix whose columns are the eigenvectors of $\mathbf{P}$ and $\mathbf{c}$ is an m-component vector of the expansion coefficients:

$$\mathbf{X} = \begin{bmatrix} \mathbf{x}_1 & \mathbf{x}_2 & \cdots & \mathbf{x}_m \end{bmatrix} \tag{3.42}$$

$$\mathbf{c} = \begin{bmatrix} c_1 & c_2 & \cdots & c_m \end{bmatrix}^t \tag{3.43}$$

We need not normalize the eigenvectors before we determine the coefficients vector $\mathbf{c}$ because any normalization constant for $\mathbf{X}$ will be accounted for while determining $\mathbf{c}$.

To find $\mathbf{s}(n)$ we will use the technique explained below. Equation (3.41) is a system of m-linear equations in m unknowns, namely the components of the column vector $\mathbf{c}$. There are many numerical techniques for finding these components like Gauss elimination, Kramer's rule, etc. MATLAB has a very simple function for finding the eigenvectors of $\mathbf{P}$:

$$\texttt{X} = \texttt{eig(P)} \tag{3.44}$$

where matrix $\mathbf{X}$ will contain the eigenvectors in its columns. To find the coefficient vector $\mathbf{c}$ we use MATLAB to solve the system of linear equations in (3.41) by typing the MATLAB command

$$\mathtt{c} = \mathtt{X} \setminus \mathtt{s(0)} \tag{3.45}$$

where the backslash operator "\" effectively solves for the unknown vector **c** using Gaussian elimination.

Note: The function `EIGPOWERS(P,s)` can be used to expand **s** in terms of the eigenvectors of **P**.

Example 14 A Markov chain has the state matrix

$$\mathbf{P} = \begin{bmatrix} 0.3 & 0.2 & 0.2 & 0.2 \\ 0.3 & 0.4 & 0.2 & 0.2 \\ 0.2 & 0.2 & 0.2 & 0.2 \\ 0.2 & 0.2 & 0.4 & 0.4 \end{bmatrix}$$

Check to see if this matrix has distinct eigenvalues. If so, expand the initial distribution vector

$$\mathbf{s}(0) = \begin{bmatrix} 1 & 0 & 0 & 0 \end{bmatrix}^t$$

in terms of the eigenvectors of **P**.

The eigenvalues of **P** are $\lambda_1 = 1$, $\lambda_2 = 0.1$, $\lambda_3 = 0.2$, and $\lambda_4 = 0$. The eigenvectors corresponding to these eigenvalues are

$$\mathbf{X} = \begin{bmatrix} 0.439 & -0.707 & 0.0 & 0.0 \\ 0.548 & 0.707 & 0.707 & 0.0 \\ 0.395 & 0.0 & 0.0 & 0.707 \\ 0.592 & 0.0 & -0.707 & -0.707 \end{bmatrix}$$

Therefore, we can expand **s**(0) in terms of the corresponding eigenvectors

$$\mathbf{s}(0) = \mathbf{X}\,\mathbf{c}$$

where **c** given by

$$\mathbf{c} = \begin{bmatrix} 0.507 & -1.1 & 0.707 & 0.283 \end{bmatrix}^t$$

■

Now let us see how to get a simple expression for evaluating $\mathbf{s}(n)$ given the expansion of $\mathbf{s}(0)$. We start by using (3.41) to find $\mathbf{s}(1)$ as

$$\begin{aligned}
\mathbf{s}(1) &= \mathbf{P}\,\mathbf{s}(0) && (3.46)\\
&= \mathbf{P}\,(c_1\mathbf{x}_1 + c_2\mathbf{x}_2 + \cdots + c_m\mathbf{x}_m) && (3.47)\\
&= c_1\lambda_1\mathbf{x}_1 + c_2\lambda_2\mathbf{x}_2 + \cdots + c_m\lambda_m\mathbf{x}_m && (3.48)
\end{aligned}$$

where λ_i is the i-th eigenvalue of $\mathbf{P}$ corresponding to the i-th eigenvector $\mathbf{x}_i$.

We can express $\mathbf{s}(1)$ in the above equation in matrix form as follows:

$$\mathbf{s}(1) = \mathbf{X}\,\mathbf{D}\,\mathbf{c} \tag{3.49}$$

where $\mathbf{X}$ was defined before and $\mathbf{D}$ is the diagonal matrix whose diagonal elements are the eigenvalues of $\mathbf{P}$.

$$\mathbf{D} = \begin{bmatrix} \lambda_1 & 0 & \cdots & 0 \\ 0 & \lambda_2 & \cdots & 0 \\ \vdots & \vdots & \ddots & \vdots \\ 0 & 0 & \cdots & \lambda_m \end{bmatrix} \tag{3.50}$$

MATLAB provides a very handy function for finding the two matrices $\mathbf{X}$ and $\mathbf{D}$ using the single command

$$[\texttt{X, D}] = \texttt{eig(P)} \tag{3.51}$$

Having found $\mathbf{s}(1)$, we now try to find $\mathbf{s}(2)$

$$\begin{aligned}
\mathbf{s}(2) &= \mathbf{P}\,\mathbf{s}(1) && (3.52)\\
&= \mathbf{P}\,\mathbf{X}\,\mathbf{D}\,\mathbf{c} && (3.53)
\end{aligned}$$

but the eigenvectors satisfy the relation

$$\mathbf{P}\mathbf{x}_i = \lambda_i\mathbf{x}_i \tag{3.54}$$

In matrix form we can write the above equation as

$$\mathbf{P}\mathbf{X} = \mathbf{X}\,\mathbf{D} \tag{3.55}$$

By substituting (3.55) into (3.53) we get

$$\mathbf{s}(2) = \mathbf{X}\,\mathbf{D}^2\,\mathbf{c} \tag{3.56}$$

where

$$\mathbf{D}^2 = \begin{bmatrix} \lambda_1^2 & 0 & \cdots & 0 \\ 0 & \lambda_2^2 & \cdots & 0 \\ \vdots & \vdots & \ddots & \vdots \\ 0 & 0 & \cdots & \lambda_m^2 \end{bmatrix} \tag{3.57}$$

In general the distribution vector at time step n is given by

$$\mathbf{s}(n) = \mathbf{X}\mathbf{D}^n\mathbf{c} \tag{3.58}$$

where

$$\mathbf{D}^n = \begin{bmatrix} \lambda_1^n & 0 & \cdots & 0 \\ 0 & \lambda_2^n & \cdots & 0 \\ \vdots & \vdots & \ddots & \vdots \\ 0 & 0 & \cdots & \lambda_m^n \end{bmatrix} \tag{3.59}$$

It is relatively easy to find the n-th power of a diagonal matrix by simply raising each diagonal element to the n-th power. Then the distribution vector at time step n is simply obtained from (3.58).

Example 15 Consider the Markov chain in Example 14. Find the values of the distribution vector at time steps 2, 5, and 20.

For the given transition matrix, we can write

$$\mathbf{s}(n) = \mathbf{X}\mathbf{D}^n\mathbf{c}$$

where the matrices **X**, **D**, and the vector **c** are given by

$$\mathbf{X} = \begin{bmatrix} 0.439 & -0.707 & 0.0 & 0.0 \\ 0.548 & 0.707 & 0.707 & 0.0 \\ 0.395 & 0.0 & 0.0 & 0.707 \\ 0.592 & 0.0 & -0.707 & -0.707 \end{bmatrix}$$

$$\mathbf{D} = \begin{bmatrix} 1 & 0 & 0 & 0 \\ 0 & 0.1 & 0 & 0 \\ 0 & 0 & 0.2 & 0 \\ 0 & 0 & 0 & 0 \end{bmatrix}$$

$$\mathbf{c} = \begin{bmatrix} 0.507 & -1.1 & 0.707 & 0.283 \end{bmatrix}^t$$

Thus we can simply write

$$\begin{aligned}
\mathbf{s}(2) &= \mathbf{X}\mathbf{D}^2\mathbf{c} \\
&= \begin{bmatrix} 0.23 & 0.29 & 0.2 & 0.28 \end{bmatrix}^t \\
\mathbf{s}(5) &= \mathbf{X}\mathbf{D}^5\mathbf{c} \\
&= \begin{bmatrix} 0.22 & 0.28 & 0.2 & 0.3 \end{bmatrix}^t \\
\mathbf{s}(20) &= \mathbf{X}\mathbf{D}^{20}\mathbf{c} \\
&= \begin{bmatrix} 0.22 & 0.28 & 0.2 & 0.3 \end{bmatrix}^t
\end{aligned}$$

We see that the distribution vector settles down to a fixed value after a few iterations. The above results could be confirmed by directly finding the distribution vectors using the usual formula

$$\mathbf{s}(n) = \mathbf{P}^n \ \mathbf{s}(0)$$

■

Sometimes two eigenvalues are the complex conjugate of each other. In that case the corresponding eigenvectors will be complex conjugate and so are the corresponding coefficient c such that the end result $\mathbf{s}(n)$ is purely real.

Example 16 A Markov chain has the transition matrix

$$\mathbf{P} = \begin{bmatrix} 0.1 & 0.4 & 0.2 \\ 0.1 & 0.4 & 0.6 \\ 0.8 & 0.2 & 0.2 \end{bmatrix}$$

Check to see if this matrix has distinct eigenvalues. If so,

(a) Expand the following initial distribution vector

$$\mathbf{s}(0) = \begin{bmatrix} 1 & 0 & 0 \end{bmatrix}^t$$

in terms of the eigenvectors of $\mathbf{P}$.

(b) Find the value of $\mathbf{s}(3)$.

The eigenvalues of $\mathbf{P}$ are

$$\begin{aligned}
\lambda_1 &= 1 \\
\lambda_2 &= -0.15 + j0.3122 \\
\lambda_3 &= -0.15 - j0.3122
\end{aligned}$$

We see that complex eigenvalues appear as complex conjugate pairs. The eigenvectors corresponding to these eigenvalues are

$$\mathbf{X} = \begin{bmatrix} -0.4234 & -0.1698 + j0.3536 & -0.1698 - j0.3536 \\ -0.6726 & -0.5095 - j0.3536 & -0.5095 + j0.3536 \\ -0.6005 & 0.6794 & 0.6794 \end{bmatrix}$$

We see that the eigenvectors corresponding to the complex eigenvalues appear also as complex conjugate pairs. Therefore, we can expand s(0) in terms of the corresponding eigenvectors.

$$\mathbf{s}(0) = \mathbf{X}\,\mathbf{c}$$

where **c** given by

$$\mathbf{c} = \begin{bmatrix} -5863 & -0.2591 - j0.9312 & -0.2591 + j0.9312 \end{bmatrix}^t$$

We see that the expansion coefficients corresponding to the complex eigenvectors appear as complex conjugate pairs also. This ensures that all the components of the distribution vector will always be real numbers.

The distribution vector at time step $n = 3$ is

$$\mathbf{s}(3) = \mathbf{X}\mathbf{D}^3\mathbf{c} = \begin{bmatrix} 0.2850 & 0.3890 & 0.3260 \end{bmatrix}^t$$

■

3.12 Finding s(n) by Diagonalizing P

According to [4], matrix **P** is diagonalizable when it has m distinct eigenvalues and we can write

$$\mathbf{P} = \mathbf{X}\mathbf{D}\mathbf{X}^{-1} \tag{3.60}$$

where **X** is the matrix whose columns are the eigenvectors of **P**, and **D** is a diagonal matrix whose diagonal elements are the eigenvalues arranged according to the ordering of the columns of **X**. MATLAB provides a very handy function for finding the matrices **X** and **D** using the single command

```
[X, D] = eig(P)
```

It does not matter here whether the columns of **X** are normalized or not since any scaling factor in **X** will be cancelled by $\mathbf{X}^{-1}$. MATLAB also offers a very simple function for inverting a matrix by using the command

```
X_inv = inv(X)
```

We can calculate $\mathbf{P}^2$ as

$$\begin{aligned}\mathbf{P}^2 &= (\mathbf{XDX}^{-1}) \times (\mathbf{XDX}^{-1}) && (3.61)\\ &= \mathbf{XD}^2\mathbf{X}^{-1} && (3.62)\end{aligned}$$

In general we have

$$\mathbf{P}^n = \mathbf{XD}^n\mathbf{X}^{-1} \tag{3.63}$$

where

$$\mathbf{D}^n = \begin{bmatrix} \lambda_1^n & 0 & 0 & \cdots & 0 \\ 0 & \lambda_2^n & 0 & \cdots & 0 \\ 0 & 0 & \lambda_3^n & \cdots & 0 \\ \vdots & \vdots & \vdots & \ddots & \vdots \\ 0 & 0 & 0 & \cdots & \lambda_m^n \end{bmatrix} \tag{3.64}$$

It is easy therefore to find $\mathbf{s}(n)$ for any value of n by simply finding $\mathbf{D}^n$ and then evaluate the simple matrix multiplication expression

$$\begin{aligned}\mathbf{s}(n) &= \mathbf{P}^n\, \mathbf{s}(0)\\ &= \mathbf{XD}^n\mathbf{X}^{-1}\, \mathbf{s}(0) && (3.65)\end{aligned}$$

3.12.1 Comparing Diagonalization with Expansion of s (0)

Diagonalizing the matrix $\mathbf{P}$ is equivalent to the previous technique of expanding $\mathbf{s}(0)$ in terms of the eigenvectors of $\mathbf{P}$. As a proof, consider calculating $\mathbf{s}(n)$ using both techniques.

Using diagonalization technique, we have

$$\begin{aligned}\mathbf{s}(n) &= \mathbf{P}^n \mathbf{s}(0) && (3.66)\\ &= \mathbf{XD}^n\mathbf{X}^{-1}\, \mathbf{s}(0) && (3.67)\end{aligned}$$

Using the expansion of $\mathbf{s}(0)$ technique in (3.41) we know that

$$\mathbf{s}(0) = \mathbf{Xc} \tag{3.68}$$

By substituting this into (3.67) we get

$$\begin{aligned}\mathbf{s}(n) &= \mathbf{XD}^n\mathbf{X}^{-1}\mathbf{X}\, \mathbf{c}\\ &= \mathbf{XD}^n\, \mathbf{c} && (3.69)\end{aligned}$$

Now the valuc for $\mathbf{c} = \mathbf{X}^{-1}\mathbf{s}(0)$ can be found to yield

$$\mathbf{s}(n) = \mathbf{X}\mathbf{D}^n\mathbf{X}^{-1}\ \mathbf{s}(0) \tag{3.70}$$

This last expression for $\mathbf{s}(n)$ is identical to (3.67). Thus we see that the two techniques are equivalent. Before we finish this section we state the following lemma which will prove useful later on.

Lemma 3 *Assume the first column of* $\mathbf{X}$ *is the normalized eigenvector corresponding to eigenvalue* $\lambda = 1$. *Then the matrix* $\mathbf{X}^{-1}$ *will have ones in its first row.*

In general, if the first column of $\mathbf{X}$ is not normalized, then the theorem would state that the matrix $\mathbf{X}^{-1}$ will have the value $1/\sigma(\mathbf{x}_1)$ in the elements in its first row.
Proof Assume $\mathbf{Y} = \mathbf{X}^{-1}$ then we can write

$$\mathbf{X} \times \mathbf{Y} = \mathbf{I} \tag{3.71}$$

where $\mathbf{I}$ is the $m \times m$ unit matrix. Let us express the above equation in terms of the elements of the two matrices $\mathbf{X}$ and $\mathbf{Y}$:

$$\begin{bmatrix} x_{11} & x_{12} & \cdots & x_{1m} \\ x_{21} & x_{22} & \cdots & x_{2m} \\ \vdots & \vdots & \ddots & \vdots \\ x_{m1} & x_{m2} & \cdots & x_{mm} \end{bmatrix} \begin{bmatrix} y_{11} & y_{12} & \cdots & y_{1m} \\ y_{21} & y_{22} & \cdots & y_{2m} \\ \vdots & \vdots & \ddots & \vdots \\ y_{m1} & y_{m2} & \cdots & y_{mm} \end{bmatrix} = \mathbf{I} \tag{3.72}$$

The element at location (i, j) is obtained from the usual formula

$$\sum_{k=1}^{m} x_{ik}\ y_{kj} = \delta(i - j) \tag{3.73}$$

where $\delta(i - j)$ is the Dirac delta function which is one only when the argument is zero, i.e. when $i = j$. The function is zero for all other values of i and j.
The sum of the j-th column on both sides of (3.71) and (3.73) is given by

$$\sum_{i=1}^{m}\sum_{k=1}^{m} x_{ik}\ y_{kj} = 1 \tag{3.74}$$

Now reverse the order of summation on the left-hand side (LHS) of the above equation

$$\sum_{k=1}^{m} y_{kj} \sum_{i=1}^{m} x_{ik} = 1 \tag{3.75}$$

Because of the properties of the eigenvectors in Theorem 3 on page 79 and Theorem 4 on page 79, the second summation becomes

$$\sum_{k=1}^{m} y_{kj}\ \delta(k-1) = 1 \tag{3.76}$$

$\delta(k-1)$ expresses the fact that only the sum of the first column of matrix $\mathbf{X}$ evaluates to 1. All other columns add up to zero. Thus the above equation becomes

$$y_{1j} = 1 \tag{3.77}$$

This proves that the elements of the first row of matrix $\mathbf{X}^{-1}$ must be all ones. ■

3.13 Expanding $\mathbf{P}^n$ in Terms of Its Eigenvalues

We shall find that expanding $\mathbf{P}^n$ in terms of the eigenvalues of $\mathbf{P}$ will give us additional insights into the transient behavior of Markov chains. Equation (3.63) related $\mathbf{P}^n$ to the n-th powers of its eigenvalues as

$$\mathbf{P}^n = \mathbf{X}\mathbf{D}^n\mathbf{X}^{-1} \tag{3.78}$$

Thus we can express $\mathbf{P}^n$ in the above equation in the form

$$\mathbf{P}^n = \lambda_1^n \mathbf{A}_1 + \lambda_2^n \mathbf{A}_2 + \cdots + \lambda_m^n \mathbf{A}_m \tag{3.79}$$

Assume that $\lambda_1 = 1$ and all other eigenvalues have magnitudes lesser than unity (fractions) because $\mathbf{P}$ is column stochastic. In that case, the above equation becomes

$$\mathbf{P}^n = \mathbf{A}_1 + \lambda_2^n \mathbf{A}_2 + \cdots + \lambda_m^n \mathbf{A}_m \tag{3.80}$$

This shows that as time progresses, n becomes large and the powers of λ_i^n will quickly decrease. The main contribution to $\mathbf{P}^n$ will be due to $\mathbf{A}_1$ only.

The matrices $\mathbf{A}_i$ can be determined from the product

$$\mathbf{A}_i = \mathbf{X}\ \mathbf{Y}_i\ \mathbf{X}^{-1} \tag{3.81}$$

where $\mathbf{Y}_i$ is the *selection matrix* which has zeros everywhere except for element $y_{ii} = 1$.

For a 3×3 matrix, we can write

$$\mathbf{P}^n = \begin{bmatrix} \mathbf{x}_1 & \mathbf{x}_2 & \mathbf{x}_3 \end{bmatrix} \begin{bmatrix} \lambda_1^n & 0 & 0 \\ 0 & \lambda_2^n & 0 \\ 0 & 0 & \lambda_3^n \end{bmatrix} \begin{bmatrix} \mathbf{x}_1 & \mathbf{x}_2 & \mathbf{x}_3 \end{bmatrix}^{-1} \tag{3.82}$$

$$\mathbf{A}_1 = \mathbf{X}\begin{bmatrix} 1 & 0 & 0 \\ 0 & 0 & 0 \\ 0 & 0 & 0 \end{bmatrix}\mathbf{X}^{-1} \quad (3.83)$$

$$\mathbf{A}_2 = \mathbf{X}\begin{bmatrix} 0 & 0 & 0 \\ 0 & 1 & 0 \\ 0 & 0 & 0 \end{bmatrix}\mathbf{X}^{-1} \quad (3.84)$$

$$\mathbf{A}_3 = \mathbf{X}\begin{bmatrix} 0 & 0 & 0 \\ 0 & 0 & 0 \\ 0 & 0 & 1 \end{bmatrix}\mathbf{X}^{-1} \quad (3.85)$$

The selection matrix effectively ensures that $\mathbf{A}_i$ is given by the product of the i-th column of $\mathbf{X}$ and the i-th row of $\mathbf{X}^{-1}$

WWW: We developed the function `SELECT(P)` that can be used to obtain the different matrices $\mathbf{A}_i$ for a diagonalizable matrix.

It will prove useful to write $\mathbf{A}_i$ explicitly in terms of the corresponding eigenvector. $\mathbf{A}_i$ can be written in the alternative form

$$\mathbf{A}_i = \mathbf{x}_i \ \mathbf{z}_i \quad (3.86)$$

where $\mathbf{x}_i$ is the i-th eigenvector of $\mathbf{P}$, which is also the i-th column of $\mathbf{X}$, and $\mathbf{z}_i$ is the m-row vector corresponding to the i-th row of $\mathbf{X}^{-1}$. Thus we have

$$\mathbf{A}_i = \begin{bmatrix} z_{i1}\mathbf{x}_i & z_{i2}\mathbf{x}_i & z_{i3}\mathbf{x}_i & \cdots & z_{im}\mathbf{x}_i \end{bmatrix} \quad (3.87)$$

The following two theorems discuss some interesting properties of the expansion matrices $\mathbf{A}_i$.

Theorem 5 *Matrix $\mathbf{A}_1$ is column stochastic and all its columns are identical.*

Proof $\mathbf{A}_1$ is found from (3.87) as

$$\mathbf{A}_1 = \begin{bmatrix} z_{11}\mathbf{x}_1 & z_{12}\mathbf{x}_1 & z_{13}\mathbf{x}_1 & \cdots & z_{1m}\mathbf{x}_1 \end{bmatrix} \quad (3.88)$$

But Lemma 3 on page 93 proved that $z_{11}, z_{12}, \ldots, z_{1m}$ are all equal to unity. Thus the above equation becomes

$$\mathbf{A}_1 = \begin{bmatrix} \mathbf{x}_1 & \mathbf{x}_1 & \cdots & \mathbf{x}_1 \end{bmatrix} \quad (3.89)$$

This proves the theorem. ∎

Theorem 5 results in the following very useful lemma.

Lemma 4 *The steady-state distribution vector $\mathbf{s}(\infty) = \mathbf{s}$ must be independent of the initial value $\mathbf{s}(0)$ and equals any column of $\mathbf{A}_1$.*

From Theorem 5 and Lemma 4 we can state the following very useful lemma.

Lemma 5 *Each column of* $\mathbf{A}_1$ *represents the steady-state distribution vector for the Markov chain.*

We define a *differential matrix* as a matrix in which the sum of each column is zero. The following theorem states that matrices $\mathbf{A}_i$ corresponding to eigenvalues $\lambda_i \neq 1$ are all differential matrices.

Theorem 6 *The expansion matrices* $\mathbf{A}_i$ *corresponding to eigenvalues* $\lambda_i \neq 1$ *are differential, i.e.* $\sigma(A_i) = 0$.

Proof The sum of column **j** in (3.80) is given by

$$\sigma_j(\mathbf{P}^n) = \sigma_j(\mathbf{A}_1) + \lambda_2^n\, \sigma_j(\mathbf{A}_2) + \lambda_3^n\, \sigma_j(\mathbf{A}_3) + \cdots \tag{3.90}$$

Lemma 1 on page 85 assures us that $\mathbf{P}^n$ is column stochastic and Theorem 5 on page 95 assures us that $\mathbf{A}_1$ is also column stochastic. Therefore, we can write the above equation as

$$1 = 1 + \lambda_2^n\, \sigma_j(\mathbf{A}_2) + \lambda_3^n\, \sigma_j(\mathbf{A}_3) + \cdots \tag{3.91}$$

Since this equation is valid for all values of λ_i and n, we must have

$$\sigma_j(\mathbf{A}_2) = \sigma_j(\mathbf{A}_3) = \cdots = \sigma_j(\mathbf{A}_m) = 0 \tag{3.92}$$

The above equations are valid for all values of $1 \leq j \leq m$. Thus all the matrices $\mathbf{A}_i$ which correspond to $\lambda_i \neq 1$ are all differential matrices. And we can write

$$\sigma(\mathbf{A}_2) = \sigma(\mathbf{A}_3) = \cdots = \sigma(\mathbf{A}_m) = 0 \tag{3.93}$$

This proves the theorem. ∎

The following theorem is related to Theorem 2 on page 77. The theorem essentially explains the effect of premultiplying any matrix by a differential matrix.

Theorem 7 *Given any matrix* **A** *and a differential matrix* **V**, *then matrix* $\mathbf{B} = \mathbf{VA}$ *will be a differential matrix.*

Proof When **A** is premultiplied by **V** matrix **B** results

$$\mathbf{B} = \mathbf{VA} \tag{3.94}$$

Element b_{ij} is given by the usual matrix product formula

$$b_{ij} = \sum_{k=1}^{m} v_{ik}\, a_{kj} \tag{3.95}$$

The sum of the j-th column of matrix $\mathbf{B}$ is denoted by $\sigma_j(\mathbf{B})$ and is given by

$$\sigma_j(\mathbf{B}) = \sum_{i=1}^{m} b_{ij} = \sum_{i=1}^{m} \sum_{k=1}^{m} v_{ik}\, a_{kj} \tag{3.96}$$

Now reverse the order of summation on the right-hand side of the above equation

$$\sigma_j(\mathbf{B}) = \sum_{i=1}^{m} b_{ij} = \sum_{k=1}^{m} a_{kj} \sum_{i=1}^{m} v_{ik} \tag{3.97}$$

Because $\mathbf{V}$ is differential we have

$$\sigma_j(\mathbf{V}) = \sum_{i=1}^{m} v_{ik} = 0 \tag{3.98}$$

Therefore, (3.97) becomes

$$\begin{aligned} \sigma_j(\mathbf{B}) &= \sum_{i=1}^{m} b_{ij} \\ &= \sum_{k=1}^{m} a_{kj} \times 0 \\ &= 0 \end{aligned} \tag{3.99}$$

Thus we proved that sum of columns of a matrix becomes zero when the matrix is premultiplied by a differential matrix. ■

Example 17 The following is a diagonalizable state matrix.

$$\mathbf{P} = \begin{bmatrix} 0.1 & 0.3 & 1 \\ 0.2 & 0.3 & 0 \\ 0.7 & 0.4 & 0 \end{bmatrix}$$

We would like to express $\mathbf{P}^n$ in the form as in (3.80).

First thing is to check that $\mathbf{P}$ is diagonalizable by checking that it has three distinct eigenvalues. Having assured ourselves that this is the case, we use the MATLAB function `select` that we developed, to find that

$$\mathbf{P}^n = \mathbf{A}_1 + \lambda_2^n \mathbf{A}_2 + \lambda_3^n \mathbf{A}_3$$

where $\lambda_1 = 1$ and

$$\mathbf{A}_1 = \begin{bmatrix} 0.476 & 0.476 & 0.476 \\ 0.136 & 0.136 & 0.136 \\ 0.388 & 0.388 & 0.388 \end{bmatrix}$$

$$\mathbf{A}_2 = \begin{bmatrix} 0.495 & 0.102 & -0.644 \\ -0.093 & -0.019 & 0.121 \\ -0.403 & -0.083 & 0.524 \end{bmatrix}$$

$$\mathbf{A}_3 = \begin{bmatrix} 0.028 & -0.578 & 0.168 \\ -0.043 & 0.883 & -0.257 \\ 0.015 & -0.305 & 0.089 \end{bmatrix}$$

Notice that matrix $\mathbf{A}_1$ is column stochastic and all its columns are equal. Notice also that the sum of columns for matrices $\mathbf{A}_2$ and $\mathbf{A}_3$ is zero. ■

Sometimes two eigenvalues are the complex conjugate of each other. In that case the corresponding eigenvectors will be complex conjugate and so are the corresponding matrices $\mathbf{A}_i$ such that the end result $\mathbf{P}^{(n)}$ is purely real.

Example 18 A Markov chain has the transition matrix

$$\mathbf{P} = \begin{bmatrix} 0.2 & 0.4 & 0.2 \\ 0.1 & 0.4 & 0.6 \\ 0.7 & 0.2 & 0.2 \end{bmatrix}$$

Check to see if this matrix has distinct eigenvalues. If so,

(a) expand the transition matrix $\mathbf{P}^n$ in terms of its eigenvalues
(b) find the value of $\mathbf{s}(3)$ using the expression in (3.80).

The eigenvalues of $\mathbf{P}$ are $\lambda_1 = 1$, $\lambda_2 = -0.1 + j0.3$, and $\lambda_3 = -0.1 - j0.3$. We see that complex eigenvalues appear as complex conjugate pairs. The eigenvectors corresponding to these eigenvalues are

$$\mathbf{X} = \begin{bmatrix} 0.476 & -0.39 - j0.122 & -0.39 + j0.122 \\ 0.660 & 0.378 - j0.523 & 0.378 + j0.523 \\ 0.581 & 0.011 + j0.645 & 0.011 - j0.645 \end{bmatrix}$$

We see that the eigenvectors corresponding to the complex eigenvalues appear also as complex conjugate pairs. Using the function `select(P)`, we express $\mathbf{P}^n$ according to (3.80)

$$\mathbf{P}^n = \mathbf{A}_1 + \lambda_2^n \mathbf{A}_2 + \lambda_3^n \mathbf{A}_3$$

where

$$\mathbf{A}_1 = \begin{bmatrix} 0.277 & 0.277 & 0.277 \\ 0.385 & 0.385 & 0.385 \\ 0.339 & 0.339 & 0.339 \end{bmatrix}$$

$$\mathbf{A}_2 = \begin{bmatrix} 0.362 + j0.008 & -0.139 - j0.159 & -0.139 + j0.174 \\ -0.192 + j0.539 & 0.308 - j0.128 & -0.192 - j0.295 \\ -0.169 - j0.546 & -0.169 + j0.287 & 0.331 + j0.121 \end{bmatrix}$$

$$\mathbf{A}_3 = \begin{bmatrix} 0.362 - j0.008 & -0.139 + j0.159 & -0.139 - j0.174 \\ -0.192 - j0.539 & 0.308 + j0.128 & -0.192 + j0.295 \\ -0.169 + j0.546 & -0.169 - j0.287 & 0.331 - j0.121 \end{bmatrix}$$

We see that the expansion coefficients corresponding to the complex eigenvectors appear as complex conjugate pairs. This ensures that all the components of the distribution vector will always be real numbers. The distribution vector at time step $n = 3$ is

$$\begin{aligned} \mathbf{s}(3) &= \mathbf{P}^3\, \mathbf{s}(0) \\ &= \left[\mathbf{A}_1 + \lambda_2^3 \mathbf{A}_2 + \lambda_3^3 \mathbf{A}_3\right] \mathbf{s}(0) \\ &= \begin{bmatrix} 0.285 & 0.389 & 0.326 \end{bmatrix}^t \end{aligned}$$

■

Example 19 Consider the on–off source example whose transition matrix was given by

$$\mathbf{P} = \begin{bmatrix} s & 1-a \\ 1-s & a \end{bmatrix} \tag{3.100}$$

Express this matrix in diagonal form and find an expression for the n-th power of $\mathbf{P}$.

Using MAPLE or MATLAB's SYMBOLIC packages, the eigenvectors for this matrix are

$$x_1 = \begin{bmatrix} (1-a)/(1-s) & 1 \end{bmatrix}^t; \quad \text{with} \lambda_0 = 1 \tag{3.101}$$

$$x_2 = \begin{bmatrix} -1 & 1 \end{bmatrix}^t; \quad \text{with} \lambda_1 = s + a - 1 \tag{3.102}$$

Thus we have

$$\mathbf{X} = \begin{bmatrix} (1-a)/(1-s) & -1 \\ 1 & 1 \end{bmatrix} \tag{3.103}$$

$$\mathbf{X}^{-1} = \frac{1}{\alpha} \begin{bmatrix} 1-s & 1-s \\ s-1 & 1-a \end{bmatrix} \tag{3.104}$$

where $\alpha = (s + a - 1)$. And $\mathbf{P}^n$ is given by

$$\mathbf{P}^n = \mathbf{X} \begin{bmatrix} 1 & 0 \\ 0 & \alpha^n \end{bmatrix} \mathbf{X}^{-1} \tag{3.105}$$

$$= \frac{1}{\alpha} \begin{bmatrix} 1 - a - \alpha^n (s - 1) & (1 - \alpha^n)(1 - a) \\ (1 - \alpha^n)(1 - s) & 1 - s + \alpha^n (1 - a) \end{bmatrix} \tag{3.106}$$

We can write $\mathbf{P}^n$ in the summation form

$$\mathbf{P}^n = \mathbf{A}_1 + \alpha^n \mathbf{A}_2 \tag{3.107}$$

where

$$\mathbf{A}_1 = \frac{1}{2 - a - s} \begin{bmatrix} 1 - a & 1 - a \\ 1 - s & 1 - s \end{bmatrix} \tag{3.108}$$

$$\mathbf{A}_2 = \frac{1}{2 - a - s} \begin{bmatrix} 1 - s & a - 1 \\ s - 1 & 1 - a \end{bmatrix} \tag{3.109}$$

The observations on the properties of the matrices $\mathbf{A}_1$ and $\mathbf{A}_2$ can be easily verified.

For large values of n, we get the simpler form

$$\lim_{n \to \infty} \mathbf{P}^n = \mathbf{A}_1 = \frac{1}{2 - a - s} \begin{bmatrix} 1 - a & 1 - a \\ 1 - s & 1 - s \end{bmatrix} \tag{3.110}$$

and in the steady-state our distribution vector becomes

$$\mathbf{s} = \frac{1}{2 - a - s} \begin{bmatrix} (1 - a) & (1 - s) \end{bmatrix}^t$$

and this is independent of the initial state of our source.

Example 20 Consider the column stochastic matrix

$$\mathbf{P} = \begin{bmatrix} 0.5 & 0.8 & 0.4 \\ 0.5 & 0 & 0.3 \\ 0 & 0.2 & 0.3 \end{bmatrix}.$$

This matrix is diagonalizable. Express it in the form given in (3.63) and obtain a general expression for the distribution vector at step n. What would be the distribution vector after 100 steps assuming an initial distribution vector $\mathbf{s}(0) = \begin{bmatrix} 1 & 0 & 0 \end{bmatrix}^t$

The eigenvalues for this matrix are $\lambda_1 = 1$, $\lambda_2 = -0.45$, $\lambda_3 = 0.25$. Since they are distinct, we know that the matrix is diagonalizable. The eigenvectors corresponding to these eigenvalues are

$$\mathbf{x}_1 = \begin{bmatrix} 0.869 & 0.475 & 0.136 \end{bmatrix}^t$$
$$\mathbf{x}_2 = \begin{bmatrix} -0.577 & 0.789 & -0.211 \end{bmatrix}^t$$
$$\mathbf{x}_3 = \begin{bmatrix} -0.577 & -0.211 & 0.789 \end{bmatrix}^t$$

and matrix $\mathbf{X}$ in (3.63) is simply $\mathbf{X} = \begin{bmatrix} \mathbf{x}_1 & \mathbf{x}_2 & \mathbf{x}_3 \end{bmatrix}$ which is given by

$$\mathbf{X} = \begin{bmatrix} 0.869 & -0.577 & -0.577 \\ 0.475 & 0.789 & -0.212 \\ 0.136 & -0.212 & 0.789 \end{bmatrix}$$

Notice that the sum of the elements in the second and third columns is exactly zero. We also need the inverse of this matrix which is

$$\mathbf{X}^{-1} = \begin{bmatrix} 0.676 & 0.676 & 0.676 \\ -0.473 & 0.894 & -0.106 \\ -0.243 & 0.123 & 1.123 \end{bmatrix}$$

Thus we can express $\mathbf{P}^n$ as

$$\begin{aligned} \mathbf{P}^n &= \mathbf{X} \begin{bmatrix} 1 & 0 & 0 \\ 0 & \lambda_2^n & 0 \\ 0 & 0 & \lambda_3^n \end{bmatrix} \mathbf{X}^{-1} \\ &= \mathbf{A}_1 + \lambda_2^n \mathbf{A}_2 + \lambda_3^n \mathbf{A}_3 \end{aligned}$$

where the three matrices $\mathbf{A}_1$, $\mathbf{A}_2$, and $\mathbf{A}_3$ are given using the following expressions.

$$\mathbf{A}_1 = \mathbf{X} \begin{bmatrix} 1 & 0 & 0 \\ 0 & 0 & 0 \\ 0 & 0 & 0 \end{bmatrix} \mathbf{X}^{-1} \tag{3.111}$$

$$\mathbf{A}_2 = \mathbf{X} \begin{bmatrix} 0 & 0 & 0 \\ 0 & 1 & 0 \\ 0 & 0 & 0 \end{bmatrix} \mathbf{X}^{-1} \tag{3.112}$$

$$\mathbf{A}_3 = \mathbf{X} \begin{bmatrix} 0 & 0 & 0 \\ 0 & 0 & 0 \\ 0 & 0 & 1 \end{bmatrix} \mathbf{X}^{-1} \tag{3.113}$$

The above formulas produce

$$\mathbf{A}_1 = \begin{bmatrix} 0.587 & 0.587 & 0.587 \\ 0.321 & 0.321 & 0.321 \\ 0.092 & 0.092 & 0.092 \end{bmatrix}$$

$$\mathbf{A}_2 = \begin{bmatrix} 0.273 & -0.516 & 0.062 \\ -0.373 & 0.705 & -0.084 \\ 0.1 & -0.189 & 0.022 \end{bmatrix}$$

$$\mathbf{A}_3 = \begin{bmatrix} 0.141 & -0.071 & -0.649 \\ 0.051 & -0.026 & -0.237 \\ -0.192 & 0.097 & 0.886 \end{bmatrix}$$

We notice that all the columns in $\mathbf{A}_1$ are identical while the sums of the columns in $\mathbf{A}_2$ and $\mathbf{A}_3$ is zero. These properties of matrices $\mathbf{A}_i$ guarantee that $\mathbf{P}^n$ remains a column stochastic matrix for all values of n between 0 and ∞.

Since each λ is a fraction, we see that the probabilities of each state consist of a steady-state solution (independent of λ) and a transient solution that contains the different $\lambda's$. In the steady state (i.e., when $n \to \infty$) we have the distribution vector that equals any column of $\mathbf{A}_1$:

$$\mathbf{s}(\infty) = \begin{bmatrix} 0.587 & 0.321 & 0.092 \end{bmatrix}^t \tag{3.114}$$

Thus we get after 100 steps

$$\begin{aligned} \mathbf{s}(100) &= \mathbf{P}^{100}\,\mathbf{s}(0) \\ &= \begin{bmatrix} 0.6 + 0.3 \times \lambda_2^{100} + 0.1 \times \lambda_3^{100} \\ 0.3 - 0.4\lambda_2^{100} \\ 0.1 + 0.1\lambda_2^{100} - 0.2\lambda_3^{100} \end{bmatrix} \\ &= \begin{bmatrix} 0.587 & 0.321 & 0.092 \end{bmatrix}^t \end{aligned}$$

Thus the distribution vector settles down to its steady-state value irrespective of the initial state. ■

3.13.1 Test for Matrix Diagonalizability

The above two techniques expanding $\mathbf{s}(0)$ and diagonalizing $\mathbf{P}$ both relied on the fact that $\mathbf{P}$ could be diagonalized. We mentioned earlier that a matrix could be diagonalized when its eigenvalues are all distinct. While this is certainly true, there is a more general test for the diagonalizability of a matrix.

A matrix is diagonalizable only when its Jordan canonic form (JCF) is diagonal [4].

We will explore this topic more in the next section.

Example 21 Indicate whether the matrix in Example 6 on page 72 is diagonalizable or not.

The eigenvalues of this matrix are $\lambda_1 = 1$ and $\lambda_2 = \lambda_3 = -1/4$. Since some eigenvalues are repeated, **P** might or might not be diagonalizable. Using Maple or the MATLAB function `JORDAN(P)`, the Jordan canonic form for this matrix is

$$\mathbf{J} = \begin{bmatrix} 1 & 0 & 0 \\ 0 & -1/4 & 1 \\ 0 & 0 & -1/4 \end{bmatrix}$$

Since **J** is not diagonal, **P** is not diagonalizable. It is as simple as that! ∎

3.14 Finding s(n) Using the Jordan Canonic Form

We saw in Section 3.12 how easy it was to find $\mathbf{s}(n)$ by diagonalizing **P**. We would like to follow the same lines of reasoning even when **P** cannot be diagonalized because one or more of its eigenvalues are repeated.

3.14.1 Jordan Canonic Form (JCF)

Any $m \times m$ matrix **P** can be transformed through a similarity transformation into its Jordan canonic form (JCF) [4]

$$\mathbf{P} = \mathbf{U}\mathbf{J}\mathbf{U}^{-1} \tag{3.115}$$

Matrix **U** is a nonsingular matrix and J is a block-diagonal matrix

$$\mathbf{J} = \begin{bmatrix} \mathbf{J}_1 & 0 & 0 & \cdots & 0 \\ 0 & \mathbf{J}_2 & 0 & \cdots & 0 \\ 0 & 0 & \mathbf{J}_3 & \cdots & 0 \\ \vdots & \vdots & \vdots & \ddots & \vdots \\ 0 & 0 & 0 & \cdots & \mathbf{J}_t \end{bmatrix} \tag{3.116}$$

where the matrix $\mathbf{J}_i$ is an $m_i \times m_i$ **Jordan block** or **Jordan box** matrix of the form

$$\mathbf{J}_i = \begin{bmatrix} \lambda_i & 1 & 0 & \cdots & 0 \\ 0 & \lambda_i & 1 & \cdots & 0 \\ 0 & 0 & \lambda_i & \cdots & 0 \\ \vdots & \vdots & \vdots & \ddots & \vdots \\ 0 & 0 & 0 & \cdots & 1 \\ 0 & 0 & 0 & \cdots & \lambda_i \end{bmatrix} \tag{3.117}$$

such that the following equation holds

$$m_1 + m_2 + \cdots + m_t = m \tag{3.118}$$

The similarity transformation performed in (3.115) above ensures that the eigenvalues of the two matrices **P** and **J** are identical. The following example proves this statement.

Example 22 Prove that if the Markov matrix **P** has Jordan canonic form matrix **J**, then the eigenvalues of **P** are identical to those of **J**. Having proved that, find the relation between the corresponding eigenvectors for both matrices.

We start first by proving that the characteristic equations for both matrices are identical. The characteristic equation or characteristic polynomial of **P** is given by

$$x(\lambda) = \det(\mathbf{P} - \lambda \mathbf{I}) \tag{3.119}$$

where det(**A**) is the determinant of the matrix **A**. The characteristic polynomial of **J** is given by

$$y(\alpha) = \det(\mathbf{J} - \alpha \mathbf{I}) \tag{3.120}$$

where α is the assumed root or eigenvalue of **J**. But (3.115) indicates that **J** can be expressed in terms of **P** and **U** as

$$\mathbf{J} = \mathbf{U}^{-1}\mathbf{P}\mathbf{U}$$

Using this, we can express (3.120) as

$$y(\alpha) = \det\left(\mathbf{U}^{-1}\mathbf{P}\mathbf{U} - \alpha\mathbf{U}^{-1}\mathbf{U}\right)$$

We can factor out the matrices $\mathbf{U}^{-1}$ and **U** to get

$$y(\alpha) = \det\left[\mathbf{U}^{-1}(\mathbf{P} - \alpha\mathbf{I})\,\mathbf{U}\right]$$

But the rules of determinants indicate that det(**AB**) = det(**A**)×det(**B**). Thus we have

$$y(\alpha) = \det\left(\mathbf{U}^{-1}\right)\ \det(\mathbf{U})\ \det(\mathbf{P} - \alpha\mathbf{I})$$

But the rules of determinants also indicate that $\det(\mathbf{A}^{-1}) \times \det(\mathbf{A}) = 1$. Thus we have

$$y(\alpha) = \det(\mathbf{P} - \alpha\mathbf{I}) \tag{3.121}$$

This proves that the polynomial $x(\lambda)$ in (3.119) and the polynomial $y(\alpha)$ in (3.121) are identical and the matrices **P** and **J** have the same roots or eigenvalues.

So far, we proved that the two matrices **P** and **J** have the same roots or eigenvalues. Now we want to prove that they have related eigenvectors. Assume **x** is an eigenvector of **P** associated with eigenvalue λ.

$$\mathbf{P}\mathbf{x} = \lambda\mathbf{x}$$

but $\mathbf{P} = \mathbf{U}\mathbf{J}\mathbf{U}^{-1}$ and we can write the above equation as

$$\mathbf{U}\mathbf{J}\mathbf{U}^{-1}\ \mathbf{x} = \lambda\mathbf{x}$$

Now we premultiply both sides of the equation by $\mathbf{U}^{-1}$ to get

$$\mathbf{J}\mathbf{U}^{-1}\ \mathbf{x} = \lambda\mathbf{U}^{-1}\mathbf{x}$$

We write $\mathbf{u} = \mathbf{U}^{-1}\mathbf{x}$ to express the above equation as

$$\mathbf{J}\ \mathbf{u} = \lambda\mathbf{u}$$

Thus the relation between the eigenvectors of **P** and **J** is

$$\mathbf{u} = \mathbf{U}^{-1}\mathbf{x}$$

This is why the transformation

$$\mathbf{P} = \mathbf{U}\mathbf{J}\mathbf{U}^{-1}$$

is called a **similarity transformation**. In other words, a similarity transformation does not change the eigenvalues and scales the eigenvectors. ■

MATLAB offers the command `[U,J] = JORDAN(P)` to obtain the Jordan canonic form as the following example shows.

Example 23 Obtain the Jordan canonic form for the transition matrix

$$\mathbf{P} = \begin{bmatrix} 0 & 0.25 & 0.25 \\ 0.75 & 0 & 0.25 \\ 0.25 & 0.75 & 0.5 \end{bmatrix}$$

Using MATLAB we are able to obtain the Jordan canonic form

```
[U, J] = jordan(P)
U =

    0.2000         0     0.8000
    0.2800    0.4000    -0.2800
    0.5200   -0.4000    -0.5200

J =

    1.0000         0          0
    0        -0.2500     1.0000
    0              0    -0.2500
```

■

3.14.2 Properties of Jordan Canonic Form

We make the following observations on the properties of the Jordan canonic form:

1. The number of Jordan blocks equals the number of linearly independent eigenvectors of **P**.
2. The elements in **P** are real but matrix **U** might have complex elements.
3. Matrix **U** can be written as

$$\mathbf{U} = \begin{bmatrix} \mathbf{U}_1 & \mathbf{U}_2 & \cdots & \mathbf{U}_t \end{bmatrix}$$

   where each matrix $\mathbf{U}_i$ is a rectangular matrix of dimension $m \times m_i$ such that

$$m_1 + m_2 + \cdots + m_t = m$$

4. Rectangular matrix $\mathbf{U}_i$ corresponds to the Jordan block $\mathbf{J}_i$ and eigenvalue λ_i.
5. Each rectangular matrix $\mathbf{U}_i$ can be decomposed into m_i column vectors:

$$\mathbf{U}_i = \begin{bmatrix} \mathbf{u}_{i,1} & \mathbf{u}_{i,2} & \cdots & \mathbf{u}_{i,m_i} \end{bmatrix} \qquad (3.122)$$

   where each column vector has m components.
6. The first column $\mathbf{u}_{i,1}$ is the eigenvector of matrix **P** and corresponds to the eigenvalue λ_i:

$$\mathbf{P}\,\mathbf{u}_{i,1} = \lambda_i\,\mathbf{u}_{i,1} \qquad (3.123)$$

7. The other column vectors of $\mathbf{U}_i$ satisfy the recursive formula

$$\mathbf{P}\,\mathbf{u}_{i,j} = \lambda_i\,\mathbf{u}_{i,j} + \mathbf{u}_{i,j-1} \qquad (3.124)$$

   where $2 \le j \le m_i$.

8. There could be one or more blocks having the same eigenvalue. In other words, we could have two Jordan blocks $\mathbf{J}_1$ and $\mathbf{J}_2$ such that both have the same eigenvalue on their main diagonals.
9. The eigenvalue λ_i is said to have algebraic multiplicity of m_i.
10. If all Jordan blocks are one-dimensional (i.e., all $m_i = 1$), then the Jordan matrix $\mathbf{J}$ becomes diagonal. In that case, matrix $\mathbf{P}$ is diagonalizable.

Example 24 Given the following Jordan matrix, identify the Jordan blocks and find the eigenvalues and the number of linearly independent eigenvectors.

$$\mathbf{J} = \begin{bmatrix} 0.5 & 1 & 0 & 0 & 0 \\ 0 & 0.5 & 0 & 0 & 0 \\ 0 & 0 & 0.2 & 1 & 0 \\ 0 & 0 & 0 & 0.2 & 0 \\ 0 & 0 & 0 & 0 & 1 \end{bmatrix}$$

We have three Jordan blocks as follows:

$$\begin{aligned} \mathbf{J}_1 &= \begin{bmatrix} 0.5 & 1 \\ 0 & 0.5 \end{bmatrix} \\ \mathbf{J}_2 &= \begin{bmatrix} 0.2 & 1 \\ 0 & 0.2 \end{bmatrix} \\ \mathbf{J}_3 &= [1] \end{aligned}$$

From the three Jordan blocks, we determine the eigenvalues as $\lambda_1 = 0.5$, $\lambda_2 = 0.2$, and $\lambda_3 = 1$. We also know that we must have three linearly independent eigenvectors. Using MATLAB, the eigenvectors of $\mathbf{J}$ are

$$\begin{aligned} \mathbf{x}_1 &= \begin{bmatrix} 1 & 0 & 0 & 0 & 0 \end{bmatrix}^t \\ \mathbf{x}_2 &= \begin{bmatrix} 0 & 0 & 1 & 0 & 0 \end{bmatrix}^t \\ \mathbf{x}_3 &= \begin{bmatrix} 0 & 0 & 0 & 0 & 1 \end{bmatrix}^t \end{aligned}$$

we notice that these eigenvectors are linearly independent because they are orthogonal to each other. ∎

3.15 Properties of Matrix U

According to Theorem 3 on page 79, Theorem 4 on page 79, and Lemma 3 on page 93, the columns of matrix $\mathbf{U}$ must satisfy the following properties:

1. The sum of the elements of the vector $\mathbf{u}$ corresponding to the eigenvalue $\lambda = 1$ is arbitrary and could be taken as unity, i.e. $\sigma(\mathbf{u}) = 1$.
2. The sum of the elements of the vectors $\mathbf{u}$ belonging to Jordan blocks with eigenvalue $\lambda \neq 1$ must be zero, i.e. $\sigma(u) = 0$.
3. Matrix $\mathbf{U}^{-1}$ has ones in its first row.

3.16 $\mathbf{P}^n$ Expressed in Jordan Canonic Form

We explained in the previous subsection that the transition matrix could be expressed in terms of its Jordan canonic form which we repeat here for convenience:

$$\mathbf{P} = \mathbf{U}\mathbf{J}\mathbf{U}^{-1} \tag{3.125}$$

where $\mathbf{U}$ is a unitary matrix. Equation 3.125 results in a very simple expression for $\mathbf{P}^n$ in (3.38). To start, we can calculate $\mathbf{P}^2$ as

$$\begin{aligned} \mathbf{P}^2 &= \left(\mathbf{U}\mathbf{J}\mathbf{U}^{-1}\right) \times \left(\mathbf{U}\mathbf{J}\mathbf{U}^{-1}\right) && (3.126) \\ &= \mathbf{U}\mathbf{J}^2\mathbf{U}^{-1} && (3.127) \end{aligned}$$

In general we have

$$\mathbf{P}^n = \mathbf{U}\mathbf{J}^n\mathbf{U}^{-1} \tag{3.128}$$

where $\mathbf{J}^n$ has the same block structure as $\mathbf{J}$

$$\mathbf{J}^n = \begin{bmatrix} \mathbf{J}_1^n & 0 & 0 & \cdots & 0 \\ 0 & \mathbf{J}_2^n & 0 & \cdots & 0 \\ 0 & 0 & \mathbf{J}_3^n & \cdots & 0 \\ \vdots & \vdots & \vdots & \ddots & \vdots \\ 0 & 0 & 0 & \cdots & \mathbf{J}_t^n \end{bmatrix} \tag{3.129}$$

In the above equation, the Jordan block $\mathbf{J}_i^n$ of dimension $m_i \times m_i$ is an upper triangular Toeplitz matrix in the form

$$\mathbf{J}_i^n = \begin{bmatrix} f_{i0}^n & f_{i1}^n & f_{i2}^n & f_{i3}^n & \cdots & f_{i,m_i-1}^n \\ 0 & f_{i0}^n & f_{i1}^n & f_{i2}^n & \cdots & f_{i,m_i-2}^n \\ 0 & 0 & f_{i0}^n & f_{i1}^n & \cdots & f_{i,m_i-3}^n \\ \vdots & \vdots & \vdots & \vdots & \ddots & \vdots \\ 0 & 0 & 0 & 0 & \cdots & f_{i0}^n \end{bmatrix} \tag{3.130}$$

where f_{ij}^n is given by

$$f_{ij}^n = \binom{n}{j} \lambda_i^{n-j} \qquad 0 \le j < m_i \tag{3.131}$$

We assumed the binomial coefficient vanishes whenever $j > n$. In fact, the term f_{ij}^n equals the j-th term in the binomial expansion

$$(\lambda_i + 1)^n = \sum_{j=0}^{m_i-1} f_{ij}^n \tag{3.132}$$

3.17 Expressing $\mathbf{P}^n$ in Terms of Its Eigenvalues

Equation (3.128) related $\mathbf{P}^n$ to the n-th power of its Jordan canonic form as

$$\mathbf{P}^n = \mathbf{U}\mathbf{J}^n\mathbf{U}^{-1} \tag{3.133}$$

Thus we can express $\mathbf{P}^n$ in the above equation in the form

$$\mathbf{P}^n = \sum_{j=0}^{m_1-1} f_{1j}^n \mathbf{A}_{1j} + \sum_{j=0}^{m_1-1} f_{2j}^n \mathbf{A}_{2j} + \sum_{j=0}^{m_2-1} f_{3j}^n \mathbf{A}_{3j} + \cdots \tag{3.134}$$

The above equation can be represented as the double summation

$$\mathbf{P}^n = \sum_{i=1}^{t} \sum_{j=0}^{m_i-1} f_{ij}^n \, \mathbf{A}_{ij} \tag{3.135}$$

where t is the number of Jordan blocks and it was assumed that f_{ij}^n is zero whenever $j > n$.

The matrices $\mathbf{A}_{ij}$ can be determined from the product

$$\mathbf{A}_{ij} = \mathbf{U} \, \mathbf{Y}_{ij} \, \mathbf{U}^{-1} \tag{3.136}$$

where $\mathbf{Y}_{ij}$ is the selection matrix which has zeros everywhere except for the elements corresponding to the superdiagonal j in Jordan block i which contains the values 1. For example, assume a 6×6 Markov matrix whose Jordan canonic form is

$$\mathbf{J} = \begin{bmatrix} 1 & 0 & 0 & 0 & 0 & 0 \\ 0 & 0.2 & 1 & 0 & 0 & 0 \\ 0 & 0 & 0.2 & 1 & 0 & 0 \\ 0 & 0 & 0 & 0.2 & 0 & 0 \\ 0 & 0 & 0 & 0 & 0.5 & 1 \\ 0 & 0 & 0 & 0 & 0 & 0.5 \end{bmatrix}$$

This Jordan canonic form has three Jordan blocks ($t = 3$) and the selection matrix $\mathbf{Y}_{21}$ indicates that we need to access the first superdiagonal of the second Jordan block:

$$\mathbf{Y}_{21} = \begin{bmatrix} 0 & 0 & 0 & 0 & 0 & 0 \\ 0 & 0 & 1 & 0 & 0 & 0 \\ 0 & 0 & 0 & 1 & 0 & 0 \\ 0 & 0 & 0 & 0 & 0 & 0 \\ 0 & 0 & 0 & 0 & 0 & 0 \\ 0 & 0 & 0 & 0 & 0 & 0 \end{bmatrix}$$

WWW: We have prepared the MATLAB function `J_POWERS(n,P)` that accepts a Markov matrix (or any square matrix) and expresses the matrix $\mathbf{P}^n$ in the form given by (3.134).

Example 25 Consider the Markov matrix

$$\mathbf{P} = \begin{bmatrix} 0 & 0.25 & 0.4 \\ 0.75 & 0 & 0.3 \\ 0.25 & 0.75 & 0.3 \end{bmatrix}$$

Use the Jordan canonic form technique to find the decomposition of $\mathbf{P}^3$ according to (3.134).

We start by finding the Jordan canonic form for the given matrix to determine whether $\mathbf{P}$ can be diagonalized or not. The Jordan decomposition produces

$$\mathbf{U} = \begin{bmatrix} 0.20 & 0.00 & 0.80 \\ 0.44 & 0.20 & -0.44 \\ 0.36 & -0.20 & -0.36 \end{bmatrix}$$

$$\mathbf{J} = \begin{bmatrix} 1 & 0 & 0 \\ 0 & -0.25 & 1 \\ 0 & 0 & -0.25 \end{bmatrix}$$

The given Markov matrix has two eigenvalues $\lambda_1 = 1$ and $\lambda_2 = -0.25$, which is repeated twice. Since the Jordan matrix is not diagonal, we cannot diagonalize the transition matrix and we need to use the Jordan decomposition techniques to find $\mathbf{P}^n$. By using the function `J_POWERS`, we get

$$\mathbf{P}^3 = \mathbf{A}_1 + f_{20}^3 \mathbf{A}_{20} + f_{21}^3 \mathbf{A}_{21}$$

where

$$f_{20}^3 = 0.0156$$
$$f_{21}^3 = 0.187$$

and the corresponding matrices are given by

$$\mathbf{A}_1 = \begin{bmatrix} 0.20 & 0.20 & 0.20 \\ 0.44 & 0.44 & 0.44 \\ 0.36 & 0.36 & 0.36 \end{bmatrix}$$

$$\mathbf{A}_{20} = \begin{bmatrix} 0.80 & -0.20 & -0.20 \\ -0.44 & 0.56 & -0.44 \\ -0.36 & -0.36 & 0.64 \end{bmatrix}$$

$$\mathbf{A}_{21} = \begin{bmatrix} 0 & 0 & 0 \\ 0.2 & -0.05 & -0.05 \\ -0.2 & 0.05 & 0.05 \end{bmatrix}$$

■

3.18 Finding $\mathbf{P}^n$ Using the Z-Transform

The z-transform technique has been proposed for finding expressions for $\mathbf{P}^n$. In our opinion, this technique is not useful for the following reasons. Obtaining the z-transform is very tedious since it involves finding the inverse of a matrix using symbolic, not numerical, techniques. This is really tough for any matrix whose dimension is above 2×2 even when symbolic arithmetic packages are used. Furthermore, the technique will not offer any new insights that have not been already covered in this chapter. For that reason, we delegate discussion of this topic to Appendix C on page xxx. The interested reader can gloss over the appendix and compare it to the techniques developed in this chapter.

3.19 Renaming the States

Sometimes we will need to rename or relabel the states of a Markov chain. When we do that, the transition matrix will assume a simple form that helps in understanding the behavior of the system.

Renaming or relabeling the states amounts to exchanging the rows and columns of the transition matrix. For example, if we exchange states s_2 and s_5, then rows 2 and 5 as well as columns 2 and 5 will be exchanged. We perform this rearranging through the elementary exchange matrix $\mathbf{E}(2, 5)$ which exchanges states 2 and 5:

$$\mathbf{E}(2,5) = \begin{bmatrix} 1 & 0 & 0 & 0 & 0 \\ 0 & 0 & 0 & 0 & 1 \\ 0 & 0 & 1 & 0 & 0 \\ 0 & 0 & 0 & 1 & 0 \\ 0 & 1 & 0 & 0 & 0 \end{bmatrix}$$

In general, the exchange matrix $\mathbf{E}(i, j)$ is similar to the identity matrix except that rows i and j of the identity matrix are exchanged. The exchange of states is achieved by pre and post multiplying the transition matrix:

$$\mathbf{P}' = \mathbf{E}(2, 5)\,\mathbf{P}\,\mathbf{E}(2, 5)$$

Assume for example that $\mathbf{P}$ is given by

$$\mathbf{P} = \begin{array}{c} 1 \\ 2 \\ 3 \\ 4 \\ 5 \end{array} \overset{\begin{array}{ccccc} 1 & 2 & 3 & 4 & 5 \end{array}}{\begin{bmatrix} 0 & 0.1 & 0 & 0.1 & 1 \\ 0 & 0.3 & 0 & 0.2 & 0 \\ 1 & 0.2 & 0 & 0.2 & 0 \\ 0 & 0.3 & 0 & 0.4 & 0 \\ 0 & 0.1 & 1 & 0.1 & 0 \end{bmatrix}}$$

where the state indices are indicated around $\mathbf{P}$ for illustration.

Exchanging states 2 and 5 results in

$$\mathbf{P}' = \begin{array}{c} 1 \\ 5 \\ 3 \\ 4 \\ 2 \end{array} \overset{\begin{array}{ccccc} 1 & 5 & 3 & 4 & 2 \end{array}}{\begin{bmatrix} 0 & 1 & 0 & 0.1 & 0.1 \\ 0 & 0 & 1 & 0.1 & 0.1 \\ 1 & 0 & 0 & 0.2 & 0.2 \\ 0 & 0 & 0 & 0.4 & 0.1 \\ 0 & 0 & 0 & 0.2 & 0.3 \end{bmatrix}}$$

Problems

Markov Chains

3.1 Consider Example 2 where the time step value is chosen as $T = 8/\lambda_a$. Estimate the packet arrival probability a.

3.2 Three workstation clusters are connected to each other using switching hubs. At steady state the following daily traffic share of each hub was observed. For hub A, 80% of its traffic is switched to its local cluster, 5% of its traffic is switched to hub B, and 15% of its traffic is switched to hub C. For hub B, 90% of its traffic is switched to its local cluster, 5% of its traffic is switched to hub A, and 5% of its traffic is switched to hub C. For hub C, 75% of its traffic is switched to its local cluster, 10% of its traffic is switched to hub A, and 15% of its traffic is switched to hub B. Assuming initially the total traffic is distributed among the three hubs as 60% in hub A, 30% in hub B, and 10% in hub C,

(a) write the initial distribution vector for the total traffic
(b) construct the transition matrix for the Markov chain that describes the traffic share of the three hubs.

3.3 In order to plan the volume of LAN traffic flow in a building, the system administrator divided the building into three floors. Traffic volume trend indicated the following hourly pattern. In the first floor, 60% of traffic is local, 30% of traffic goes to second floor, 10% of traffic goes to third floor. In the second floor, 30% of traffic is local, 40% of traffic goes to first floor, 30% of traffic goes to third floor. In the third floor, 60% of traffic is local, 30% of traffic goes to first floor, 10% of traffic goes to second floor. Assuming initially traffic volume is distributed as 10% in first floor, 40% in second floor, and 50% in third floor,

(a) write the initial distribution vector for the total traffic
(b) construct the transition matrix for the Markov chain that describes the traffic share of the three floors.

3.4 The transition matrix for a Markov chain is given by

$$\mathbf{P} = \begin{bmatrix} 0.3 & 0.6 \\ 0.7 & 0.4 \end{bmatrix}$$

What does each entry represent?

3.5 A traffic data generator could be either idle or is generating data at five different rates $\lambda_1 < \lambda_2 < \cdots < \lambda_5$. When idle, the source could equally likely remain idle or it could start transmitting at the lowest rate λ_1. When in the highest rate state λ_5, the source could equally likely remain in that state or it could switch to the next lower rate λ_4. When in the other states, the source is equally likely to remain in its present state or it could start transmitting at the next lower or higher rate. Identify the system states and write down the transition matrix.

3.6 Repeat the above problem when transitions between the different states is equally likely.

3.7 The market over reaction theory proposes that stocks with low return (called "losers") subsequently outperform stocks with high return (called "winners") over some observation period. The rest of the market share is stocks with medium return (called "medium"). It was observed that winners split according to the following ratios: 70% become losers, 25% become medium, and 5% stay winners. Medium stocks split according to the following ratios: 5% become losers, 90% stay medium, and 5% become winers. Losers split according to the following ratios: 80% stay losers, 5% become medium, and 15% become winners. The Markov chain representing the state of a stock is defined as follows: s_0 represents loser stock, s_1 represent medium stock, and s_2 represent winner stocks. Assuming an aggressive manager's portfolio is initially split among the stocks in the following percentages, 5% losers, 70% medium, and 25% winners,

(a) write the initial distribution vector for the portfolio
(b) construct the transition matrix for the Markov chain that describes the stock share of the portfolio.

3.8 Consider Example 8, but this time the source is transmitting packets having random lengths. Assume for simplicity that the transmitted packets can be one out of three lengths selected at random with probability l_i ($i = 1, 2, 3$).

(a) Identify the different states of the system.
(b) Draw the corresponding Markov state transition diagram.
(c) Write down the transition matrix.

Time Step Selection

3.9 Develop the proper packet arrival statistics for the case considered in Section 3.3.1 when the line is sampled at a rate r while the packets arrive at an average rate λ_a.

Markov Transition Matrices

3.10 Explain what is meant by a homogeneous Markov chain.

3.11 By Assuming **P** is a transition matrix, prove that the unit row vector

$$\mathbf{u} = \begin{bmatrix} 1 & 1 & 1 & \cdots & 1 \end{bmatrix}$$

is a left eigenvector of the matrix and its associated eigenvalue is $\lambda = 1$.

3.12 Prove (3.11) on page 70 which states that at any time step n, the sum of the components of the distribution vector $\mathbf{s}(n)$ equals 1. Do your proof by proceeding as follows:

(a) Start with an initial distribution vector $\mathbf{s}(0)$ of an arbitrary dimension m such that $\sum_{i=1}^{m} s_i(0) = 1$.
(b) Prove that $\mathbf{s}(1)$ satisfies (3.11).
(c) Prove that $\mathbf{s}(2)$ also satisfies (3.11) and so on.

3.13 Prove the properties stated in Section 3.9.1 on page 85.

3.14 Prove Lemma 1 on page 85.

3.15 Given a column stochastic matrix **P** with an eigenvector **x** that corresponds to the eigenvalue $\lambda = -1$, prove that $\sigma(x) = 0$ in accordance with Theorem 4.

3.16 In problems 3.17–3.26 determine which of the given matrices are Markov matrices and justify your answer. For the Markov matrices, determine the eigenvalues, the corresponding eigenvectors, and the rank.

3.17

$$\begin{bmatrix} 0.4 & 0.4 \\ 0.6 & 0.3 \end{bmatrix}$$

3.18

$$\begin{bmatrix} 0.8 & 0.5 \\ 0.3 & 0.5 \end{bmatrix}$$

3.19

$$\begin{bmatrix} -0.1 & 0.8 \\ -0.9 & 0.2 \end{bmatrix}$$

3.20

$$\begin{bmatrix} 1.2 & 0.8 \\ -0.2 & 0.2 \end{bmatrix}$$

3.21

$$\begin{bmatrix} 1 & 0 & 0 \\ 0 & 1 & 0 \\ 0 & 0 & 1 \end{bmatrix}$$

3.22

$$\begin{bmatrix} 0.4 & 0.4 \\ 0.6 & 0.3 \end{bmatrix}$$

3.23

$$\begin{bmatrix} 0.1 & 0.3 \\ 0.2 & 0.5 \\ 0.7 & 0.2 \end{bmatrix}$$

3.24

$$\begin{bmatrix} 0 & 1 & 0 \\ 1 & 0 & 0 \\ 0 & 0 & 1 \end{bmatrix}$$

3.25

$$\begin{bmatrix} 0.8 & 0.3 & 0.5 \\ 0.2 & 0.7 & 0.5 \end{bmatrix}$$

3.26

$$\begin{bmatrix} 0.7 & 0.3 & 0.5 & 0.3 \\ 0.1 & 0.1 & 0.2 & 0.3 \\ 0.1 & 0.3 & 0.1 & 0.3 \\ 0.1 & 0.3 & 0.2 & 0.3 \end{bmatrix}$$

3.27 Choose any column stochastic matrix from the matrices in the above problems, or choose one of your own, then reduce the value of one of its nonzero elements slightly (keeping the matrix nonnegative of course). In that way, the matrix will not be a column stochastic matrix any longer. Observe the change in the maximum eigenvalue and the corresponding eigenvector.

3.28 Assume a source is sending packets on a wireless channel. The source could be in one of three states: (1) idle state. (2) successful transmission state where source is active and transmitted packet is received without errors. (3) erroneous transmission state where source is active and transmitted packet is received with errors.

Assume the probability the source switches from idle to active is a and the probability that the source successfully transmits a packet is s. Draw a state transition diagram indicating the transition probabilities between states and find the transition matrix.

Transient Behavior

3.29 For problem 3.2 how much share of the traffic will be maintained by each hub after one day and after two days?

3.30 For problem 3.3 how much share of the traffic will be maintained by each floor after one hour and after two hours?

3.31 The transition matrix for a Markov chain is given by

$$\mathbf{P} = \begin{bmatrix} 0.8 & 0.1 \\ 0.2 & 0.9 \end{bmatrix}$$

(a) Given that the system is in state s_0, what is the probability the next state will be s_1?

(b) For the initial distribution vector $\mathbf{s}(0)$ find $\mathbf{s}(1)$:

$$\mathbf{s}(0) = \begin{bmatrix} 0.4 & 0.6 \end{bmatrix}^t$$

3.32 The transition matrix for a Markov chain is given by

$$\mathbf{P} = \begin{bmatrix} 0.5 & 0.3 & 0.5 \\ 0 & 0.3 & 0.25 \\ 0.5 & 0.4 & 0.25 \end{bmatrix}$$

(a) What does the entry p_{01} represent?
(b) Given that the system is in state s_0, what is the probability the next state will be s_1?
(c) For the initial distribution vector $\mathbf{s}(0)$

$$\mathbf{s}(0) = \begin{bmatrix} 0.4 & 0.6 & 0 \end{bmatrix}^t$$

find $\mathbf{s}(1)$.

3.33 Given a transition matrix

$$\mathbf{P} = \begin{bmatrix} 0.2 & 0.7 \\ 0.8 & 0.3 \end{bmatrix}$$

what is the probability of making a transition to state s_0 given that we are in state s_1?

3.34 Assume a hypothetical city where yearly computer buying trends indicate that 95% of the people who own a desktop computer will purchase a desktop and the rest will switch to laptops. On the other hand, 60% of laptop owners will continue to buy laptops and the rest will switch to desktops. At the beginning of the year 65% of the computer owners had desktops. What will be the percentages of desktop and laptop owners aftcr onc, two, and ten years?

3.35 Assume the state of a typical winter day in Cairo to be sunny, bright, or partly cloudy. Observing the weather pattern reveals the following. When today is sunny, tomorrow will be bright with probability 80%, and partly cloudy with probability 20%. When today is bright, tomorrow will be sunny with probability 60%, bright with probability 30%, and partly cloudy with probability 10%. When today is partly cloudy, tomorrow will be sunny with probability 30%, bright with probability 40%, and partly cloudy with probability 30%. Assume state 1 represents a sunny day, state 2 represents a bright day, and state 3 represents a partly cloudy day.

(a) Construct a state transition matrix.
(b) What is the probability that it will be partly cloudy tomorrow given that it is sunny today?
(c) What is the probability that it will be bright day after tomorrow given that is bright today?

3.36 Assume a gambler plays double or nothing game using a fair coin and starting with one dollar.

(a) Draw a state diagram for the amount of money with the gambler and explain how much money corresponds to each state.
(b) Derive the transition matrix.
(c) What is the probability that the gambler will have more than $500 after playing the game for 10 tosses of the coin?

3.37 Suppose you play the following game with a friend, both of you start with $2. The game starts when the coin is filipped. If the coin comes up heads, you win $1. If the coin comes up tails, you lose $1. The game ends when either of you do not have anymore money.

(a) Construct a Markov transition diagram and transition matrix for this game.
(b) Find the eigenvectors and eigenvalues for this matrix.
(c) What is the initial probability vector when you start the game?

3.38 Assume you are playing a truncated form of the snakes and ladder game using a fair coin instead of the dice. The number of squares is assumed to be 10, to make things simple, and each player starts at the first square (we label it square 1 and the last square is labeled 10). Tails mean the player advances one square and heads mean the player advances by two squares. To make the game interesting, some squares have special transitions according to the following rules which indicate the address on the next square upon the flip of the coin.

Square	Heads	Tails
2	4	2
3	6	1
5	7	2
8	9	4

Write the initial distribution vector and the transition matrix. What will be the distribution vector be after 5 flips of the coin? What are the chances of a player winning the game after 10 flips?

3.39 Assume a particle is allowed to move on a one-dimensional grid and starts at the middle. The probability of the particle moving to the right is p and to the left is q, where $p+q=1$. Assume the size of the gird to extend from 1 to N, with N assumed odd. Assume that at the end points of the grid, the particle is reflected back with probability 1. Draw a Markov transition diagram and write down the corresponding transition matrix. Assume $p=0.6$ and $q=0.4$, and $N=7$. Plot the most probable position for the particle versus time.

3.40 A parrot breeder has birds of two colors, blue and green. She finds that 60% of the males are blue if the father was blue and 80% of the males are green if the father was green. Write down the transition matrix for the males parrot. What is the probability that a blue male has a blue male after two and three generations?

3.41 A virus can mutate between N different states (typically $N = 20$). In each generation it either stays the same or mutates to another sate. Assume that the virus is equally likely to change state to any of the N states. You can reduce the size of the transition matrix $\mathbf{P}$ from $N \times N$ to $2{\times}2$ only by studying two sates: original state and "others" state which contains all other mutations. Construct the transition matrix describing the two-state Markov chain. What is the probability that in the n-th generation it will return to its original state?

3.42 Consider the stock portfolio Problem 3.7.

(a) What will be the performance of the portfolio after one, two, and ten years?
(b) Investigate the performance of the conservative portfolio that starts with the following percentage distribution of stocks $\mathbf{s}(0) = \begin{bmatrix} 0.5 & 0.5 & 0 \end{bmatrix}^t$ over the same period of time as in (a) above.
(c) Investigate the performance of a "very aggressive" portfolio that starts with the following percentage distribution of stocks $\mathbf{s}(0) = \begin{bmatrix} 0 & 0 & 1 \end{bmatrix}^t$ over the same period of time as in (a) above.
(d) Compare the long-term performance of the conservative and aggressive portfolios.

3.43 A hidden Markov chain model can be used as a waveform generator. Your task is to generate random waveforms using the following procedure.

(1) Define a set of quantization levels $Q_1, Q_2, \cdots, Q_m$ for the signal values.
(2) Define a $m \times m$ Markov transition matrix for the system. Choose your own transition probabilities for the matrix.
(3) Choose an initial state vector from the set with only one nonzero entry chosen at random from the set of quantization levels

$$\mathbf{s}(0) = \begin{bmatrix} 0 & \cdots & 0 & 1 & 0 & \cdots & 0 \end{bmatrix}^t$$

If that single element is in position k, then the corresponding initial output value is Q_k.

(4) Evaluate the next state vector using the iteration

$$\mathbf{s}(i) = \mathbf{P}\,\mathbf{s}(i-1)$$

(5) Generate the cumulative function

$$F_j(i) = \sum_{k=1}^{j} s_k(i)$$

(6) Generate a random variable x using the uniform distribution and estimate the index j for the output Q_j that satisfies the inequality

$$F_{j-1}(i) < x \le F_j(i)$$

(7) Repeat 4.

Finding P^n by Expanding s(0)

3.44 In Section 3.11 we expressed **s**(0) in terms of the eigenvectors of the transition matrix according to (3.41). Prove that if a pair of the eigenvectors is a complex conjugate pair, then the corresponding coefficients of **c** are also a complex conjugate pair.

3.45 The given transition matrix

$$\mathbf{P} = \begin{bmatrix} 0.1 & 0.4 & 0.6 \\ 0.1 & 0.4 & 0.2 \\ 0.8 & 0.2 & 0.2 \end{bmatrix}$$

has distinct eigenvalues. Express the initial state vector

$$\mathbf{s}(0) = \begin{bmatrix} 0 & 1 & 0 \end{bmatrix}^t$$

in terms of its eigenvectors, then find the distribution vector for $n = 4$.

3.46 The given transition matrix

$$\mathbf{P} = \begin{bmatrix} 0.1 & 0.4 & 0.4 & 0.1 \\ 0.1 & 0.4 & 0.2 & 0.1 \\ 0.1 & 0.1 & 0.2 & 0.7 \\ 0.7 & 0.1 & 0.2 & 0.1 \end{bmatrix}$$

has distinct, but complex, eigenvalues. Express the initial state vector

$$\mathbf{s}(0) = \begin{bmatrix} 0 & 1 & 0 & 0 \end{bmatrix}^t$$

in terms of its eigenvectors, then find the distribution vector for $n = 5$.

3.47 The given transition matrix

$$\mathbf{P} = \begin{bmatrix} 0.1 & 0.4 & 0.1 & 0.8 \\ 0.1 & 0.4 & 0.5 & 0 \\ 0.3 & 0.1 & 0.4 & 0.1 \\ 0.5 & 0.1 & 0.0 & 0.1 \end{bmatrix}$$

has distinct, but complex, eigenvalues. Express the initial state vector

$$\mathbf{s}(0) = \begin{bmatrix} 0 & 1 & 0 & 0 \end{bmatrix}^t$$

in terms of its eigenvectors, then find the distribution vector for $n = 5$.

3.48 The given transition matrix

$$\mathbf{P} = \begin{bmatrix} 0.5 & 0.3 & 0.5 \\ 0.5 & 0.3 & 0 \\ 0 & 0.4 & 0.5 \end{bmatrix}$$

has distinct, but complex, eigenvalues. Express the initial state vector

$$\mathbf{s}(0) = \begin{bmatrix} 0 & 1 & 0 \end{bmatrix}^t$$

in terms of its eigenvectors, then find the distribution vector for $n = 7$.

3.49 The given transition matrix

$$\mathbf{P} = \begin{bmatrix} 0.5 & 0.75 \\ 0.5 & 0.25 \end{bmatrix}$$

has distinct, and real, eigenvalues. Express the initial state vector

$$\mathbf{s}(0) = \begin{bmatrix} 0 & 1 \end{bmatrix}^t$$

in terms of its eigenvectors, then find the distribution vector for $n = 7$.

3.50 The given transition matrix

$$\mathbf{P} = \begin{bmatrix} 0.9 & 0.75 \\ 0.1 & 0.25 \end{bmatrix}$$

has distinct, and real, eigenvalues. Express the initial state vector

$$\mathbf{s}(0) = \begin{bmatrix} 0 & 1 \end{bmatrix}^t$$

in terms of its eigenvectors, then find the distribution vector for $n = 7$.

References

1. J. Warland and R. Varaiya *High-Performance Communication Networks*, Moragan Kaufmann, San Francisco, 2000.
2. M. E. Woodward, *Communication and Computer Networks*, IEEE Computer Society Press, Los Alamitos, CA, 1994.
3. W.J. Stewart, *Introduction to Numerical Solutions of Markov Chains*, Princeton University Press, Princeton, New Jersey, 1994.
4. R. Horn and C. R. Johnson, *Matrix Analysis*, Cambridge University Press, Cambridge, England, 1985.
5. M. Gardner, *Aha! Insight*, Scientific American, Inc./W.H.Freeman and Company, New York City, 1978.

Chapter 4
Markov Chains at Equilibrium

4.1 Introduction

In this chapter we will study the long-term behavior of Markov chains. In other words, we would like to know the distribution vector $\mathbf{s}(n)$ when $n \to \infty$. The state of the system at equilibrium or steady state can then be used to obtain performance parameters such as throughput, delay, loss probability, etc.

4.2 Markov Chains at Equilibrium

Assume a Markov chain in which the transition probabilities are not a function of time t or n, for the continuous-time or discrete-time cases, respectively. This defines a *homogeneous* Markov chain. At steady state as $n \to \infty$ the distribution vector $\mathbf{s}$ settles down to a unique value and satisfies the equation

$$\mathbf{P}\,\mathbf{s} = \mathbf{s} \tag{4.1}$$

This is because the distribution vector value does not vary from one time instant to another at steady state. We immediately recognize that $\mathbf{s}$ in that case is an eigenvector for $\mathbf{P}$ with corresponding eigenvalue $\lambda = 1$. We say that the Markov chain has reached its steady state when the above equation is satisfied.

4.3 Significance of s at "Steady State"

Equation (4.1) indicates that if $\mathbf{s}$ is the present value of the distribution vector, then after one time step the distribution vector will be $\mathbf{s}$ still. The system is now in equilibrium or steady state. The reader should realize that we are talking about probabilities here.

At steady state the system will not settle down to one particular state, as one might suspect. Steady state means that the probability of being in any state will not change with time. The probabilities, or components, of the vector $\mathbf{s}$ are the ones that

F. Gebali, *Analysis of Computer and Communication Networks*,
DOI: 10.1007/978-0-387-74437-7_4, © Springer Science+Business Media, LLC 2008

are in steady state. The components of the transition matrix $\mathbf{P}^n$ will also reach their steady state. The system is then said to be in steady state.

Assume a five-state system whose equilibrium or steady-state distribution vector is

$$\mathbf{s} = \begin{bmatrix} s_1 & s_2 & s_3 & s_4 & s_5 \end{bmatrix}^t \quad (4.2)$$

$$= \begin{bmatrix} 0.2 & 0.1 & 0.4 & 0.1 & 0.2 \end{bmatrix}^t \quad (4.3)$$

Which state would you think the system will be in at equilibrium? The answer is, the system is in state s_1 **with probability** 20%, or the system is in state s_2 with probability 10%, etc. However, we can say that at steady state the system is most probably in state s_3 since it has the highest probability value.

4.4 Finding Steady-State Distribution Vector s

The main goal of this chapter is to find **s** for a given **P**. Knowledge of this vector helps us find many performance measures for our system such as packet loss probability, throughput, delay, etc. The technique we choose for finding **s** depends on the size and structure of **P**.

Since the steady-state distribution is independent of the initial distribution vector **s** (0), we conclude therefore that $\mathbf{P}^n$ approaches a special structure for large values of n. In this case we find that the columns of $\mathbf{P}^n$, for large values of n, will all be identical and equal to the steady-state distribution vector **s**. We could see that in Examples 3.11 on page 82 and Example 3.20 on page 100.

Example 1 Find the steady-state distribution vector for the given transition matrix by

(a) calculating higher powers for the matrix $\mathbf{P}^n$
(b) calculating the eigenvectors for the matrix.

$$\mathbf{P} = \begin{bmatrix} 0.2 & 0.4 & 0.5 \\ 0.8 & 0 & 0.5 \\ 0 & 0.6 & 0 \end{bmatrix}$$

The given matrix is column stochastic and hence could describe a Markov chain. Repeated multiplication shows that the entries for $\mathbf{P}^n$ settle down to their stable values.

$$\mathbf{P}^2 = \begin{bmatrix} 0.36 & 0.38 & 0.30 \\ 0.16 & 0.62 & 0.40 \\ 0.48 & 0 & 0.30 \end{bmatrix}$$

$$\mathbf{P}^5 = \begin{bmatrix} 0.3648 & 0.3438 & 0.3534 \\ 0.4259 & 0.3891 & 0.3970 \\ 0.2093 & 0.2671 & 0.2496 \end{bmatrix}$$

$$\mathbf{P}^{10} = \begin{bmatrix} 0.3535 & 0.3536 & 0.3536 \\ 0.4042 & 0.4039 & 0.4041 \\ 0.2424 & 0.2426 & 0.2423 \end{bmatrix}$$

$$\mathbf{P}^{20} = \begin{bmatrix} 0.3535 & 0.3535 & 0.3535 \\ 0.4040 & 0.4040 & 0.4040 \\ 0.2424 & 0.2424 & 0.2424 \end{bmatrix}$$

The entries for $\mathbf{P}^{20}$ all reached their stable values. Since all the columns of $\mathbf{P}^{20}$ are identical, the stable-distribution vector is independent of the initial distribution vector. (Could you prove that? It is rather simple.) Furthermore, any column of $\mathbf{P}^{20}$ gives us the value of the equilibrium distribution vector.

The eigenvector corresponding to unity eigenvalue is found to be

$$\mathbf{s} = \begin{bmatrix} 0.3535 & 0.4040 & 0.2424 \end{bmatrix}^t$$

Notice that the equilibrium distribution vector is identical to the columns of the transition matrix $\mathbf{P}^{20}$. ■

4.5 Techniques for Finding s

We can use one of the following approaches for finding the steady-state distribution vector **s**. Some approaches are algebraic while the others rely on numerical techniques.

1. Repeated multiplication of **P** to obtain $\mathbf{P}^n$ for high values of n.
2. Eigenvector corresponding to eigenvalue $\lambda = 1$ for **P**.
3. Difference equations.
4. Z-transform (generating functions).
5. Direct numerical techniques for solving a system of linear equations.
6. Iterative numerical techniques for solving a system of linear equations.
7. Iterative techniques for expressing the states of **P** in terms of other states.

Which technique is easier depends on the structure of **P**. Some rough guidelines follow.

1. The repeated multiplication technique in 1 is prone to numerical roundoff errors and one has to use repeated trials until the matrix entries stop changing for increasing values of n.

2. The eigenvector technique (2) is used when **P** is expressed numerically and its size is reasonable so that any mathematical package could easily find the eigenvector. Some communication systems are described by a small 2×2 transition matrix and it is instructive to get a closed-form expression for **s**. We shall see this for the case of packet generators.
3. The difference equations technique (3) is used when **P** is banded with few subdiagonals. Again, many communication systems have banded transition matrices. We shall see many examples throughout this book about such systems.
4. The z-transform technique (4) is used when **P** is lower triangular or lower Hessenberg such that each diagonal has identical elements. Again, some communication systems have this structure and we will discuss many of them throughout this book.
5. The direct technique (5) is used when **P** is expressed numerically and **P** has no particular structure. Furthermore, the size of **P** is not too large such that rounding or truncation noise is insignificant. Direct techniques produce results with accuracies dependent on the machine precision and the number of calculations involved.
6. The iterative numerical technique (6) is used when **P** is expressed numerically and **P** has no particular structure. The size of **P** has little effect on truncation noise because iterative techniques produce results with accuracies that depend only on the machine precision and independent of the number of calculations involved.
7. The iterative technique (7) for expressing the states of **P** in terms of other states is illustrated in Section 9.3.2 on page 312.

We illustrate these approaches in the following sections.

4.6 Finding s Using Eigenvector Approach

In this case we are interested in finding the eigenvector **s** which satisfies the condition

$$\mathbf{P}\,\mathbf{s} = \mathbf{s} \tag{4.4}$$

MATLAB and other mathematical packages such as Maple and Mathematica have commands for finding that eigenvector as is explained in Appendix E. This technique is useful only if **P** is expressed numerically. Nowadays, those mathematical packages can also do symbolic computations and can produce an answer for **s** when **P** is expressed in symbols. However, symbolic computations demand that the the size of **P** must be small, in the range of 2–5, at the most to get any useful data.

Having found a numeric or symbolic answer, we must normalize **s** to ensure that

$$\sum_i s_i = 1 \tag{4.5}$$

Example 2 Find the steady-state distribution vector for the following transition matrix.

$$\mathbf{P} = \begin{bmatrix} 0.8 & 0.7 & 0.5 \\ 0.15 & 0.2 & 0.3 \\ 0.05 & 0.1 & 0.2 \end{bmatrix}$$

We use MATLAB to find the eigenvectors and eigenvalues for $\mathbf{P}$:

$$\mathbf{s}_1 = \begin{bmatrix} 0.9726 & 0.2153 & .0877 \end{bmatrix}^t \qquad \leftrightarrow \lambda_1 = 1$$
$$\mathbf{s}_2 = \begin{bmatrix} 0.8165 & -0.4882 & -0.4082 \end{bmatrix}^t \qquad \leftrightarrow \lambda_2 = 0.2$$
$$\mathbf{s}_3 = \begin{bmatrix} 0.5345 & -0.8018 & 0.2673 \end{bmatrix}^t \qquad \leftrightarrow \lambda_3 = 0$$

The steady-state distribution vector $\mathbf{s}$ corresponds to $\mathbf{s}_1$ and we have to normalize it. We have

$$\sum_i s_i = 1.2756$$

Dividing $\mathbf{s}_1$ by this value we get the steady-state distribution vector as

$$\mathbf{s}_1 = \begin{bmatrix} 0.7625 & 0.1688 & 0.0687 \end{bmatrix}^t$$

■

4.7 Finding s Using Difference Equations

This technique for finding $\mathbf{s}$ is useful only when the state transition matrix $\mathbf{P}$ is banded. Consider the Markov chain representing a simple discrete-time birth–death process whose state transition diagram is shown in Fig. 4.1. For example, each state might correspond to the number of packets in a buffer whose size grows by one or decreases by one at each time step. The resulting state transition matrix $\mathbf{P}$ is tridiagonal with each subdiagonal composed of identical elements.

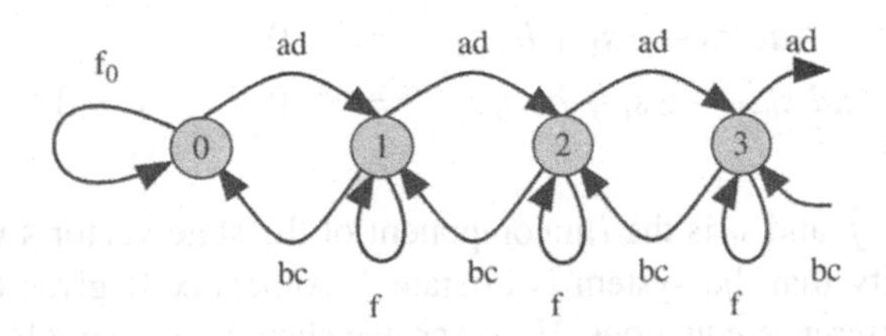

Fig. 4.1 State transition diagram for a discrete-time birth–death Markov chain

We make the following assumptions for the Markov chain.

1. The state of the Markov chain corresponds to the number of packets in the buffer or queue. s_i is the probability that i packets are in the buffer.
2. The size of the buffer or queue is assumed unrestricted.
3. The probability of a packet arriving to the system is a at a particular time, and the probability that a packet does not arrive is $b = 1 - a$.
4. The probability of a packet departing the system is c at a particular time, and the probability that a packet does not depart is $d = 1 - c$.
5. When a packet arrives it could be serviced at the same time step and it could leave the queue, at that time step, with probability c.

From the transition diagram, we write the state transition matrix as

$$\mathbf{P} = \begin{bmatrix} f_0 & bc & 0 & 0 & \cdots \\ ad & f & bc & 0 & \cdots \\ 0 & ad & f & bc & \cdots \\ 0 & 0 & ad & f & \cdots \\ \vdots & \vdots & \vdots & \vdots & \ddots \end{bmatrix} \tag{4.6}$$

where $f_0 = ac + b$ and $f = ac + bd$. For example, starting with state 1, the system goes to state 0 when a packet does not arrive and the packet that was in the buffer departs. This is represented by the term bc at location (1,2) of the matrix.

Using this matrix, or the transition diagram, we can arrive at difference equations relating the equilibrium distribution vector components as follows. We start by writing the equilibrium equation

$$\mathbf{P}\,\mathbf{s} = \mathbf{s} \tag{4.7}$$

Equating corresponding elements on both sides of the equation, we get the following equations.

$$ad\ s_0 - bc\ s_1 = 0 \tag{4.8}$$

$$ad\ s_0 - g\ s_1 + bc\ s_2 = 0 \tag{4.9}$$

$$ad\ s_{i-1} - g\ s_i + bc\ s_{i+1} = 0 \qquad i > 0 \tag{4.10}$$

where $g = 1 - f$ and s_i is the ith component of the state vector $\mathbf{s}$ which is equal to the probability that the system is in state i. Appendix B gives techniques for solving such difference equations. However, we show here a simple method based on iterations. From (4.8), (4.9), and (4.10) we can write the expressions

$$s_1 = \left(\frac{a\,d}{b\,c}\right) s_0$$

$$s_2 = \left(\frac{a\,d}{b\,c}\right)^2 s_0$$

$$s_3 = \left(\frac{a\,d}{b\,c}\right)^3 s_0$$

and in general

$$s_i = \left(\frac{ad}{bc}\right)^i s_0 \qquad i \geq 0 \tag{4.11}$$

It is more convenient to write s_i in the form

$$s_i = \rho^i\, s_0 \qquad i \geq 0 \tag{4.12}$$

where

$$\rho = \frac{a\,d}{b\,c} < 1 \tag{4.13}$$

In that sense ρ can be thought of as *distribution index* that dictates the magnitude of the distribution vector components. The complete solution is obtained from the above equations, plus the condition

$$\sum_{i=0}^{\infty} s_i = 1 \tag{4.14}$$

By substituting the expressions for each s_i in the above equation, we get

$$s_0 \sum_{i=0}^{\infty} \rho^i = 1 \tag{4.15}$$

Thus we obtain

$$\frac{s_0}{1-\rho} = 1 \tag{4.16}$$

from which we obtain the probability that the system is in state 0 as

$$s_0 = 1 - \rho \tag{4.17}$$

and the components of the equilibrium distribution vector is given from (4.12) by

$$s_i = (1-\rho)\rho^i \qquad i \geq 0 \tag{4.18}$$

For the system to be stable we must have $\rho < 1$. If we are interested in situations where $\rho \geq 1$ then we must deal with finite-sized systems where the highest state could be s_B and the system could exist in one of $B + 1$ states only. This situation will be treated more fully in Section 7.6 on page 233.

Example 3 Consider the transition matrix for the discrete-time birth–death process that describes single-arrival, single-departure queue with the following parameters $a = 0.4$ and $c = 0.6$. Construct the transition matrix and find the equilibrium distribution vector.

The transition matrix becomes

$$\mathbf{P} = \begin{bmatrix} 0.84 & 0.36 & 0 & 0 & \cdots \\ 0.16 & 0.48 & 0.36 & 0 & \cdots \\ 0 & 0.16 & 0.48 & 0.36 & \cdots \\ 0 & 0 & 0.16 & 0.48 & \cdots \\ \vdots & \vdots & \vdots & \vdots & \ddots \end{bmatrix}$$

Using (4.13), the distribution index is equal to

$$\rho = 0.4444$$

From (4.17) we have

$$s_0 = 0.5556$$

and from (4.18) we have

$$s_i = (1 - \rho)\rho^i = (0.5556) \times 0.4444^i$$

The distribution vector at steady state is

$$\mathbf{s} = \begin{bmatrix} 0.5556 & 0.2469 & 0.1097 & 0.0488 & 0.0217 & \cdots \end{bmatrix}^t$$

■

4.8 Finding s Using Z-Transform

This technique is useful to find the steady-state distribution vector **s** only if the state transition matrix **P** is lower Hessenberg such that elements on one diagonal are mostly identical. This last restriction will result in difference equations with constant coefficients for most of the elements of **s**. This type of transition matrix occurs naturally in multiple-arrival, single-departure ($M^m/M/1$) queues. This queue will be discussed in complete detail in Section 7.6 on page 233.

Let us consider a Markov chain where we can move from state j to any state i where $i \geq j-1$. This system corresponds to a buffer, or queue, whose size can increase by more than one due to multiple packet arrivals at any time, but the size of the queue can only decrease by one due to the presence of a single server. Assume the probability of K arrivals at instant n is given by

$$p(K \text{ arrivals}) = a_K \qquad K = 0, 1, 2, \ldots \tag{4.19}$$

for all time instants $n = 0, 1, 2, \ldots$ Assume that the probability that a packet is able to leave the queue is c and the probability that it is not able to leave the queue is $d = 1 - c$.

The condition for the stability of the system is

$$\sum_{K=0}^{\infty} K\, a_K < c \tag{4.20}$$

which indicates that the average number of arrivals at a given time is less than the average number of departures from the system. The state transition matrix will be lower Hessenberg:

$$\mathbf{P} = \begin{bmatrix} a_0 & b_0 & 0 & 0 & \cdots \\ a_1 & b_1 & b_0 & 0 & \cdots \\ a_2 & b_2 & b_1 & b_0 & \cdots \\ a_3 & b_3 & b_2 & b_1 & \cdots \\ \vdots & \vdots & \vdots & \vdots & \ddots \end{bmatrix} \tag{4.21}$$

where $b_i = a_i c + a_{i-1} d$ and we assumed

$$a_i = 0 \qquad \text{when } i < 0$$

Note that the sum of each column is unity, as required from the definition of a Markov chain transition matrix. At equilibrium we have

$$\mathbf{P}\,\mathbf{s} = \mathbf{s} \tag{4.22}$$

The general expression for the equilibrium equations for the states is given by the i-th term in the above equation

$$s_i = a_i s_0 + \sum_{j=1}^{i+1} b_{i-j+1}\, s_j \tag{4.23}$$

This analysis is a modified and simpler version of the one given in reference [1]. Define the z-transform for the state transition probabilities a_i and b_i as

$$A(z) = \sum_{i=0}^{\infty} a_i \, z^{-i} \quad (4.24)$$

$$B(z) = \sum_{i=0}^{\infty} b_i \, z^{-i} \quad (4.25)$$

One observation worth mentioning here for later use is that since all entries in the transition matrix are positive, all the coefficients of $A(z)$ and $B(z)$ are positive. We define the z-transform for the equilibrium distribution vector **s** as

$$S(z) = \sum_{i=0}^{\infty} s_i \, z^{-i} \quad (4.26)$$

where s_i are the components of the distribution vector. From (4.23) we can write

$$S(z) = s_0 \sum_{i=0}^{\infty} a_i \, z^{-i} + \sum_{i=0}^{\infty} \sum_{j=1}^{i+1} b_{i-j+1} \, s_j \, z^{-i} \quad (4.27)$$

$$= s_0 A(z) + \sum_{i=0}^{\infty} \sum_{j=1}^{\infty} b_{i-j+1} \, s_j \, z^{-i} \quad (4.28)$$

we were able to change the upper limit for j, in the above equation, by assuming

$$b_i = 0 \qquad \text{when } i < 0 \quad (4.29)$$

Now we change the order of the summation

$$S(z) = s_0 A(z) + \sum_{j=1}^{\infty} \sum_{i=0}^{\infty} b_{i-j+1} \, s_j \, z^{-i} \quad (4.30)$$

Making use of the assumption in (4.29), we can change the limits of the summation for i

$$S(z) = s_0 A(z) + \left[\sum_{j=1}^{\infty} s_j \, z^{-j} \right] \times \sum_{i=j-1}^{\infty} b_{i-j+1} \, z^{-(i-j)} \quad (4.31)$$

We make use of the definition of $S(z)$ to change the term in the square brackets:

$$S(z) = s_0 A(z) + [S(z) - s_0] \times \sum_{i=j-1}^{\infty} b_{i-j+1} \, z^{-(i-j)} \quad (4.32)$$

We now change the summation symbol i by using the new variable $m = i - j + 1$

$$\begin{aligned} S(z) =& s_0 A(z) + [S(z) - s_0] \times z \sum_{m=0}^{\infty} b_m z^{-m} \\ =& s_0 A(z) + z\,[S(z) - s_0]\,B(z) \end{aligned} \tag{4.33}$$

We finally get

$$S(z) = s_0 \times \frac{z^{-1} A(z) - B(z)}{z^{-1} - B(z)} \tag{4.34}$$

Below we show how we can obtain a numerical value for s_0. Assuming for the moment that s_0 is found, MATLAB allows us to find the inverse z-transform of $S(z)$ using the command `RESIDUE(a,b)` where a and b are the coefficients of $A(z)$ and $B(z)$, respectively, in descending powers of z^{-1}. The function `RESIDUE` returns the column vectors r, p, and c which give the residues, poles, and direct terms, respectively.

The solution for s_i is given by the expression

$$s_i = c_i + \sum_{j=1}^{m} r_j (p_j)^{(i-1)} \qquad i > 0 \tag{4.35}$$

where m is the number of elements in r or p vectors. The examples below show how this procedure is done.

When $z = 1$, the z-transforms for s, a_i, and b_i are

$$S(1) = \sum_{j=1}^{\infty} s_j = 1 \tag{4.36}$$

$$A(1) = B(1) = 1 \tag{4.37}$$

Thus we can put $z = 1$ in (4.34) and use L'Hospital's rule to get

$$s_0 = \frac{1 + B'(1)}{1 + B'(1) - A'(1)} \tag{4.38}$$

where

$$A'(1) = \left. \frac{dA(z)}{dz} \right|_{z=1} < 0 \tag{4.39}$$

$$B'(1) = \left. \frac{dB(z)}{dz} \right|_{z=1} < 0 \tag{4.40}$$

Since all the coefficients of $A(z)$ and $B(z)$ are positive, all the coefficients of A' and B' are negative and the two numbers $A'(1)$ and $B'(1)$ are smaller than zero. As a result of this observation, it is guaranteed that $s_0 < 1$ as expected for the probability that the queue is empty.

We should note that the term $-A'(1)$ represents the average number of packet arrivals per time step when the queue is empty. Similarly, $-B'(1)$ represents the average number of packet arrivals per time step when the queue is not empty.

Having found a numerical value for s_0, we use (4.34) to obtain the inverse z-transform of $S(z)$ and get expressions for the steady-state distribution vector

$$\mathbf{s} = \begin{bmatrix} s_0 & s_1 & s_2 & \cdots \end{bmatrix} \tag{4.41}$$

Example 4 Use the z-transform technique to find the equilibrium distribution vector for the Markov chain whose transition matrix is

$$\mathbf{P} = \begin{bmatrix} 0.84 & 0.36 & 0 & 0 & \cdots \\ 0.16 & 0.48 & 0.36 & 0 & \cdots \\ 0 & 0.16 & 0.48 & 0.36 & \cdots \\ 0 & 0 & 0.16 & 0.48 & \cdots \\ \vdots & \vdots & \vdots & \vdots & \ddots \end{bmatrix}$$

The transition probabilities are given by

$$\begin{aligned} a_0 &= 0.84 \\ a_1 &= 0.16 \\ a_i &= 0 \qquad \text{when } i > 1 \\ b_0 &= 0.36 \\ b_1 &= 0.48 \\ b_2 &= 0.16 \\ b_i &= 0 \qquad \text{when } i > 2 \end{aligned}$$

We have the z-transforms of $A(z)$ and $B(z)$ as

$$\begin{aligned} A(z) &= 0.84 + 0.16z^{-1} \\ B(z) &= 0.36 + 0.48z^{-1} + 0.16z^{-2} \end{aligned}$$

Differentiation of the above two expressions gives

$$\begin{aligned} A(z)' &= -0.16z^{-2} \\ B(z)' &= -0.48z^{-2} - 0.32z^{-3} \end{aligned}$$

By substituting $z^{-1} = 1$ in the above expressions, we get

$$A(1)' = -0.16$$
$$B(1)' = -0.8$$

By using (4.38) we get the probability that the queue is empty

$$s_0 = \frac{1 + B'(1)}{1 + B'(1) - A'(1)} = 0.5556$$

From (4.34) we can write

$$S(z) = \frac{0.2 - 0.2z^{-1}}{0.36 - 0.52z^{-1} + 0.16z^{-2}}$$

We can convert the polynomial expression for $S(z)$ into a partial-fraction expansion (residues) using the MATLAB command RESIDUE:

```
b = [0.2, -0.2]; a = [0.36,
-0.52, 0.16]; [r,p,c] = residue(b,a)
r =
   0.0000
   0.2469
p =
   1.0000
   0.4444
c =
   0.5556
```

where the column vectors r, p, and c give the residues, poles, and direct terms, respectively. Thus we have

$$s_0 = c = 0.5555$$

which confirms the value obtained earlier. For $i \geq 1$ we have

$$s_i = \sum_j r_j \, (p_j)^{i-1}$$

Thus the distribution vector at steady state is given by

$$\mathbf{s} = \begin{bmatrix} 0.5556 & 0.2469 & 0.1097 & 0.0488 & 0.0217 & \cdots \end{bmatrix}^t$$

Note that this is the same distribution vector that was obtained for the same matrix using the difference equations approach in Example 3.

As as a check, we generated the first 50 components of s and ensured that their sum equals unity. ■

Example 5 Use the z-transform technique to find the equilibrium distribution vector for the Markov chain whose transition matrix is

$$\mathbf{P} = \begin{bmatrix} 0.5714 & 0.4082 & 0 & 0 & 0 & \cdots \\ 0.2857 & 0.3673 & 0.4082 & 0 & 0 & \cdots \\ 0.1429 & 0.1837 & 0.3673 & 0.4082 & 0 & \cdots \\ 0 & 0.0408 & 0.1837 & 0.3673 & 0.4082 & \cdots \\ 0 & 0 & 0.0408 & 0.1837 & 0.3673 & \cdots \\ \vdots & \vdots & \vdots & \vdots & \vdots & \ddots \end{bmatrix}$$

We have the z-transforms of $A(z)$ and $B(z)$ as

$$\begin{aligned} A(z) &= 0.5714 + 0.2857z^{-1} + 0.1429z^{-2} \\ B(z) &= 0.4082 + 0.3673z^{-1} + 0.1837z^{-2} + 0.0408z^{-3} \end{aligned}$$

Differentiation of the above two expressions gives

$$\begin{aligned} A(z)' &= -0.2857z^{-2} - 0.2857z^{-3} \\ B(z)' &= -0.3673z^{-2} - 0.3673z^{-3} - 0.1224z^{-4} \end{aligned}$$

By substituting $z^{-1} = 1$ in the above expressions, we get

$$\begin{aligned} A(1)' &= -0.5714 \\ B(1)' &= -0.8571 \end{aligned}$$

By using (4.38) we get the probability that the queue is empty

$$s_0 = \frac{1 + B'(1)}{1 + B'(1) - A'(1)} = 0.2$$

From (4.34) we can write

$$S(z) = \frac{0.0816 - 0.0408z^{-1} - 0.0204z^{-2} - 0.0204z^{-3}}{0.4082 - 0.6326z^{-1} + 0.1837z^{-2} + 0.0408z^{-3}}$$

We can convert the polynomial expression for $S(z)$ into a partial-fraction expansion (residues) using the MATLAB command `residue`:

```
b = [0.0816, -0.0408, -0.0204, -0.0204];
a = [0.4082, -0.6327, 0.1837, 0.0408];
[r,p,c] = residue(b,a)
r =
    0.0000
    0.2574
   -0.0474
```

```
p =
     1.0000
     0.6941
    -0.0474
c =
     0.2000
```

where the column vectors r, p, and c give the residues, poles, and direct terms, respectively. Thus we have

$$s_0 = c = 0.2$$

which confirms the value obtained earlier. For $i \geq 1$ we have

$$s_i = \sum_j r_j \, (p_j)^{i-1}$$

Thus the distribution vector at steady state is given by

$$\mathbf{s} = \left[\begin{array}{cccccc} 0.2100 & 0.1855 & 0.1230 & 0.0860 & 0.0597 & \cdots \end{array} \right]^t$$

As as a check, we generated the first 50 components of **s** and ensured that their sum equals unity. ■

4.9 Finding s Using Forward- or Back-Substitution

This technique is useful when the transition matrix **P** is a lower Hessenberg matrix and the elements in each diagonal are not equal. In such matrices the elements $p_{ij} = 0$ for $j > i+1$. The following example shows a lower Hessenberg matrix of order 6:

$$\mathbf{P} = \begin{bmatrix} h_{11} & h_{12} & 0 & 0 & 0 & 0 \\ h_{21} & h_{22} & h_{23} & 0 & 0 & 0 \\ h_{31} & h_{32} & h_{33} & h_{34} & 0 & 0 \\ h_{41} & h_{42} & h_{43} & h_{44} & h_{45} & 0 \\ h_{51} & h_{52} & h_{53} & h_{54} & h_{55} & h_{56} \\ h_{61} & h_{62} & h_{63} & h_{64} & h_{65} & h_{66} \end{bmatrix} \quad (4.42)$$

This matrix describes the $M^m/M/1$ queue in which a maximum of m packets may arrive as will be explained in Chapter 7. At steady state the distribution vector **s** satisfies

$$\mathbf{Ps} = \mathbf{s} \quad (4.43)$$

and when **P** is lower Hessenberg we have

$$\begin{bmatrix} h_{11} & h_{12} & 0 & 0 & \cdots \\ h_{21} & h_{22} & h_{23} & 0 & \cdots \\ h_{31} & h_{32} & h_{33} & h_{34} & \cdots \\ \vdots & \vdots & \vdots & \vdots & \ddots \end{bmatrix} \begin{bmatrix} s_1 \\ s_2 \\ s_3 \\ \vdots \end{bmatrix} = \begin{bmatrix} s_1 \\ s_2 \\ s_3 \\ \vdots \end{bmatrix} \quad (4.44)$$

where we assumed our states are indexed as $s_1, s_2, \ldots$ Forward substitution starts with estimating a value for s_1 then proceeding to find s_2, s_3, etc. The first row gives

$$s_1 = h_{11}\, s_1 + h_{12}\, s_2 \quad (4.45)$$

We assume an arbitrary value for $s_1 = 1$. Thus the above equation gives us a value for s_2

$$s_2 = (1 - h_{11}) / h_{12} \quad (4.46)$$

We remind the reader again that $s_1 = 1$ by assumption. The second row gives

$$s_2 = h_{21}\, s_1 + h_{22}\, s_2 + h_{23}\, s_3 \quad (4.47)$$

By substituting the values we have so far for s_1 and s_2, we get

$$s_3 = (1 - h_{11})(1 - h_{22}) / (h_{12}\, h_{23}) - h_{21} / h_{23} \quad (4.48)$$

Continuing in this fashion, we can find all the states s_i where $i > 1$.
To get the true value for the distribution vector **s**, we use the normalizing equation

$$\sum_{i=1}^{m} s_i = 1 \quad (4.49)$$

Let us assume that the sum of the components that we obtained for the vector **s** gives

$$\sum_{i=1}^{m} s_i = b \quad (4.50)$$

then we must divide each value of **s** by b to get the true normalized vector that we desire.

Backward substitution is similar to forward substitution but starts by assuming a value for s_m then we estimate s_{m-1}, s_{m-2}, etc. Obviously, backward substitution applies only to finite matrices.

Example 6 Use forward substitution to find the equilibrium distribution vector **s** for the Markov chain with transition matrix given by

$$\mathbf{P} = \begin{bmatrix} 0.4 & 0.2 & 0 & 0 & 0 & 0 \\ 0.3 & 0.35 & 0.2 & 0 & 0 & 0 \\ 0.2 & 0.25 & 0.35 & 0.2 & 0 & 0 \\ 0.1 & 0.15 & 0.25 & 0.35 & 0.2 & 0 \\ 0 & 0.05 & 0.15 & 0.25 & 0.35 & 0.2 \\ 0 & 0 & 0.05 & 0.2 & 0.45 & 0.8 \end{bmatrix}$$

Assume $s_1 = 1$. The distribution vector must satisfy the equation

$$\mathbf{s} = \begin{bmatrix} 0.4 & 0.2 & 0 & 0 & 0 & 0 \\ 0.3 & 0.35 & 0.2 & 0 & 0 & 0 \\ 0.2 & 0.25 & 0.35 & 0.2 & 0 & 0 \\ 0.1 & 0.15 & 0.25 & 0.35 & 0.2 & 0 \\ 0 & 0.05 & 0.15 & 0.25 & 0.35 & 0.2 \\ 0 & 0 & 0.05 & 0.2 & 0.7 & 0.8 \end{bmatrix} \begin{bmatrix} 1 \\ s_2 \\ s_3 \\ s_4 \\ s_5 \\ s_6 \end{bmatrix} = \begin{bmatrix} 1 \\ s_2 \\ s_3 \\ s_4 \\ s_5 \\ s_6 \end{bmatrix}$$

The first row gives us a value for $s_2 = 3$. Continuing, with successive rows, we get

$$s_3 = 8.25$$
$$s_4 = 22.0625$$
$$s_5 = 58.6406$$
$$s_6 = 156.0664$$

Summing the values of all the components gives us

$$\sum_{j=1}^{6} s_j = 249.0195 \quad (4.51)$$

Thus the normalized distribution vector is

$$\mathbf{s} = \begin{bmatrix} 0.0040 & 0.0120 & 0.0331 & 0.0886 & 0.2355 & 0.6267 \end{bmatrix}^t \quad (4.52)$$

■

4.10 Finding s Using Direct Techniques

Direct techniques are useful when the transition matrix **P** has no special structure but its size is small such that rounding errors [1] are below a specified maximum level. In that case we start with the equilibrium equation

$$\mathbf{P}\,\mathbf{s} = \mathbf{s} \tag{4.53}$$

where **s** is the unknown n-component distribution vector. This can be written as

$$(\mathbf{P} - \mathbf{I})\,\mathbf{s} = \mathbf{0} \tag{4.54}$$

$$\mathbf{A}\,\mathbf{s} = \mathbf{0} \tag{4.55}$$

which describes a *homogeneous system of linear equations* with $\mathbf{A} = \mathbf{P} - \mathbf{I}$. The rank of **A** is $n - 1$ since the sum of the columns must be zero. Thus, there are many possible solutions to the system and we need an extra equation to get a unique solution.

The extra equation that is required is the normalizing condition

$$\sum_{i=1}^{m} s_i = 1 \tag{4.56}$$

where we assumed our states are indexed as $s_1, s_2, \ldots, s_m$. We can delete any row matrix **A** in (4.55) and replace it with (4.56). Let us replace the last row with (4.56). In that case we have the system of linear equations

$$\begin{bmatrix} a_{11} & a_{12} & \cdots & a_{1m} \\ a_{21} & a_{22} & \cdots & a_{2m} \\ \vdots & \vdots & \ddots & \vdots \\ a_{m-1,1} & a_{m-1,2} & \cdots & a_{m-1,m} \\ 1 & 1 & \cdots & 1 \end{bmatrix} \begin{bmatrix} s_1 \\ s_2 \\ \vdots \\ s_{m-1} \\ s_m \end{bmatrix} = \begin{bmatrix} 0 \\ 0 \\ \vdots \\ 0 \\ 1 \end{bmatrix} \tag{4.57}$$

This gives us a system of linear equations whose solution is the desired steady-state distribution vector.

Example 7 Find the steady-state distribution vector for the state transition matrix

$$\mathbf{P} = \begin{bmatrix} 0.4 & 0.2 & 0 \\ 0.1 & 0.5 & 0.6 \\ 0.5 & 0.3 & 0.4 \end{bmatrix}$$

---

[1] Rounding errors occur due to finite word length in computers. In floating-point arithmetic, rounding errors occur whenever addition or multiplication operations are performed. In fixed-point arithmetic, rounding errors occur whenever multiplication or shift operations are performed.

First, we have to obtain matrix $\mathbf{A} = \mathbf{P} - \mathbf{I}$

$$\mathbf{A} = \begin{bmatrix} -0.6 & 0.2 & 0 \\ 0.1 & -0.5 & 0.6 \\ 0.5 & 0.3 & -0.6 \end{bmatrix}$$

Now we replace the last row in **A** with all ones to get

$$\mathbf{A} = \begin{bmatrix} -0.6 & 0.2 & 0 \\ 0.1 & -0.5 & 0.6 \\ 1 & 1 & 1 \end{bmatrix}$$

The system of linear equations we have to solve is

$$\begin{bmatrix} -0.6 & 0.2 & 0 \\ 0.1 & -0.5 & 0.6 \\ 1 & 1 & 1 \end{bmatrix} \begin{bmatrix} s_1 \\ s_2 \\ s_3 \end{bmatrix} = \begin{bmatrix} 0 \\ 0 \\ 1 \end{bmatrix}$$

The solution for **s** is

$$\mathbf{s} = \begin{bmatrix} 0.1579 & 0.4737 & 0.3684 \end{bmatrix}^t$$

■

4.11 Finding s Using Iterative Techniques

Iterative techniques are useful when the transition matrix **P** has no special structure and its size is large such that direct techniques will produce too much rounding errors. Iterative techniques obtain a solution to the system of linear equations without arithmetic rounding noise. The accuracy of the results is limited only by machine precision. We enumerate below three techniques for doing the iterations. These techniques are explained in more detail in Appendix D.

1. Successive overrelaxation
2. Jacobi iterations
3. Gauss-Seidel iterations.

Basically, the solution is obtained by first assuming a trial solution, then this is improved through successive iterations. Each iteration improves the guess solution and an answer is obtained when successive iterations do not result in significant changes in the answer.

4.12 Balance Equations

In steady state the probability of finding ourselves in state s_i is given by

$$s_i = \sum_j p_{ij} \, s_j \tag{4.58}$$

The above equation is called the *balance equation* because it provides an expression for each state of the queue at steady state.

From the definition of transition probability, we can write

$$\sum_j p_{ji} = 1 \tag{4.59}$$

which is another way of saying that the sum of all probabilities of leaving state i is equal to one. From the above two equations we can write

$$s_i \sum_j p_{ji} = \sum_j p_{ij} \, s_j \tag{4.60}$$

Since s_i is independent of the index of summation on the LHS, we can write

$$\sum_j p_{ji} \, s_i = \sum_j p_{ij} \, s_j \tag{4.61}$$

Now the LHS represents all the probabilities of *flowing out* of state i. The RHS represents all the probabilities of *flowing into* state i. The above equation describes the *flow balance* for state i.

Thus we proved that in steady state, the probability of moving out of a state equals the probability of moving into the same state. This conclusion will help us derive the steady-state distributions in addition to the other techniques we have discussed above.

Problems

Finding s *Using Eigenvectors*

4.1 Assume **s** is the eigenvector corresponding to unity eigenvalue for matrix **P**. Prove that this vector cannot have a zero component in it if **P** does not have any zero elements.

4.2 Find the steady-state distribution vector corresponding to the unity eigenvalue for the following transition matrix.

$$\mathbf{P} = \begin{bmatrix} 0.45 & 0.2 & 0.5 \\ 0.5 & 0.2 & 0.3 \\ 0.05 & 0.6 & 0.2 \end{bmatrix}$$

4.3 Find the steady-state distribution vector corresponding to the unity eigenvalue for the following transition matrix.

$$\mathbf{P} = \begin{bmatrix} 0.29 & 0.46 & 0.4 \\ 0.4 & 0.45 & 0.33 \\ 0.31 & 0.09 & 0.27 \end{bmatrix}$$

4.4 Find the steady-state distribution vector corresponding to the unity eigenvalue for the following transition matrix.

$$\mathbf{P} = \begin{bmatrix} 0.33 & 0.48 & 0.41 \\ 0.3 & 0.01 & 0.48 \\ 0.37 & 0.51 & 0.11 \end{bmatrix}$$

4.5 Find the steady-state distribution vector corresponding to the unity eigenvalue for the following transition matrix.

$$\mathbf{P} = \begin{bmatrix} 0.33 & 0.51 & 0.12 \\ 0.24 & 0.17 & 0.65 \\ 0.43 & 0.32 & 0.23 \end{bmatrix}$$

4.6 Find the steady-state distribution vector corresponding to the unity eigenvalue for the following transition matrix.

$$\mathbf{P} = \begin{bmatrix} 0.03 & 0.19 & 0.07 \\ 0.44 & 0.17 & 0.53 \\ 0.53 & 0.64 & 0.4 \end{bmatrix}$$

4.7 Find the steady-state distribution vector corresponding to the unity eigenvalue for the following transition matrix.

$$\mathbf{P} = \begin{bmatrix} 0.56 & 0.05 & 0.2 \\ 0.14 & 0.57 & 0.24 \\ 0.3 & 0.38 & 0.56 \end{bmatrix}$$

4.8 Find the steady-state distribution vector corresponding to the unity eigenvalue for the following transition matrix.

$$\mathbf{P} = \begin{bmatrix} 0.08 & 0.19 & 0.07 \\ 0.04 & 0.17 & 0.53 \\ 0.18 & 0.17 & 0.53 \\ 0.7 & 0.64 & 0.4 \end{bmatrix}$$

4.9 Find the steady-state distribution vector corresponding to the unity eigenvalue for the following transition matrix.

$$\mathbf{P} = \begin{bmatrix} 0.12 & 0.06 & 0.42 & 0.1 & 0.09 \\ 0.18 & 0.14 & 0.03 & 0.14 & 0.01 \\ 0.23 & 0.33 & 0.17 & 0.14 & 0.32 \\ 0.26 & 0.32 & 0.38 & 0.43 & 0.18 \\ 0.21 & 0.15 & 0 & 0.19 & 0.4 \end{bmatrix}$$

Finding s *by Difference Equations*

4.10 A queuing system is described by the following transition matrix:

$$\mathbf{P} = \begin{bmatrix} 0.8 & 0.5 & 0 & 0 & 0 \\ 0.2 & 0.3 & 0.5 & 0 & 0 \\ 0 & 0.2 & 0.3 & 0.5 & 0 \\ 0 & 0 & 0.2 & 0.3 & 0.5 \\ 0 & 0 & 0 & 0.2 & 0.5 \end{bmatrix}$$

(a) Find the steady-state distribution vector using the difference equations approach.
(b) What is the probability that the queue is full?

4.11 A queuing system is described by the following transition matrix.

$$\mathbf{P} = \begin{bmatrix} 0.3 & 0.2 & 0 & 0 & 0 \\ 0.7 & 0.1 & 0.2 & 0 & 0 \\ 0 & 0.7 & 0.1 & 0.2 & 0 \\ 0 & 0 & 0.7 & 0.1 & 0.2 \\ 0 & 0 & 0 & 0.7 & 0.8 \end{bmatrix}$$

(a) Find the steady-state distribution vector using the difference equations approach.
(b) What is the probability that the queue is full?

4.12 A queuing system is described by the following transition matrix.

$$\mathbf{P} = \begin{bmatrix} 0.9 & 0.1 & 0 & 0 & 0 \\ 0.1 & 0.8 & 0.1 & 0 & 0 \\ 0 & 0.1 & 0.8 & 0.1 & 0 \\ 0 & 0 & 0.1 & 0.8 & 0.1 \\ 0 & 0 & 0 & 0.1 & 0.9 \end{bmatrix}$$

(a) Find the steady-state distribution vector using the difference equations approach.
(b) What is the probability that the queue is full?

4.13 A queuing system is described by the following transition matrix.

$$\mathbf{P} = \begin{bmatrix} 0.75 & 0.25 & 0 & 0 & 0 \\ 0.25 & 0.5 & 0.25 & 0 & 0 \\ 0 & 0.25 & 0.5 & 0.25 & 0 \\ 0 & 0 & 0.25 & 0.5 & 0.25 \\ 0 & 0 & 0 & 0.25 & 0.75 \end{bmatrix}$$

(a) Find the steady-state distribution vector using the difference equations approach.
(b) What is the probability that the queue is full?

4.14 A queuing system is described by the following transition matrix.

$$\mathbf{P} = \begin{bmatrix} 0.6 & 0.2 & 0 & 0 & 0 \\ 0.2 & 0.4 & 0.2 & 0 & 0 \\ 0.2 & 0.2 & 0.4 & 0.2 & 0 \\ 0 & 0.2 & 0.2 & 0.4 & 0.2 \\ 0 & 0 & 0.2 & 0.4 & 0.8 \end{bmatrix}$$

(a) Find the steady-state distribution vector using the difference equations approach.
(b) What is the probability that the queue is full?

4.15 A queuing system is described by the following transition matrix.

$$\mathbf{P} = \begin{bmatrix} 0.8 & 0.5 & 0 & \cdots \\ 0.2 & 0.3 & 0.5 & \cdots \\ 0 & 0.2 & 0.3 & \cdots \\ 0 & 0 & 0.2 & \cdots \\ \vdots & \vdots & \vdots & \ddots \end{bmatrix}$$

Find the steady-state distribution vector using the difference equations approach.

4.16 A queuing system is described by the following transition matrix.

$$\mathbf{P} = \begin{bmatrix} 0.8 & 0.7 & 0 & \cdots \\ 0.2 & 0.1 & 0.7 & \cdots \\ 0 & 0.2 & 0.1 & \cdots \\ 0 & 0 & 0.2 & \cdots \\ \vdots & \vdots & \vdots & \ddots \end{bmatrix}$$

Find the steady-state distribution vector using the difference equations approach.

4.17 A queuing system is described by the following transition matrix.

$$\mathbf{P} = \begin{bmatrix} 0.9 & 0.2 & 0 & \cdots \\ 0.1 & 0.7 & 0.2 & \cdots \\ 0 & 0.1 & 0.7 & \cdots \\ 0 & 0 & 0.1 & \cdots \\ \vdots & \vdots & \vdots & \ddots \end{bmatrix}$$

Find the steady-state distribution vector using the difference equations approach.

4.18 A queuing system is described by the following transition matrix.

$$\mathbf{P} = \begin{bmatrix} 0.85 & 0.35 & 0 & \cdots \\ 0.15 & 0.5 & 0.35 & \cdots \\ 0 & 0.15 & 0.5 & \cdots \\ 0 & 0 & 0.15 & \cdots \\ \vdots & \vdots & \vdots & \ddots \end{bmatrix}$$

Find the steady-state distribution vector using the difference equations approach.

4.19 A queuing system is described by the following transition matrix.

$$\mathbf{P} = \begin{bmatrix} 0.7 & 0.6 & 0 & 0 & \cdots \\ 0.2 & 0.1 & 0.6 & 0 & \cdots \\ 0.1 & 0.2 & 0.1 & 0.6 & \cdots \\ 0 & 0.1 & 0.2 & 0.1 & \cdots \\ 0 & 0 & 0.1 & 0.2 & \cdots \\ \vdots & \vdots & \vdots & \vdots & \ddots \end{bmatrix}$$

Find the steady-state distribution vector using the difference equations approach.

Finding s *Using Z-Transform*

4.20 Given the following state transition matrix, find the first 10 components of the equilibrium distribution vector using the z-transform approach.

$$\mathbf{P} = \begin{bmatrix} 0.8 & 0.3 & 0 & \cdots \\ 0.2 & 0.5 & 0.3 & \cdots \\ 0 & 0.2 & 0.5 & \cdots \\ 0 & 0 & 0.2 & \cdots \\ \vdots & \vdots & \vdots & \ddots \end{bmatrix}$$

4.21 Given the following state transition matrix, find the first 10 components of the equilibrium distribution vector using the z-transform approach.

$$\mathbf{P} = \begin{bmatrix} 0.95 & 0.45 & 0 & \cdots \\ 0.05 & 0.5 & 0.45 & \cdots \\ 0 & 0.05 & 0.5 & \cdots \\ 0 & 0 & 0.05 & \cdots \\ \vdots & \vdots & \vdots & \ddots \end{bmatrix}$$

4.22 Given the following state transition matrix, find the first 10 components of the equilibrium distribution vector using the z-transform approach.

$$\mathbf{P} = \begin{bmatrix} 0.95 & 0.45 & 0 & \cdots \\ 0.05 & 0.5 & 0.45 & \cdots \\ 0 & 0.05 & 0.5 & \cdots \\ 0 & 0 & 0.05 & \cdots \\ \vdots & \vdots & \vdots & \ddots \end{bmatrix}$$

4.23 Given the following state transition matrix, find the first 10 components of the equilibrium distribution vector using the z-transform approach.

$$\mathbf{P} = \begin{bmatrix} 0.86 & 0.24 & 0 & \cdots \\ 0.14 & 0.62 & 0.24 & \cdots \\ 0 & 0.14 & 0.62 & \cdots \\ 0 & 0 & 0.14 & \cdots \\ \vdots & \vdots & \vdots & \ddots \end{bmatrix}$$

4.24 Given the following state transition matrix, find the first 10 components of the equilibrium distribution vector using the z-transform approach.

$$\mathbf{P} = \begin{bmatrix} 0.93 & 0.27 & 0 & \cdots \\ 0.07 & 0.66 & 0.27 & \cdots \\ 0 & 0.07 & 0.66 & \cdots \\ 0 & 0 & 0.07 & \cdots \\ \vdots & \vdots & \vdots & \ddots \end{bmatrix}$$

4.25 Given the following state transition matrix, find the first 10 components of the equilibrium distribution vector using the z-transform approach.

$$\mathbf{P} = \begin{bmatrix} 0.512 & 0.3584 & 0 & 0 & \cdots \\ 0.384 & 0.4224 & 0.3584 & 0 & \cdots \\ 0.096 & 0.1824 & 0.4224 & 0.3584 & \cdots \\ 0.008 & 0.0344 & 0.1824 & 0.4224 & \cdots \\ 0 & 0.0024 & 0.0344 & 0.1824 & \cdots \\ 0 & 0 & 0.0024 & 0.0344 & \cdots \\ \vdots & \vdots & \vdots & \vdots & \ddots \end{bmatrix}$$

4.26 Given the following state transition matrix, find the first 10 components of the equilibrium distribution vector using the z-transform approach.

$$\mathbf{P} = \begin{bmatrix} 0.720 & 0.4374 & 0 & 0 & \cdots \\ 0.243 & 0.4374 & 0.4374 & 0 & \cdots \\ 0.027 & 0.1134 & 0.4374 & 0.4374 & \cdots \\ 0.001 & 0.0114 & 0.1134 & 0.4374 & \cdots \\ 0 & 0.0004 & 0.0114 & 0.1134 & \cdots \\ 0 & 0 & 0.0004 & 0.0114 & \cdots \\ \vdots & \vdots & \vdots & \vdots & \ddots \end{bmatrix}$$

4.27 Given the following state transition matrix, find the first 10 components of the equilibrium distribution vector using the z-transform approach.

$$\mathbf{P} = \begin{bmatrix} 0.9127 & 0.3651 & 0 & 0 & \cdots \\ 0.0847 & 0.5815 & 0.3651 & 0 & \cdots \\ 0.0026 & 0.0519 & 0.5815 & 0.3651 & \cdots \\ 0 & 0.0016 & 0.0519 & 0.5815 & \cdots \\ 0 & 0 & 0.0016 & 0.0519 & \cdots \\ 0 & 0 & 0 & 0.0016 & \cdots \\ \vdots & \vdots & \vdots & \vdots & \ddots \end{bmatrix}$$

4.28 Given the following state transition matrix, find the first 10 components of the equilibrium distribution vector using the z-transform approach.

$$\mathbf{P} = \begin{bmatrix} 0.6561 & 0.5249 & 0 & 0 & \cdots \\ 0.2916 & 0.3645 & 0.5249 & 0 & \cdots \\ 0.0486 & 0.0972 & 0.3645 & 0.5249 & \cdots \\ 0.0036 & 0.0126 & 0.0972 & 0.3645 & \cdots \\ 0.0001 & 0.0008 & 0.0126 & 0.0972 & \cdots \\ 0 & 0 & 0.0008 & 0.0126 & \cdots \\ \vdots & \vdots & \vdots & \vdots & \ddots \end{bmatrix}$$

4.29 Given the following state transition matrix, find the first 10 components of the equilibrium distribution vector using the z-transform approach.

$$\mathbf{P} = \begin{bmatrix} 0.512 & 0.4096 & 0 & 0 & \cdots \\ 0.384 & 0.4096 & 0.4096 & 0 & \cdots \\ 0.096 & 0.1536 & 0.4096 & 0.4096 & \cdots \\ 0.008 & 0.0256 & 0.1536 & 0.4096 & \cdots \\ 0.0001 & 0.0016 & 0.0256 & 0.1536 & \cdots \\ 0 & 0 & 0.0016 & 0.0256 & \cdots \\ \vdots & \vdots & \vdots & \vdots & \ddots \end{bmatrix}$$

References

1. M.E. Woodward, *Communication and Computer Networks*, IEEE Computer Society Press, Los Alamitos, CA, 1994.

Chapter 5
Reducible Markov Chains

5.1 Introduction

Reducible Markov chains describe systems that have particular states such that once we visit one of those states, we cannot visit other states. An example of systems that can be modeled by reducible Markov chains is games of chance where once the gambler is broke, the game stops and the casino either kicks him out or gives him some compensation (comp). The gambler moved from being in a state of play to being in a comp state and the game stops there. Another example of reducible Markov chains is studying the location of a fish swimming in the ocean. The fish is free to swim at any location as dictated by the currents, food, or presence of predators. Once the fish is caught in a net, it cannot escape and it has limited space where it can swim.

Consider the transition diagram in Fig. 5.1(a). Starting at any state, we are able to reach any other state in the diagram directly, in one step, or indirectly, through one or more intermediate states. Such a Markov chain is termed *irreducible Markov chain* for reasons that will be explained shortly. For example, starting at s_1, we can directly reach s_2 and we can indirectly reach s_3 through either of the intermediate s_2 or s_5. We encounter irreducible Markov chains in systems that can operate for long periods of time such as the state of the lineup at a bank, during business hours. The number of customers lined up changes all the time between zero and maximum. Another example is the state of buffer occupancy in a router or a switch. The buffer occupancy changes between being completely empty and being completely full depending on the arriving traffic pattern.

Consider now the transition diagram in Fig. 5.1(b). Starting from any state, we might not be able to reach other states in the diagram, directly or indirectly. Such a Markov chain is termed *reducible Markov chain* for reasons that will be explained shortly. For example, if we start at s_1, we can never reach any other state. If we start at state s_4, we can only reach state s_5. If we start at state s_3, we can reach all other states. We encounter reducible Markov chains in systems that have terminal conditions such as most games of chance like gambling. In that case, the player keeps on playing till she loses all her money or wins. In either cases, she leaves the game. Another example is the game of snakes and ladders where the player keeps

F. Gebali, *Analysis of Computer and Communication Networks*,
DOI: 10.1007/978-0-387-74437-7_5,

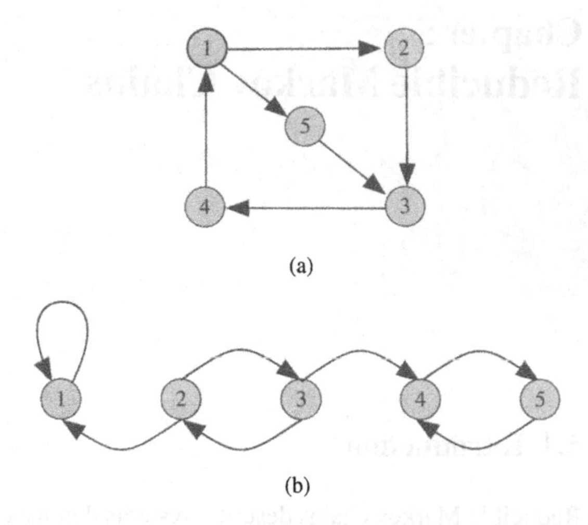

Fig. 5.1 State transition diagrams. (**a**) Irreducible Markov chain. (**b**) Reducible Markov chain

on playing but cannot go back to the starting position. Ultimately, the player reaches the final square and could not go back again to the game.

5.2 Definition of Reducible Markov Chain

The traditional way to define a reducible Markov chain is as follows.

A Markov chain is irreducible if there is some integer $k > 1$ such that all the elements of $\mathbf{P}^k$ are nonzero.

What is the value of k? No one seems to know; the only advice is to keep on multiplying till the conditions are satisfied or computation noise overwhelms us!

This chapter is dedicated to shed more light on this situation and introduce, for the first time, a simple and rigorous technique for identifying a reducible Markov chain. As a bonus, the states of the Markov chain will be simply classified too without too much effort on our part.

5.3 Closed and Transient States

We defined an irreducible (or regular) Markov chain as one in which every state is reachable from every other state either directly or indirectly. We also defined a reducible Markov chain as one in which some states cannot reach other states. Thus the states of a reducible Markov chain are divided into two sets: closed state (C) and transient state (T). Figure 5.2 shows the two sets of states and the directions of transitions between the two sets of states.

When the system is in T, it can make a transition to either T or C. However, once our system is in C, it can never make a transition to T again no matter how long we iterate. In other words, the probability of making a transition from a closed state to a transient state is exactly zero.

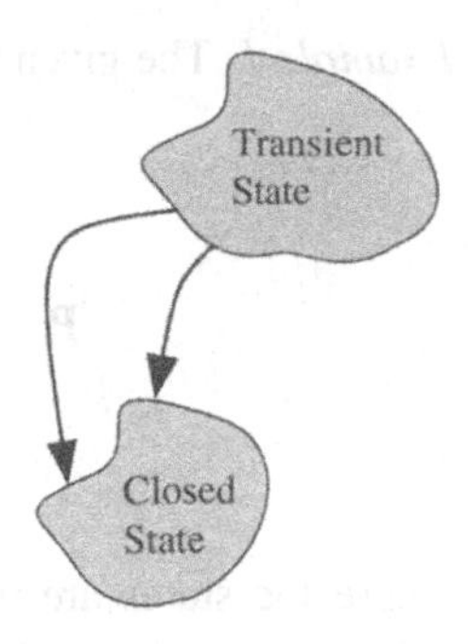

Fig. 5.2 Reducible Markov chain with two sets of states. There are no transitions from the closed states to the transient states as shown.

When C consists of only one state, then that state is called an *absorbing* state. When s_i is an absorbing state, we would have $p_{ii} = 1$. Thus inspection of the transition matrix quickly informs us of the presence of any absorbing states since the diagonal element for that state will be 1.

5.4 Transition Matrix of Reducible Markov Chains

Through proper state assignment, the transition matrix **P** for a reducible Markov chain could be partitioned into the *canonic form*

$$\mathbf{P} = \begin{bmatrix} \mathbf{C} & \mathbf{A} \\ \mathbf{0} & \mathbf{T} \end{bmatrix} \tag{5.1}$$

where

$\mathbf{C}$ = square column stochastic matrix
$\mathbf{A}$ = rectangular nonnegative matrix
$\mathbf{T}$ = square column substochastic matrix

Appendix D defines the meaning of nonnegative and substochastic matrices. The matrix **C** is a column stochastic matrix that can be studied separately from the rest of the transition matrix **P**. In fact, the eigenvalues and eigenvectors of **C** will be used to define the behavior of the Markov chain at equilibrium.

The states of the Markov chain are now partitioned into two mutually exclusive subsets as shown below.

C = set of closed states belonging to matrix **C**
T = set of transient states belonging to matrix **T**

The following equation explicitly shows the partitioning of the states into two sets, closed state C and transient state T.

$$\mathbf{P} = \begin{array}{c} \\ C \\ T \end{array} \begin{array}{c} \begin{array}{cc} C & T \end{array} \\ \begin{bmatrix} \mathbf{C} & \mathbf{A} \\ \mathbf{0} & \mathbf{T} \end{bmatrix} \end{array} \tag{5.2}$$

Example 1 The given transition matrix represents a reducible Markov chain.

$$\mathbf{P} = \begin{array}{c} s_1 \\ s_2 \\ s_3 \\ s_4 \end{array} \overset{\begin{array}{cccc} s_1 & s_2 & s_3 & s_4 \end{array}}{\begin{bmatrix} 0.8 & 0 & 0.1 & 0.1 \\ 0 & 0.5 & 0 & 0.2 \\ 0.2 & 0.2 & 0.9 & 0 \\ 0 & 0.3 & 0 & 0.7 \end{bmatrix}}$$

where the states are indicated around **P** for illustration. Rearrange the rows and columns to express the matrix in the canonic form in (5.1) or (5.2) and identify the matrices **C**, **A**, and **T**. Verify the assertions that **C** is column stochastic, **A** is nonnegative, and **T** is column substochastic.

After exploring a few possible transitions starting from any initial state, we see that if we arrange the states in the order 1, 3, 2, 4 then the following state matrix is obtained

$$\mathbf{P} = \begin{array}{c} s_1 \\ s_3 \\ s_2 \\ s_4 \end{array} \overset{\begin{array}{cccc} s_1 & s_2 & s_3 & s_4 \end{array}}{\begin{bmatrix} 0.8 & 0.1 & 0 & 0.1 \\ 0.2 & 0.9 & 0.2 & 0 \\ 0 & 0 & 0.5 & 0.2 \\ 0 & 0 & 0.3 & 0.7 \end{bmatrix}}$$

We see that the matrix exhibits the reducible Markov chain structure and matrices **C**, **A**, and **T** are

$$\mathbf{C} = \begin{bmatrix} 0.8 & 0.1 \\ 0.2 & 0.9 \end{bmatrix}$$

$$\mathbf{A} = \begin{bmatrix} 0 & 0.1 \\ 0.2 & 0 \end{bmatrix}$$

$$\mathbf{T} = \begin{bmatrix} 0.5 & 0.2 \\ 0.3 & 0.7 \end{bmatrix}$$

The sum of each column of **C** is exactly 1, which indicates that it is column stochastic. The sum of columns of **T** is less than 1, which indicates that it is column substochastic.

The set of closed states is $C = \{1,\ 3\}$ and the set of transient states is $T = \{2,\ 4\}$. Starting in state s_2 or s_4, we will ultimately go to states s_1 and s_3. Once we are there, we can never go back to state s_2 or s_4 because we entered the closed states. ■

Example 2 Consider the reducible Markov chain of the previous example. Assume that the system was initially in state s_3. Find the distribution vector at 20 time step intervals.

We do not have to rearrange the transition matrix to do this example. We have

$$\mathbf{P} = \begin{bmatrix} 0.8 & 0 & 0.1 & 0.1 \\ 0 & 0.5 & 0 & 0.2 \\ 0.2 & 0.2 & 0.9 & 0 \\ 0 & 0.3 & 0 & 0.7 \end{bmatrix}$$

The initial distribution vector is

$$\mathbf{s} = \begin{bmatrix} 0 & 0 & 1 & 0 \end{bmatrix}^t$$

The distribution vector at 20 time step intervals is

$$\mathbf{s}(20) = \begin{bmatrix} 0.3208 & 0.0206 & 0.6211 & 0.0375 \end{bmatrix}^t$$
$$\mathbf{s}(40) = \begin{bmatrix} 0.3327 & 0.0011 & 0.6642 & 0.0020 \end{bmatrix}^t$$
$$\mathbf{s}(60) = \begin{bmatrix} 0.3333 & 0.0001 & 0.6665 & 0.0001 \end{bmatrix}^t$$
$$\mathbf{s}(80) = \begin{bmatrix} 0.3333 & 0.0000 & 0.6667 & 0.0000 \end{bmatrix}^t$$

We note that after 80 time steps, the probability of being in the transient state s_2 or s_4 is nil. The system will definitely be in the closed set composed of states s_1 and s_3. ■

5.5 Composite Reducible Markov Chains

In the general case, the reducible Markov chain could be composed of two or more sets of closed states. Figure 5.3 shows a reducible Markov chain with two sets of closed states. If the system is in the transient state T, it can move to either sets of closed states, C_1 or C_2. However, if the system is in state C_1, it cannot move to T or C_2. Similarly, if the system is in state C_2, it cannot move to T or C_1. In that case, the canonic form for the transition matrix $\mathbf{P}$ for a reducible Markov chain could be expanded into several subsets of noncommunicating closed states

$$\mathbf{P} = \begin{bmatrix} \mathbf{C}_1 & \mathbf{0} & \mathbf{A}_1 \\ \mathbf{0} & \mathbf{C}_2 & \mathbf{A}_2 \\ \mathbf{0} & \mathbf{0} & \mathbf{T} \end{bmatrix} \tag{5.3}$$

where

$\mathbf{C}_1$ and $\mathbf{C}_2$ = square column stochastic matrices
$\mathbf{A}_1$ and $\mathbf{A}_2$ = rectangular nonnegative matrices
$\mathbf{T}$ = square column substochastic matrix

Since the transition matrix contains two-column stochastic matrices $\mathbf{C}_1$ and $\mathbf{C}_2$, we expect to get two eigenvalues $\lambda_1 = 1$ and $\lambda_2 = 1$ also. And we will be getting

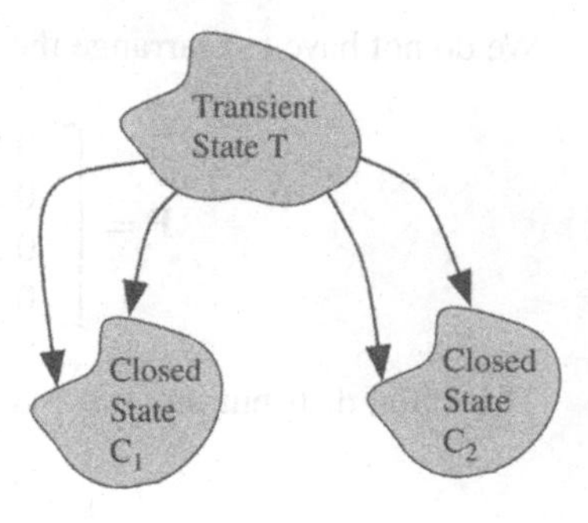

Fig. 5.3 A reducible Markov chain with two sets of closed states

two possible steady-state distributions based on the initial value of the distribution vector **s**(0)—more on that in Sections 5.7 and 5.9.

The states of the Markov chain are now divided into three mutually exclusive sets as shown below.

1. C_1 = set of closed states belonging to matrix $\mathbf{C}_1$
2. C_2 = set of closed states belonging to matrix $\mathbf{C}_2$
3. T = set of transient states belonging to matrix $\mathbf{T}$

The following equation explicitly shows the partitioning of the states.

$$\mathbf{P} \quad = \quad \begin{matrix} \\ C_1 \\ C_2 \\ T \end{matrix} \begin{matrix} \begin{matrix} C_1 & C_2 & T \end{matrix} \\ \begin{bmatrix} \mathbf{C}_1 & \mathbf{0} & \mathbf{A}_1 \\ \mathbf{0} & \mathbf{C}_2 & \mathbf{A}_2 \\ \mathbf{0} & \mathbf{0} & \mathbf{T} \end{bmatrix} \end{matrix} \tag{5.4}$$

Notice from the structure of **P** in (5.4) that if we were in the first set of closed state C_1, then we cannot escape that set to visit C_2 or T. Similarly, if we were in the second set of closed state C_2, then we cannot escape that set to visit C_1 or T. On the other hand, if we were in the set of transient states T, then we can not stay in that set since we will ultimately fall into C_1 or C_2.

Example 3 You play a coin tossing game with a friend. The probability that one player winning \$1 is p, and the probability that he loses \$1 is $q = 1 - p$. Assume the combined assets of both players is \$6 and the game ends when one of the players is broke. Define a Markov chain whose state s_i means that you have \$$i$ and construct the transition matrix. If the Markov chain is reducible, identify the closed and transient states and rearrange the matrix to conform to the structure of (5.3) or (5.4).

Since this is a gambling game, we suspect that we have a reducible Markov chain with closed states where one player is the winner and the other is the loser.

A player could have \$0, \$1, $\cdots$, or \$6. Therefore, the transition matrix is of dimension 7×7 as shown in (5.5).

$$\mathbf{P} = \begin{bmatrix} 1 & q & 0 & 0 & 0 & 0 & 0 \\ 0 & 0 & q & 0 & 0 & 0 & 0 \\ 0 & p & 0 & q & 0 & 0 & 0 \\ 0 & 0 & p & 0 & q & 0 & 0 \\ 0 & 0 & 0 & p & 0 & q & 0 \\ 0 & 0 & 0 & 0 & p & 0 & 0 \\ 0 & 0 & 0 & 0 & 0 & p & 1 \end{bmatrix} \tag{5.5}$$

Notice that states 0 and 6 are absorbing states since $p_{00} = p_{66} = 1$. The set $T = \{ s_1, \quad s_2, \quad \cdots, \quad c_5 \}$ is the set of transient states. We could rearrange our transition matrix such that states s_0 and s_6 are adjacent as shown below.

$$\mathbf{P} = \begin{array}{c} s_0 \\ s_6 \\ \\ s_1 \\ s_2 \\ s_3 \\ s_4 \\ s_5 \end{array} \overset{\begin{array}{cccccccc} s_0 & s_6 & & s_1 & s_2 & s_3 & s_4 & s_5 \end{array}}{\left[\begin{array}{cc|ccccc} 1 & 0 & q & 0 & 0 & 0 & 0 \\ 0 & 1 & 0 & 0 & 0 & 0 & p \\ \hline 0 & 0 & 0 & q & 0 & 0 & 0 \\ 0 & 0 & p & 0 & q & 0 & 0 \\ 0 & 0 & 0 & p & 0 & q & 0 \\ 0 & 0 & 0 & 0 & p & 0 & q \\ 0 & 0 & 0 & 0 & 0 & p & 0 \end{array} \right]}$$

We have added spaces between the elements of the matrix to show the outline of the component matrices $\mathbf{C}_1$, $\mathbf{C}_2$, $\mathbf{A}_1$, $\mathbf{A}_2$, and $\mathbf{T}$. In that case, each closed matrix corresponds to a single absorbing state (s_0 and s_6), while the transient states correspond to a 5×5 matrix. ■

5.6 Transient Analysis

We might want to know how a reducible Markov chain varies with time n since this leads to useful results such as the probability of visiting a certain state at any given time value. In other words, we want to find $\mathbf{s}(n)$ from the expression

$$\mathbf{s}(n) = \mathbf{P}^n \ \mathbf{s}(0) \tag{5.6}$$

Without loss of generality we assume the reducible transition matrix to be given in the form

$$\mathbf{P} = \begin{bmatrix} \mathbf{C} & \mathbf{A} \\ \mathbf{0} & \mathbf{T} \end{bmatrix} \tag{5.7}$$

After n time steps the transition matrix of a reducible Markov chain will still be reducible and will have the form

$$\mathbf{P}^n = \begin{bmatrix} \mathbf{C}^n & \mathbf{Y}^n \\ \mathbf{0} & \mathbf{T}^n \end{bmatrix} \tag{5.8}$$

where matrix $\mathbf{Y}^n$ is given by

$$\mathbf{Y}^n = \sum_{i=0}^{n-1} \mathbf{C}^{n-i-1}\, \mathbf{A}\, \mathbf{T}^i \tag{5.9}$$

We can always find $\mathbf{C}^n$ and $\mathbf{T}^n$ using the techniques discussed in Chapter 3 such as diagonalization, finding the Jordan canonic form, or even repeated multiplications. The stochastic matrix $\mathbf{C}^n$ can be expressed in terms of its eigenvalues using (3.80) on page 94.

$$\mathbf{C}^n = \mathbf{C}_1 + \lambda_2^n \mathbf{C}_2 + \lambda_3^n \mathbf{C}_3 + \cdots \tag{5.10}$$

where it was assumed that $\mathbf{C}_1$ is the expansion matrix corresponding to the eigenvalue $\lambda_1 = 1$ and $\mathbf{C}$ is assumed to be of dimension $m_c \times m_c$. Similarly, the substochastic matrix $\mathbf{T}^n$ can be expressed in terms of its eigenvalues using (3.80) on page 94.

$$\mathbf{T}^n = \lambda_1^n \mathbf{T}_1 + \lambda_2^n \mathbf{T}_2 + \lambda_3^n \mathbf{T}_3 + \cdots \tag{5.11}$$

We should note here that all the magnitudes of the eigenvalues in the above equation are less than unity. Equation (5.9) can then be expressed in the form

$$\mathbf{Y}^n = \sum_{j=1}^{m} \mathbf{C}_j\, \mathbf{A} \sum_{i=0}^{n-1} \lambda_j^{n-i-1}\, \mathbf{T}^i \tag{5.12}$$

After some algebraic manipulations, we arrive at the form

$$\mathbf{Y}^n = \sum_{j=1}^{m} \lambda_j^{n-1} \mathbf{C}_j\, \mathbf{A} \left[\mathbf{I} - \left(\frac{\mathbf{T}}{\lambda_j} \right)^n \right] \left(\mathbf{I} - \frac{\mathbf{T}}{\lambda_j} \right)^{-1} \tag{5.13}$$

This can be written in the form

$$
\begin{aligned}
\mathbf{Y}^n &= \mathbf{C}_1\mathbf{A}(\mathbf{I}-\mathbf{T})^{-1}\left[\mathbf{I}-\mathbf{T}^n\right] + \\
& \lambda_2^{n-1}\mathbf{C}_2\mathbf{A}\left(\mathbf{I}-\frac{1}{\lambda_2}\mathbf{T}\right)^{-1}\left[\mathbf{I}-\frac{1}{\lambda_2^n}\mathbf{T}^n\right] + \\
& \lambda_3^{n-1}\mathbf{C}_3\mathbf{A}\left(\mathbf{I}-\frac{1}{\lambda_3}\mathbf{T}\right)^{-1}\left[\mathbf{I}-\frac{1}{\lambda_3^n}\mathbf{T}^n\right] + \cdots
\end{aligned} \tag{5.14}
$$

If some of the eigenvalues of $\mathbf{C}$ are repeated, then the above formula has to be modified as explained in Section 3.14 on page 103. Problem 5.25 discusses this situation.

Example 4 A reducible Markov chain has the transition matrix

$$
\mathbf{P} = \begin{bmatrix}
0.5 & 0.3 & 0.1 & 0.3 & 0.1 \\
0.5 & 0.7 & 0.2 & 0.1 & 0.3 \\
0 & 0 & 0.2 & 0.2 & 0.1 \\
0 & 0 & 0.1 & 0.3 & 0.1 \\
0 & 0 & 0.4 & 0.1 & 0.4
\end{bmatrix}
$$

Find the value of $\mathbf{P}^{20}$ and from that find the probability of making the following transitions:

(a) From s_3 to s_2.
(b) From s_2 to s_2.
(c) From s_4 to s_1.
(d) From s_3 to s_4.

The components of the transition matrix are

$$
\begin{aligned}
\mathbf{C} &= \begin{bmatrix} 0.5 & 0.3 \\ 0.5 & 0.7 \end{bmatrix} \\
\mathbf{A} &= \begin{bmatrix} 0.1 & 0.3 & 0.1 \\ 0.2 & 0.1 & 0.3 \end{bmatrix} \\
\mathbf{T} &= \begin{bmatrix} 0.2 & 0.2 & 0.1 \\ 0.1 & 0.3 & 0.1 \\ 0.4 & 0.1 & 0.4 \end{bmatrix}
\end{aligned}
$$

We use the MATLAB function `EIGPOWERS`, which we developed to expand matrix $\mathbf{C}$ in terms of its eigenpowers, and we have $\lambda_1 = 1$ and $\lambda_2 = 0.2$. The corresponding matrices according to (3.80) are

$$\mathbf{C}_1 = \begin{bmatrix} 0.375 & 0.375 \\ 0.625 & 0.625 \end{bmatrix}$$
$$\mathbf{C}_2 = \begin{bmatrix} 0.625 & -0.375 \\ -0.625 & 0.375 \end{bmatrix}$$

We could now use (5.13) to find $\mathbf{P}^{20}$ but instead we use repeated multiplication here

$$\mathbf{P}^{20} = \begin{bmatrix} 0.375 & 0.375 & 0.375 & 0.375 & 0.375 \\ 0.625 & 0.625 & 0.625 & 0.625 & 0.625 \\ 0 & 0 & 0 & 0 & 0 \\ 0 & 0 & 0 & 0 & 0 \\ 0 & 0 & 0 & 0 & 0 \end{bmatrix}$$

(a) $p_{32} = 0.625$
(b) $p_{22} = 0.625$
(c) $p_{14} = 0.375$
(d) $p_{43} = 0$

■

Example 5 Find an expression for the transition matrix at times $n = 4$ and $n = 20$ for the reducible Markov chain characterized by the transition matrix

$$\mathbf{P} = \begin{bmatrix} 0.9 & 0.3 & 0.3 & 0.3 & 0.2 \\ 0.1 & 0.7 & 0.2 & 0.1 & 0.3 \\ 0 & 0 & 0.2 & 0.2 & 0.1 \\ 0 & 0 & 0.1 & 0.3 & 0.1 \\ 0 & 0 & 0.2 & 0.1 & 0.2 \end{bmatrix}$$

The components of the transition matrix are

$$\mathbf{C} = \begin{bmatrix} 0.9 & 0.3 \\ 0.1 & 0.7 \end{bmatrix}$$
$$\mathbf{A} = \begin{bmatrix} 0.3 & 0.1 & 0.3 \\ 0.2 & 0.1 & 0.3 \end{bmatrix}$$
$$\mathbf{T} = \begin{bmatrix} 0.2 & 0.2 & 0.1 \\ 0.3 & 0.3 & 0.1 \\ 0.2 & 0.1 & 0.4 \end{bmatrix}$$

$\mathbf{C}^n$ is expressed in terms of its eigenvalues as

$$\mathbf{C}^n = \lambda_1^n \mathbf{C}_1 + \lambda_2^n \mathbf{C}_2$$

where $\lambda_1 = 1$ and $\lambda_2 = 0.6$ and

$$\mathbf{C}_1 = \begin{bmatrix} 0.75 & 0.75 \\ 0.25 & 0.25 \end{bmatrix}$$

$$\mathbf{C}_2 = \begin{bmatrix} 0.25 & -0.75 \\ -0.25 & 0.75 \end{bmatrix}$$

At any time instant n the matrix $\mathbf{Y}^n$ has the value

$$\mathbf{Y}^n = \mathbf{C}_1\mathbf{A}(\mathbf{I}-\mathbf{T})^{-1}\left[\mathbf{I}-\mathbf{T}^n\right] + (0.6)^{n-1}\mathbf{C}_2\mathbf{A}\left(\mathbf{I}-\frac{1}{0.6}\mathbf{T}\right)^{-1}\left[\mathbf{I}-\frac{1}{0.6^n}\mathbf{T}^n\right]$$

By substituting $n = 4$, we get

$$\mathbf{P}^4 = \begin{bmatrix} 0.7824 & 0.6528 & 0.6292 & 0.5564 & 0.6318 \\ 0.2176 & 0.3472 & 0.2947 & 0.3055 & 0.3061 \\ 0 & 0 & 0.0221 & 0.0400 & 0.0180 \\ 0 & 0 & 0.0220 & 0.0401 & 0.0180 \\ 0 & 0 & 0.0320 & 0.0580 & 0.0261 \end{bmatrix}$$

By substituting $n = 20$, we get

$$\mathbf{P}^{20} = \begin{bmatrix} 0.7500 & 0.7500 & 0.7500 & 0.7500 & 0.7500 \\ 0.2500 & 0.2500 & 0.2500 & 0.2500 & 0.2500 \\ 0 & 0 & 0 & 0 & 0 \\ 0 & 0 & 0 & 0 & 0 \\ 0 & 0 & 0 & 0 & 0 \end{bmatrix}$$

We see that all the columns of $\mathbf{P}^{20}$ are identical, which indicates that the steady-state distribution vector is independent of its initial value. ■

5.7 Reducible Markov Chains at Steady-State

Assume we have a reducible Markov chain with transition matrix **P** that is expressed in the canonic form

$$\mathbf{P} = \begin{bmatrix} \mathbf{C} & \mathbf{A} \\ \mathbf{0} & \mathbf{T} \end{bmatrix} \tag{5.15}$$

According to (5.8), after n time steps the transition matrix will have the form

$$\mathbf{P}^n = \begin{bmatrix} \mathbf{C}^n & \mathbf{Y}^n \\ \mathbf{0} & \mathbf{T}^n \end{bmatrix} \tag{5.16}$$

where matrix $\mathbf{Y}^n$ is given by

$$\mathbf{Y}^n = \sum_{i=0}^{n-1} \mathbf{C}^{n-i-1} \mathbf{A} \mathbf{T}^i \tag{5.17}$$

To see how $\mathbf{P}^n$ will be like when $n \to \infty$, we express the matrices $\mathbf{C}$ and $\mathbf{T}$ in terms of their eigenvalues as in (5.10) and (5.11).

When $n \to \infty$, matrix $\mathbf{Y}^\infty$ becomes

$$\mathbf{Y}^\infty = \mathbf{C}^\infty \mathbf{A} \sum_{i=0}^{\infty} \mathbf{T}^i \tag{5.18}$$

$$= \mathbf{C}_1 \mathbf{A} (\mathbf{I} - \mathbf{T})^{-1} \tag{5.19}$$

where $\mathbf{I}$ is the unit matrix whose dimensions match that of $\mathbf{T}$.

We used the following matrix identity to derive the above equation

$$(\mathbf{I} - \mathbf{T})^{-1} = \sum_{i=0}^{\infty} \mathbf{T}^i \tag{5.20}$$

Finally, we can write the steady-state expression for the transition matrix of a reducible Markov chain as

$$\mathbf{P}^\infty = \begin{bmatrix} \mathbf{C}_1 & \mathbf{Y}^\infty \\ \mathbf{0} & \mathbf{0} \end{bmatrix} \tag{5.21}$$

$$= \begin{bmatrix} \mathbf{C}_1 & \mathbf{C}_1 \mathbf{A} (\mathbf{I} - \mathbf{T})^{-1} \\ \mathbf{0} & \mathbf{0} \end{bmatrix} \tag{5.22}$$

The above matrix is column stochastic since it represents a transition matrix. We can prove that the columns of the matrix $\mathbf{C}_1 \mathbf{A} (\mathbf{I} - \mathbf{T})^{-1}$ are all identical and equal to the columns of $\mathbf{C}_1$. This is left as an exercise (see Problem 5.16). Since all the columns of $\mathbf{P}$ at steady-state are equal, all we have to do to find $\mathbf{P}^\infty$ is to find one column only of $\mathbf{C}_1$. The following examples show this.

Example 6 Find the steady-state transition matrix for the reducible Markov chain characterized by the transition matrix

$$\mathbf{P} = \begin{bmatrix} 0.8 & 0.4 & 0 & 0.3 & 0.1 \\ 0.2 & 0.6 & 0.2 & 0.2 & 0.3 \\ 0 & 0 & 0.2 & 0.2 & 0.1 \\ 0 & 0 & 0 & 0.3 & 0.1 \\ 0 & 0 & 0.6 & 0 & 0.4 \end{bmatrix}$$

The components of the transition matrix are

$$\mathbf{C} = \begin{bmatrix} 0.8 & 0.4 \\ 0.2 & 0.6 \end{bmatrix}$$
$$\mathbf{A} = \begin{bmatrix} 0 & 0.3 & 0.1 \\ 0.2 & 0.2 & 0.3 \end{bmatrix}$$
$$\mathbf{T} = \begin{bmatrix} 0.2 & 0.2 & 0.1 \\ 0 & 0.3 & 0.1 \\ 0.6 & 0 & 0.4 \end{bmatrix}$$

The steady-state value of $\mathbf{C}$ is

$$\mathbf{C}^{\infty} = \mathbf{C}_1 = \begin{bmatrix} 0.6667 & 0.6667 \\ 0.3333 & 0.3333 \end{bmatrix}$$

The matrix $\mathbf{Y}^{\infty}$ has the value

$$\mathbf{Y}^{\infty} = \mathbf{C}_1 \ \mathbf{A} \ (\mathbf{I} - \mathbf{T})^{-1} = \begin{bmatrix} 0.6667 & 0.6667 \\ 0.3333 & 0.3333 \end{bmatrix}$$

Thus the steady state value of $\mathbf{P}$ is

$$\mathbf{P}^{\infty} = \begin{bmatrix} 0.6667 & 0.6667 & 0.6667 & 0.6667 & 0.6667 \\ 0.3333 & 0.3333 & 0.3333 & 0.3333 & 0.3333 \\ 0 & 0 & 0 & 0 & 0 \\ 0 & 0 & 0 & 0 & 0 \\ 0 & 0 & 0 & 0 & 0 \end{bmatrix}$$

The first thing we notice about the steady-state value of the transition matrix is that all columns are identical. This is exactly the same property for the transition matrix of an irreducible Markov chain. The second observation we can make about the transition matrix at steady-state is that there is no possibility of moving to a transient state irrespective of the value of the initial distribution vector. The third observation we can make is that no matter what the initial distribution vector was, we will always wind up in the same steady-state distribution. ■

Example 7 Find the steady-state transition matrix for the reducible Markov chain characterized by the transition matrix

$$\mathbf{P} = \begin{bmatrix} 0.5 & 0.3 & 0.1 & 0.3 & 0.1 \\ 0.5 & 0.7 & 0.2 & 0.1 & 0.3 \\ 0 & 0 & 1 & 0.2 & 0.1 \\ 0 & 0 & 0 & 0.3 & 0.1 \\ 0 & 0 & 0 & 0.1 & 0.4 \end{bmatrix}$$

The components of the transition matrix are

$$\mathbf{C} = \begin{bmatrix} 0.5 & 0.3 \\ 0.5 & 0.7 \end{bmatrix}$$

$$\mathbf{A} = \begin{bmatrix} 0.1 & 0.3 & 0.1 \\ 0.2 & 0.1 & 0.3 \end{bmatrix}$$

$$\mathbf{T} = \begin{bmatrix} 0.2 & 0.2 & 0.1 \\ 0.1 & 0.3 & 0.1 \\ 0.4 & 0.1 & 0.4 \end{bmatrix}$$

The steady-state value of **C** is

$$\mathbf{C}^{\infty} = \mathbf{C}_1 = \begin{bmatrix} 0.375 & 0.375 \\ 0.625 & 0.625 \end{bmatrix}$$

The matrix $\mathbf{Y}^{\infty}$ has the value

$$\mathbf{Y}^{\infty} = \mathbf{C}_1 \; \mathbf{A} \; (\mathbf{I} - \mathbf{T})^{-1} = \begin{bmatrix} 0.375 & 0.375 & 0.375 \\ 0.625 & 0.625 & 0.625 \end{bmatrix}$$

Thus the steady-state value of **P** is

$$\mathbf{P}^{\infty} = \begin{bmatrix} 0.375 & 0.375 & 0.375 & 0.375 & 0.375 \\ 0.625 & 0.625 & 0.625 & 0.625 & 0.625 \\ 0 & 0 & 0 & 0 & 0 \\ 0 & 0 & 0 & 0 & 0 \\ 0 & 0 & 0 & 0 & 0 \end{bmatrix}$$

■

5.8 Reducible Composite Markov Chains at Steady-State

In this section, we will study the steady-state behavior of reducible composite Markov chains. In the general case, the reducible Markov chain could be composed of two or more closed states. Figure 5.4 shows a reducible Markov chain with two sets of closed states. If the system is in the transient state T, it can move to either sets of closed states, C_1 or C_2. However, if the system is in state C_1, it cannot move to T or C_2. Similarly, if the system is in state C_2, it can not move to T or C_1.

Assume the transition matrix is given by the canonic form

$$\mathbf{P} = \begin{bmatrix} \mathbf{C}_1 & \mathbf{0} & \mathbf{A}_1 \\ \mathbf{0} & \mathbf{C}_2 & \mathbf{A}_2 \\ \mathbf{0} & \mathbf{0} & \mathbf{T} \end{bmatrix} \tag{5.23}$$

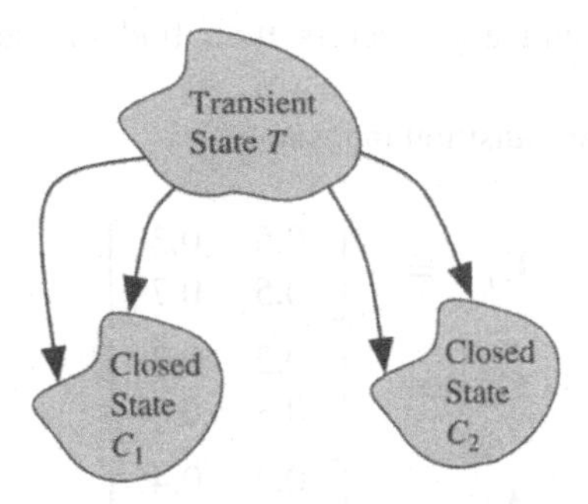

Fig. 5.4 A reducible Markov chain with two sets of closed states

where

$$\begin{aligned} \mathbf{C}_1 \text{ and } \mathbf{C}_2 &= \text{square column stochastic matrices} \\ \mathbf{A}_1 \text{ and } \mathbf{A}_2 &= \text{rectangular nonnegative matrices} \\ \mathbf{T} &= \text{square column substochastic matrix} \end{aligned}$$

It is easy to verify that the steady-state transition matrix for such a system will be

$$\mathbf{P}^{\infty} = \begin{bmatrix} \mathbf{C}_1' & \mathbf{0} & \mathbf{Y}_1 \\ \mathbf{0} & \mathbf{C}_1'' & \mathbf{Y}_2 \\ \mathbf{0} & \mathbf{0} & \mathbf{0} \end{bmatrix} \tag{5.24}$$

where

$$\mathbf{C}_1' = \mathbf{C}_1^{\infty} \tag{5.25}$$

$$\mathbf{C}_1'' = \mathbf{C}_2^{\infty} \tag{5.26}$$

$$\mathbf{Y}_1 = \mathbf{C}_1'\mathbf{A}_1 (\mathbf{I} - \mathbf{T})^{-1} \tag{5.27}$$

$$\mathbf{Y}_2 = \mathbf{C}_1''\mathbf{A}_2 (\mathbf{I} - \mathbf{T})^{-1} \tag{5.28}$$

Essentially, $\mathbf{C}_1'$ is the matrix that is associated with $\lambda = 1$ in the expansion of $\mathbf{C}_1$ in terms of its eigenvalues. The same also applies to $\mathbf{C}_1''$, which is the matrix that is associated with $\lambda = 1$ in the expansion of $\mathbf{C}_2$ in terms of its eigenvalues.

We observe that each column of matrix $\mathbf{Y}_1$ is a scaled copy of the columns of $\mathbf{C}_1'$. Also, the sum of each column of $\mathbf{Y}_1$ is lesser than one. We can make the same observations about matrix $\mathbf{Y}_2$.

Example 8 The given transition matrix corresponds to a composite reducible Markov chain.

$$\mathbf{P} = \begin{bmatrix} 0.5 & 0.3 & 0 & 0 & 0.1 & 0.4 \\ 0.5 & 0.7 & 0 & 0 & 0.3 & 0.1 \\ 0 & 0 & 0.2 & 0.7 & 0.1 & 0.2 \\ 0 & 0 & 0.8 & 0.3 & 0.1 & 0.1 \\ 0 & 0 & 0 & 0 & 0.1 & 0.2 \\ 0 & 0 & 0 & 0 & 0.3 & 0 \end{bmatrix}$$

Find its eigenvalues and eigenvectors then find the steady-state distribution vector.

The components of the transition matrix are

$$\mathbf{C}_1 = \begin{bmatrix} 0.5 & 0.3 \\ 0.5 & 0.7 \end{bmatrix}$$

$$\mathbf{C}_2 = \begin{bmatrix} 0.2 & 0.7 \\ 0.8 & 0.3 \end{bmatrix}$$

$$\mathbf{A}_1 = \begin{bmatrix} 0.2 & 0.4 \\ 0.2 & 0.1 \end{bmatrix}$$

$$\mathbf{A}_2 = \begin{bmatrix} 0.1 & 0.2 \\ 0.1 & 0.1 \end{bmatrix}$$

$$\mathbf{T} = \begin{bmatrix} 0.1 & 0.2 \\ 0.3 & 0 \end{bmatrix}$$

The steady-state value of $\mathbf{C}_1$ is

$$\mathbf{C}_1' = \begin{bmatrix} 0.375 & 0.375 \\ 0.625 & 0.625 \end{bmatrix}$$

The matrix $\mathbf{Y}_1$ has the value

$$\mathbf{Y}_1 = \mathbf{C}_1' \; \mathbf{A}_1 \; (\mathbf{I} - \mathbf{T})^{-1} = \begin{bmatrix} 0.2455 & 0.2366 \\ 0.4092 & 0.3943 \end{bmatrix}$$

The steady-state value of $\mathbf{C}_2$ is

$$\mathbf{C}_1'' = \begin{bmatrix} 0.4667 & 0.4667 \\ 0.5333 & 0.5333 \end{bmatrix}$$

The matrix $\mathbf{Y}_2$ has the value

$$\mathbf{Y}_2 = \mathbf{C}_1'' \; \mathbf{A}_2 \; (\mathbf{I} - \mathbf{T})^{-1} = \begin{bmatrix} 0.1611 & 0.1722 \\ 0.1841 & 0.1968 \end{bmatrix}$$

Thus the steady-state value of $\mathbf{P}$ is

$$\mathbf{P}^{\infty} = \begin{bmatrix} 0.375 & 0.375 & 0 & 0 & 0.2455 & 0.2366 \\ 0.625 & 0.625 & 0 & 0 & 0.4092 & 0.3943 \\ 0 & 0 & 0.4667 & 0.4667 & 0.1611 & 0.1722 \\ 0 & 0 & 0.5333 & 0.5333 & 0.1841 & 0.1968 \\ 0 & 0 & 0 & 0 & 0 & 0 \\ 0 & 0 & 0 & 0 & 0 & 0 \end{bmatrix}$$

We notice that all columns are identical for the closed state matrices. However, the columns for the matrices corresponding to the transient states ($\mathbf{Y}_1$ and $\mathbf{Y}_2$) are not. The second observation we can make about the transition matrix at steady-state is that there is no possibility of moving to a transient state irrespective of the value of the initial distribution vector. The third observation we can make is that no matter what the initial distribution vector was, we will always wind up in the same steady-state distribution. ■

5.9 Identifying Reducible Markov Chains

We saw above that reducible Markov chains have a transition matrix that can be expressed, by proper reordering of the states, into the canonic form

$$\mathbf{P} = \begin{bmatrix} \mathbf{C} & \mathbf{A} \\ \mathbf{0} & \mathbf{T} \end{bmatrix} \tag{5.29}$$

This rearranged matrix allowed us to determine the closed and transient states. We want to show in this section how to easily identify a reducible Markov chain and how to find its closed and transient states without having to rearrange the matrix. The following theorem helps us to determine if our Markov chain is reducible or not by observing the structure of its eigenvector corresponding to the eigenvalue $\lambda = 1$.

Theorem 1 *Let* $\mathbf{P}$ *be the transition matrix of a Markov chain whose eigenvalue* $\lambda = 1$ *corresponds to an eigenvector* $\mathbf{s}$*. Then this chain is reducible if and only if* $\mathbf{s}$ *has one or more zero elements.*

Proof. We start by assuming that the eigenvector $\mathbf{s}$ has k nonzero elements and $m - k$ zero elements where m is the number of rows and columns of $\mathbf{P}$. Without loss of generality we can write $\mathbf{s}$ in the canonic form

$$\mathbf{s} = \begin{bmatrix} \mathbf{a} \\ \mathbf{0} \end{bmatrix} \tag{5.30}$$

where the vector $\mathbf{a}$ has k elements none of which is zero such that $0 < k < m$. Partition $\mathbf{P}$ into the form

$$\mathbf{P} = \begin{bmatrix} \mathbf{A} & \mathbf{B} \\ \mathbf{C} & \mathbf{D} \end{bmatrix} \tag{5.31}$$

where $\mathbf{A}$ is a square $k \times k$ matrix, $\mathbf{D}$ is a square $(m - k) \times (m - k)$ matrix, and the other two matrices are rectangular with the proper dimensions. Since $\mathbf{s}$ is the eigenvector corresponding to $\lambda = 1$, we can write

$$\mathbf{P}\,\mathbf{s} = \mathbf{s} \tag{5.32}$$

or

$$\begin{bmatrix} \mathbf{A} & \mathbf{B} \\ \mathbf{C} & \mathbf{D} \end{bmatrix} \begin{bmatrix} \mathbf{a} \\ \mathbf{0} \end{bmatrix} = \begin{bmatrix} \mathbf{a} \\ \mathbf{0} \end{bmatrix} \tag{5.33}$$

This equation results in

$$\mathbf{A}\,\mathbf{a} = \mathbf{a} \tag{5.34}$$

and

$$\mathbf{C}\,\mathbf{a} = \mathbf{0} \tag{5.35}$$

Having $\mathbf{a} = \mathbf{0}$ is contrary to our assumptions. Since the above two equations are valid for any nonzero value of $\mathbf{a}$, we conclude that $\mathbf{A}$ is column stochastic and $\mathbf{C} = \mathbf{0}$.

Thus the transition matrix reduces to the form

$$\mathbf{P} = \begin{bmatrix} \mathbf{A} & \mathbf{B} \\ \mathbf{0} & \mathbf{D} \end{bmatrix} \tag{5.36}$$

This is the general canonic form for a reducible Markov chain and this completes one part of the proof.

Now let us assume that $\mathbf{P}$ corresponds to a reducible Markov chain. In that case, we can write $\mathbf{P}$ in the canonic form

$$\mathbf{P} = \begin{bmatrix} \mathbf{C} & \mathbf{A} \\ \mathbf{0} & \mathbf{T} \end{bmatrix} \tag{5.37}$$

There are two cases to consider here: $\mathbf{A} = \mathbf{0}$ and $\mathbf{A} \neq \mathbf{0}$.

Case 1: $\mathbf{A} = \mathbf{0}$

This is the case when we have

$$\mathbf{P} = \begin{bmatrix} \mathbf{C} & \mathbf{0} \\ \mathbf{0} & \mathbf{T} \end{bmatrix} \tag{5.38}$$

We have in reality two independent and noncommunicating Markov systems. Assume vector $\mathbf{s}$ is the distribution vector associated with the unity eigenvalue for matrix $\mathbf{C}$. In that case, we can express $\mathbf{s}$ as

$$\mathbf{s} = \begin{bmatrix} \mathbf{a} & \mathbf{b} \end{bmatrix}^t \tag{5.39}$$

and $\mathbf{s}$ satisfies the equations

$$\begin{bmatrix} \mathbf{C} & \mathbf{0} \\ \mathbf{0} & \mathbf{T} \end{bmatrix} \begin{bmatrix} \mathbf{a} \\ \mathbf{b} \end{bmatrix} = \begin{bmatrix} \mathbf{a} \\ \mathbf{b} \end{bmatrix} \tag{5.40}$$

We can write

$$\mathbf{C\,a} = \mathbf{a} \tag{5.41}$$

$$\mathbf{T\,b} = \mathbf{b} \tag{5.42}$$

The first equation indicates that **a** is an eigenvector of **C** and it should be a valid distribution vector. Since the sum of the components of **a** must be unity, the sum of the components of **b** must be zero, which is possible only when

$$\mathbf{b} = \mathbf{0} \tag{5.43}$$

This completes the second part of the proof for Case 1.

The same is true for the eigenvector corresponding to unity eigenvalue for matrix **T**. In that case, **a** will be null and **b** will be the valid distribution vector. Either way, this completes the second part of the proof for Case 1.

Case 2: $\mathbf{A} \neq \mathbf{0}$
This is the case when we have

$$\mathbf{P} = \begin{bmatrix} \mathbf{C} & \mathbf{A} \\ \mathbf{0} & \mathbf{T} \end{bmatrix} \tag{5.44}$$

In that case, **T** is a substochastic matrix and $\mathbf{T}^{\infty} = 0$. Now for large time values $(n \to \infty)$ we have

$$\mathbf{P}^{\infty} = \begin{bmatrix} \mathbf{X} \\ \mathbf{0} \end{bmatrix} \tag{5.45}$$

But we can also write

$$\mathbf{P}^{\infty}\,\mathbf{s} = \mathbf{s} \tag{5.46}$$

We partition **s** into the form

$$\mathbf{s} = \begin{bmatrix} \mathbf{a} \\ \mathbf{b} \end{bmatrix} \tag{5.47}$$

Substitute the above equation into (5.46) to get

$$\begin{bmatrix} \mathbf{X} \\ \mathbf{0} \end{bmatrix} \begin{bmatrix} \mathbf{a} \\ \mathbf{b} \end{bmatrix} = \begin{bmatrix} \mathbf{a} \\ \mathbf{b} \end{bmatrix} \tag{5.48}$$

And we get the two equations

$$\mathbf{X\,a} = \mathbf{a} \tag{5.49}$$

and

$$\mathbf{b} = \mathbf{0} \tag{5.50}$$

Thus the distribution vector corresponding to the eigenvalue $\lambda = 1$ will have the form

$$\mathbf{s} = \begin{bmatrix} \mathbf{a} \\ \mathbf{0} \end{bmatrix} \tag{5.51}$$

This completes the proof of the theorem for Case 2.

■

Example 9 Prove that the given transition matrix corresponds to a reducible Markov chain.

$$\mathbf{P} = \begin{bmatrix} 0 & 0.7 & 0.1 & 0 & 0 \\ 0.3 & 0 & 0.1 & 0 & 0 \\ 0.1 & 0.1 & 0.2 & 0 & 0 \\ 0.4 & 0 & 0.1 & 0.6 & 0.7 \\ 0.2 & 0.2 & 0.5 & 0.4 & 0.3 \end{bmatrix}$$

We calculate the eigenvalues and eigenvectors for the transition matrix. The distribution vector associated with the eigenvalue $\lambda = 1$ is

$$\mathbf{s} = \begin{bmatrix} 0 & 0 & 0 & 0.6364 & 0.3636 \end{bmatrix}^t$$

Since we have zero elements, we conclude that we have a reducible Markov chain.

5.9.1 Determining Closed and Transient States

Now that we know how to recognize a reducible Markov chain, we need to know how to recognize its closed and transient states. The following theorem provides the answer.

Theorem 2 *Let* $\mathbf{P}$ *be the transition matrix of a reducible Markov chain whose eigenvalue* $\lambda = 1$ *corresponds to an eigenvector* $\mathbf{s}$*. The closed states of the chain correspond to the nonzero elements of* $\mathbf{s}$ *and the transient states of the chain correspond to the zero elements of* $\mathbf{s}$.

Proof. Since we are dealing with a reducible Markov chain, then without loss of generality, the transition matrix can be arranged in the canonic form

$$\mathbf{P} = \begin{bmatrix} \mathbf{C} & \mathbf{A} \\ \mathbf{0} & \mathbf{T} \end{bmatrix} \tag{5.52}$$

where it is assumed that $\mathbf{C}$ is a $k \times k$ matrix and $\mathbf{T}$ is a $(m-k) \times (m-k)$ matrix. The first k states correspond to closed states and the last $m-k$ states correspond to transient states.

Assume the eigenvector $\mathbf{s}$ is expressed in the form

$$\mathbf{s} = \begin{bmatrix} \mathbf{a} \\ \mathbf{b} \end{bmatrix}^t \tag{5.53}$$

where some of the elements of $\mathbf{s}$ are zero according to Theorem 1. Since this is the eigenvector corresponding to unity eigenvalue, we must have

$$\begin{bmatrix} \mathbf{C} & \mathbf{A} \\ \mathbf{0} & \mathbf{T} \end{bmatrix} \begin{bmatrix} \mathbf{a} \\ \mathbf{b} \end{bmatrix} = \begin{bmatrix} \mathbf{a} \\ \mathbf{b} \end{bmatrix} \tag{5.54}$$

And we get the two equations

$$\mathbf{C}\,\mathbf{a} + \mathbf{A}\,\mathbf{b} = \mathbf{a} \tag{5.55}$$

and

$$\mathbf{T}\,\mathbf{b} = \mathbf{b} \tag{5.56}$$

The above equation seems to indicate that $\mathbf{T}$ has an eigenvector $\mathbf{b}$ with unity eigenvalue. However, this is a contradiction since $\mathbf{T}$ is column substochastic and it cannot have a unity eigenvalue. The absolute values of all the eigenvalues of $\mathbf{T}$ are less than unity [1]. For such a matrix, we say that its spectral radius cannot equal unity.[1] The above equation is satisfied only if

$$\mathbf{b} = \mathbf{0} \tag{5.57}$$

In that case, (5.55) becomes

$$\mathbf{C}\,\mathbf{a} = \mathbf{a} \tag{5.58}$$

Thus the eigenvector $\mathbf{s}$ will have the form

$$\mathbf{s} = \begin{bmatrix} \mathbf{a} \\ \mathbf{0} \end{bmatrix}^t \tag{5.59}$$

where $\mathbf{a}$ is a k-distribution vector corresponding to unity eigenvalue of $\mathbf{C}$.

We can therefore associate the closed states with the nonzero components of $\mathbf{s}$ and associate the transient states with the zero components of $\mathbf{s}$.

---

[1] Spectral radius equals the largest absolute value of the eigenvalues of a matrix.

So far we have proven that **s** has the form given in (5.59). We must prove now that *all* the components of **a** are nonzero. This will allow us to state with certainty that any zero component of **s** belongs solely to a transient state.

We prove this by proving that a contradiction results if **a** is assumed to have one or more zero components in it. Assume that **a** has one or more zero components. We have proven, however, that **a** satisfies the equation

$$\mathbf{C}\,\mathbf{a} = \mathbf{a} \tag{5.60}$$

where **C** is a nonreducible matrix. Applying Theorem 1, on page 167, to the above equation would indicate that **C** is reducible. This is a contradiction since **C** is a nonreducible matrix.

Thus the k closed states correspond to the nonzero elements of **s** and the transient states of the chain correspond to the zero elements of **s**. This proves the theorem. ∎

Example 10 Prove that the given transition matrix corresponds to a reducible Markov chain.

$$\mathbf{P} = \begin{bmatrix} 0.3 & 0.2 & 0.3 & 0.4 & 0.1 & 0.2 \\ 0 & 0.1 & 0.2 & 0 & 0 & 0 \\ 0 & 0.2 & 0.1 & 0 & 0 & 0 \\ 0.4 & 0.1 & 0.2 & 0.1 & 0.2 & 0.3 \\ 0.1 & 0.1 & 0.2 & 0.2 & 0.3 & 0.4 \\ 0.2 & 0.3 & 0 & 0.3 & 0.4 & 0.1 \end{bmatrix}$$

We calculate the eigenvalues and eigenvectors for the transition matrix. The distribution vector associated with the eigenvalue $\lambda = 1$ is

$$\mathbf{s} = \begin{bmatrix} 0.25 & 0 & 0 & 0.25 & 0.25 & 0.25 \end{bmatrix}^t$$

Since we have zero elements, we conclude that we have a reducible Markov chain. The zero elements identify the transient states and the nonzero elements identify the closed states.

Closed States	Transient States
1, 4, 5, 6	2, 3

5.10 Identifying Reducible Composite Matrices

We can generalize Theorem 2 as follows. Let **P** be the transition matrix of a composite reducible Markov chain with u mutually exclusive closed states corresponding to the sets $C_1, C_2, \ldots, C_u$. The canonic form for the transition matrix of such a system will be

$$\mathbf{P} = \begin{bmatrix} \mathbf{C}_1 & \mathbf{0} & \cdots & \mathbf{0} & \mathbf{A}_1 \\ \mathbf{0} & \mathbf{C}_1 & \cdots & \mathbf{0} & \mathbf{A}_2 \\ \vdots & \vdots & \ddots & \vdots & \vdots \\ \mathbf{0} & \mathbf{0} & \cdots & \mathbf{C}_u & \mathbf{A}_u \\ \mathbf{0} & \mathbf{0} & \cdots & \mathbf{0} & \mathbf{T} \end{bmatrix} \tag{5.61}$$

The eigenvalue $\lambda = 1$ corresponds to the eigenvectors $\mathbf{s}_1, \mathbf{s}_2, \ldots, \mathbf{s}_u$ such that

$$\mathbf{P}\,\mathbf{s}_1 = \mathbf{s}_1 \tag{5.62}$$
$$\mathbf{P}\,\mathbf{s}_2 = \mathbf{s}_2 \tag{5.63}$$
$$\vdots$$
$$\mathbf{P}\,\mathbf{s}_u = \mathbf{s}_u \tag{5.64}$$

The eigenvectors also satisfy the equations

$$\mathbf{C}_1\,\mathbf{s}_1 = \mathbf{s}_1 \tag{5.65}$$
$$\mathbf{C}_2\,\mathbf{s}_2 = \mathbf{s}_2 \tag{5.66}$$
$$\vdots$$
$$\mathbf{C}_u\,\mathbf{s}_u = \mathbf{s}_u \tag{5.67}$$

We can in fact write each eigenvector $\mathbf{s}_i$ in block form as

$$\mathbf{s}_1 = \begin{bmatrix} \mathbf{a}_1 & \mathbf{0} & \mathbf{0} & \cdots & \mathbf{0} & \mathbf{0} \end{bmatrix}^t \tag{5.68}$$
$$\mathbf{s}_2 = \begin{bmatrix} \mathbf{0} & \mathbf{a}_2 & \mathbf{0} & \cdots & \mathbf{0} & \mathbf{0} \end{bmatrix}^t \tag{5.69}$$
$$\vdots$$
$$\mathbf{s}_u = \begin{bmatrix} \mathbf{0} & \mathbf{0} & \mathbf{0} & \cdots & \mathbf{a}_u & \mathbf{0} \end{bmatrix}^t \tag{5.70}$$

where each $\mathbf{a}_i$ is a nonzero vector whose dimension matches $\mathbf{C}_i$ such that

$$\mathbf{C}_1\,\mathbf{a}_1 = \mathbf{a}_1 \tag{5.71}$$
$$\mathbf{C}_2\,\mathbf{a}_2 = \mathbf{a}_2 \tag{5.72}$$
$$\vdots$$
$$\mathbf{C}_u\,\mathbf{a}_u = \mathbf{a}_u \tag{5.73}$$

which means that $\mathbf{a}_i$ is a distribution (sum of its components is unity).

Vector $\mathbf{s}_i$ corresponds to the set of closed states C_i and the transient states of the chain correspond to the zero elements common to all the vectors $\mathbf{s}_i$.

Example 11 Assume a composite reducible transition matrix where the number of closed states is $u = 3$ such that the partitioned matrices are

$$
\begin{aligned}
\mathbf{C}_1 &= \begin{bmatrix} 0.3 & 0.6 \\ 0.7 & 0.4 \end{bmatrix} \\
\mathbf{C}_2 &= \begin{bmatrix} 0.5 & 0.1 \\ 0.5 & 0.9 \end{bmatrix} \\
\mathbf{C}_3 &= \begin{bmatrix} 0.2 & 0.3 \\ 0.8 & 0.7 \end{bmatrix} \\
\mathbf{A}_1 &= \begin{bmatrix} 0.1 & 0 \\ 0.1 & 0.1 \end{bmatrix} \\
\mathbf{A}_2 &= \begin{bmatrix} 0.3 & 0.1 \\ 0.2 & 0.1 \end{bmatrix} \\
\mathbf{A}_3 &= \begin{bmatrix} 0 & 0.2 \\ 0.1 & 0 \end{bmatrix} \\
\mathbf{T} &= \begin{bmatrix} 0.1 & 0.2 \\ 0.1 & 0.3 \end{bmatrix}
\end{aligned}
$$

Determine the eigenvectors corresponding to the eigenvalue $\lambda = 1$ and identify the closed and transient states with the elements of those eigenvectors.

The composite transition matrix $\mathbf{P}$ is given by

$$
\mathbf{P} = \begin{bmatrix} \mathbf{C}_1 & \mathbf{0} & \mathbf{0} & \mathbf{A}_1 \\ \mathbf{0} & \mathbf{C}_2 & \mathbf{0} & \mathbf{A}_2 \\ \mathbf{0} & \mathbf{0} & \mathbf{C}_3 & \mathbf{A}_3 \\ \mathbf{0} & \mathbf{0} & \mathbf{0} & \mathbf{T} \end{bmatrix}
$$

Let us find the eigenvalues and eigenvectors for $\mathbf{P}$. The eigenvalues are

$$
\begin{aligned}
\lambda_1 &= -0.3 \\
\lambda_2 &= 1 \\
\lambda_3 &= 0.4 \\
\lambda_4 &= 1 \\
\lambda_5 &= -0.1 \\
\lambda_6 &= 1 \\
\lambda_7 &= 0.0268 \\
\lambda_8 &= 0.3732
\end{aligned}
$$

The eigenvectors corresponding to unity eigenvalue (after normalization so their sums is unity) are

$$\begin{aligned}
\mathbf{s}_1 &= \begin{bmatrix} 0.4615 & 0.5385 & 0 & 0 & 0 & 0 & 0 & 0 \end{bmatrix}^t \\
\mathbf{s}_2 &= \begin{bmatrix} 0 & 0 & 0.1667 & 0.8333 & 0 & 0 & 0 & 0 \end{bmatrix}^t \\
\mathbf{s}_3 &= \begin{bmatrix} 0 & 0 & 0 & 0 & 0.2727 & 0.7273 & 0 & 0 \end{bmatrix}^t
\end{aligned}$$

The sets of closed and transient states are as follows.

Set	States
C_1	1, 2
C_2	3, 4
C_3	5, 6
T	7, 8

Problems

Reducible Markov Chains

For Problems 5.1–5.8: (a) Determine whether the given Markov matrices have absorbing or closed states; (b) express such matrices in the form given in (5.2) or (5.3); (c) identify the component matrices **C**, **A**, and **T**; and (d) identify the closed and transient states.

5.1

$$\mathbf{P} = \begin{bmatrix} 0.3 & 0 \\ 0.7 & 1 \end{bmatrix}$$

5.2

$$\mathbf{P} = \begin{bmatrix} 0.5 & 0 & 0.2 \\ 0.3 & 1 & 0.3 \\ 0.2 & 0 & 0.5 \end{bmatrix}$$

5.3

$$\mathbf{P} = \begin{bmatrix} 0.5 & 0.5 & 0 \\ 0.3 & 0.5 & 0 \\ 0.2 & 0 & 1 \end{bmatrix}$$

5.4

$$\mathbf{P} = \begin{bmatrix} 0.7 & 0.5 & 0.1 \\ 0.3 & 0.5 & 0 \\ 0 & 0 & .9 \end{bmatrix}$$

5.5

$$\mathbf{P} = \begin{bmatrix} 0.2 & 0 & 0 \\ 0.3 & 1 & 0 \\ 0.5 & 0 & 1 \end{bmatrix}$$

5.6

$$\mathbf{P} = \begin{bmatrix} 0.1 & 0 & 0.5 & 0 \\ 0.2 & 1 & 0 & 0 \\ 0.3 & 0 & 0.5 & 0 \\ 0.4 & 0 & 0 & 1 \end{bmatrix}$$

5.7

$$\mathbf{P} = \begin{bmatrix} 1 & 0 & 0 & 0 \\ 0 & 0.5 & 0 & 0.2 \\ 0 & 0.2 & 1 & 0 \\ 0 & 0.3 & 0 & 0.8 \end{bmatrix}$$

5.8

$$\mathbf{P} = \begin{bmatrix} 1 & 0 & 0 & 0 \\ 0 & 0.1 & 0 & 0.5 \\ 0 & 0.2 & 1 & 0 \\ 0 & 0.7 & 0 & 0.5 \end{bmatrix}$$

Composite Reducible Markov Chains

The transition matrices in Problems 5.9–5.12 represent composite reducible Markov chains. Identify the sets of closed and transient states, find the eigenvalues and eigenvectors for the matrices, and find the value of each matrix for large values of n, say when $n = 50$.

5.9

$$\mathbf{P} = \begin{bmatrix} 1 & 0 & 0 & 0.3 \\ 0 & 0.5 & 0 & 0 \\ 0 & 0.5 & 1 & 0.6 \\ 0 & 0 & 0 & 0.1 \end{bmatrix}$$

5.10

$$\mathbf{P} = \begin{bmatrix} 0.7 & 0 & 0 & 0 \\ 0.1 & 0.9 & 0.2 & 0 \\ 0.1 & 0.1 & 0.8 & 0 \\ 0.1 & 0 & 0 & 1 \end{bmatrix}$$

5.11

$$\mathbf{P} = \begin{bmatrix} 0.4 & 0 & 0 & 0 & 0 \\ 0.2 & 0.5 & 0.8 & 0 & 0 \\ 0.1 & 0.5 & 0.2 & 0 & 0 \\ 0.1 & 0 & 0 & 0.7 & 0.9 \\ 0.2 & 0 & 0 & 0.3 & 0.1 \end{bmatrix}$$

5.12

$$\mathbf{P} = \begin{bmatrix} 0.2 & 0 & 0 & 0.2 & 0 & 0.6 \\ 0 & 0.3 & 0 & 0.3 & 0 & 0 \\ 0 & 0 & 0.2 & 0.2 & 0.3 & 0 \\ 0 & 0.3 & 0 & 0.3 & 0 & 0 \\ 0 & 0.2 & 0.8 & 0 & 0.7 & 0 \\ 0.8 & 0.2 & 0 & 0 & 0 & 0.4 \end{bmatrix}$$

Transient Analysis

5.13 Check whether the given transitions matrix is reducible or irreducible. Identify the closed and transient states and express the matrix in the form of (5.2) or (5.3) and identify the component matrices **C**, **A**, and **T**.

$$\mathbf{P} = \begin{bmatrix} 0.5 & 0 & 0 & 0 & 0.5 \\ 0 & 0.5 & 0 & 0.25 & 0 \\ 0 & 0 & 1 & 0.25 & 0 \\ 0 & 0.5 & 0 & 0.25 & 0 \\ 0.5 & 0 & 0 & 0.25 & 0.5 \end{bmatrix}$$

Find the value of $\mathbf{P}^{10}$ using (5.8) and (5.9) on page 158 and verify your results using repeated multiplications.

5.14 Assume the transition matrix $\mathbf{P}$ has the structure given in (5.1) or (5.2). Prove that $\mathbf{P}^n$ also possesses the same structure as the original matrix and prove also that the component matrices $\mathbf{C}$, $\mathbf{A}$, and $\mathbf{T}$ have the same properties as the original component matrices.

5.15 Find an expression for the transition matrix using (5.13) at time $n = 4$ and $n = 20$ for the reducible Markov chain characterized by the transition matrix

$$\mathbf{P} = \begin{bmatrix} 0.7 & 0.9 & 0.3 & 0.3 & 0.2 \\ 0.3 & 0.1 & 0.2 & 0.1 & 0.3 \\ 0 & 0 & 0.2 & 0.2 & 0 \\ 0 & 0 & 0.2 & 0.2 & 0.3 \\ 0 & 0 & 0.1 & 0.2 & 0.2 \end{bmatrix}$$

Find the value of $\mathbf{P}^{10}$ using (5.8) and verify your results using repeated multiplications.

Reducible Markov Chains at Steady-State

5.16 In Section 5.7, it was asserted that the transition matrix for a reducible Markov chain will have the form of (5.22) where all the columns of the matrix are identical. Prove that assertion knowing that

(a) all the columns of $\mathbf{C}_1$ are all identical
(b) matrix $\mathbf{C}_1$ is column stochastic
(c) matrix $\mathbf{Y}^\infty$ is column stochastic
(d) the columns of $\mathbf{Y}^\infty$ are identical to the columns of $\mathbf{C}_1$.

5.17 Find the steady-state transition matrix and distribution vector for the reducible Markov chain characterized by the matrix

$$\mathbf{P} = \begin{bmatrix} 0.9 & 0.2 & 0.5 & 0.1 & 0.1 \\ 0.1 & 0.8 & 0.1 & 0.2 & 0.3 \\ 0 & 0 & 0.2 & 0.2 & 0.1 \\ 0 & 0 & 0.2 & 0.2 & 0.1 \\ 0 & 0 & 0 & 0.3 & 0.4 \end{bmatrix}$$

5.18 Find the steady-state transition matrix and distribution vector for the reducible Markov chain characterized by the matrix

$$\mathbf{P} = \begin{bmatrix} 0.9 & 0.2 & 0.5 & 0.1 \\ 0.1 & 0.8 & 0.1 & 0.2 \\ 0 & 0 & 0.4 & 0.2 \\ 0 & 0 & 0 & 0.5 \end{bmatrix}$$

5.19 Find the steady-state transition matrix and distribution vector for the reducible Markov chain characterized by the matrix

$$\mathbf{P} = \begin{bmatrix} 0.1 & 0.2 & 0.5 & 0.1 & 0.1 \\ 0.5 & 0.7 & 0.1 & 0.2 & 0.3 \\ 0.4 & 0.1 & 0.4 & 0.2 & 0.1 \\ 0 & 0 & 0 & 0.2 & 0.1 \\ 0 & 0 & 0 & 0.3 & 0.4 \end{bmatrix}$$

5.20 Find the steady-state transition matrix and distribution vector for the reducible Markov chain characterized by the matrix

$$\mathbf{P} = \begin{bmatrix} 1 & 0.2 & 0.2 & 0.1 & 0.1 \\ 0 & 0.3 & 0.1 & 0.2 & 0.3 \\ 0 & 0.1 & 0.4 & 0.2 & 0.1 \\ 0 & 0.3 & 0.1 & 0.2 & 0.1 \\ 0 & 0.1 & 0.2 & 0.3 & 0.4 \end{bmatrix}$$

5.21 Find the steady-state transition matrix and distribution vector for the reducible Markov chain characterized by the matrix

$$\mathbf{P} = \begin{bmatrix} 1 & 0 & 0.2 & 0.1 & 0.1 \\ 0 & 1 & 0.1 & 0.2 & 0.3 \\ 0 & 0 & 0.4 & 0.2 & 0.1 \\ 0 & 0 & 0.1 & 0.2 & 0.1 \\ 0 & 0 & 0.2 & 0.3 & 0.4 \end{bmatrix}$$

Note that this matrix has two absorbing states.

5.22 Consider the state transition matrix

$$\mathbf{P} = \begin{bmatrix} 0 & 1 & 0 \\ 1-p & 0 & q \\ p & 0 & 1-q \end{bmatrix}$$

(a) Can this matrix represent a reducible Markov chain?
(b) Find the distribution vector at equilibrium.
(c) What values of p and q give $s_0 = s_1 = s_2$?

5.23 Consider a discrete-time Markov chain in which the transition probabilities are given by

$$p_{ij} = q^{|i-j|} p$$

For a 3×3 case, what are the values of p and q to make this a reducible Markov chain? What are the values of p and q to make this an irreducible Markov chain and find the steady-state distribution vector.

5.24 Consider the coin-tossing Example 3 on page 156. Derive the equilibrium distribution vector and comment on it for the cases $p < q$, $p = q$, and $p > q$.

5.25 Rewrite (5.10) on page 158 to take into account the fact that some of the eigenvalues of $\mathbf{C}$ might be repeated using the results of Section 3.14 on page 103.

Identification of Reducible Markov Chains

Use the results of Section 5.9 to verify that the transition matrices in the following problems correspond to reducible Markov chains and identify the closed and transient states. Rearrange each matrix to the standard form as in (5.2) or (5.3).

5.26

$$\mathbf{P} = \begin{bmatrix} 0.3 & 0.1 & 0.4 \\ 0 & 0.1 & 0 \\ 0.7 & 0.8 & 0.6 \end{bmatrix}$$

5.27

$$\mathbf{P} = \begin{bmatrix} 0.4 & 0.6 & 0 \\ 0.4 & 0.3 & 0 \\ 0.2 & 0.1 & 1 \end{bmatrix}$$

5.28

$$\mathbf{P} = \begin{bmatrix} 0.5 & 0.3 & 0 & 0 \\ 0.1 & 0.1 & 0 & 0 \\ 0.3 & 0.4 & 0.2 & 0.9 \\ 0.1 & 0.2 & 0.8 & 0.1 \end{bmatrix}$$

5.29

$$\mathbf{P} = \begin{bmatrix} 0.1 & 0 & 0 & 0 \\ 0.3 & 0.1 & 0.5 & 0.3 \\ 0.4 & 0.3 & 0.4 & 0.2 \\ 0.2 & 0.6 & 0.1 & 0.5 \end{bmatrix}$$

5.30

$$\mathbf{P} = \begin{bmatrix} 0.2 & 0.3 & 0 & 0.3 & 0 \\ 0.2 & 0.4 & 0 & 0 & 0 \\ 0.3 & 0 & 0.5 & 0.1 & 0.2 \\ 0.2 & 0 & 0 & 0.4 & 0 \\ 0.1 & 0.3 & 0.5 & 0.2 & 0.8 \end{bmatrix}$$

5.31

$$\mathbf{P} = \begin{bmatrix} 0.2 & 0.1 & 0.3 & 0.1 & 0.6 \\ 0 & 0.3 & 0 & 0.2 & 0 \\ 0.2 & 0.2 & 0.4 & 0.1 & 0.3 \\ 0 & 0.1 & 0 & 0.4 & 0 \\ 0.6 & 0.3 & 0.3 & 0.2 & 0.1 \end{bmatrix}$$

5.32

$$\mathbf{P} = \begin{bmatrix} 0.8 & 0.4 & 0 & 0.5 & 0 \\ 0 & 0.1 & 0 & 0 & 0 \\ 0 & 0.1 & 0.8 & 0 & 0.3 \\ 0.2 & 0.2 & 0 & 0.5 & 0 \\ 0 & 0.2 & 0.2 & 0 & 0.7 \end{bmatrix}$$

5.33

$$\mathbf{P} = \begin{bmatrix} 0.1 & 0 & 0.2 & 0 & 0 & 0.3 \\ 0.2 & 0.5 & 0.1 & 0 & 0.4 & 0.1 \\ 0.1 & 0 & 0.3 & 0 & 0 & 0.3 \\ 0.2 & 0 & 0.1 & 1 & 0 & 0 \\ 0.3 & 0.5 & 0.2 & 0 & 0.6 & 0.1 \\ 0.1 & 0 & 0.1 & 0 & 0 & 0.2 \end{bmatrix}$$

5.34

$$\mathbf{P} = \begin{bmatrix} 0.1 & 0.5 & 0 & 0.1 & 0 & 0 & 0.6 \\ 0.7 & 0.3 & 0 & 0.2 & 0 & 0 & 0.3 \\ 0 & 0 & 0.5 & 0.3 & 0.1 & 0 & 0 \\ 0 & 0 & 0 & 0.2 & 0 & 0.3 & 0 \\ 0 & 0 & 0.5 & 0 & 0.9 & 0.2 & 0 \\ 0 & 0 & 0 & 0.2 & 0 & 0.4 & 0 \\ 0.2 & 0.2 & 0 & 0 & 0 & 0.1 & 0.1 \end{bmatrix}$$

References

1. R.A. Horn and C.R. Johnson, *Matrix Analysis, Cambridge University Press, Cambridge, 1985.*

Chapter 6
Periodic Markov Chains

6.1 Introduction

From Chapter 4, we infer that a Markov chain settles down to a steady-state distribution vector **s** when $n \to \infty$. This is true for most transition matrices representing most Markov chains we studied. However, there are other times when the Markov chain never settles down to an equilibrium distribution vector, no matter how long we iterate. So this chapter will illustrate *periodic Markov chains* whose distribution vector $\mathbf{s}(n)$ repeats its values at regular intervals of time and never settles down to an equilibrium value no matter how long we iterate.

Periodic Markov chains could be found in systems that show repetitive behavior or task sequences. An intuitive example of a periodic Markov chain is the population of wild salmon. In that fish species, we can divide the life cycle as eggs, hatchlings, subadults, and adults. Once the adults reproduce, they die, and the resulting eggs hatch and repeat the cycle as shown in Fig. 6.1. Fluctuations in the salmon population can thus be modeled as a periodic Markov chain. It is interesting to note that other fishes that lay eggs without dying can be modeled as nonperiodic Markov chains.

Another classic example from nature where periodic Markov chains apply is the predator–prey relation—where the population of deer, say, is related to the population of wolves. When deer numbers are low, the wolf population is low. This results in more infant deer survival rate, and the deer population grows during the next year. When this occurs, the wolves start having more puppies and the wolf population also increases. However, the large number of wolves results in more deer kills, and the deer population diminishes. The reduced number of deer results in wolf starvation, and the number of wolves also decreases. This cycle repeats as discussed in Problem 6.14.

Another example for periodic Markov chains in communications is data transmission. In such a system, first, data are packetized to be transmitted, and then the packets are sent over a channel. The received packets are then analyzed for the presence of errors. Based on the number of bits in error, the receiver is able to correct for the errors or inform the transmitter to retransmit, perhaps even with a higher level of data redundancy. Figure 6.2 shows these states which are modeled as a periodic

F. Gebali, *Analysis of Computer and Communication Networks*,
DOI: 10.1007/978-0-387-74437-7_6,

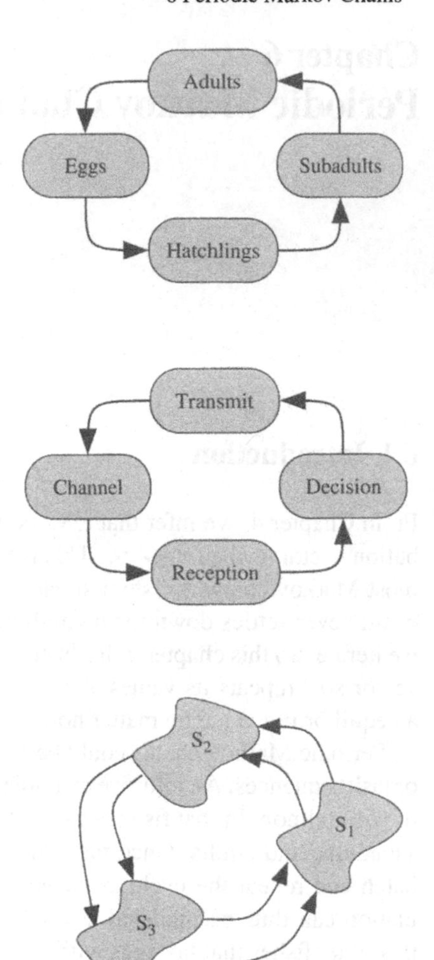

Fig. 6.1 Wild salmon can be modeled as a periodic Markov chain with the states representing the number of each phase of the fish life cycle

Fig. 6.2 Data communication over a noisy channel can be modeled as a periodic Markov chain with the states representing the state of each phase

Fig. 6.3 A periodic Markov chain where states are divided into groups and allowed transitions occur only between adjacent groups

Markov chain. In each transmission phase, there could be several states indicating the level of decoding required or the number of random errors introduced.

Consider the abstract transition diagram shown in Fig. 6.3, where the states of the Markov chain are divided into groups and allowed transitions occur only between adjacent groups. The sets of states S_1, S_2, $\cdots$ are called *periodic classes* of the Markov chain. A state in set S_1 is allowed to make a transition to any other state in set S_2 only. Thus, the states in the set S_1 cannot make a transition to any state in S_1 or S_3. A similar argument applies to the states in sets S_2 or S_3.

6.2 Definition

A *periodic Markov chain* has the property that the number of single-step transitions that must be made after leaving a state to return to that state is a multiple of some integer $\gamma > 1$ [1]. This definition implies that the distribution vector never settles down to a fixed steady-state value no matter how long we iterate.

Having mentioned that Markov chains could be periodic, we naturally want to know how to recognize that the transition matrix $\mathbf{P}$ represents a periodic Markov chain. In addition, we will also want to know the period of such a system. The theorems presented here will specify the properties of the transition matrix when the Markov chain is periodic.

Of course, a Markov chain that is not periodic would be called nonperiodic. A *nonperiodic Markov chain* does not show repetitive behavior as time progresses, and its distribution vector settles down to a fixed steady-state value.

6.3 Types of Periodic Markov Chains

There are two types of periodic Markov chains

1. *Strongly periodic Markov chains* the distribution vector repeats its values with a period $\gamma > 1$. In the next section, we will find out that the state transition matrix satisfies the relation

$$\mathbf{P}^{\gamma} = \mathbf{I}. \tag{6.1}$$

   In other words, in a strongly periodic Markov chain, the probability of returning to the starting state after γ time steps is unity for all states of the system.
2. *Weakly periodic Markov chains* the system shows periodic behavior only when $n \to \infty$. In other words, the distribution vector repeats its values with a period $\gamma > 1$ only when $n \to \infty$. We will find out in Section 6.12 that the state transition matrix satisfies the relation

$$\mathbf{P}^{\gamma} \neq \mathbf{I} \tag{6.2}$$

   In other words, in a weakly periodic Markov chain, the probability of returning to the starting state after γ time steps is less than unity for some or all states of the system.

In both cases, there is no equilibrium distribution vector. Strongly periodic Markov chains are not encountered in practice since they are a special case of the more widely encountered weakly periodic Markov chains. We start this chapter, however, with the strongly periodic Markov chains since they are easier to study, and they will pave the ground for studying the weakly periodic type.

6.4 The Transition Matrix

Let us start by making general observations on the transition matrix of a strongly periodic Markov chain. Assume that somehow, we have a strongly periodic Markov chain with period γ. Then the probability of making a transition from state j to state i at time instant n will repeat its value at instant $n+\gamma$. This is true for all valid values of i and j:

$$p_{ij}(n+\gamma) = p_{ij}(n) \tag{6.3}$$

It is remarkable that this equation is valid for all valid values of n, i, and j. But then again, this is what a strongly periodic Markov chain does.

We can apply the above equation to the components of the distribution vector and write

$$\mathbf{s}(n+\gamma) = \mathbf{s}(n) \tag{6.4}$$

But the two distribution vectors are also related to each other through the transition matrix

$$\mathbf{s}(n+\gamma) = \mathbf{P}^{\gamma}\ \mathbf{s}(n) \tag{6.5}$$

From the above two equations, we can write

$$\mathbf{P}^{\gamma}\ \mathbf{s}(n) = \mathbf{s}(n) \tag{6.6}$$

or

$$(\mathbf{P}^{\gamma} - \mathbf{I})\ \mathbf{s}(n) = \mathbf{0} \tag{6.7}$$

and this equation is valid for all values of n and $\mathbf{s}(\mathbf{n})$. The solution to the above equations is

$$\mathbf{P}^{\gamma} = \mathbf{I} \tag{6.8}$$

where $\mathbf{I}$ is the unit matrix and $\gamma > 1$. The case is trivial when $\gamma = 1$, which indicates that $\mathbf{P}$ is the identity matrix.

Example 1 The following transition matrix corresponds to a strongly periodic Markov chain. Estimate the period of the chain

$$\mathbf{P} = \begin{bmatrix} 0 & 1 \\ 1 & 0 \end{bmatrix}$$

We start by performing repeated multiplications to see when $\mathbf{P}^k = \mathbf{I}$ for some value of k. We are lucky since

$$\mathbf{P}^2 = \mathbf{I}$$

The period of this Markov chain is $\gamma = 2$. The given transition matrix is also known as a circulant matrix where the adjacent rows or columns advance by one position. The matrix $\mathbf{P}$ could also be considered a permutation matrix or exchange matrix where rows 1 and 2 are exchanged after premultiplying any $2 \times m$ matrix [2]. ■

6.5 The Transition Matrix Determinant

This section provides one specification on the transition matrix of a strongly periodic Markov chain. Theorem 1 indicates the allowed values of the determinant of the transition matrix.

Theorem 1 *Let* **P** *be the transition matrix of a strongly periodic Markov chain. The determinant of* **P** *will be given by [3]*

$$\Delta = \pm 1$$

Proof We start by assuming that the Markov chain is strongly periodic. We have from the assumptions

$$\mathbf{P}^{\gamma} = \mathbf{I} \tag{6.9}$$

Equate the determinants of both sides

$$\Delta^{\gamma} = 1 \tag{6.10}$$

where Δ is the determinant of $\mathbf{P}$ and Δ^{γ} is the determinant of $\mathbf{P}^{\gamma}$.

Taking the γ-root of the above equation, we find that Δ is the γ-root of unity:

$$\Delta = \exp\left(j2\pi \times \frac{k}{\gamma}\right) \quad k = 1, 2, \cdots, \gamma \tag{6.11}$$

But Δ must be real since the components of $\mathbf{P}$ are all real. Thus, the only possible values for Δ are ± 1. This proves the theorem. ■

From the properties of the determinant of the transition matrix of a strongly periodic Markov chain, we conclude that $\mathbf{P}$ must have the following equivalent properties:

1. The $m \times m$ transition matrix **P** of a strongly periodic Markov chain is full rank, i.e., rank(**P**) $= m$.
2. The rows and columns of the transition matrix **P** of a strongly periodic Markov chain are linearly independent.
3. $\lambda < 1$ can never be an eigenvalue for the transition matrix **P** of a strongly periodic Markov chain. Any value of $\lambda < 1$ would produce a determinant that is not equal to ± 1.
4. All eigenvalues must obey the relation $|\lambda| = 1$. This will be proved in the next section.

6.6 Transition Matrix Diagonalization

The following theorem indicates that the transition matrix of a strongly periodic Markov chain can be diagonalized. This fact naturally leads to a great simplification in the study of strongly periodic Markov chains.

Theorem 2 *Let* **P** *be the transition matrix of a strongly periodic Markov chain with period* $\gamma > 1$. *Then* **P** *is diagonalizable.*

Proof If **P** is diagonalizable, then its Jordan canonic form will turn into a diagonal matrix. Let us assume that **P** is not diagonalizable. In that case, **P** is similar to its Jordan canonic form

$$\mathbf{P} = \mathbf{U}\mathbf{J}\mathbf{U}^{-1} \tag{6.12}$$

Since **P** is periodic, we must have

$$\mathbf{P}^{\gamma} = \mathbf{U}\mathbf{J}^{\gamma}U^{-1} = \mathbf{I} \tag{6.13}$$

Multiplying both sides of the equation from the right by **U**, we get

$$\mathbf{U}\mathbf{J}^{\gamma} = \mathbf{I}\mathbf{U} = \mathbf{U}\mathbf{I} \tag{6.14}$$

This implies that we must have

$$\mathbf{J}^{\gamma} = \mathbf{I} \tag{6.15}$$

The above equation states that the matrix $\mathbf{J}^{\gamma}$ is equal to the diagonal matrix **I**. However, **J** can never be diagonal if it is not already so. Thus the above equation is only possible when the Jordan canonic form **J** is diagonal, which happens when **P** is diagonalizable, and the theorem is proved. ■

Example 2 Verify that the following transition matrix is diagonalizable.

$$\mathbf{P} = \begin{bmatrix} 0 & 0 & 1 & 0 \\ 1 & 0 & 0 & 0 \\ 0 & 1 & 0 & 0 \\ 0 & 0 & 0 & 1 \end{bmatrix}$$

We start by finding the Jordan canonic form for the matrix **P**.

```
[V,J] = jordan(P)

V =
 0.3333     0.3333                      0.3333                0
 0.3333    -0.1667   - 0.2887j         -0.1667   + 0.2887j              0
 0.3333    -0.1667   + 0.2887j         -0.1667   - 0.2887j              0
 1.0000          0                           0                     1.0000

J =

   1.0000          0                          0                         0
        0    -0.5000 + 0.8660j                0                         0
        0          0                    -0.5000    - 0.8660j            0
        0          0                          0                    1.0000
```

Thus we see that **P** is diagonalizable. We also see that all the eigenvalues lie on the unit circle.

$$\lambda_1 = \exp\left(j2\pi \times \frac{1}{3}\right)$$

$$\lambda_2 = \exp\left(j2\pi \times \frac{2}{3}\right)$$

$$\lambda_3 = \exp\left(j2\pi \times \frac{3}{3}\right)$$

$$\lambda_4 = \exp\left(j2\pi \times \frac{3}{3}\right)$$

where $j = \sqrt{-1}$. ■

The following theorem will add one more specification on the transition matrix of a strongly periodic Markov chain.

Theorem 3 *Let* **P** *be the transition matrix of a strongly periodic Markov chain. Then* **P** *is unitary (orthogonal)*

Proof Assume **x** is an eigenvector of the transition matrix **P**, then we can write

$$\mathbf{P}\,\mathbf{x} = \lambda\,\mathbf{x} \tag{6.16}$$

Transposing and taking the complex conjugate of both sides of the above equation, we get

$$\mathbf{x}^H \, \mathbf{P}^H = \lambda^* \, \mathbf{x}^H \tag{6.17}$$

where the symbol H indicates complex conjugate of the transposed matrix or vector, and λ^* is the complex conjugate of λ.

Now multiply the corresponding sides of Equations (6.16) and (6.17) to get

$$\mathbf{x}^H \, \mathbf{P}^H \, \mathbf{P} \, \mathbf{x} = \lambda^* \, \lambda \, \mathbf{x}^H \, \mathbf{x} \tag{6.18}$$

or

$$\mathbf{x}^H \, \mathbf{P}^H \, \mathbf{P} \, \mathbf{x} = |\lambda|^2 \; |\mathbf{x}|^2 \tag{6.19}$$

where

$$|\mathbf{x}|^2 = x_1^* \, x_1 + x_2^* \, x_2 + \cdots + x_m^* \, x_m \tag{6.20}$$

We know from Theorem 5 that the eigenvalues of $\mathbf{P}$ lie on the unit circle, and (6.19) can be written as

$$\mathbf{x}^H \, \mathbf{Y} \, \mathbf{x} = |\mathbf{x}|^2 \tag{6.21}$$

where $\mathbf{Y} = \mathbf{P}^H \, \mathbf{P}$. The above equation can be written as

$$\sum_{i=1}^{m} \sum_{j=1}^{m} x_i^* \, y_{ij} \, x_j = \sum_{i=1}^{m} x_i^* \, x_i \tag{6.22}$$

This equation can only be satisfied for arbitrary values of the eigenvectors when

$$\sum_{j=1}^{m} x_i^* \, y_{ij} \, x_j = x_i^* \, x_i \tag{6.23}$$

Similarly, this equation can only be satisfied for arbitrary values of the eigenvectors when

$$y_{ij} = \delta_{ij} \tag{6.24}$$

where δ_{ij} is the Kronecker delta which satisfies the equation

$$\delta_{ij} = \begin{cases} 1 & \text{when } i = j \\ 0 & \text{when } i \neq j \end{cases} \tag{6.25}$$

We conclude therefore that

$$\mathbf{P}^H \mathbf{P} = \mathbf{I} \tag{6.26}$$

Thus **P** is a unitary matrix whose inverse equals the complex conjugate of its transpose. Since **P** is real, the unitary matrix is usually called orthogonal matrix. This proves the theorem. ■

6.7 Transition Matrix Eigenvalues

So far, we have discovered several restrictions on the transition matrix of a strongly periodic Markov chain. The following theorem specifies the allowed magnitudes of the eigenvalues for strongly periodic Markov chains.

Theorem 4 *Let* **P** *be the transition matrix of a Markov chain. The Markov chain will be strongly periodic if and only if all the eigenvalues of* **P** *lie on the unit circle.*

Proof Let us start by assuming that the Markov chain is strongly periodic. According to Theorem 1, the determinant of the transition matrix is

$$\Delta = \pm 1$$

The determinant can be written as the product of all the eigenvalues of the transition matrix

$$\Delta = \prod_i \lambda_i \tag{6.27}$$

but we know that $\Delta = \pm 1$ and we can write

$$\prod_i \lambda_i = \pm 1 \tag{6.28}$$

Since **P** is column stochastic, all its eigenvalues must satisfy the inequality

$$|\lambda_i| \leq 1 \tag{6.29}$$

The above inequality together with (6.28) imply that

$$|\lambda_i| = 1 \tag{6.30}$$

This proves one side of the theorem.

Let us now assume that all the eigenvalues of **P** lie on the unit circle. In that case, we can write the eigenvalues as

$$\lambda_i = \exp\left(\frac{j2\pi}{\gamma_i}\right) \tag{6.31}$$

Therefore, we can write the transition matrix determinant as

$$\Delta = \prod_i \lambda_i \tag{6.32}$$

Raising the above equation to some power γ, we get

$$\Delta^\gamma = \prod_i \lambda_i^\gamma \tag{6.33}$$

Thus we have

$$\Delta^\gamma = \exp\left[j2\pi\left(\frac{\gamma}{\gamma_1} + \frac{\gamma}{\gamma_2} + \cdots + \frac{\gamma}{\gamma_m}\right)\right] \tag{6.34}$$

If γ is chosen to satisfy the equation

$$\gamma = \mathrm{lcm}(\gamma_i) \tag{6.35}$$

where lcm is the least common multiple, then we can write

$$\Delta^\gamma = 1 \tag{6.36}$$

According to Theorem 1, this proves that the Markov chain is strongly periodic. This proves the other part of the theorem. ∎

Figure 6.4 shows the locations of the eigenvalues of **P** in the complex plane. We see that all the eigenvalues of a strongly periodic Markov chain lie on the unit circle as indicated by the $\times$s. Since **P** is column stochastic, the eigenvalue $\lambda = 1$ must be present as is also indicated. The figure also indicates that complex eigenvalues appear in complex conjugate pairs.

Example 3 Consider the following transition matrix

$$\mathbf{P} = \begin{bmatrix} 0 & 0 & 1 & 0 \\ 1 & 0 & 0 & 0 \\ 0 & 1 & 0 & 0 \\ 0 & 0 & 0 & 1 \end{bmatrix}$$

Verify that this matrix is strongly periodic and find its period.

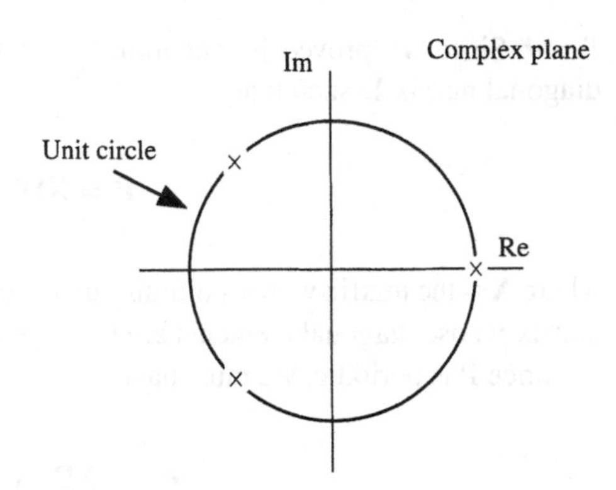

Fig. 6.4 All the eigenvalues of a strongly periodic Markov chain lie on the unit circle in the complex plane

The determinant is det(**P**) $= 1$, and the rank of the transition matrix is rank (**P**) $= 4$.

Performing repeated multiplications, we find that

$$\mathbf{P}^3 = \mathbf{I}$$

The period of the given transition matrix is $\gamma = 3$.

The eigenvalues of the transition matrix are

$$\lambda_1 = \exp\left(j2\pi \times \frac{1}{3}\right)$$

$$\lambda_2 = \exp\left(j2\pi \times \frac{2}{3}\right)$$

$$\lambda_3 = \exp\left(j2\pi \times \frac{3}{3}\right)$$

$$\lambda_4 = \exp\left(j2\pi \times \frac{3}{3}\right)$$

Thus all the eigenvalues lie on the unit circle. ■

The following theorem defines the allowed eigenvalues for the transition matrix of a strongly periodic Markov chain.

Theorem 5 *Let* **P** *be the transition matrix of an irreducible strongly periodic Markov chain with period* $\gamma > 1$*. Then the eigenvalues of* **P** *are the* γ*-roots of unity, i.e.,*

$$\lambda_i = \exp\left(j2\pi \times \frac{i}{\gamma}\right) \qquad i = 1, 2, \cdots, \gamma \tag{6.37}$$

where $j = \sqrt{-1}$.

Proof Since we proved in Theorem 2 that **P** is diagonalizable, it is similar to a diagonal matrix **D** such that

$$\mathbf{P} = \mathbf{X}\mathbf{D}\mathbf{X}^{-1} \tag{6.38}$$

where **X** is the matrix whose columns are the eigenvectors of **P**, and **D** is the diagonal matrix whose diagonal elements are the eigenvalues of **P**.

Since **P** is periodic, we must have

$$\mathbf{P}^{\gamma} = \mathbf{X}\mathbf{D}^{\gamma}\mathbf{X}^{-1} = \mathbf{I} \tag{6.39}$$

Multiplying both sides from the right by **X**, we get

$$\mathbf{X}\mathbf{D}^{\gamma} = \mathbf{I}\mathbf{X} = \mathbf{X}\mathbf{I} \tag{6.40}$$

Therefore, we must have

$$\mathbf{D}^{\gamma} = \mathbf{I} \tag{6.41}$$

This implies that any diagonal element d_i^{γ} of $\mathbf{D}^{\gamma}$ must obey the relation

$$d_i^{\gamma} = 1 \tag{6.42}$$

Therefore, the eigenvalues of **P** are the γ-root of unity and are given by the equation

$$d_i = \exp\left(j2\pi \times \frac{i}{\gamma}\right) \qquad i = 1, 2, \cdots, m \tag{6.43}$$

This proves the theorem. Note that the theorem does not preclude repeated eigenvalues having the same value as long as **P** can be diagonalized. ■

Example 4 The given transition matrix corresponds to a strongly periodic Markov chain. Confirm the conclusions of Theorems 1, 2, and 5

$$\mathbf{P} = \begin{bmatrix} 0 & 1 & 0 & 0 \\ 0 & 0 & 1 & 0 \\ 0 & 0 & 0 & 1 \\ 1 & 0 & 0 & 0 \end{bmatrix}$$

The period of the given matrix is 4 since $\mathbf{P}^4 = \mathbf{I}$. Using MATLAB, we find that the determinant of $\mathbf{P}$ is $\Delta = -1$. The eigenvalues of $\mathbf{P}$ are

$$\begin{aligned} \lambda_1 &= \ \ 1 = \exp\left(j2\pi \times \frac{4}{4}\right) \\ \lambda_2 &= \ \ j = \exp\left(j2\pi \times \frac{1}{4}\right) \\ \lambda_3 &= -j = \exp\left(j2\pi \times \frac{3}{4}\right) \\ \lambda_4 &= -1 = \exp\left(j2\pi \times \frac{2}{4}\right) \end{aligned}$$

The matrix can be diagonalized as

$$\mathbf{P} = \mathbf{X}\mathbf{D}\mathbf{X}^{-1}$$

where

$$\mathbf{X} = \begin{bmatrix} 0.5 & -0.49 - 0.11j & -0.49 + 0.11j & -0.5 \\ -0.5 & 0.11 - 0.40j & 0.11 + 0.49j & -0.5 \\ 0.5 & 0.40 + 0.11j & 0.49 - 0.11j & -0.5 \\ -0.5 & -0.11 + 0.40j & -0.11 - 0.49j & -0.5 \end{bmatrix}$$

and

$$\mathbf{D} = \begin{bmatrix} -1 & 0 & 0 & 0 \\ 0 & j & 0 & 0 \\ 0 & 0 & -j & 0 \\ 0 & 0 & 0 & 1 \end{bmatrix}$$

We see that Theorems 1, 2, and 5 are verified.

6.8 Transition Matrix Elements

The following theorem imposes surprising restrictions on the values of the elements of the transition matrix for a strongly periodic Markov chain.

Theorem 6 *Let* $\mathbf{P}$ *be the* $m \times m$ *transition matrix of a Markov chain. The Markov chain is strongly periodic if and only if the elements of* $\mathbf{P}$ *are all zeros except for* m *elements that have 1s arranged such that each column and each row contains only a single 1 entry in a unique location.*

Proof We begin by assuming that the transition matrix $\mathbf{P}$ corresponds to a strongly periodic Markov chain. We proved that $\mathbf{P}$ must be unitary which implies that the rows and columns are orthonormal. This implies that no two rows shall have nonzero elements at the same location since all elements are nonnegative. Similarly, no two columns shall have nonzero elements at the same location. This is only possible if all rows and columns are zero except for a single element at a unique location in each row or column.

Further, since we are dealing with a Markov matrix, these nonzero elements should be equal to 1. This completes the first part of the proof.

Now let us assume that the transition matrix **P** is all zeros except for m elements that have 1s arranged such that each column and each row of **P** contains only a single 1 entry in a unique location.

The unique arrangement of 1s implies that **P** is column stochastic. The unique arrangement of 1s also implies that the determinant of **P** must be ± 1. Thus the product of the eigenvalues of **P** is given by

$$\prod_i \lambda_i = \pm 1 \tag{6.44}$$

Since **P** was proved to be the column stochastic, we have

$$|\lambda_i| \leq 1 \tag{6.45}$$

The above equation together with (6.44) imply that

$$|\lambda_i| = 1 \tag{6.46}$$

Thus we proved that all the eigenvalues of **P** lie on the unit circle of the complex plane when **P** has a unique distribution of 1s among its rows and columns. According to Theorem 4, this implies that the Markov chain is strongly periodic. This proves the theorem. ■

6.9 Canonic Form for P

A strongly periodic Markov chain will have its $m \times m$ transition matrix expressed in the canonic form

$$\mathbf{P} = \begin{bmatrix} 0 & 0 & 0 & \cdots & 0 & 0 & 1 \\ 1 & 0 & 0 & \cdots & 0 & 0 & 0 \\ 0 & 1 & 0 & \cdots & 0 & 0 & 0 \\ \vdots & \vdots & \vdots & \ddots & \vdots & \vdots & \vdots \\ 0 & 0 & 0 & \cdots & 0 & 0 & 0 \\ 0 & 0 & 0 & \cdots & 1 & 0 & 0 \\ 0 & 0 & 0 & \cdots & 0 & 1 & 0 \end{bmatrix} \tag{6.47}$$

This matrix can be obtained by proper ordering of the states and will have a period $\gamma = m$, which can be easily proved (see Problem 5).This matrix is also known as a *circulant matrix* since multiplying any matrix by this matrix will shift the rows or columns by one location.

6.10 Transition Diagram

Based on the above theorems, specifying the structure of a strongly periodic Markov chain, we find that the transition diagram for a strongly periodic Markov chain of period $\gamma = 3$ is as shown in Fig. 6.5. We see that each set of periodic classes consists of one state only, and the number of states equals the period of the Markov chain.

Example 5 Prove that the given transition matrix is periodic and determine the period.

$$\mathbf{P} = \begin{bmatrix} 0 & 1 & 0 & 0 & 0 \\ 0 & 0 & 0 & 0 & 1 \\ 1 & 0 & 0 & 0 & 0 \\ 0 & 0 & 1 & 0 & 0 \\ 0 & 0 & 0 & 1 & 0 \end{bmatrix}$$

The given matrix is column stochastic, full rank, and all its rows and columns are zeros except for five elements that contain 1 at unique locations in each row and column. Thus **P** is periodic according to Theorem 6. From MATLAB, the eigenvalues of **P** are given by

$$\lambda_1 = 1\angle 144^\circ \quad = \exp\left(j2\pi \times \tfrac{2}{5}\right)$$

$$\lambda_2 = 1\angle -144^\circ = \exp\left(j2\pi \times \tfrac{3}{5}\right)$$

$$\lambda_3 = 1\angle 72^\circ \quad = \exp\left(j2\pi \times \tfrac{1}{5}\right)$$

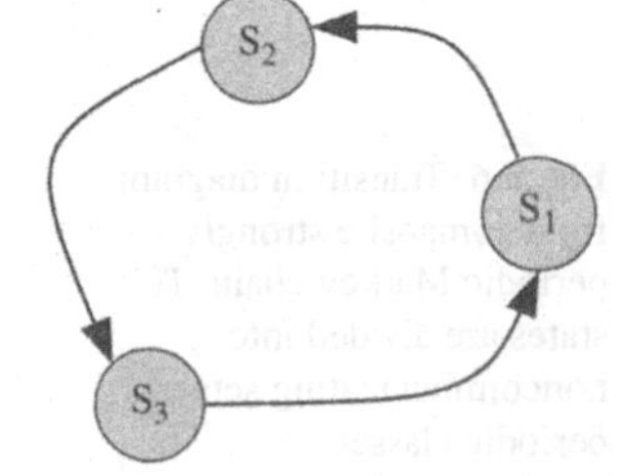

Fig. 6.5 Transition diagram for a strongly periodic Markov chain with period $\gamma = 3$

$$\lambda_4 = 1\angle -72^\circ = \exp\left(j2\pi \times \tfrac{4}{5}\right)$$

$$\lambda_5 = 1\angle 0^\circ \quad = \exp\left(j2\pi \times \tfrac{5}{5}\right)$$

Thus the period is $\gamma = 5$. As a verification, MATLAB assures us that $\mathbf{P}^5 = \mathbf{I}$.

6.11 Composite Strongly Periodic Markov Chains

In general, a composite strongly periodic Markov chain can be expressed, through proper ordering of states, in the canonic form

$$\mathbf{P} = \begin{bmatrix} \mathbf{C}_1 & \mathbf{0} & \mathbf{0} & \cdots & \mathbf{0} & \mathbf{0} \\ \mathbf{0} & \mathbf{C}_2 & \mathbf{0} & \cdots & \mathbf{0} & \mathbf{0} \\ \mathbf{0} & \mathbf{0} & \mathbf{C}_3 & \cdots & \mathbf{0} & \mathbf{0} \\ \vdots & \vdots & \vdots & \ddots & \vdots & \vdots \\ \mathbf{0} & \mathbf{0} & \mathbf{0} & \cdots & \mathbf{C}_{h-1} & \mathbf{0} \\ \mathbf{0} & \mathbf{0} & \mathbf{0} & \cdots & \mathbf{0} & \mathbf{C}_h \end{bmatrix} \tag{6.48}$$

where $\mathbf{C}_i$ is an $m_i \times m_i$ circulant matrix whose period is $\gamma_i = m_i$. The period of the composite Markov chain is given by the equation

$$\gamma = \operatorname{lcm}(\gamma_1, \gamma_2, \cdots, \gamma_h) \tag{6.49}$$

Figure 6.6 shows the periodic classes of a composite strongly periodic Markov chain. The states are divided into noncommunicating sets of periodic classes.

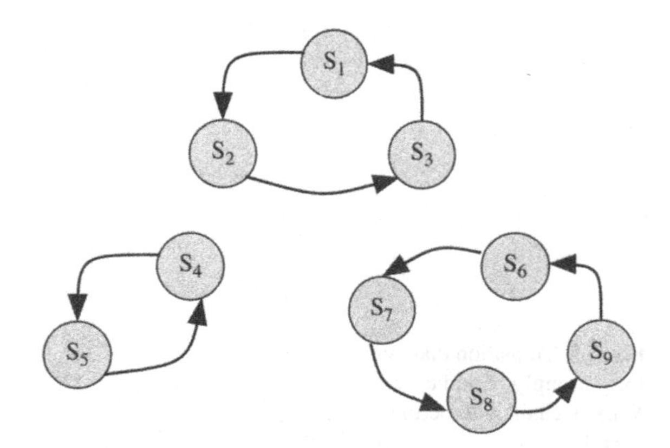

Fig. 6.6 Transition diagram for a composite strongly periodic Markov chain. The states are divided into noncommunicating sets of periodic classes

Example 6 Find the period of the following periodic transition matrix

$$\mathbf{P} = \begin{bmatrix} 0 & 0 & 0 & 1 & 0 & 0 & 0 \\ 1 & 0 & 0 & 0 & 0 & 0 & 0 \\ 0 & 1 & 0 & 0 & 0 & 0 & 0 \\ 0 & 0 & 1 & 0 & 0 & 0 & 0 \\ 0 & 0 & 0 & 0 & 0 & 0 & 1 \\ 0 & 0 & 0 & 0 & 1 & 0 & 0 \\ 0 & 0 & 0 & 0 & 0 & 1 & 0 \end{bmatrix}$$

We identify $\mathbf{C}_1$ as a circulant 4×4 matrix with period $\gamma_1 = 4$, and we identify $\mathbf{C}_2$ as a circulant 3×3 matrix with period $\gamma_2 = 3$. Indeed, the eigenvalues of $\mathbf{P}$ are

$$\lambda_1 = 1\angle 180^\circ = \exp\left(j2\pi \times \frac{2}{4}\right)$$

$$\lambda_2 = 1\angle 90^\circ = \exp\left(j2\pi \times \frac{1}{4}\right)$$

$$\lambda_3 = 1\angle -90^\circ = \exp\left(j2\pi \times \frac{3}{4}\right)$$

$$\lambda_4 = 1\angle 0^\circ = \exp\left(j2\pi \times \frac{4}{4}\right)$$

$$\lambda_5 = 1\angle 120^\circ = \exp\left(j2\pi \times \frac{1}{3}\right)$$

$$\lambda_6 = 1\angle -120^\circ = \exp\left(j2\pi \times \frac{2}{3}\right)$$

$$\lambda_7 = 1\angle 0^\circ = \exp\left(j2\pi \times \frac{3}{3}\right)$$

We expect $\mathbf{P}$ to have a period equal to the least common multiple of 3 and 4, which is 12. Indeed, this is verified by MATLAB where the smallest power k for which $\mathbf{P}^k = \mathbf{I}$ is when $k = 12$. ■

Example 7 Find the period of the following transition matrix for a strongly periodic Markov chain and express $\mathbf{P}$ in the form given by (6.48).

$$\mathbf{P} = \begin{bmatrix} 0 & 0 & 0 & 1 & 0 & 0 & 0 & 0 & 0 \\ 0 & 0 & 0 & 0 & 0 & 0 & 0 & 1 & 0 \\ 1 & 0 & 0 & 0 & 0 & 0 & 0 & 0 & 0 \\ 0 & 0 & 1 & 0 & 0 & 0 & 0 & 0 & 0 \\ 0 & 0 & 0 & 0 & 0 & 0 & 1 & 0 & 0 \\ 0 & 0 & 0 & 0 & 0 & 0 & 0 & 0 & 1 \\ 0 & 0 & 0 & 0 & 0 & 1 & 0 & 0 & 0 \\ 0 & 1 & 0 & 0 & 0 & 0 & 0 & 0 & 0 \\ 0 & 0 & 0 & 0 & 1 & 0 & 0 & 0 & 0 \end{bmatrix}$$

By inspection, we know that the given matrix is periodic since the elements are either 1 or 0 and each row or column has a single 1 at a unique location.

Indeed, all the eigenvalues of $\mathbf{P}$ lie on the unit circle

$$\begin{aligned}
\lambda_1 &= 1\angle 120^\circ = \exp\left(j2\pi \times \frac{1}{3}\right) \\
\lambda_2 &= 1\angle -120^\circ = \exp\left(j2\pi \times \frac{2}{3}\right) \\
\lambda_3 &= 1\angle 0^\circ = \exp\left(j2\pi \times \frac{3}{3}\right) \\
\lambda_4 &= 1\angle 0^\circ = \exp\left(j2\pi \times \frac{2}{2}\right) \\
\lambda_5 &= 1\angle 180^\circ = \exp\left(j2\pi \times \frac{1}{2}\right) \\
\lambda_6 &= 1\angle 180^\circ = \exp\left(j2\pi \times \frac{2}{4}\right) \\
\lambda_7 &= 1\angle 90^\circ = \exp\left(j2\pi \times \frac{1}{4}\right) \\
\lambda_8 &= 1\angle -90^\circ = \exp\left(j2\pi \times \frac{3}{4}\right) \\
\lambda_9 &= 1\angle 0^\circ = \exp\left(j2\pi \times \frac{4}{4}\right)
\end{aligned}$$

Using MATLAB, we find that $\mathbf{P}^3 \neq \mathbf{I}$, $\mathbf{P}^2 \neq \mathbf{I}$, and $\mathbf{P}^4 \neq \mathbf{I}$. However, the LCM of 2, 3, and 4 is 12 and $\mathbf{P}^{12} = \mathbf{I}$.

We notice that there are three eigenvalues equal to 1; mainly

$$\lambda_3 = \lambda_4 = \lambda_9 = 1$$

The eigenvectors corresponding to these eigenvalues give three sets of periodic classes:

$$C_1 = \{1,\ 3,\ 4\}$$
$$C_2 = \{2,\ 8\}$$
$$C_3 = \{5,\ 6,\ 7,\ 9\}$$

Each set is identified by the nonzero components of the corresponding eigenvector.

To group each set of periodic states together, we exchange states 2 and 4 so that C_1 will contain the new states 1, 3, and 4. This is done using the elementary exchange matrix

$$\mathbf{E}(2,4) = \begin{bmatrix} 1 & 0 & 0 & 0 & 0 & 0 & 0 & 0 & 0 \\ 0 & 0 & 0 & 1 & 0 & 0 & 0 & 0 & 0 \\ 0 & 0 & 1 & 0 & 0 & 0 & 0 & 0 & 0 \\ 0 & 1 & 0 & 0 & 0 & 0 & 0 & 0 & 0 \\ 0 & 0 & 0 & 0 & 1 & 0 & 0 & 0 & 0 \\ 0 & 0 & 0 & 0 & 0 & 1 & 0 & 0 & 0 \\ 0 & 0 & 0 & 0 & 0 & 0 & 1 & 0 & 0 \\ 0 & 0 & 0 & 0 & 0 & 0 & 0 & 1 & 0 \\ 0 & 0 & 0 & 0 & 0 & 0 & 0 & 0 & 1 \end{bmatrix}$$

The new matrix will be obtained with the operations

$$\mathbf{P}' = \mathbf{E}(2,4) \times \mathbf{P} \times \mathbf{E}(2,4)$$

Next, we group the states of C_2 together by exchanging states 5 and 8 so that C_2 will contain the new states 2 and 8, and C_3 will contain the states 5, 6, 7, and 9. The rearranged matrix will be

$$\mathbf{P} = \begin{bmatrix} 0 & 1 & 0 & 0 & 0 & 0 & 0 & 0 & 0 \\ 0 & 0 & 1 & 0 & 0 & 0 & 0 & 0 & 0 \\ 1 & 0 & 0 & 0 & 0 & 0 & 0 & 0 & 0 \\ 0 & 0 & 0 & 0 & 1 & 0 & 0 & 0 & 0 \\ 0 & 0 & 0 & 1 & 0 & 0 & 0 & 0 & 0 \\ 0 & 0 & 0 & 0 & 0 & 0 & 0 & 0 & 1 \\ 0 & 0 & 0 & 0 & 0 & 1 & 0 & 0 & 0 \\ 0 & 0 & 0 & 0 & 0 & 0 & 1 & 0 & 0 \\ 0 & 0 & 0 & 0 & 0 & 0 & 0 & 1 & 0 \end{bmatrix}$$

From the structure of the matrix, we see that the periods of the three diagonal circulant matrices is $\gamma_1 = 4$, $\gamma_2 = 2$, and $\gamma_3 = 3$. The period of the matrix is $\gamma = \text{lcm}(4, 2, 3) = 12$. As a verification, we find that the first time $\mathbf{P}^n$ becomes the identity matrix when $n = 12$. ■

6.12 Weakly Periodic Markov Chains

Periodic behavior can sometimes be observed in Markov chains even when some of the eigenvalues of **P** lie on the unit circle while other eigenvalues lie inside the unit circle. In spite of that, periodic behavior is observed because the structure of the matrix is closely related to the canonic form for a periodic Markov chain in (6.47) on page 196. To generalize the structure of a circulant matrix, we replace each "1" with a block matrix and obtain the *canonic* form for a weakly periodic Markov chain:

$$\mathbf{P} = \begin{bmatrix} \mathbf{0} & \mathbf{0} & \mathbf{0} & \cdots & \mathbf{0} & \mathbf{0} & \mathbf{W}_h \\ \mathbf{W}_1 & \mathbf{0} & \mathbf{0} & \cdots & \mathbf{0} & \mathbf{0} & \mathbf{0} \\ \mathbf{0} & \mathbf{W}_2 & \mathbf{0} & \cdots & \mathbf{0} & \mathbf{0} & \mathbf{0} \\ \vdots & \vdots & \vdots & \ddots & \vdots & \vdots & \vdots \\ \mathbf{0} & \mathbf{0} & \mathbf{0} & \cdots & \mathbf{0} & \mathbf{0} & \mathbf{0} \\ \mathbf{0} & \mathbf{0} & \mathbf{0} & \cdots & \mathbf{W}_{h-2} & \mathbf{0} & \mathbf{0} \\ \mathbf{0} & \mathbf{0} & \mathbf{0} & \cdots & \mathbf{0} & \mathbf{W}_{h-1} & \mathbf{0} \end{bmatrix} \tag{6.50}$$

where the block-diagonal matrices are square zero matrices and the nonzero matrices $\mathbf{W}_i$ could be rectangular but the sum of each of their columns is unity since **P** is column stochastic. Such a matrix will exhibit periodic behavior with a period $\gamma = h$ where h is the number of **W** blocks.

As an example, consider the following transition matrix

$$\mathbf{P} = \begin{bmatrix} 0 & 0 & 0.1 & 0.6 & 0.5 \\ 0 & 0 & 0.9 & 0.4 & 0.5 \\ 0.5 & 0.2 & 0 & 0 & 0 \\ 0.1 & 0.4 & 0 & 0 & 0 \\ 0.4 & 0.4 & 0 & 0 & 0 \end{bmatrix} \equiv \begin{bmatrix} \mathbf{0} & \mathbf{W}_2 \\ \mathbf{W}_1 & \mathbf{0} \end{bmatrix} \tag{6.51}$$

We know this matrix does not correspond to a strongly periodic Markov chain because it is not a 0–1 matrix whose elements are 0 or 1. However, let us look at the eigenvalues of this matrix:

$$\lambda_1 = -1$$
$$\lambda_2 = 1$$

$$
\begin{aligned}
\lambda_3 &= 0.3873j \\
\lambda_4 &= -0.3873j \\
\lambda_5 &= 0
\end{aligned}
$$

Some of the eigenvalues are inside the unit circle and represent decaying modes, but two eigenvalues lie on the unit circle. It is this extra eigenvalue on the unit circle that is responsible for the periodic behavior.

Let us now see the long-term behavior of the matrix. When $n > 25$, the contribution of the decaying modes will be $< 10^{-10}$. So let us see how $\mathbf{P}^n$ behaves when $n > 25$.

$$
\mathbf{P}^{25} = \begin{bmatrix} 0 & 0 & 0.4 & 0.4 & 0.4 \\ 0 & 0 & 0.6 & 0.6 & 0.6 \\ 0.32 & 0.32 & 0 & 0 & 0 \\ 0.28 & 0.28 & 0 & 0 & 0 \\ 0.40 & 0.40 & 0 & 0 & 0 \end{bmatrix} \tag{6.52}
$$

$$
\mathbf{P}^{26} = \begin{bmatrix} 0.28 & 0.28 & 0 & 0 & 0 \\ 0.40 & 0.40 & 0 & 0 & 0 \\ 0 & 0 & 0.4 & 0.4 & 0.4 \\ 0 & 0 & 0.6 & 0.6 & 0.6 \\ 0.32 & 0.32 & 0 & 0 & 0 \end{bmatrix} \tag{6.53}
$$

$$
\mathbf{P}^{27} = \begin{bmatrix} 0 & 0 & 0.4 & 0.4 & 0.4 \\ 0 & 0 & 0.6 & 0.6 & 0.6 \\ 0.32 & 0.32 & 0 & 0 & 0 \\ 0.28 & 0.28 & 0 & 0 & 0 \\ 0.40 & 0.40 & 0 & 0 & 0 \end{bmatrix} \tag{6.54}
$$

We see that $\mathbf{P}^n$ repeats its structure every two iterations.

We can make several observations about this transition matrix:

1. The transition matrix displays periodic behavior for large value of n.
2. $\mathbf{P}^n$ has a block structure that does not disappear. The blocks just move vertically at different places after each iteration.
3. The columns of each block in $\mathbf{P}^n$ are identical and the distribution vector will be independent of its initial value.
4. The period of $\mathbf{P}^n$ is 2.

Let us see now the distribution vector values for $n = 25$, 26, and 27. The initial value of the distribution vector is not important, and we arbitrarily pick

$$
\mathbf{s}(0) = \begin{bmatrix} 0.2 & 0.2 & 0.2 & 0.2 & 0.2 \end{bmatrix}^t \tag{6.55}
$$

We get

$$
\begin{aligned}
\mathbf{s}(25) = \mathbf{P}^{25}\mathbf{s}(0) &= \begin{bmatrix} 0.240 & 0.360 & 0.128 & 0.112 & 0.160 \end{bmatrix}^t && (6.56)\\
\mathbf{s}(26) &= \begin{bmatrix} 0.160 & 0.240 & 0.192 & 0.168 & 0.240 \end{bmatrix}^t && (6.57)\\
\mathbf{s}(27) &= \begin{bmatrix} 0.240 & 0.360 & 0.128 & 0.112 & 0.160 \end{bmatrix}^t && (6.58)
\end{aligned}
$$

We notice that the distribution vector repeats its value for every two iterations. Specifically, we see that $\mathbf{s}(25) = \mathbf{s}(27)$, and so on.

Example 8 The following transition matrix can be expressed in the form of (6.50). Find this form and estimate the period of the Markov chain.

$$
\mathbf{P} = \begin{bmatrix}
0 & 0.5 & 0.1 & 0 & 0 & 0 \\
0 & 0.3 & 0.5 & 0 & 0 & 0 \\
0 & 0.2 & 0.4 & 0 & 0 & 0 \\
0.9 & 0 & 0 & 0 & 0 & 0 \\
0.1 & 0 & 0 & 0 & 0 & 0 \\
0 & 0 & 0 & 1 & 1 & 1
\end{bmatrix}
$$

Our strategy for rearranging the states is to make the diagonal zero matrices appear in ascending order. The following ordering of states gives the desired result:

$$
1 \leftrightarrow 6 \quad 2 \leftrightarrow 5 \quad 3 \leftrightarrow 4
$$

The rearranged matrix will be

$$
\mathbf{P} = \begin{bmatrix}
0 & 0 & 0 & 1 & 1 & 1 \\
0.1 & 0 & 0 & 0 & 0 & 0 \\
0.9 & 0 & 0 & 0 & 0 & 0 \\
0 & 0.2 & 0.4 & 0 & 0 & 0 \\
0 & 0.3 & 0.5 & 0 & 0 & 0 \\
0 & 0.5 & 0.1 & 0 & 0 & 0
\end{bmatrix}
$$

The structure of the matrix is now seen to be in the form

$$
\mathbf{P} = \begin{bmatrix}
\mathbf{0} & \mathbf{0} & \mathbf{W}_3 \\
\mathbf{W}_1 & \mathbf{0} & \mathbf{0} \\
\mathbf{0} & \mathbf{W}_2 & \mathbf{0}
\end{bmatrix}
$$

where

$$\mathbf{W}_1 = \begin{bmatrix} 0.1 \\ 0.9 \end{bmatrix}$$
$$\mathbf{W}_2 = \begin{bmatrix} 0.2 & 0.4 \\ 0.3 & 0.5 \\ 0.5 & 0.1 \end{bmatrix}$$
$$\mathbf{W}_3 = \begin{bmatrix} 1 & 1 & 1 \end{bmatrix}$$

The eigenvalues for $\mathbf{P}$ are

$$\lambda_1 = \exp\left(j2\pi \times \frac{1}{3}\right)$$
$$\lambda_2 = \exp\left(j2\pi \times \frac{2}{3}\right)$$
$$\lambda_3 = 1 = \exp\left(j2\pi \times \frac{3}{3}\right)$$
$$\lambda_4 = 0$$
$$\lambda_5 = 0$$
$$\lambda_6 = 0$$

The period of $\mathbf{P}$ is 3. As a verification, we chose

$$\mathbf{s}(0) = \begin{bmatrix} 0.2 & 0.2 & 0.2 & 0.2 & 0.2 & 0 \end{bmatrix}^t$$

and found the following distribution vectors

$$\mathbf{s}(3) = \begin{bmatrix} 0.2000 & 0.0400 & 0.3600 & 0.1520 & 0.1920 & 0.0560 \end{bmatrix}^t$$
$$\mathbf{s}(4) = \begin{bmatrix} 0.4000 & 0.0200 & 0.1800 & 0.1520 & 0.1920 & 0.0560 \end{bmatrix}^t$$
$$\mathbf{s}(5) = \begin{bmatrix} 0.4000 & 0.0400 & 0.3600 & 0.0760 & 0.0960 & 0.0280 \end{bmatrix}^t$$
$$\mathbf{s}(6) = \begin{bmatrix} 0.2000 & 0.0400 & 0.3600 & 0.1520 & 0.1920 & 0.0560 \end{bmatrix}^t$$
$$\mathbf{s}(7) = \begin{bmatrix} 0.4000 & 0.0200 & 0.1800 & 0.1520 & 0.1920 & 0.0560 \end{bmatrix}^t$$

We see that the distribution vector repeats itself over a period of three iterations. Specifically, we see that $\mathbf{s}(3) = \mathbf{s}(6)$, $\mathbf{s}(4) = \mathbf{s}(7)$, and so on.

Example 9 The following transition matrix has the form of (6.50). Estimate the period of the Markov chain and study the distribution vector after the transients have decayed.

$$\mathbf{P} = \begin{bmatrix} 0 & 0 & 0 & 0 & 0.3 & 1 \\ 0 & 0 & 0 & 0 & 0.7 & 0 \\ 0.9 & 0.6 & 0 & 0 & 0 & 0 \\ 0.1 & 0.4 & 0 & 0 & 0 & 0 \\ 0 & 0 & 0.5 & 0.9 & 0 & 0 \\ 0 & 0 & 0.5 & 0.1 & 0 & 0 \end{bmatrix}$$

The eigenvalues for this matrix are

$$\lambda_1 = \exp\left(j2\pi \times \frac{1}{3}\right)$$
$$\lambda_2 = \exp\left(j2\pi \times \frac{2}{3}\right)$$
$$\lambda_3 = 1 = \exp\left(j2\pi \times \frac{3}{3}\right)$$
$$\lambda_4 = 0.438 \exp\left(j2\pi \times \frac{1}{3}\right)$$
$$\lambda_5 = 0.438 \exp\left(j2\pi \times \frac{2}{3}\right)$$
$$\lambda_6 = 0.438 \exp\left(j2\pi \times \frac{3}{3}\right)$$

The period of this matrix is $\gamma = 3$.

The transients would die away after about 30 iterations since the value of the eigenvalues within the unit circle would be

$$\lambda^{30} = 0.438^{30} = 1.879 \times 10^{-11}$$

The value of $\mathbf{P}^n$ at high values for n would start to approach an equilibrium pattern

$$\mathbf{P}^n = \begin{bmatrix} 0.5873 & 0.5873 & 0 & 0 & 0 & 1 \\ 0.4127 & 0.4127 & 0 & 0 & 0 & 0 \\ 0 & 0 & 0.7762 & 0.7762 & 0 & 0 \\ 0 & 0 & 0.2238 & 0.2238 & 0 & 0 \\ 0 & 0 & 0 & 0 & 0.5895 & 0.5895 \\ 0 & 0 & 0 & 0 & 0.4105 & 0.4105 \end{bmatrix}$$

where $n > 30$. As a verification, we chose

$$\mathbf{s}(0) = \begin{bmatrix} 0.2 & 0.2 & 0.2 & 0.2 & 0.2 & 0 \end{bmatrix}^t$$

and found the following distribution vectors

$$\mathbf{s}(30) = \begin{bmatrix} 0.2349 & 0.1651 & 0.3105 & 0.0895 & 0.1179 & 0.0821 \end{bmatrix}^t$$
$$\mathbf{s}(31) = \begin{bmatrix} 0.1175 & 0.0825 & 0.3105 & 0.0895 & 0.2358 & 0.1642 \end{bmatrix}^t$$
$$\mathbf{s}(32) = \begin{bmatrix} 0.2349 & 0.1651 & 0.1552 & 0.0448 & 0.2358 & 0.1642 \end{bmatrix}^t$$
$$\mathbf{s}(33) = \begin{bmatrix} 0.2349 & 0.1651 & 0.3105 & 0.0895 & 0.1179 & 0.0821 \end{bmatrix}^t$$
$$\mathbf{s}(34) = \begin{bmatrix} 0.1175 & 0.0825 & 0.3105 & 0.0895 & 0.2358 & 0.1642 \end{bmatrix}^t$$

We see that the distribution vector repeats itself with a period of three iterations. Specifically, we see that $\mathbf{s}(30) = \mathbf{s}(33)$, $\mathbf{s}(31) = \mathbf{s}(34)$, and so on. ■

Example 10 Assume a wireless packet transmission system that employs an adaptive forward error correction scheme. There are two levels of data forward error correction that could be employed depending on the number of errors detected in the received packet as shown in Fig. 6.7. The upper row of states corresponds to lower error levels and the lower row of states corresponds to higher error levels.

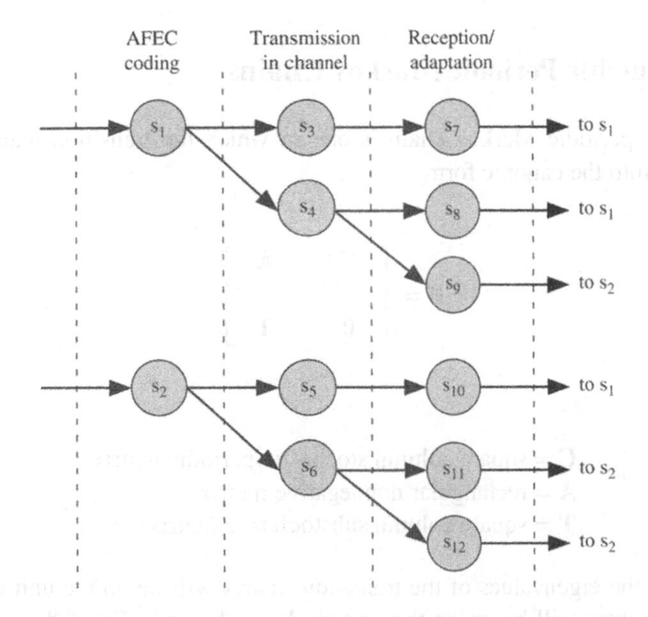

Fig. 6.7 An adaptive forward error correction scheme that uses two levels of encoding depending on the number of errors introduced in the channel

For example, if the coder is in state s_1 and errors occur during transmission, we move to state s_4. If the receiver is able to correct the errors, we move to state s_8. We then conclude that the error correction coding is adequate, and we move to state s_1 for the next transmission. If the errors cannot be corrected, we conclude that the error coding is not adequate and we move from state s_9 to s_2 at the next transmission.

Write down the transition matrix and show that it corresponds to a weakly periodic Markov chain.

The sets of periodic states are identified at the bottom of the figure. We can see that we have three sets of states such that the states in each set make transitions only to the next set. This seems to imply a weakly periodic Markov chain. As a verification, we construct the transition matrix and see its structure.

$$\mathbf{P} = \begin{bmatrix} 0 & 0 & 0 & 0 & p_{15} & p_{16} \\ 0 & 0 & 0 & 0 & p_{25} & p_{26} \\ p_{31} & p_{32} & 0 & 0 & 0 & 0 \\ p_{41} & p_{42} & 0 & 0 & 0 & 0 \\ 0 & 0 & p_{53} & p_{54} & 0 & 0 \\ 0 & 0 & p_{63} & p_{64} & 0 & 0 \end{bmatrix}$$

We see that the transition matrix has the same structure as a weakly periodic Markov chain with period $\gamma = 3$. The exact values of the transition probabilities will depend on the details of the system being investigated. ■

6.13 Reducible Periodic Markov Chains

A reducible periodic Markov chain is one in which the transition matrix can be partitioned into the canonic form

$$\mathbf{P} = \begin{bmatrix} \mathbf{C} & \mathbf{A} \\ \mathbf{0} & \mathbf{T} \end{bmatrix} \tag{6.59}$$

where

$\mathbf{C}$ = square column stochastic periodic matrix
$\mathbf{A}$ = rectangular nonnegative matrix
$\mathbf{T}$ = square column substochastic matrix

Some of the eigenvalues of the transition matrix will lie on the unit circle. The other eigenvalues will be inside the unit circle as shown in Fig. 6.8. Note that the periodic matrix $\mathbf{C}$ could be strongly periodic or could be weakly periodic.

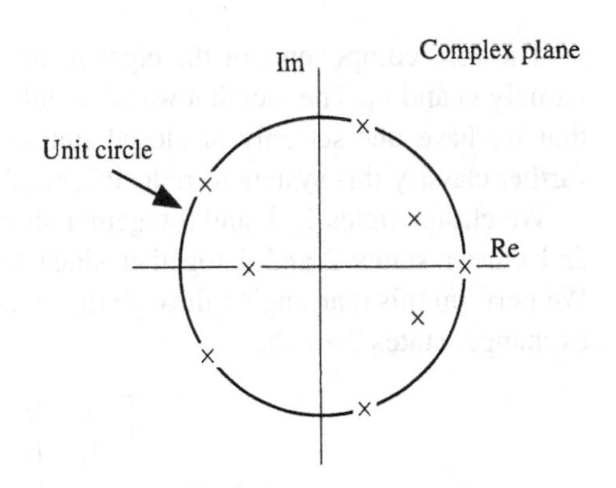

Fig. 6.8 The eigenvalues of a reducible periodic Markov chain. Some of the eigenvalues lie on the unit circle in the complex plane and some lie inside the unit circle

Example 11 Check to see if the given matrix below corresponds to a reducible periodic Markov chain.

$$\mathbf{P} = \begin{array}{c} \\ 1 \\ 2 \\ 3 \\ 4 \\ 5 \end{array} \begin{array}{c} \begin{array}{ccccc} 1 & 2 & 3 & 4 & 5 \end{array} \\ \begin{bmatrix} 0 & 0.1 & 0 & 0.1 & 1 \\ 0 & 0.3 & 0 & 0.2 & 0 \\ 1 & 0.2 & 0 & 0.2 & 0 \\ 0 & 0.3 & 0 & 0.4 & 0 \\ 0 & 0.1 & 1 & 0.1 & 0 \end{bmatrix} \end{array}$$

where the state indices are indicated around **P** for illustration. Rearrange the rows and columns to express the matrix in the form of (6.59) and identify the matrices **C**, **A**, and **T**. Verify the assertions that **C** is column stochastic, **A** is nonnegative, and **T** is column substochastic.

The best way to study a Markov chain is to explore its eigenvalues.

$$\lambda_1 = \exp\left(j2\pi \times \frac{1}{3}\right)$$

$$\lambda_2 = \exp\left(j2\pi \times \frac{2}{3}\right)$$

$$\lambda_3 = \exp\left(j2\pi \times \frac{3}{3}\right)$$

$$\lambda_4 = 0.6 \exp(j2\pi)$$

$$\lambda_5 = 0.1 \exp(j2\pi)$$

Thus we see that we have two decaying modes, but three other eigenvalues lie on the unit circle. We classify this system as a weakly periodic Markov chain.

The vector corresponding to the unity eigenvalue is given by

$$\mathbf{x} = \begin{bmatrix} 0.5774 & 0 & 0.5774 & 0 & 0.5774 \end{bmatrix}^t$$

The zero components of the eigenvector indicate that we have transient states, namely s_2 and s_4. The fact that we have only one eigenvalue, that is unity, indicates that we have one set only of closed states: $C = 1$, 3, and 5. Based on that, we further classify this system as reducible weakly periodic Markov chain.

We cluster states 1, 3, and 5 together since they correspond to the closed states, and cluster states 2 and 4 together since they correspond to the transient states. We perform this rearranging through the elementary exchange matrix $\mathbf{E}(2, 5)$ which exchanges states 2 and 5:

$$\mathbf{E}(2,5) = \begin{bmatrix} 1 & 0 & 0 & 0 & 0 \\ 0 & 0 & 0 & 0 & 1 \\ 0 & 0 & 1 & 0 & 0 \\ 0 & 0 & 0 & 1 & 0 \\ 0 & 1 & 0 & 0 & 0 \end{bmatrix}$$

The exchange of states is achieved by pre- and post-multiplying the transition matrix:

$$\mathbf{P}' = \mathbf{E}(2,5)\,\mathbf{P}\,\mathbf{E}(2,5)$$

This results in

$$\mathbf{P}' = \begin{array}{c} \\ 1 \\ 5 \\ 3 \\ 4 \\ 2 \end{array} \begin{array}{c} \begin{array}{ccccc} 1 & 5 & 3 & 4 & 2 \end{array} \\ \left[\begin{array}{ccccc} 0 & 1 & 0 & 0.1 & 0.1 \\ 0 & 0 & 1 & 0.1 & 0.1 \\ 1 & 0 & 0 & 0.2 & 0.2 \\ 0 & 0 & 0 & 0.4 & 0.1 \\ 0 & 0 & 0 & 0.2 & 0.3 \end{array}\right] \end{array}$$

We see that the transition matrix represents a reducible periodic Markov chain, and matrices $\mathbf{C}$, $\mathbf{A}$, and $\mathbf{T}$ are

$$\mathbf{C} = \begin{bmatrix} 0 & 1 & 0 \\ 0 & 0 & 1 \\ 1 & 0 & 0 \end{bmatrix}$$

$$\mathbf{A} = \begin{bmatrix} 0.1 & 0.1 \\ 0.1 & 0.1 \\ 0.2 & 0.2 \end{bmatrix}$$

$$\mathbf{T} = \begin{bmatrix} 0.5 & 0.1 \\ 0.2 & 0.3 \end{bmatrix}$$

The sum of each column of $\mathbf{C}$ is exactly 1, which indicates that it is column stochastic and strongly periodic. The sum of columns of $\mathbf{T}$ is less than 1, which indicates that it is column substochastic.

The set of closed periodic states is $C = \{1, 3, 5\}$, and the set of transient states is $T = \{2, 4\}$

Starting in state 2 or 4, we will ultimately go to states 1, 3, or 5. Once we are there, we cannot ever go back to state 2 or 4 because we entered the closed periodic states. ■

6.14 Transient Analysis

After n time steps, the transition matrix of a reducible Markov chain will still be reducible and will have the form

$$\mathbf{P}^n = \begin{bmatrix} \mathbf{C}^n & \mathbf{Y}^n \\ \mathbf{0} & \mathbf{T}^n \end{bmatrix} \tag{6.60}$$

where matrix $\mathbf{Y}^n$ is given by

$$\mathbf{Y}^n = \sum_{i=0}^{n-1} \mathbf{C}^{n-i-1} \mathbf{A} \mathbf{T}^i \tag{6.61}$$

We can always find $\mathbf{C}^n$ and $\mathbf{T}^n$ using the techniques discussed in Chapter 3 such as diagonalization, finding the Jordan canonic form, or even repeated multiplications.

The stochastic matrix $\mathbf{C}^n$ can be expressed in terms of its eigenvalues using (3.80) on page 94.

$$\mathbf{C}^n = \mathbf{C}_1 + \lambda_2^n \mathbf{C}_2 + \lambda_3^n \mathbf{C}_3 + \cdots \tag{6.62}$$

where it was assumed that $\mathbf{C}_1$ is the expansion matrix corresponding to the eigenvalue $\lambda_1 = 1$.

Similarly, the substochastic matrix $\mathbf{T}^n$ can be expressed in terms of its eigenvalues using (3.80) on page 94.

$$\mathbf{T}^n = \lambda_1^n \mathbf{T}_1 + \lambda_2^n \mathbf{T}_2 + \lambda_3^n \mathbf{T}_3 + \cdots \tag{6.63}$$

Equation (6.61) can then be expressed in the form

$$\mathbf{Y}^n = \sum_{j=1}^{m} \mathbf{C}_j \mathbf{A} \sum_{i=0}^{n-1} \lambda_j^{n-i-1} \mathbf{T}^i \tag{6.64}$$

After some algebraic manipulations, we arrive at the form

$$\mathbf{Y}^n = \sum_{j=1}^{m} \lambda_j^{n-1} \mathbf{C}_j \mathbf{A} \left[\mathbf{I} - \left(\frac{\mathbf{T}}{\lambda_j} \right)^n \right] \left(\mathbf{I} - \frac{\mathbf{T}}{\lambda_j} \right)^{-1} \tag{6.65}$$

This can be written in the form

$$\mathbf{Y}^n = \mathbf{C}_1\mathbf{A}(\mathbf{I}-\mathbf{T})^{-1}\left[\mathbf{I}-\mathbf{T}^n\right] + \lambda_2^{n-1}\mathbf{C}_2\mathbf{A}\left(\mathbf{I}-\frac{1}{\lambda_2}\mathbf{T}\right)^{-1}\left[\mathbf{I}-\frac{1}{\lambda_2^n}\mathbf{T}^n\right] + \lambda_3^{n-1}\mathbf{C}_3\mathbf{A}\left(\mathbf{I}-\frac{1}{\lambda_3}\mathbf{T}\right)^{-1}\left[\mathbf{I}-\frac{1}{\lambda_3^n}\mathbf{T}^n\right] + \cdots \tag{6.66}$$

If some of the eigenvalues of $\mathbf{C}$ are repeated, then the above formula has to be modified as explained in Section 3.14 on page 103.

Example 12 Consider the reducible weakly periodic Markov chain of the previous example. Assume that the system was initially in state 4. Explore how the distribution vector changes as time progresses.

We do not have to rearrange the transition matrix to do this example. We have

$$\mathbf{P} = \begin{bmatrix} 0 & 0.1 & 0 & 0.1 & 1 \\ 0 & 0.3 & 0 & 0.2 & 0 \\ 1 & 0.2 & 0 & 0.2 & 0 \\ 0 & 0.3 & 0 & 0.4 & 0 \\ 0 & 0.1 & 1 & 0.1 & 0 \end{bmatrix}$$

The eigenvalues for this matrix are

$$\lambda_1 = \exp\left(j2\pi \times \frac{1}{3}\right)$$
$$\lambda_2 = \exp\left(j2\pi \times \frac{2}{3}\right)$$
$$\lambda_3 = \exp\left(j2\pi \times \frac{3}{3}\right)$$
$$\lambda_4 = 0.1\exp\left(j2\pi \times \frac{3}{3}\right)$$

We see that the period of this system is $\gamma = 3$. The initial distribution vector is

$$\mathbf{s} = \begin{bmatrix} 0 & 0 & 1 & 0 \end{bmatrix}^t$$

The distribution vector at the start is given by

$$\begin{aligned}
\mathbf{s}(0) &= \begin{bmatrix} 0 & 0 & 0 & 1 & 0 \end{bmatrix}^t \\
\mathbf{s}(1) &= \begin{bmatrix} 0.1000 & 0.2000 & 0.2000 & 0.4000 & 0.1000 \end{bmatrix}^t \\
\mathbf{s}(2) &= \begin{bmatrix} 0.1600 & 0.1400 & 0.2200 & 0.2200 & 0.2600 \end{bmatrix}^t
\end{aligned}$$

$$
\begin{aligned}
\mathbf{s}(3) &= \begin{bmatrix} 0.2960 & 0.0860 & 0.2320 & 0.1300 & 0.2560 \end{bmatrix}^t \\
\mathbf{s}(4) &= \begin{bmatrix} 0.2776 & 0.0518 & 0.3392 & 0.0778 & 0.2536 \end{bmatrix}^t \\
\mathbf{s}(5) &= \begin{bmatrix} 0.2666 & 0.0311 & 0.3035 & 0.0467 & 0.3522 \end{bmatrix}^t
\end{aligned}
$$

Continuing the iterations, the distribution vector settles down to the following sequence.

$$
\begin{aligned}
\mathbf{s}(19) &= \begin{bmatrix} 0.3265 & 0.0000 & 0.3775 & 0.0000 & 0.2959 \end{bmatrix}^t \\
\mathbf{s}(20) &= \begin{bmatrix} 0.2959 & 0.0000 & 0.3265 & 0.0000 & 0.3775 \end{bmatrix}^t \\
\mathbf{s}(21) &= \begin{bmatrix} 0.3775 & 0.0000 & 0.2959 & 0.0000 & 0.3265 \end{bmatrix}^t \\
\mathbf{s}(22) &= \begin{bmatrix} 0.3265 & 0.0000 & 0.3775 & 0.0000 & 0.2959 \end{bmatrix}^t \\
\mathbf{s}(23) &= \begin{bmatrix} 0.2959 & 0.0000 & 0.3265 & 0.0000 & 0.3775 \end{bmatrix}^t \\
\mathbf{s}(24) &= \begin{bmatrix} 0.3775 & 0.0000 & 0.2959 & 0.0000 & 0.3265 \end{bmatrix}^t
\end{aligned}
$$

We note that after about 20 time steps, the probability of being in the transient state 2 or 4 is nil. The system will definitely be in the closed set composed of states 1, 3, or 5. The distribution vector will show periodic behavior with a period $\gamma = 3$. ■

6.15 Asymptotic Behavior

Assume we have a reducible periodic Markov chain with transition matrix **P** that is expressed in the form

$$
\mathbf{P} = \begin{bmatrix} \mathbf{C} & \mathbf{A} \\ \mathbf{0} & \mathbf{T} \end{bmatrix} \tag{6.67}
$$

According to (6.60), after n time steps, the transition matrix will have the form

$$
\mathbf{P}^n = \begin{bmatrix} \mathbf{C}^n & \mathbf{Y}^n \\ \mathbf{0} & \mathbf{T}^n \end{bmatrix} \tag{6.68}
$$

where matrix $\mathbf{Y}^n$ is given by (6.65) in the form

$$
\mathbf{Y}^n = \sum_{j=1}^{m} \lambda_j^{n-1} \mathbf{C}_j \, \mathbf{A} \left[\mathbf{I} - \left(\frac{\mathbf{T}}{\lambda_j} \right)^n \right] \left(\mathbf{I} - \frac{\mathbf{T}}{\lambda_j} \right)^{-1} \tag{6.69}
$$

when $n \rightarrow \infty$; $\mathbf{T}^n$ will become zero since it is column substochastic. Furthermore, the eigenvalues of **C** satisfy the following equations because of the periodicity of **C**.

$$|\lambda_i| = 1 \qquad 1 \le i \le K \tag{6.70}$$

$$|\lambda_i| < 1 \qquad K < i \le m \tag{6.71}$$

The eigenvalues that lie in the unit circle will have no contribution at large values of n, and matrix $\mathbf{P}^\infty$ becomes

$$\mathbf{P}^\infty = \begin{bmatrix} \mathbf{E} & \mathbf{F} \\ \mathbf{0} & \mathbf{0} \end{bmatrix} \tag{6.72}$$

The matrices **E** and **F** are given by

$$\mathbf{E} = \sum_{k=1}^{K} \lambda_k^i \mathbf{C}_k \tag{6.73}$$

$$\mathbf{F} = \left[\sum_{i=1}^{\gamma} \sum_{k=1}^{K} \lambda_k^{i-1} \mathbf{C}_k \, \mathbf{A} \, \mathbf{T}^{i-1} \right] (\mathbf{I} - \mathbf{T}^\gamma)^{-1} \tag{6.74}$$

where **I** is the unit matrix whose dimensions match that of **T**.

Example 13 Find the asymptotic transition matrix for the reducible weakly periodic Markov chain characterized by the transition matrix

$$\mathbf{P} = \begin{bmatrix} 0 & 0 & 0.9 & 0.1 & 0.1 & 0.3 & 0.1 \\ 0 & 0 & 0.1 & 0.9 & 0.2 & 0.1 & 0.3 \\ 0.2 & 0.8 & 0 & 0 & 0 & 0.1 & 0 \\ 0.8 & 0.2 & 0 & 0 & 0.1 & 0 & 0.1 \\ 0 & 0 & 0 & 0 & 0.2 & 0.2 & 0.1 \\ 0 & 0 & 0 & 0 & 0.1 & 0.1 & 0 \\ 0 & 0 & 0 & 0 & 0.3 & 0.2 & 0.4 \end{bmatrix}$$

The components of the transition matrix are

$$\mathbf{C} = \begin{bmatrix} 0 & 0 & 0.9 & 0.1 \\ 0 & 0 & 0.1 & 0.9 \\ 0.2 & 0.8 & 0 & 0 \\ 0.8 & 0.2 & 0 & 0 \end{bmatrix}$$

$$\mathbf{A} = \begin{bmatrix} 0.1 & 0.3 & 0.1 \\ 0.2 & 0.1 & 0.3 \\ 0 & 0.1 & 0 \\ 0.1 & 0 & 0.1 \end{bmatrix}$$

$$\mathbf{T} = \begin{bmatrix} 0.2 & 0.2 & 0.1 \\ 0.1 & 0.1 & 0 \\ 0.3 & 0.2 & 0.4 \end{bmatrix}$$

C is a weakly periodic Markov chain whose eigenvalues are

$$\lambda_1 = 1$$
$$\lambda_2 = -1$$
$$\lambda_3 = 0.6928j$$
$$\lambda_3 = -0.6928j$$

and $\mathbf{C}^i$ can be decomposed into the form

$$\mathbf{C}^i = \mathbf{C}_1 + (-1)^i \mathbf{C}_2$$

where

$$\mathbf{C}_1 = \begin{bmatrix} 0.25 & 0.25 & 0.25 & 0.25 \\ 0.25 & 0.25 & 0.25 & 0.25 \\ 0.25 & 0.25 & 0.25 & 0.25 \\ 0.25 & 0.25 & 0.25 & 0.25 \end{bmatrix}$$

$$\mathbf{C}_2 = \begin{bmatrix} 0.25 & 0.25 & -0.25 & -0.25 \\ 0.25 & 0.25 & -0.25 & -0.25 \\ -0.25 & -0.25 & 0.25 & 0.25 \\ -0.25 & -0.25 & 0.25 & 0.25 \end{bmatrix}$$

According to (6.72), the matrix **E** has the value

$$\mathbf{E} = \begin{bmatrix} 0.2742 & 0.3044 & 0.3018 \\ 0.2742 & 0.3044 & 0.3018 \\ 0.2258 & 0.1956 & 0.1982 \\ 0.2258 & 0.1956 & 0.1982 \end{bmatrix}$$

According to (6.72), the matrix **F** has the value

$$\mathbf{F} = \begin{bmatrix} 0.2742 & 0.3044 & 0.3018 \\ 0.2742 & 0.3044 & 0.3018 \\ 0.2258 & 0.1956 & 0.1982 \\ 0.2258 & 0.1956 & 0.1982 \end{bmatrix}$$

The asymptotic value of **P** is given by the two values

$$\mathbf{P}_1 = \begin{bmatrix} 0.5 & 0.5 & 0 & 0 & 0.2258 & 0.1956 & 0.1982 \\ 0.5 & 0.5 & 0 & 0 & 0.2258 & 0.1956 & 0.1982 \\ 0 & 0 & 0.5 & 0.5 & 0.2742 & 0.3044 & 0.3018 \\ 0 & 0 & 0.5 & 0.5 & 0.2742 & 0.3044 & 0.3018 \\ 0 & 0 & 0 & 0 & 0 & 0 & 0 \\ 0 & 0 & 0 & 0 & 0 & 0 & 0 \\ 0 & 0 & 0 & 0 & 0 & 0 & 0 \end{bmatrix}$$

$$\mathbf{P}_2 = \begin{bmatrix} 0 & 0 & 0.5 & 0.5 & 0.2742 & 0.3044 & 0.3018 \\ 0 & 0 & 0.5 & 0.5 & 0.2742 & 0.3044 & 0.3018 \\ 0.5 & 0.5 & 0 & 0 & 0.2258 & 0.1956 & 0.1982 \\ 0.5 & 0.5 & 0 & 0 & 0.2258 & 0.1956 & 0.1982 \\ 0 & 0 & 0 & 0 & 0 & 0 & 0 \\ 0 & 0 & 0 & 0 & 0 & 0 & 0 \\ 0 & 0 & 0 & 0 & 0 & 0 & 0 \end{bmatrix}$$

At steady state, the two matrices are related by

$$\mathbf{P}_1 = \mathbf{P} \quad \mathbf{P}_2$$
$$\mathbf{P}_1 = \mathbf{P} \quad \mathbf{P}_1$$

Therefore, at steady state, the system oscillates between two state transition matrices and periodic behavior is observed.

We can make several observations on the asymptotic value of the transition matrix.

(a) The columns are not identical as in nonperiodic Markov chains.
(b) There is no possibility of moving to a transient state irrespective of the value of the initial distribution vector.
(c) The asymptotic values of the transition matrix has two forms P_1 and P_2 such that the system makes periodic transitions between the states, and a steady-state value can never be reached. ■

6.16 Identification of Markov Chains

e are now able to discuss how we can determine if the given transition matrix corresponds to a periodic Markov chain or not and to determine if the chain is strongly or weakly periodic. Further, we are also able to tell if the Markov chain is reducible or irreducible and to identify the transient and closed states.

Inspection of the elements of the transition matrix does not help much except for the simplest case when we are dealing with a strongly periodic Markov chain. In that case, the transition matrix will be a 0–1 matrix. However, determining the period of the chain is not easy since we have to rearrange the matrix to correspond to the form (6.48) on page 198.

A faster and more direct way to classify a Markov chain is to simply study its eigenvalues and eigenvector corresponding to $\lambda = 1$. The following subsections summarize the different cases that can be encountered.

6.16.1 Nonperiodic Markov chain

This is the case when only one eigenvalue is 1 and all other eigenvalues lie inside the unit circle:

$$|\lambda_i| < 1 \tag{6.75}$$

For large values of the time index $n \to \infty$, all modes will decay except the one corresponding to $\lambda_1 = 1$, which gives us the steady-state distribution vector.

6.16.2 Strongly periodic Markov chain

This is the case when all the eigenvalues of the transition matrix lie on the unit circle:

$$\lambda_i = \exp\left(j2\pi \times \frac{i}{\gamma}\right) \tag{6.76}$$

where $1 \leq i \leq \gamma$.

For all values of the time index, the distribution vector will exhibit periodic behavior, and the period of the system is γ.

6.16.3 Weakly periodic Markov chain

This is the case when γ eigenvalues of the transition matrix lie on the unit circle, and the rest of the eigenvalues lie inside the unit circle. Thus we can write

$$|\lambda_i| = 1 \qquad \text{when } 1 \leq i \leq \gamma \tag{6.77}$$

$$|\lambda_i| < 1 \qquad \text{when } \gamma < i \leq m \tag{6.78}$$

The eigenvalues that lie on the unit circle will be given by

$$\lambda_i = \exp\left(j2\pi \times \frac{i}{\gamma}\right) \tag{6.79}$$

where $1 \leq i \leq \gamma$.

For large values of the time index $n \to \infty$, some of the modes will decay but γ of them will not, and the distribution vector will never settle down to a stable value. The period of the system is γ.

6.17 Problems

Strongly Periodic Markov Chains

6.1 The following matrix is a circulant matrix of order 3, we denote it by $\mathbf{C}_3$. If it corresponds to a transition matrix, then the resulting Markov chain is strongly periodic. Find the period of the Markov chain.

$$\mathbf{P} = \begin{bmatrix} 0 & 0 & 1 \\ 1 & 0 & 0 \\ 0 & 1 & 0 \end{bmatrix}$$

6.2 Prove Theorem 5 using the result of Theorem 1 and the fact that all eigenvalues of a column stochastic matrix are in the range $0 \leq |\lambda| \leq 1$.

6.3 Does the following transition matrix represent a strongly periodic Markov chain?

$$\mathbf{P} = \begin{bmatrix} 0 & 0 & 0 & 1 & 0 & 0 & 0 \\ 0 & 1 & 0 & 0 & 0 & 0 & 0 \\ 0 & 0 & 0 & 0 & 1 & 0 & 0 \\ 0 & 0 & 0 & 0 & 0 & 0 & 1 \\ 0 & 0 & 1 & 0 & 0 & 0 & 0 \\ 1 & 0 & 0 & 0 & 0 & 0 & 0 \\ 0 & 0 & 0 & 0 & 0 & 1 & 0 \end{bmatrix}$$

6.4 Verify that the following transition matrix corresponds to a periodic Markov chain and find the period. Find also the values of all the eigenvalues.

$$\mathbf{P} = \begin{bmatrix} 0 & 0 & 0 & 0 & 0 & 0 & 1 \\ 0 & 0 & 0 & 1 & 0 & 0 & 0 \\ 0 & 0 & 0 & 0 & 1 & 0 & 0 \\ 0 & 0 & 0 & 0 & 0 & 1 & 0 \\ 0 & 0 & 1 & 0 & 0 & 0 & 0 \\ 1 & 0 & 0 & 0 & 0 & 0 & 0 \\ 0 & 1 & 0 & 0 & 0 & 0 & 0 \end{bmatrix}$$

6.5 Prove that the $m \times m$ circulant matrix in (6.47) has a period $\gamma = m$. You can do that by observing the effect of premultiplying any $m \times k$ matrix by this matrix.

From that, you will be able to find a pattern for the repeated multiplication of the transition matrix by itself.

Weakly Periodic Markov Chains

6.6 Does the following transition matrix represent a periodic Markov chain?

$$\mathbf{P} = \begin{bmatrix} 1/3 & 0 & 0 & 0 & 0 & 1/3 & 1/3 \\ 1/3 & 1/3 & 0 & 0 & 0 & 0 & 1/3 \\ 1/3 & 1/3 & 1/3 & 0 & 0 & 0 & 0 \\ 0 & 1/3 & 1/3 & 1/3 & 0 & 0 & 0 \\ 0 & 0 & 1/3 & 1/3 & 1/3 & 0 & 0 \\ 0 & 0 & 0 & 1/3 & 1/3 & 1/3 & 0 \\ 0 & 0 & 0 & 0 & 1/3 & 1/3 & 1/3 \end{bmatrix}$$

6.7 The weather in a certain island in the middle of nowhere is either sunny or rainy. A recent shipwreck survivor found that if the day is sunny, then the next day will be sunny with 80% probability. If the day is rainy, then the next day will be rainy with 70% probability. Find the asymptotic behavior of the weather on this island.

6.8 A traffic source is modeled as a periodic Markov chain with three stages. Each stage has three states: silent, transmitting at rate λ_1, and transmitting at rate λ_2.

(a) Draw the periodic transition diagram.
(b) Write down the transition matrix.
(c) Assign transition probabilities to the system and find the asymptotic behavior of the source.
(d) Plot the state of the system versus time by choosing the state with the highest probability at each time instant. Can you see any periodic behavior in the traffic pattern?

6.9 A computer goes through the familiar fetch, decode, and execute stages for instruction execution. Assume that the fetch stage has three states depending on the location of the operands in cache, RAM, or in main memory. The decode stage has three states depending on the instruction type. The execute state also has three states depending on the length of the instruction.

(a) Draw the periodic transition diagram.
(b) Write down the transition matrix.
(c) Assign transition probabilities to the system and find the asymptotic behavior of the program being run on the machine.

6.10 Table look up for packet routing can be divided into three phases: database search, packet classification, and packet processing. Assume that the database search phase consists of four states depending on the location of the data among the different storage modules. The packet classification phase consists of three states depending on the type of packet received. The packet processing phase consists of four different states depending on the nature of the processing being performed.

(a) Draw the periodic transition diagram.
(b) Write down the transition matrix.
(c) Assign transition probabilities to the system and find the asymptotic behavior of the lookup table operation.

6.11 A bus company did a study on commuter habits during a typical week. Essentially, the company wanted to know the percentage of commuters who use the bus, their own car, or stay at home each day of the week. Based on this study, the company can plan ahead and assign more buses or even bigger buses during heavy usage days.

(a) Draw the periodic transition diagram.
(b) Write down the transition matrix.
(c) Assign transition probabilities to the system and find the asymptotic behavior of the commuter patterns.

6.12 Assume a certain species of wild parrot to have two colors: red and blue. It was found that when a red–red pair breeds, their offspring are red with a 90% probability due to some obscure reasons related to recessed genes and so on. When a blue–blue pair breeds, their offspring are blue with a 70% probability. When a red–blue pair breeds, their offspring are blue with a 50% probability.

(a) Draw the periodic transition diagram.
(b) Write down the transition matrix.
(c) Assign transition probabilities to the system and find the asymptotic behavior of the parrot colors in the wild.

6.13 A packet source is modeled using periodic Markov chain. The source is assumed to go through four repeating phases and each phase has two states: idle and active. Transitions from phase 1 to phase 2 are such that the source switches to the other state with a probability 0.05. Transitions from phase 2 to phase 3 are such that the source switches to the other state with a probability 0.1. Transitions from phase 3 to phase 4 are such that the source switches to the other state with a probability 0.2. Transitions from phase 4 to phase 1 are such that the source switches to the other state with a probability 0.90.

(a) Draw the periodic transition diagram.
(b) Write down the transition matrix.
(c) Assign transition probabilities to the system and find the asymptotic behavior of the source.

6.14 The predator and prey populations are related as was explained at the start of this chapter. Assume that the states of the predator–prey population are as follows.

State	Predator polulation	Prey population
s_0	Low	Low
s_1	Low	High
s_2	High	Low
s_3	High	High

The transitions between the states of the system take place once each year. The transition probabilities are left for the reader to determine or assume.

1. Construct state transition diagram and a Markov transition matrix.
2. Justify the entries you choose for the matrix.
3. Study the periodicity of the system.

References

1. W.J. Stewart, *Introduction to Numerical Solutions of Markov Chains*, Princeton University Press, Princeton, New Jersey, 1994.
2. R. Horn and C.R. Johnson, *Matrix Analysis*, Cambridge University Press, Cambridge, England, 1985.
3. Private discussions with Dr. W.-S. Lu of the Department of Electrical & Computer Engineering, University of Victoria, 2001.

Chapter 7
Queuing Analysis

7.1 Introduction

Queuing analysis is one of the most important tools for studying communication systems. The analysis allows us to answer endless questions about the system performance. This chapter explains that queuing analysis is a special case of Markov chains. Some examples of queues are as follows:

- The number of patients in a doctor's waiting room
- The number of customers in a store checkout line
- The number of packets stored in a router's buffer
- The number of print jobs present in a printer's queue
- The number of workstations requesting access to the LAN

Without being specific to a certain system, we can state that *queuing analysis* deals with *queues* where *customers* compete to be processed by shared *servers*. The *queue size* is the waiting room provided for the customers that have not been served yet plus the customers that are being served. We will talk about packets instead of customers in this chapter since most of networking analysis deals with transmitting and processing packets.

The objective of queuing analysis is to predict the system performance such as how many customers get processed per time step, the average delay a customer endures before being served, and the size of the queue or waiting room required. These performance measures have obvious applications in telecommunication systems and the design of hardware for such systems.

We list here some typical examples of queues and point out the customers and servers in each.

- People lining up at a bank where the bank teller is the server and the bank patrons are of course the customers.
- Workstations connected in a local-area network (LAN) where the communication medium (e.g., Ethernet cable) represents the shared resource while the communicating applications represent the customers and the server is the media access control (MAC) protocol that enables access to the medium.

F. Gebali, *Analysis of Computer and Communication Networks*,
DOI: 10.1007/978-0-387-74437-7_7,

- A parallel processing system in which a common shared memory is accessed by all the computers. The computers, or rather the memory requests, represent the customers and the arbitration protocol that resolves memory access conflicts represents the server.

Most of the literature and textbooks deal with continuous-time systems. In these systems, only single customer arrival or departure takes place at a given time instant. However, analyzing the systems for general arrival or departure statistics proved to be difficult such that most textbooks simply provide tables of performance formulas for the most common situations. Subsequent researchers simply adopt these formulas without any form of adaptation or innovation.

Most of computer and communication systems, however, are migrating to the digital domain. In this domain, time is measured in discrete units or steps of finite size. As a consequence, there could be multiple arrivals or departures at a given time step. We will find that discrete-time systems are simple to analyze compared to continuous-time systems. Adapting discrete-time systems to a large variety of situations is really simple using the techniques we provide in this book.

In this chapter, we study different types of discrete-time queues characterized by the following attributes:

1. The total number of customers in the system.
2. The number of customers that could possibly request service at a given time step.
3. The arrival process statistics for the customers.
4. The number of servers which dictate how many customers can leave the queue at a given time step.
5. The service discipline for deciding which customer or customers are to be served. Examples of service disciplines are first-come-first-serve (FIFO), random selection, polling, and priority [1].
6. The size of the queue to accommodate customers waiting for service.

7.1.1 Kendall's Notation

Kendall's notation is frequently used to succinctly describe a queuing system. This notation is represented as $A/B/c/n/p$, where [2]

A	Arrival statistics
B	Service or departure statistics
c	Number of servers
n	Queue size
p	Customer population size

The final two fields are optional and are assumed infinite if they are omitted [3]. The letters A and B denoting arrival and server statistics are given the following notations:

D	*Deterministic*, process has fixed arrival or service rates
M	*Markovian*, process is Poisson or binomial
G	*General* or constant time

A common practice is to attach a superscript to the letters A and B to denote multiple arrivals or "batch service". Using this notation, the discrete-time $M/M/1$ queue has binomial, or Poisson, arrivals and departures. At a given time step at most one customer arrives and at most one customer departs. The queue has infinite buffer size and the population size is also infinite.

The queue $M/M/1/B$ has binomial, or Poisson, arrivals and departures. At a given time step, at most one customer arrives and at most one customer departs. The queue has a finite buffer of size B and the population size is infinite. This queue is frequently encountered when the time step is so short that only one customer can arrive or depart in that time.

The queue $M/M/J/B$ has binomial, or Poisson, arrivals and departures. At a given time step, at most one customer arrives and at most one customer could depart through one of the J available servers. The queue has a finite buffer of size B and the population size is infinite. This type of queue might be encountered when a server might be busy serving a customer at a given time step and the remaining servers become available to serve other customers in the queue.

The queue $M^m/M/1/B$ has binomial, or Poisson, arrivals and departures. There is one server in the queue, and at a given time step, at most m customers arrive and at most one customer departs because there is one server. The queue has a finite buffer of size B and the population size is infinite.

The queue $M/M^m/1/B$ has binomial, or Poisson, arrivals and departures. There is one server in the queue, and at a given time step, at most one customer arrives and at most J customers depart because the server could handle J customers in one time step. The queue has a finite buffer of size B and the population size is infinite.

The queues we shall deal with here will be one of the following types:

1. Single arrival, single departure infinite-sized queues in which the transition matrix **P** is tridiagonal. Such a queue will be denoted by the symbols $M/M/1$.
2. Single arrival, single departure finite-sized queues in which **P** is tridiagonal. Such a queue will be denoted by the symbols $M/M/1/B$.
3. Multiple arrival, single departure finite queues in which **P** is lower Hessenberg. Such a queue will be denoted by the symbols $M^m/M/1/B$.
4. Single arrival, multiple departure finite-sized queues in which **P** is upper Hessenberg. Such a queue will be denoted by the symbols $M/M^m/1/B$.

7.2 Queue Throughput (Th)

Most often we are interested in estimating the rate of customers leaving the queue; which is expressed as customers per time step or customers per second. We call this rate the *average output traffic* $N_a(\text{out})$, or *throughput* (Th) of a queue. The throughput is given by

$$\text{Th} = \text{output data rate} = N_a(\text{out}) \quad (7.1)$$

The units of Th in the above expression are packets/time step. Notice that this definition implies that Th could never be negative.

When we talk about throughput of packets, we usually mean *goodput* which represents the packets that got through intact without collisions or other problems. Sometimes authors talk about throughput to include good and corrupted packets. The difference between throughput and goodput is seldom discussed and the reader is advised to make certain which quantity is being dealt with in any discussion. Throughout this book, we shall use the term throughput to imply good packets that got sent without corruption.

7.3 Efficiency (η) or Access Probability (p_a)

The *efficiency* (η) of the queue or its *access probability* (p_a) essentially gives the percentage of customers or packets transmitted in one time step through the system relative to the total number of arriving customers or packets in one time step also. We shall use the term efficiency when we talk about queues and switches and will use the term access probability when we talk about interconnection networks. Efficiency or access probability essentially measures the effectiveness of the queue at processing data present at the input.

We define the *access probability* (p_a) or *efficiency* η as the ratio of the average output traffic relative to the average input traffic:

$$p_a = \eta = \frac{N_a(\text{out})}{N_a(\text{in})} \quad (7.2)$$

This can be expressed in terms of the throughput

$$p_a = \eta = \frac{N_a(\text{out})}{N_a(\text{in})} = \frac{\text{Th}}{N_a(\text{in})} \leq 1 \quad (7.3)$$

Notice that the access probability or efficiency could never be negative and could never be more than one.

If customers or packets are created within the queue, then we have to modify our definitions of both the input and output traffic to ensure that the efficiency never exceeds unity. This situation could in fact happen. For example, consider a packet switch that receives packets at its inputs and then routes these packets to the output ports. We expect that the number of packets leaving the switch per unit time will be smaller than or equal to the number of packets coming to the switch per unit time. The output traffic could be smaller than the input traffic when packets are lost within the switch when its internal buffers become full. However, the switch itself might generate its own packets to communicate with other switches or routers in the network. In that case, the traffic at the input could in fact be smaller than the output traffic due to the internally generated traffic. Let us leave this point to the problems at the end of the chapter.

Throughput and access probability are very useful in describing the behavior of a system. The two concepts are not equivalent and in fact we will see that the curves showing variation of efficiency with the queue parameters are not scaled version of throughput variation with queue parameters. This point is clearly illustrated in Fig. 14.4 on page 515, or Fig. 14.20 on page 538.

If the queue produces three customers on the average per time step while five customers could arrive, then the throughput is Th $= 3$ while its efficiency is $\eta = 3/5 = 60\%$. On the other hand, if the queue produces four customers on the average per time step while a maximum of six customers could potentially leave, then the throughput is Th $= 4$ and its efficiency is $\eta = 4/6 = 66.67\%$.

7.4 Traffic Conservation

When the queue reaches steady state, the incoming customers or packets have two options when they arrive at a queue: either they get processed and move through the queue or they get lost. We can therefore write the traffic conservation as

$$N_a(\text{in}) = N_a(\text{out}) + N_a(\text{lost}) \tag{7.4}$$

where $N_a(\text{lost})$ is the average number of lost traffic or customers per unit time. The above equation is valid at steady state since the number of packets in the queue will be constant.

Dividing the above equation by $N_a(\text{in})$ to normalize, we get

$$\eta + L = 1 \tag{7.5}$$

where L is the customer, or traffic loss probability,

$$L = 1 - \eta \tag{7.6}$$

Systems that have high efficiency will have low loss probability and vice versa. This situation is very similar to mechanical or electrical energy conversion systems characterized by input power, output power, and lost power due to friction or resistive effects.

7.5 M/M/1 Queue

In an $M/M/1$ queue at any time step, at most one customer could arrive and at most one customer could leave. An example of an $M/M/1$ queue is a first-in-first-out buffer of a communication system. This gives rise to simplest of queues in queuing analysis. The queue size here is assumed infinite. This is the discrete time equivalent of the famous $M/M/1$ queue for the continuous-time case. At a certain time step, the probability of packet arrival is a, which is equivalent to a birth event or increase in the queue population. The probability that a packet did not arrive is $b = 1 - a$. The probability that a packet leaves the queue is c, which is equivalent to death or reduction in the queue population. The probability that a packet does not leave the queue is $d = 1 - c$. The probability c is representative of the server's ability to process the customers or packets in the queue in one time step.

The number of customers or packets stored in the queue is the state of our system. Thus, the queue is in state s_i when there are i customers or packets in the queue. The state transition diagram for the discrete-time M/M/1 queue is shown in Fig. 7.1. Changes in the queue size occur by at most one, i.e., only one packet could arrive or depart. Thus, we expect the transition matrix to be tridiagonal.

The future state of the queue depends only on its current state. Thus, we can model the queue as a discrete-time Markov chain. Since packet arrivals and departures are independent of the time index value, we have a homogeneous Markov chain. We assume that when a packet arrives, it could be serviced at the same time step and it could leave the queue with probability c. This results in the transition matrix given by

$$\mathbf{P} = \begin{bmatrix} f_0 & bc & 0 & 0 & \cdots \\ ad & f & bc & 0 & \cdots \\ 0 & ad & f & bc & \cdots \\ 0 & 0 & ad & f & \cdots \\ \vdots & \vdots & \vdots & \vdots & \ddots \end{bmatrix} \tag{7.7}$$

where $b = 1 - a$, $d = 1 - c$, $f_0 = 1 - ad$, and $f = ac + bd$.

For example, starting with an empty queue (state s_0), Fig. 7.1 indicates that we move to state s_1 only when a packet arrives and no packet can depart, which is term ad in the diagram. This transition is also indicated in the above transition matrix as the term at location (2,1) of the transition matrix.

In order for the queue to be stable, the arrival probability must be smaller than the departure probability ($a < c$).

Since the dimension of $\mathbf{P}$ is infinite, we are going to obtain an expression for the distribution vector $\mathbf{s}$ using difference equation techniques instead of studying the eigenvectors of the matrix. The difference equations for the steady-state distribution vector are obtained from the equation

$$\mathbf{P}\,\mathbf{s} = \mathbf{s} \tag{7.8}$$

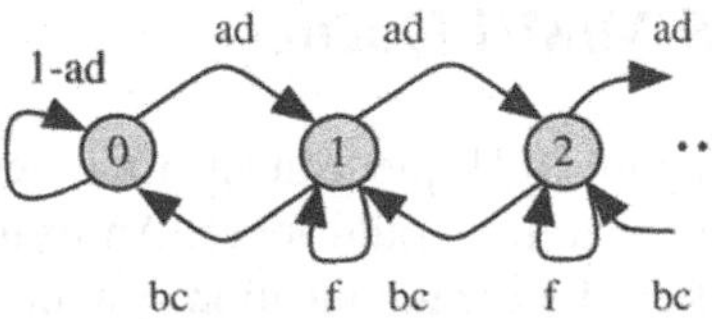

Fig. 7.1 State transition diagram for the discrete-time M/M/1 queue

which produces the following difference equations

$$ad\ s_0 - bc\ s_1 = 0 \tag{7.9}$$

$$ad\ s_0 - g\ s_1 + bc\ s_2 = 0 \tag{7.10}$$

$$ad\ s_{i-1} - g\ s_i + bc\ s_{i+1} = 0 \qquad i > 0 \tag{7.11}$$

where $g = 1 - f$ and s_i is the probability that the system is in state i. For an infinite-sized queue, we have $i \geq 0$.

The solution to the above equations is given as

$$s_1 = \left(\frac{a\,d}{b\,c}\right) s_0$$

$$s_2 = \left(\frac{a\,d}{b\,c}\right)^2 s_0$$

$$s_3 = \left(\frac{a\,d}{b\,c}\right)^3 s_0$$

and in general,

$$s_i = \left(\frac{a\,d}{b\,c}\right)^i s_0 \qquad i \geq 0$$

It is more convenient to write s_i in the form

$$s_i = \rho^i\ s_0 \qquad i \geq 0 \tag{7.12}$$

where ρ is the *distribution index*

$$\rho = \frac{a\,d}{b\,c} < 1 \tag{7.13}$$

The value of the distribution index will affect the component values of the distribution vector.

The complete solution is obtained from the above equations plus the condition $\sum_{i=0}^{\infty} s_i = 1$.

$$s_0 \sum_{i=0}^{\infty} \rho^i = 1 \tag{7.14}$$

Thus, we obtain

$$\frac{s_0}{1 - \rho} = 1 \tag{7.15}$$

from which we obtain

$$s_0 = 1 - \rho \tag{7.16}$$

and the components of the equilibrium distribution vector are given from (7.12) by

$$s_i = (1 - \rho)\rho^i \qquad i \geq 0 \tag{7.17}$$

It is interesting to compare this expression with the continuous-time $M/M/1$ queue. The components of the distribution vector for a continuous-time queue are given by

$$s_i = (1 - \rho)\rho^i \qquad i \geq 0 \tag{7.18}$$

where ρ for the continuous-time queue is called the *traffic intensity* and equals the ratio of the arrival rate to the service rate. The two expressions are identical for the simple $M/M/1$ queue.

7.5.1 $M/M/1$ Queue Performance

Once $\mathbf{s}$ is found, we can find the queue performance such as the throughput, average queue size, and packet delay.

The output traffic or average number of packets leaving the queue per time step is given by

$$N_a(\text{out}) = ac\, s_0 + \sum_{i=1}^{\infty} c\, s_i \tag{7.19}$$

The first term on the RHS is the number of packets leaving the queue, given the queue is empty. The second term on the RHS is the average number of packets leaving the queue when it is not empty. Simplifying, we get

$$\begin{aligned} N_a(\text{out}) &= a\, c\, s_0 + c\,(1 - s_0) \\ &= c - bc\, s_0 \\ &= a \end{aligned} \tag{7.20}$$

The units of $N_a(\text{out})$ are packets/time step.

The throughput for the $M/M/1$ queue is given by

$$\text{Th} = N_a(\text{out}) = a \tag{7.21}$$

The throughput is measured in units of packets/time step. To obtain the throughput in units of packets/s, we use the time step value:

$$\text{Th}' = \frac{\text{Th}}{T} \tag{7.22}$$

The input traffic or average number of packets entering the queue per time step is given by

$$N_{\text{a}}(\text{in}) = 1 \times a + 0 \times b = a \tag{7.23}$$

This output traffic is measured in units of packets/time step.

The efficiency of the $M/M/1$ queue is given by

$$\eta = \frac{N_{\text{a}}(\text{out})}{N_{\text{a}}(\text{in})} = 1 \tag{7.24}$$

The $M/M/1$ queue is characterized by maximum efficiency since input and output data rates are equal. There is no chance for packets to be lost since the infinite queue does not ever fill up.

The average queue size is given by the equation

$$Q_{\text{a}} = \sum_{i=0}^{\infty} i \; s_i \tag{7.25}$$

Using (7.17) we get

$$Q_{\text{a}} = \frac{\rho}{1-\rho} \tag{7.26}$$

The queue size is measured in units of packets or customers.

Figure 7.2 shows the exponential growth of the average queue size as the distribution index increases. A semilog plot was chosen here to show in more detail the size of the queue.

We can invoke Little's result to estimate the *wait time*, which is the average number of time steps a packet spends in the queue before it is routed, as

$$Q_{\text{a}} = W \times \text{Th} \tag{7.27}$$

where W is the average number of time steps that a packet spends in the queue. Thus, W is given by

$$W = \frac{\rho}{a(1-\rho)} \tag{7.28}$$

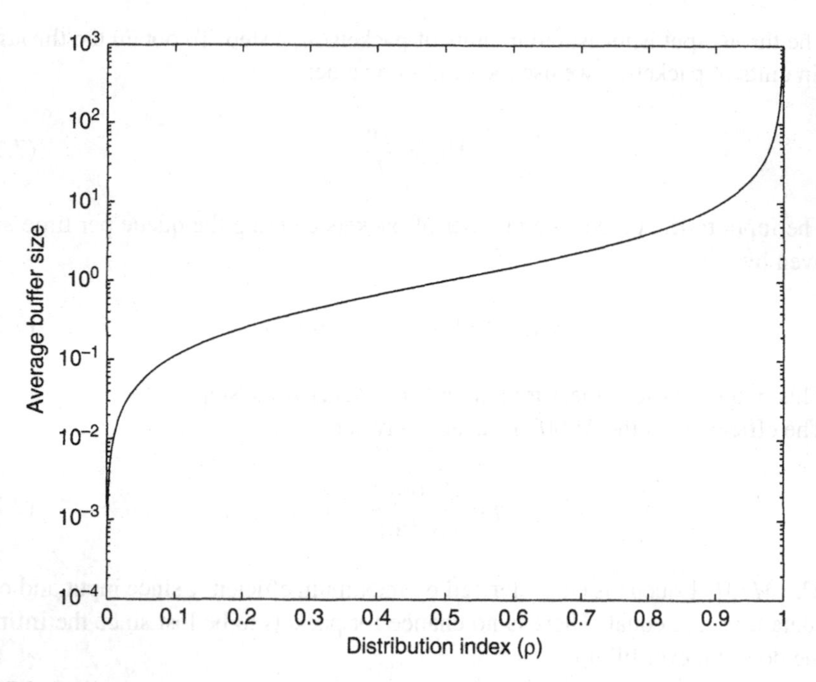

Fig. 7.2 Average queue size versus the distribution index ρ for the $M/M/1$ queue

This wait time is measured in units of time steps. The wait time in units of seconds is given by the unnormalized version of Little's result:

$$W' = \frac{Q_a}{\text{Th}'} \tag{7.29}$$

Example 1 Consider the $M/M/1$ queue with the following parameters $a = 0.6$ and $c = 0.8$. Find the equilibrium distribution vector and the queue performance.

From (7.7), the transition matrix is

$$\mathbf{P} = \begin{bmatrix} 0.88 & 0.32 & 0 & 0 & 0 & \cdots \\ 0.12 & 0.56 & 0.32 & 0 & 0 & \cdots \\ 0 & 0.12 & 0.56 & 0.32 & 0 & \cdots \\ 0 & 0 & 0.12 & 0.56 & 0.32 & \cdots \\ 0 & 0 & 0 & 0.12 & 0.56 & \cdots \\ 0 & 0 & 0 & 0 & 0.12 & \cdots \\ \vdots & \vdots & \vdots & \vdots & \vdots & \ddots \end{bmatrix}$$

The steady-state distribution vector is found using (7.17):

$$\mathbf{s} = \begin{bmatrix} 0.6250 & 0.2344 & 0.0879 & 0.0330 & 0.0124 & 0.0046 & 0.0017 & \cdots \end{bmatrix}^t$$

The probability of being in state i decreases exponentially as i increases.

The queue performance is as follows:

$$\begin{aligned} \text{Th} &= 0.6 && \text{packets/time step} \\ \eta &= 1 && \\ Q_a &= 0.6 && \text{packets} \\ W &= 1 && \text{time steps} \end{aligned}$$

■

Example 2 Investigate the queue in the previous example when the arrival probability is very close to the departure probability.

For the queue to remain stable, we must have $a < c$. Let us try $a = 0.6$ and $c = a + 0.01$. The steady-state distribution vector is found using (7.17):

$$\mathbf{s} = \begin{bmatrix} 0.0410 & 0.0393 & 0.0377 & 0.0361 & 0.0347 & 0.0332 & 0.0319 & \cdots \end{bmatrix}^t$$

Comparing this distribution vector with its counterpart in the previous example, we see that the probability of being in state i is increased for $i > 0$. This is an indication that the queue is getting close to being unstable.

The queue performance is as follows:

$$\begin{aligned} \text{Th} &= 0.6 && \text{packets/time step} \\ \eta &= 1 && \\ Q_a &= 23.4 && \text{packets} \\ W &= 39 && \text{time steps} \end{aligned}$$

We see that the throughput is increased since the probability that the queue is empty (state 0) is decreased. ■

7.6 M/M/1/B Queue

This queue is similar to the discrete-time $M/M/1$ queue except that the queue has finite size B. The state transition diagram is shown in Fig. 7.3.

Since packet arrivals and departures are independent of the time index value, we have a homogeneous Markov chain. We assume that when a packet arrives, it could

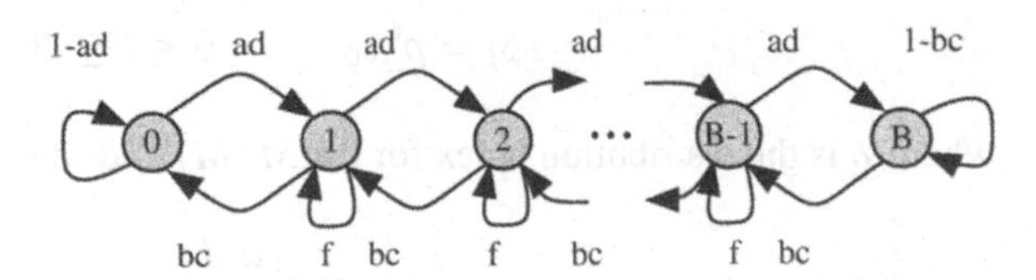

Fig. 7.3 State transition diagram for the discrete-time $M/M/1/B$ queue

be serviced at the same time step and it could leave the queue with probability c. This results in the transition matrix given by

$$\mathbf{P} = \begin{bmatrix} f_0 & bc & 0 & \cdots & 0 & 0 & 0 \\ ad & f & bc & \cdots & 0 & 0 & 0 \\ 0 & ad & f & \cdots & 0 & 0 & 0 \\ \vdots & \vdots & \vdots & \ddots & \vdots & \vdots & \vdots \\ 0 & 0 & 0 & \cdots & f & bc & 0 \\ 0 & 0 & 0 & \cdots & ad & f & bc \\ 0 & 0 & 0 & \cdots & 0 & ad & 1-bc \end{bmatrix} \tag{7.30}$$

where $f_0 = 1 - ad$ and $f = ac + bd$.

Since the dimension of $\mathbf{P}$ is arbitrary, we are going to obtain an expression for the equilibrium distribution vector $\mathbf{s}$ using difference equations techniques. The difference equations for the state probability vector are given by

$$ad\ s_0 - bc\ s_1 = 0 \tag{7.31}$$

$$ad\ s_0 - g\ s_1 + bc\ s_2 = 0 \tag{7.32}$$

$$ad\ s_{i-1} - g\ s_i + bc\ s_{i+1} = 0 \qquad 0 < i < B \tag{7.33}$$

where $g = ad + bc$ and s_i is the component of the distribution vector corresponding to state i.

The solution to the above equations is given as

$$s_1 = \left(\frac{a\,d}{b\,c}\right) s_0$$

$$s_2 = \left(\frac{a\,d}{b\,c}\right)^2 s_0$$

$$s_3 = \left(\frac{a\,d}{b\,c}\right)^3 s_0$$

and in general,

$$s_i = \left(\frac{a\,d}{b\,c}\right)^i s_0 \qquad i \geq 0$$

It is more convenient to write s_i in the form

$$s_i = \rho^i\ s_0 \qquad 0 \leq i \leq B \tag{7.34}$$

where ρ is the distribution index for the $M/M/1/B$ queue:

$$\rho = \frac{a\,d}{b\,c}$$

The complete solution is obtained from the above equations plus the condition $\sum_{i=0}^{B} s_i = 1$, which gives

$$s_0 \sum_{i=0}^{B} \rho^i = 1 \tag{7.35}$$

from which we obtain s_0, which is the probability that the queue is empty,

$$s_0 = \frac{1-\rho}{1-\rho^{B+1}} \tag{7.36}$$

and the equilibrium distribution for the other states is given from (7.34) by

$$s_i = \frac{(1-\rho)\rho^i}{1-\rho^{B+1}} \qquad 0 \le i \le B \tag{7.37}$$

Note that ρ for the finite-sized queue *can* be more than one. In that case, the queue will not be stable in the following sense:

$$s_0 < s_1 < s_2 \cdots < s_B \tag{7.38}$$

This indicates that the probability that the queue is full (s_B) is bigger than the probability that it is empty (s_0).

7.6.1 $M/M/1/B$ Queue Performance

The throughput or output traffic for the $M/M/1/B$ queue is given by

$$\begin{aligned} \text{Th} &= N_\text{a}(\text{out}) \\ &= ac\, s_0 + \sum_{i=1}^{B} c\, s_i \\ &= ac\, s_0 + c\,(1 - s_0) \\ &= c\,(1 - b\, s_0) \end{aligned} \tag{7.39}$$

This throughput is measured in units of packets/time step. The throughput in units of packets/s is

$$\text{Th}' = \frac{\text{Th}}{T} \tag{7.40}$$

The input traffic is given by

$$N_\text{a}(\text{in}) = 1 \times a + 0 \times b = a \tag{7.41}$$

Input traffic is measured in units of packets/time step.

The efficiency of the $M/M/1/B$ queue is given by

$$\begin{aligned}\eta &= \frac{N_\mathrm{a}(\mathrm{out})}{N_\mathrm{a}(\mathrm{in})} \\ &= \frac{\mathrm{Th}}{a} \\ &= \frac{c\,(1 - b\, s_0)}{a} \end{aligned} \tag{7.42}$$

Data are lost in the $M/M/1/B$ queue when it is full and packets arrive but does not leave. The average lost traffic $N_\mathrm{a}(\mathrm{lost})$ is given by

$$N_\mathrm{a}(\mathrm{lost}) = s_B\; a\; d \tag{7.43}$$

The above equation is simply the probability that a packet is lost which equals the probability that the queue is full, and a packet arrives, and no packets can leave.

Lost traffic is measured in units of packets/time step. The average lost traffic per second is given by

$$N'_\mathrm{a}(\mathrm{lost}) = \frac{N_\mathrm{a}(\mathrm{lost})}{T} \tag{7.44}$$

The packet loss probability L is the ratio of lost traffic relative to the input traffic

$$L = \frac{N_\mathrm{a}(\mathrm{lost})}{N_\mathrm{a}(\mathrm{in})} = s_B\; d \tag{7.45}$$

The average queue size is given by the equation

$$Q_\mathrm{a} = \sum_{i=0}^{B} i\; s_i \tag{7.46}$$

Queue size is measured in units of packets. Using (7.37), the average queue size is given by

$$Q_\mathrm{a} = \frac{\rho \times \left[1 - (B+1)\rho^B + B\rho^{B+1}\right]}{(1-\rho) \times \left(1 - \rho^{B+1}\right)} \tag{7.47}$$

Figure 7.4 shows the exponential growth of the average queue size as the distribution index increases ($B = 10$ in that case). The solid line is for the $M/M/1/B$ queue and the dotted line is for the $M/M/1$ queue for comparison. We see that Q_a for the finite-sized queue grows at a slower rate with increasing distribution index compared to the infinite-sized queue. Furthermore, ρ for the infinite-sized queue could increase beyond unity value.

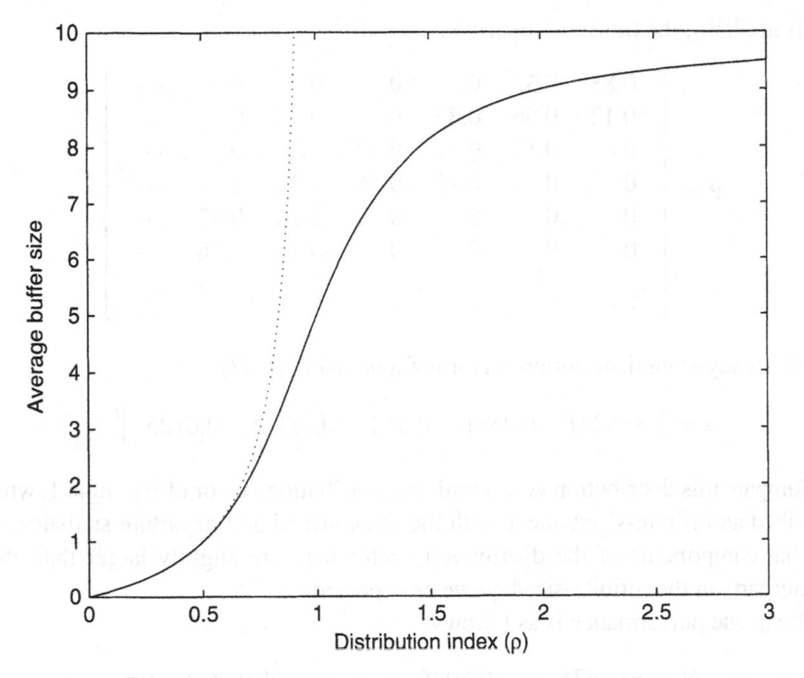

Fig. 7.4 Average queue size versus the distribution index ρ for the $M/M/1/B$ queue when $B = 10$ (*solid line*). The *dotted line* is average queue size for an infinite-sized $M/M/1$ queue

We can invoke Little's result to estimate the average number of time steps a packet spends in the queue before it is routed as

$$Q_{\mathrm{a}} = W \times \mathrm{Th} \tag{7.48}$$

where W is the wait time, or average number of time steps, that a packet spends in the queue. The throughput in the above expression for the wait time must be in units of packets time step. The wait time is simply

$$W = \frac{Q_{\mathrm{a}}}{\mathrm{Th}} \tag{7.49}$$

Wait time is measured in units of time steps. The wait time in units of seconds is given by

$$W' = \frac{Q_{\mathrm{a}}}{\mathrm{Th}'} \qquad \text{seconds} \tag{7.50}$$

Example 3 Consider the $M/M/1/B$ queue with the following parameters $a = 0.6$, $c = 0.8$ and $B = 4$. Find the equilibrium distribution vector and the queue performance.

From (7.30), the transition matrix is

$$\mathbf{P} = \begin{bmatrix} 0.88 & 0.32 & 0 & 0 & 0 & 0 & \cdots \\ 0.12 & 0.56 & 0.32 & 0 & 0 & 0 & \cdots \\ 0 & 0.12 & 0.56 & 0.32 & 0 & 0 & \cdots \\ 0 & 0 & 0.12 & 0.56 & 0.32 & 0 & \cdots \\ 0 & 0 & 0 & 0.12 & 0.56 & 0.32 & \cdots \\ 0 & 0 & 0 & 0 & 0.12 & 0.56 & \cdots \\ \vdots & \vdots & \vdots & \vdots & \vdots & \vdots & \ddots \end{bmatrix}$$

The steady-state distribution vector is found using (7.37)

$$\mathbf{s} = \begin{bmatrix} 0.6297 & 0.2361 & 0.0885 & 0.0332 & 0.0125 \end{bmatrix}^t$$

Compare this distribution vector with the distribution vector of Example 1, which described an infinite-sized queue with the same arrival and departure statistics. We see that components of the distribution vector here are slightly larger than their counterparts in the infinite-sized queue as expected.

The queue performance is as follows:

$$\begin{aligned} N_a(\text{out}) = \text{Th} &= 0.5985 && \text{packets/time step} \\ \eta &= 0.9975 && \\ N_a(\text{lost}) &= 1.5 \times 10^{-3} && \text{packets/time step} \\ L &= 0.0025 && \\ Q_a &= 0.5626 && \text{packets} \\ W &= 0.9401 && \text{time steps} \end{aligned}$$

We note that the $M/M/1/B$ queue has smaller average size Q_a and smaller wait time W compared to the $M/M/1$ queue with the same arrival and departure statistics. As expected, we have

$$N_a(\text{out}) + N_a(\text{lost}) = N_a(\text{in})$$

■

Example 4 Find the performance of the queue in the previous example when the queue size becomes $B = 20$.

The queue performance is as follows:

$$\begin{aligned} N_a(\text{out}) = \text{Th} &= 0.6 && \text{packets/time step} \\ \eta &= 1 && \\ N_a(\text{lost}) &= 2.2682 \times 10^{-10} && \text{packets/time step} \\ L &= 3.7804 \times 10^{-10} && \\ Q_a &= 0.6 && \text{packets} \\ W &= 1 && \text{time steps} \end{aligned}$$

We see that increasing the queue size exponentially decreases the loss probability. The throughput is not changed by much, but the wait time is slightly increased due to the increased average queue size. ■

Example 5 Plot the $M/M/1/B$ performance when the input traffic varies between $0 \le a \le 1$ for $B = 10$ and $c = 0.5$.

Figure 7.5 shows the throughput, efficiency, loss probability, and delay to plot these quantities versus input traffic.

The important things to note from this example are as follows:

1. The throughput of the queue could not exceed the maximum value for the average output traffic. Section 7.6.2 below will prove that this maximum value is simply c.
2. The efficiency of the queue is very close to 100% until the input traffic approaches the maximum output traffic c.
3. Packet loss probability is always present but starts to increase when the input traffic approaches the packet maximum output traffic c.
4. Packet delay increases sharply when the input traffic approaches the packet maximum output traffic c.

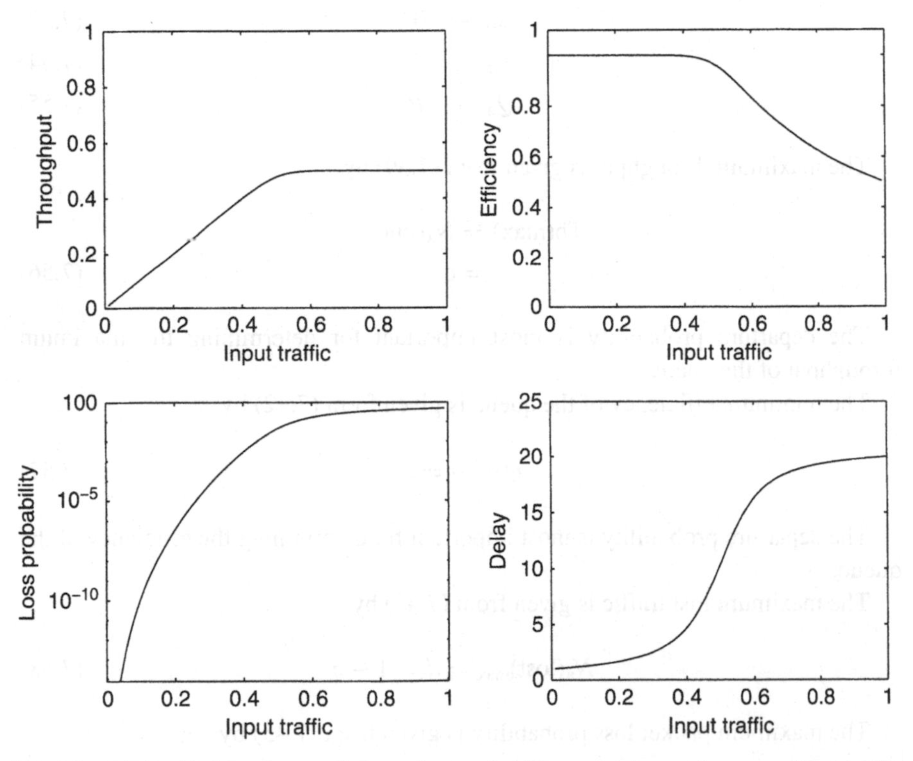

Fig. 7.5 $M/M/1/B$ throughput, efficiency, loss probability, and delay to plot versus input traffic when $B = 10$ and $c = 0.5$

5. Congestion conditions occur as soon as the input traffic exceeds the maximum output traffic c. Congestion is characterized by decreased efficiency, increased packet loss, and increased delay.
6. The delay reaches a maximum value determined by the maximum size of the queue and the maximum output traffic c. The maximum delay could be approximated as

$$\text{Maximum Delay} = \frac{B}{c} = 20 \qquad \text{time steps}$$

■

7.6.2 Performance Bounds on $M/M/1/B$ Queue

The previous example helped us get some rough estimates for the performance bounds of the $M/M/1/B$ queue. This section formalizes these estimates.

Under full load conditions, the $M/M/1/B$ become full and we can assume

$$a \to 1 \tag{7.51}$$

$$b \to 0 \tag{7.52}$$

$$s_0 \to 0 \tag{7.53}$$

$$s_B \to 1 \tag{7.54}$$

$$Q_a \to B \tag{7.55}$$

The maximum throughput is given from (7.39) by

$$\begin{aligned} \text{Th}(\max) &= N_a(\text{out})_{\max} \\ &= c \end{aligned} \tag{7.56}$$

The departure probability is most important for determining the maximum throughput of the queue.

The minimum efficiency of the queue is given from (7.42) by

$$\eta(\min) = c \tag{7.57}$$

The departure probability is most important for determining the efficiency of the queue.

The maximum lost traffic is given from (7.43) by

$$N_a(\text{lost})_{\max} = d = 1 - c \tag{7.58}$$

The maximum packet loss probability is given from (7.45) by

$$L(\max) = 1 - c \tag{7.59}$$

The maximum wait time is given by the approximate formula

$$W(\max) = \frac{B}{c} \tag{7.60}$$

Larger queues result in larger wait times as expected.

7.7 $M^m/M/1/B$ Queue

In an $M^m/M/1$ queue at any time step, at most m customers could arrive and at most one customer could leave. We shall encounter this type of queue when we study network switches where first-in-first-out (FIFO) queues exist at each output port. For each queue, a maximum of m packets arrive at the queue input but only one packet can leave the queue. Therefore, the queue size can increase by more than one, but can only decrease by one in each time step. Assume that the binomial probability of k arrivals at instant n is given by

$$a_k = \binom{m}{k} a^k \, b^{m-k} \tag{7.61}$$

where a is the probability that a packet arrives, $b = 1 - a$, and m is the maximum number of packets that could arrive at the queue input. The queue size can only decrease by at most one at any instant with probability c. The probability that no packet leaves the queue is $d = 1 - c$. We assume that when a packet arrives, it could be serviced at the same time step and it could leave the queue with probability c.

The condition for the stability of the queue is

$$\sum_{k=0}^{m} k \, a_k = a \, m < c \tag{7.62}$$

which indicates that the average number of arrivals at a given time is less than the average number of departures from the system. The resulting state transition matrix is a lower $(B+1) \times (B+1)$ Hessenberg matrix in which all the elements $p_{ij} = 0$ for $j > i + 1$:

$$\mathbf{P} = \begin{bmatrix} x & y_0 & 0 & \cdots & 0 & 0 \\ y_2 & y_1 & y_0 & \cdots & 0 & 0 \\ y_3 & y_2 & y_1 & \cdots & 0 & 0 \\ \vdots & \vdots & \vdots & \ddots & \vdots & \vdots \\ y_B & y_{B-1} & y_{B-2} & \cdots & y_1 & y_0 \\ z_B & z_{B-1} & z_{B-2} & \cdots & z_1 & z_0 \end{bmatrix} \tag{7.63}$$

where

$$x = a_1 \, c + a_0 \tag{7.64}$$
$$y_i = a_i \, c + a_{i-1} \, d \tag{7.65}$$
$$z_i = a_i \, d + \sum_{k=i+1}^{m} a_k \tag{7.66}$$

where we assumed

$$a_k = 0 \qquad k < 0 \tag{7.67}$$
$$a_k = 0 \qquad k > m \tag{7.68}$$

The above transition matrix has m subdiagonals when $m \leq B$.

7.7.1 $M^m/M/1/B$ Queue Performance

To calculate the throughput of the $M^m/M/1/B$, we need to consider the queue in two situations: when it is empty and when it is not.

The throughput of the $M^m/M/1/B$ queue when the queue is in state s_0 is given by

$$\mathrm{Th}_0 = (1 - a_0) \, c \tag{7.69}$$

This is simply the probability that one or more packets arrive and one packet leaves the queue. When the queue is in any other state, the throughput is given by

$$\mathrm{Th}_i = c \qquad 1 \leq i \leq B \tag{7.70}$$

This is simply the probability that a packet leaves the queue. The average throughput is estimated as

$$\begin{aligned}\mathrm{Th} &= \sum_{i=0}^{B} \mathrm{Th}_i \; s_i \\ &= c \; (1 - a_0 \, s_0)\end{aligned} \tag{7.71}$$

The throughput is measured in units of packets/time step. The throughput in units of packets/s is

$$\mathrm{Th}' = \frac{\mathrm{Th}}{T} \tag{7.72}$$

The input traffic is given by

$$N_a(\text{in}) = \sum_{i=0}^{m} i\ a_i = m\ a \tag{7.73}$$

The efficiency of the $M^m/M/1/B$ queue is given by

$$\eta = \frac{N_a(\text{out})}{N_a(\text{in})} = \frac{\text{Th}}{m\ a} = \frac{c\ (1 - a_0\ s_0)}{m\ a} \tag{7.74}$$

Data are lost in the $M^m/M/1/B$ queue when it becomes full and packets arrive but does not leave. However, due to multiple arrivals, packets could be lost even when the queue is not completely full. For example, we could still have one location left in the queue but three customers arrive. Definitely packets will be lost then. Therefore, we conclude that average lost traffic $N_a(\text{lost})$ is a bit difficult to obtain. However, the traffic conservation principle is useful in getting a simple expression for lost traffic. The average lost traffic $N_a(\text{lost})$ is given by

$$N_a(\text{lost}) = N_a(\text{in}) - N_a(\text{out}) = m\ a - c\ (1 - a_0\ s_0) \tag{7.75}$$

The lost traffic is measured in units of packets per time step. The average lost traffic measured in packets per second is given by

$$N_a'(\text{lost}) = \frac{N_a(\text{lost})}{T} \tag{7.76}$$

The packet loss probability L is the ratio of lost traffic relative to the input traffic:

$$L = \frac{N_a(\text{lost})}{N_a(\text{in})} = 1 - \eta = 1 - \frac{c\ (1 - a_0\ s_0)}{m\ a} \tag{7.77}$$

The average queue size is given by the equation

$$Q_a = \sum_{i=0}^{B} i\ s_i \tag{7.78}$$

We can invoke Little's result to estimate the *wait time*, which is the average number of time steps a packet spends in the queue before it is routed, as

$$Q_a = W \times \mathrm{Th} \tag{7.79}$$

where W is the average number of time steps that a packet spends in the queue. Thus, W is given by

$$W = \frac{Q_a}{\mathrm{Th}} \tag{7.80}$$

The wait time is measured in units of time steps. The wait time in units of seconds is given by the unnormalized version of Little's result:

$$W' = \frac{Q_a}{\mathrm{Th}'} \tag{7.81}$$

Example 6 Consider the $M^m/M/1/B$ queue with the following parameters $a = 0.04$, $m = 2$, $c = 0.1$, and $B = 5$. Check its stability condition and find the equilibrium distribution vector and queue performance.

The packet arrival probability is

$$a_k = \binom{2}{k}(0.04)^k(0.96)^{2-k}$$

The stability condition is found from (7.62):

$$\sum_{k=0}^{2} k\, a_k = 0.08 < c$$

The queue is stable and the transition matrix is

$$\mathbf{P} = \begin{bmatrix} 0.9293 & 0.0922 & 0 & 0 & 0 & 0 \\ 0.0693 & 0.8371 & 0.0922 & 0 & 0 & 0 \\ 0.0014 & 0.0693 & 0.8371 & 0.0922 & 0 & 0 \\ 0 & 0.0014 & 0.0693 & 0.8371 & 0.0922 & 0 \\ 0 & 0 & 0.0014 & 0.0693 & 0.8371 & 0.0922 \\ 0 & 0 & 0 & 0.0014 & 0.0707 & 0.9078 \end{bmatrix}$$

The transition matrix has two subdiagonals because $m = 2$.

The steady-state distribution vector is the eigenvector of $\mathbf{P}$ that corresponds to unity eigenvalue. We have

$$\mathbf{s} = \begin{bmatrix} 0.2843 & 0.2182 & 0.1719 & 0.1353 & 0.1065 & 0.0838 \end{bmatrix}^t$$

We see that the most probable state for the queue is state s_0 which occurs with probability 28.43%.

The queue performance is as follows:

$$
\begin{aligned}
\text{Th} &= 0.0738 && \text{packets/time step} \\
\eta &= 0.9225 && \\
N_a(\text{lost}) &= 6.2 \times 10^{-3} && \text{packets/time step} \\
L &= 0.0775 && \\
Q_a &= 1.813 && \text{packets} \\
W &= 24.5673 && \text{time steps}
\end{aligned}
$$

Flow conservation is verified since the sum of the throughput and the lost traffic equals the input traffic $N_a(\text{in}) = m\, a$. ■

Example 7 Find the performance of the queue in the previous example when the queue size becomes $B = 20$. The queue performance is as follows:

$$
\begin{aligned}
\text{Th} &= 0.0799 && \text{packets/time step} \\
\eta &= 0.9984 && \\
N_a(\text{lost}) &= 1.3167 \times 10^{-4} && \text{packets/time step} \\
L &= 0.0016 && \\
Q_a &= 1.813 && \text{packets} \\
W &= 44.3432 && \text{time steps}
\end{aligned}
$$

We see that increasing the queue size decreases the loss probability which results in only a slight increase in the throughput. The wait time is almost doubled. ■

In Chapter 4, we explored different techniques for finding the equilibrium distribution for the distribution vector **s**. For simple situations when the value of m is small (1 or 2), we can use the difference equation approach. When m is large, we can use the z-transform technique as in Section 4.8.

Example 8 Plot the $M^m/M/1/B$ performance when $m = 2$ and the input traffic varies between $0 \le a \le 1$ for $B = 10$ and $c = 0.5$.

Figure 7.6 shows the variation of the throughput, efficiency, loss probability, and delay versus the average input traffic.

The important things to note from this example are as follows:

1. The throughput of the queue could not exceed the maximum output traffic c.
2. The efficiency of the queue is very close to 100% until the input traffic approaches the maximum output traffic c.
3. Packet loss probability is always present but starts to increase when the input traffic approaches the packet maximum output traffic c.

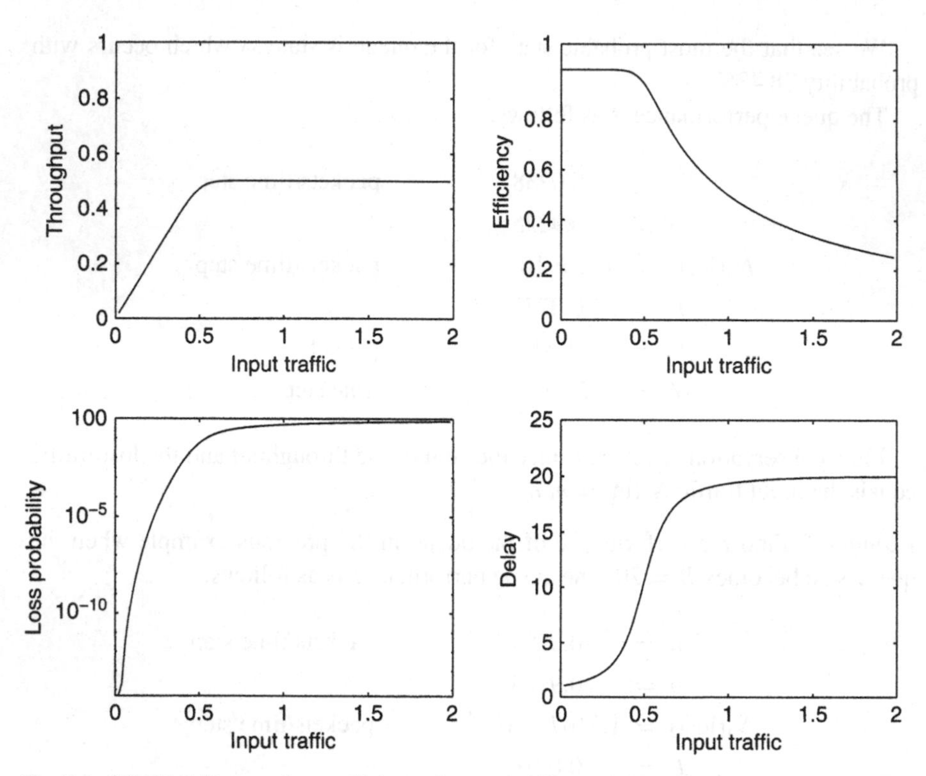

Fig. 7.6 $M^m/M/1/B$ throughput, efficiency, loss probability, and delay to plot versus input traffic when $m = 2$, $B = 10$ and $c = 0.5$

4. Packet delay increases sharply when the input traffic approaches the packet maximum output traffic c.
5. Congestion conditions occur as soon as the input traffic exceeds the maximum output traffic c. Congestion is characterized by decreased efficiency, increased packet loss, and increased delay.

■

7.7.2 Performance Bounds on $M^m/M/1/B$ Queue

The previous example helped us get some rough estimates for the performance bounds of the $M^m/M/1/B$ queue. This section formalizes these estimates.

Under full load conditions, the $M^m/M/1/B$ becomes full and we can assume

$$a \rightarrow 1 \tag{7.82}$$

$$b \rightarrow 0 \tag{7.83}$$

$$s_0 \rightarrow 0 \tag{7.84}$$

$$s_B \rightarrow 1 \tag{7.85}$$

$$Q_\mathrm{a} \rightarrow B \tag{7.86}$$

The maximum value for the throughput from (7.71) becomes

$$\text{Th}\,(\text{max}) = c \tag{7.87}$$

The packet departure probability determines the maximum throughput of the queue.

The minimum efficiency of the $M^m/M/1/B$ queue is given from (7.74) by

$$\eta(\text{min}) = \frac{c}{m} \tag{7.88}$$

The maximum lost traffic is given from (7.77) by

$$N_a(\text{lost})_{\text{max}} = m - c \tag{7.89}$$

The maximum packet loss probability is given from (7.77) by

$$L(\text{max}) = 1 - \frac{c}{m} \tag{7.90}$$

When more customers arrive per time step (large m), the probability of loss increases. Maximum loss probability increases when packet departure probability decreases and when the number of arriving customers increases.

The maximum delay is given from (7.80) by

$$\begin{aligned} W(\text{max}) &= \frac{B}{\text{Th}(\text{max})} \\ &= \frac{B}{c} \end{aligned} \tag{7.91}$$

7.7.3 Alternative Solution Method

When B is large, it is better to use numerical techniques such as forward- or backward substitution using Givens rotations.[1] [5]

At steady state, we can write

$$\mathbf{P}\,\mathbf{s} = \mathbf{s} \tag{7.92}$$

We can use the technique explained in Section 4.10 on page 140 to construct a system of linear equations that can be solved using any of the specialized software designed to solve large systems of linear equations.

---

[1] Another useful technique for triangularizing a matrix is to use Householder transformation. However, we prefer Givens rotation due to its numerical stability. Alston Householder once commented that he would never fly in an airplane that was designed with the help of a computer using floating point arithmetic [4].

7.8 $M/M^m/1/B$ Queue

In an $M/M^m/1$ queue at any time step, at most one customer could arrive and at most m customers could leave. We shall encounter this type of queue when we study network switches where first-in-first-out (FIFO) queues exist at each input port. For each queue, one packet arrives at the input line to the storage buffer, but a maximum of m packets can leave the queue destined to the different switch outputs. Therefore, the queue size can increase by one, but can decrease by more than one. The probability that a packet arrives is a and $b = 1 - a$ is the probability that a packet does not arrive at a time step. Define $c_{i,j}$ as the probability that j customers leave the queue when there are i customers in the queue.

$$c_{i,j} = \binom{i}{j} c^j \, d^{i-j} \tag{7.93}$$

where c is the probability that a packet departs and $d = 1 - c$. The state transition matrix is an upper $(B+1) \times (B+1)$ Hessenberg matrix in which all the elements of subdiagonals 2, 3, ... are zero—i.e., $p_{ij} = 0$ for $i > j + 1$. The matrix has only m superdiagonals. If we assume an arriving packet is served at the same time step, then the transition matrix for the case when $B = 6$ and $m = 3$ will have the form

$$\mathbf{P} = \begin{bmatrix} q_0 & r_1 & s_2 & t_3 & 0 & 0 & 0 \\ p_0 & q_1 & r_2 & s_3 & t_4 & 0 & 0 \\ 0 & p_1 & q_2 & r_3 & s_4 & t_5 & 0 \\ 0 & 0 & p_2 & q_3 & r_4 & s_5 & y \\ 0 & 0 & 0 & p_3 & q_4 & r_5 & x_2 \\ 0 & 0 & 0 & 0 & p_4 & q_5 & x_1 \\ 0 & 0 & 0 & 0 & 0 & p_5 & x_0 \end{bmatrix} \tag{7.94}$$

where the matrix elements are given by the general expressions

$$p_i = a\, c_{i+1,0} \tag{7.95}$$

$$q_i = a\, c_{i+1,1} + b\, c_{i,0} \tag{7.96}$$

$$r_i = a\, c_{i+1,2} + b\, c_{i,1} \tag{7.97}$$

$$s_i = a \sum_{j=m}^{i+1} c_{i+1,j} + b\, c_{i,2} \tag{7.98}$$

$$t_i = b \sum_{j=m}^{i} c_{i,j} \tag{7.99}$$

$$x_i = c_{6,i} \tag{7.100}$$

$$y = \sum_{j=m}^{B} c_{B,j} \tag{7.101}$$

The condition for the stability of the queue is when average traffic at the queue input is smaller than the average traffic at the queue output:

$$a < m\,c \tag{7.102}$$

7.8.1 $M/M^m/1/B$ Queue Performance

The average input traffic for the $M/M^m/1/B$ queue is obtained simply as

$$N_a(\text{in}) = a \tag{7.103}$$

Traffic is lost in the $M/M^m/1/B$ queue when it becomes full and a packet arrives while no packets leave. The average lost traffic $N_a(\text{lost})$ is expressed simply as

$$N_a(\text{lost}) = a\; s_B\; c_{B,0} \tag{7.104}$$

The packet loss probability L is the ratio of lost traffic relative to the input traffic:

$$L = \frac{N_a(\text{lost})}{N_a(\text{in})} = s_B\; c_{B,0} \tag{7.105}$$

The throughput of the queue is obtained using the traffic conservation principle:

$$\begin{aligned} \text{Th} &= N_a(\text{in}) - N_a(\text{lost}) \\ &= a\,(1 - s_B) \end{aligned} \tag{7.106}$$

The efficiency of the $M/M^m/1/B$ queue is given by

$$\begin{aligned} \eta &= 1 - L \\ &= 1 - s_B\; c_{B,0} \end{aligned} \tag{7.107}$$

The average queue size is given by the equation

$$Q_a = \sum_{i=0}^{B} i\; s_i \tag{7.108}$$

We can invoke Little's result to estimate the *wait time*, which is the average number of time steps a packet spends in the queue before it is routed:

$$Q_a = W \times \text{Th} \tag{7.109}$$

where W is the average number of time steps that a packet spends in the queue. Thus, W is give by

$$W = \frac{Q_a}{\text{Th}} \tag{7.110}$$

The wait time is measured in units of time steps. The wait time in units of seconds is given by the unnormalized version of Little's result.

$$W' = \frac{Q_a}{Th'} \tag{7.111}$$

Example 9 Consider the $M/M^m/1/B$ queue with the following parameters $a = 0.1$, $c = 0.07$, $m = 2$, and $B = 5$. Check its stability condition and find the equilibrium distribution vector and the queue performance.

Using (7.94), the transition matrix is

$$\mathbf{P} = \begin{bmatrix} 0.9070 & 0.0635 & 0.0044 & 0 & 0 & 0 \\ 0.0930 & 0.8500 & 0.1186 & 0.0126 & 0 & 0 \\ 0 & 0.0865 & 0.7966 & 0.1661 & 0.0241 & 0 \\ 0 & 0 & 0.0804 & 0.7464 & 0.2069 & 0.0425 \\ 0 & 0 & 0 & 0.0748 & 0.6994 & 0.2618 \\ 0 & 0 & 0 & 0 & 0.0696 & 0.6957 \end{bmatrix}$$

The transition matrix has two superdiagonals because $m = 2$.

The steady-state distribution vector is the eigenvector of $\mathbf{P}$ that corresponds to unity eigenvalue. We have

$$\mathbf{s} = \begin{bmatrix} 0.2561 & 0.3584 & 0.2414 & 0.1043 & 0.0324 & 0.0074 \end{bmatrix}^t$$

We see that the most probable state for the queue is state s_1 which occurs with probability 35.8%.

The queue performance is as follows:

$$\begin{aligned} N_a(\text{lost}) &= 6.4058 \times 10^{-4} && \text{packets/time step} \\ \text{Th} &= 0.0994 && \text{packets/time step} \\ L &= 0.0064 \times 10^{-2} && \\ \eta &= 0.9936 && \\ Q_a &= 1.3205 && \text{packets} \\ W &= 13.2906 && \text{time steps} \end{aligned}$$

■

Example 10 Find the performance of the queue in the previous example when the queue size becomes $B = 10$.

The queue performance is as follows:

$$
\begin{aligned}
N_a(\text{lost}) &= 2.6263 \times 10^{-8} && \text{packets/time step} \\
\text{Th} &= 0.1 && \text{packets/time step} \\
L &= 2.6263 \times 10^{-7} \\
\eta &= 1 \\
Q_a &= 1.3286 && \text{packets} \\
W &= 13.2862 && \text{time steps}
\end{aligned}
$$

We see that increasing the queue size exponentially decreases the loss probability. The throughput is not changed by much. ■

Example 11 Plot the $M/M^m/1/B$ performance when $m = 2$ and the input traffic varies between $0 \leq a \leq 1$ for $B = 5$ and $c = 0.05$.

Figure 7.7 shows the throughput, efficiency, loss probability, and delay to plot these quantities versus input traffic.

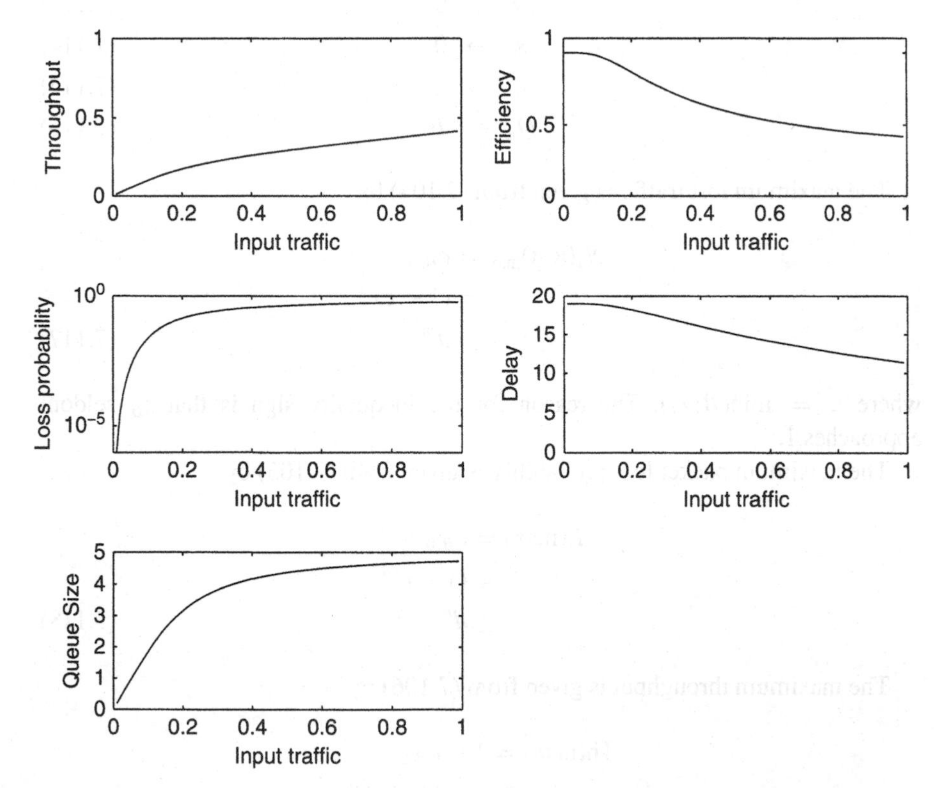

Fig. 7.7 $M/M^m/1/B$ throughput, efficiency, loss probability and delay versus input traffic when $m = 2$, $B = 5$, and $c = 0.05$

The important things to note from this example are as follows:

1. The throughput of the queue increases with increasing input traffic but shows slight decrease when the input traffic approaches the value $m\,c$.
2. The efficiency of the queue is very close to 100% until the input traffic approaches the maximum output traffic $m\,c$.
3. Packet loss probability is always present but starts to increase when the input traffic approaches the value $m\,c$.
4. Packet decreases when the input traffic approaches the value $m\,c$. This is due to the queue size becoming constant while the throughput keeps increasing.

■

7.8.2 Performance Bounds on $M/M^m/1/B$ Queue

The previous examples help us get some rough estimates for the performance bounds of the $M/M/^m1/B$ queue.

Under full load conditions, the $M/M^m/1/B$ becomes full and we can assume

$$a \to 1 \tag{7.112}$$
$$b \to 0 \tag{7.113}$$
$$s_0 \to 0 \tag{7.114}$$
$$s_B \to 1 \tag{7.115}$$
$$Q_a \to B \tag{7.116}$$

The maximum lost traffic is given from (7.104) by

$$\begin{aligned} N_a(\text{lost})_{\max} &= c_{B,0} \\ &= (1-c)^m \\ &\le d^m \end{aligned} \tag{7.117}$$

where $n = \min(B, m)$. The reason for the inequality sign is that s_B seldom approaches 1.

The maximum packet loss probability is given from (7.105) by

$$\begin{aligned} L(\max) &= c_{B,0} \\ &= (1-c)^n \\ &\le d^n \end{aligned} \tag{7.118}$$

The maximum throughput is given from (7.106) by

$$\begin{aligned} \text{Th}(\max) &= 1 - c_{B,0} \\ &\ge 1-(1-c)^m \\ &= 1-d^m \end{aligned} \tag{7.119}$$

The minimum efficiency of the $M/M^m/1/B$ queue is given from (7.107) by

$$\eta(\text{min}) \geq 1 - d^m \tag{7.120}$$

The maximum delay is given by the approximate formula

$$\begin{aligned} W(\text{max}) &\leq \frac{B}{\text{Th(max)}} \\ &\leq \frac{B}{1 - d^m} \\ &\leq \frac{B}{1 - d^m} \end{aligned} \tag{7.121}$$

7.8.3 Alternative Solution Method

When B is large, it is better to use numerical techniques such as forward- or backward substitution using Givens rotations.

At steady state, we can write

$$\mathbf{P}\,\mathbf{s} = \mathbf{s} \tag{7.122}$$

We can use the technique explained in Section 4.10 on page 140 to construct a system of linear equations that can be solved using any of the specialized software designed to solve large systems of linear equations.

7.9 The $D/M/1/B$ Queue

In the $D/M/1/B$ queue, packets arrive at a fixed rate but leave the queue in a random fashion. Let us assume that the time step is chosen such that exactly one packet arrives at the nth time step. Assume also that c is the probability that a packet leaves the queue during one time step. We also assume that at most one packet leaves the queue in one time step. $d = 1 - c$ has the usual meaning. Figure 7.8 shows the state transition diagram for such queue for the case when $n = 4$ and $B = 4$ also. The number of rows correspond to the number of time steps between packet arrivals. The last row corresponds to the states when a packet arrives. The number of columns corresponds to the size of the queue such that each column corresponds to a particular state of queue occupancy. For example, the leftmost column corresponds to the case when the queue is empty. The rightmost column corresponds to the case when the queue is full. The state of occupancy of the queue is indicated in Table 7.1. In that case, we know that a packet arrives when the queue is in one of the bottom states $s_{3,0}$, $s_{3,1}$, $s_{3,2}$, $s_{3,3}$, or $s_{3,4}$.

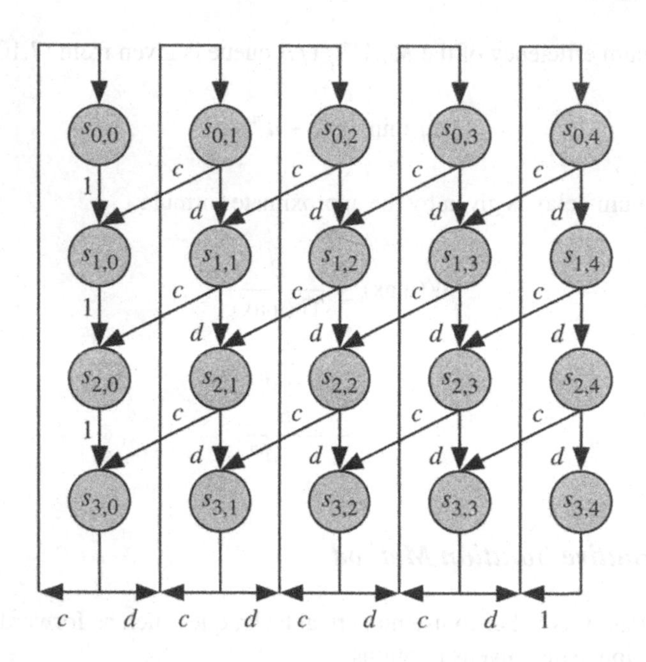

Fig. 7.8 State transition diagram for the discrete-time $D/M/1/B$ queue

Table 7.1 Relation of the queue states and the $D/M/1/B$ queue occupancy

States	Queue occupancy
$s_{0,0}$–$s_{3,0}$	Queue empty
$s_{0,1}$–$s_{3,1}$	One customer in queue
$s_{0,2}$–$s_{3,2}$	Two customers in queue
$\vdots$	$\vdots$
$s_{0,j}$–$s_{3,j}$	j customers in queue
$\vdots$	$\vdots$

The state vector **s** can be grouped into B subvectors

$$\mathbf{s} = \begin{bmatrix} \mathbf{s}_0 & \mathbf{s}_1 & \cdots & \mathbf{s}_B \end{bmatrix}^t \tag{7.123}$$

where the subvector $\mathbf{s}_j$ correspond to the jth column in Fig. 7.8 and is given by

$$\mathbf{s}_j = \begin{bmatrix} s_{0,j} & s_{1,j} & \cdots & s_{n-1,j} \end{bmatrix}^t \tag{7.124}$$

which corresponds to the case when there are j customers in the queue. The state transition matrix **P** corresponding to the state vector **s** will be of dimension $n(B+1) \times n(B+1)$. We describe the transition matrix **P** as a composite matrix of size $(B+1) \times (B+1)$ as follows:

$$
\mathbf{P} = \begin{bmatrix}
\mathbf{A} & \mathbf{C} & \mathbf{0} & \cdots & \mathbf{0} & \mathbf{0} & \mathbf{0} \\
\mathbf{B} & \mathbf{D} & \mathbf{C} & \cdots & \mathbf{0} & \mathbf{0} & \mathbf{0} \\
\mathbf{0} & \mathbf{B} & \mathbf{D} & \cdots & \mathbf{0} & \mathbf{0} & \mathbf{0} \\
\vdots & \vdots & \vdots & \ddots & \vdots & \vdots & \vdots \\
\mathbf{0} & \mathbf{0} & \mathbf{0} & \cdots & \mathbf{D} & \mathbf{C} & \mathbf{0} \\
\mathbf{0} & \mathbf{0} & \mathbf{0} & \cdots & \mathbf{B} & \mathbf{D} & \mathbf{C} \\
\mathbf{0} & \mathbf{0} & \mathbf{0} & \cdots & \mathbf{0} & \mathbf{B} & \mathbf{E}
\end{bmatrix} \tag{7.125}
$$

where all the matrices **A**, **B**, **C**, **D**, and **E** are of dimension $n \times n$. For the case $n = 4$, we can write

$$
\mathbf{A} = \begin{bmatrix} 0 & 0 & 0 & c \\ 1 & 0 & 0 & 0 \\ 0 & 1 & 0 & 0 \\ 0 & 0 & 1 & 0 \end{bmatrix}, \quad
\mathbf{B} = \begin{bmatrix} 0 & 0 & 0 & d \\ 0 & 0 & 0 & 0 \\ 0 & 0 & 0 & 0 \\ 0 & 0 & 0 & 0 \end{bmatrix}, \quad
\mathbf{C} = \begin{bmatrix} 0 & 0 & 0 & 0 \\ c & 0 & 0 & 0 \\ 0 & c & 0 & 0 \\ 0 & 0 & c & 0 \end{bmatrix}
$$

$$
\mathbf{D} = \begin{bmatrix} 0 & 0 & 0 & c \\ d & 0 & 0 & 0 \\ 0 & d & 0 & 0 \\ 0 & 0 & d & 0 \end{bmatrix}, \quad
\mathbf{E} = \begin{bmatrix} 0 & 0 & 0 & 1 \\ d & 0 & 0 & 0 \\ 0 & d & 0 & 0 \\ 0 & 0 & d & 0 \end{bmatrix}
$$

Having found the state transition matrix, we are now able to find the steady-state value for the distribution vector of the $D/M/1/B$ queue. At steady state, the distribution vector **s** is derived from the two equations

$$
\mathbf{P}\,\mathbf{s} = \mathbf{s} \tag{7.126}
$$

$$
\mathbf{1}\,\mathbf{s} = 1 \tag{7.127}
$$

where **1** is a row vector whose components are all *1*s.

We can find the vector **s** by iterations as follows. We start by assuming a value for element $s_{0,0} = 1$. As a consequence, all the elements of the vector $\mathbf{s}_0$ can be found as follows

$$
s_{1,0} = s_{0,0} = 1 \tag{7.128}
$$

$$
s_{2,0} = s_{1,0} = 1 \tag{7.129}
$$

$$
s_{3,0} = s_{2,0} = 1 \tag{7.130}
$$

$$
\vdots
$$

Thus, we know that $\mathbf{s}_0$ is assumed to be a vector whose components are all *1*s. To find $\mathbf{s}_1$ we use the equation

$$
\mathbf{s}_0 = \mathbf{A}\mathbf{s}_0 + \mathbf{C}\mathbf{s}_1 \tag{7.131}
$$

or

$$\mathbf{s}_1 = \mathbf{C}^{-1}(\mathbf{I} - \mathbf{A})\mathbf{s}_0 \tag{7.132}$$

Since we have an initial assumed value for $\mathbf{s}_0$, we now know $\mathbf{s}_1$. In general, we can write the iterative expressions

$$\mathbf{s}_i = \mathbf{C}^{-1}(\mathbf{I} - \mathbf{A})\mathbf{s}_{i-1} \quad 1 \le i \le B \tag{7.133}$$

Having found all the vectors $\mathbf{s}_i$, we obtain the normalized distribution vector $\mathbf{s}'$ as

$$\mathbf{s}' = s \Bigg/ \sum_{i=0}^{n-1} \sum_{j=0}^{B} s_{i,j} \tag{7.134}$$

7.9.1 *Performance of the D/M/1/B Queue*

The average input traffic $N_a(\text{in})$ is needed to estimate the efficiency of the queue and queue delay. Since we get one packet every n time steps, $N_a(\text{in})$ is given by

$$N_a(\text{in}) = \frac{1}{n} \tag{7.135}$$

The throughput of the queue is given by

$$\begin{aligned} \text{Th} &= c s_{n-1,0} + c \sum_{i=0}^{n-1} \sum_{j=1}^{B} s_{i,j} \\ &= c \left(1 - \sum_{i=0}^{n-2} s_{i,0}\right) \end{aligned} \tag{7.136}$$

The efficiency of the $D/M/1/B$ queue is given by

$$\eta = \frac{\text{Th}}{N_a(\text{in})} = cn \left(1 - \sum_{i=0}^{n-2} s_{i,0}\right) \tag{7.137}$$

Packets are lost in the $D/M/1B$ queue when the queue is full and a packet arrives but does not leave. The average lost traffic is given by

$$N_a(\text{lost}) = d s_{n-1,B} \tag{7.138}$$

The packet loss probability L is given by

$$L = \frac{N_a(\text{lost})}{N_a(\text{in})} = dns_{n-1,B} \tag{7.139}$$

The average queue size is given by

$$Q_a = \sum_{i=0}^{n-1} \sum_{j=0}^{B} i \ s_{i,j} \tag{7.140}$$

7.10 The $M/D/1/B$ Queue

In the $M/D/1/B$ queue, packets arrive in a random fashion but leave the queue at a fixed rate. Let us assume that the time step is chosen such that a packet leaves at the nth time step. Assume also that a is the probability that it arrives at queue during one time step. $b = 1 - a$ has the usual meaning. We also assume that at most one packet arrives in the queue in one time step. Figure 7.9 shows the state transition diagram for such queue for the case when $n = 4$ and $B = 4$ also. The number of rows corresponds to the number of time steps between packet departures. The last row corresponds to the states when a packet leaves. The number of columns corresponds to the size of the queue such that each column corresponds to a particular state of queue occupancy. For example, the leftmost column corresponds to the case when the queue is empty. The rightmost column corresponds to the case when the queue

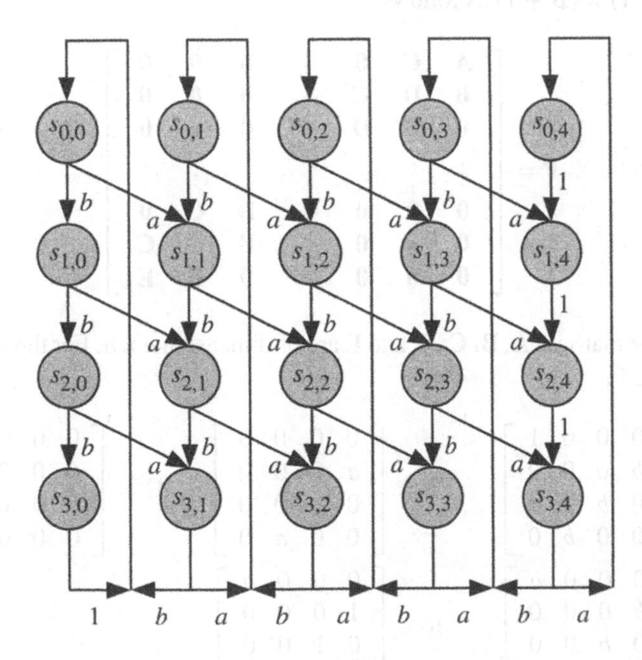

Fig. 7.9 State transition diagram for the discrete-time $M/D/1/B$ queue

Table 7.2 Relation of the queue states and the $M/D/1/B$ queue occupancy

States	Queue occupancy
$s_{0,0}$–$s_{3,0}$	Queue empty
$s_{0,1}$–$s_{3,1}$	One customer in queue
$s_{0,2}$–$s_{3,2}$	Two customers in queue
$\vdots$	$\vdots$
$s_{0,j}$–$s_{3,j}$	j customers in queue
$\vdots$	$\vdots$

is full. The state of occupancy of the queue is indicated in Table 7.2. In that case, we know that a packet leaves when the queue is in one of the bottom states $s_{3,1}$, $s_{3,2}$, $s_{3,3}$, or $s_{3,4}$.

The state vector **s** can be grouped into B sub vectors

$$\mathbf{s} = \begin{bmatrix} \mathbf{s}_0 & \mathbf{s}_1 & \cdots & \mathbf{s}_B \end{bmatrix}^t \tag{7.141}$$

where the subvector $\mathbf{s}_j$ corresponds to the jth column in Fig. 7.8 and is given by

$$\mathbf{s}_j = \begin{bmatrix} s_{0,j} & s_{1,j} & \cdots & s_{n-1,j} \end{bmatrix}^t \tag{7.142}$$

corresponds to the case when there are j customers in the queue.

The state transition matrix **P** corresponding to the state vector **s** will be of dimension $n(B+1) \times n(B+1)$. We describe the transition matrix **P** as a composite matrix of size $(B+1) \times (B+1)$ as follows:

$$\mathbf{P} = \begin{bmatrix} \mathbf{A} & \mathbf{C} & \mathbf{0} & \cdots & \mathbf{0} & \mathbf{0} & \mathbf{0} \\ \mathbf{B} & \mathbf{D} & \mathbf{C} & \cdots & \mathbf{0} & \mathbf{0} & \mathbf{0} \\ \mathbf{0} & \mathbf{B} & \mathbf{D} & \cdots & \mathbf{0} & \mathbf{0} & \mathbf{0} \\ \vdots & \vdots & \vdots & \ddots & \vdots & \vdots & \vdots \\ \mathbf{0} & \mathbf{0} & \mathbf{0} & \cdots & \mathbf{D} & \mathbf{C} & \mathbf{0} \\ \mathbf{0} & \mathbf{0} & \mathbf{0} & \cdots & \mathbf{B} & \mathbf{D} & \mathbf{C} \\ \mathbf{0} & \mathbf{0} & \mathbf{0} & \cdots & \mathbf{0} & \mathbf{B} & \mathbf{E} \end{bmatrix} \tag{7.143}$$

where all the matrices **A**, **B**, **C**, **D**, and **E** are of dimension $n \times n$. For the case $n = 4$, we can write

$$\mathbf{A} = \begin{bmatrix} 0 & 0 & 0 & 1 \\ b & 0 & 0 & 0 \\ 0 & b & 0 & 0 \\ 0 & 0 & b & 0 \end{bmatrix}, \quad \mathbf{B} = \begin{bmatrix} 0 & 0 & 0 & 0 \\ a & 0 & 0 & 0 \\ 0 & a & 0 & 0 \\ 0 & 0 & a & 0 \end{bmatrix}, \quad \mathbf{C} = \begin{bmatrix} 0 & 0 & 0 & b \\ 0 & 0 & 0 & 0 \\ 0 & 0 & 0 & 0 \\ 0 & 0 & 0 & 0 \end{bmatrix}$$

$$\mathbf{D} = \begin{bmatrix} 0 & 0 & 0 & a \\ b & 0 & 0 & 0 \\ 0 & b & 0 & 0 \\ 0 & 0 & b & 0 \end{bmatrix}, \quad \mathbf{E} = \begin{bmatrix} 0 & 0 & 0 & a \\ 1 & 0 & 0 & 0 \\ 0 & 1 & 0 & 0 \\ 0 & 0 & 1 & 0 \end{bmatrix}$$

Having found the state transition matrix, we are now able to find the steady-state value for the distribution vector of the $M/D/1/B$ queue. At steady state, the distribution vector $\mathbf{s}$ is derived from the two equations

$$\mathbf{P}\,\mathbf{s} = \mathbf{s} \tag{7.144}$$

$$\mathbf{1}\,\mathbf{s} = 1 \tag{7.145}$$

where $\mathbf{1}$ is a row vector whose components are all *1*s.

We can find the vector $\mathbf{s}$ by iterations as follows. We start by assuming a value for element $s_{0,0} = 1$. As a consequence, all the elements of the vector $\mathbf{s}_0$ can be found as follows:

$$s_{1,0} = b s_{0,0} = b \tag{7.146}$$

$$s_{2,0} = b s_{1,0} = b^2 \tag{7.147}$$

$$s_{3,0} = b s_{2,0} = b^3 \tag{7.148}$$

$$\vdots$$

Thus we know $\mathbf{s}_0$. To find $\mathbf{s}_1$, we use the equation

$$\mathbf{s}_0 = \mathbf{A}\mathbf{s}_0 + \mathbf{C}\mathbf{s}_1 \tag{7.149}$$

or

$$\mathbf{s}_1 = \mathbf{C}^{-1}(\mathbf{I} - \mathbf{A})\mathbf{s}_0 \tag{7.150}$$

Since we have an initial assumed value for $\mathbf{s}_0$, we now know $\mathbf{s}_1$. In general, we can write the iterative expressions

$$\mathbf{s}_i = \mathbf{C}^{-1}(\mathbf{I} - \mathbf{A})\mathbf{s}_{i-1} \quad 1 \le i \le B \tag{7.151}$$

Having found all the vectors $\mathbf{s}_i$, we obtain the normalized distribution vector $\mathbf{s}'$ as

$$\mathbf{s}' = s \Bigg/ \sum_{i=0}^{n-1} \sum_{j=0}^{B} s_{i,j} \tag{7.152}$$

7.10.1 Performance of the $M/D/1/B$ Queue

The average input traffic $N_a(\text{in})$ is needed to estimate the efficiency of the queue and queue delay. $N_a(\text{in})$ is given by

$$N_a(\text{in}) = N\ a \tag{7.153}$$

The throughput of the queue is given by

$$\text{Th} = a \sum_{j=0}^{B} s_{n-1,j} \tag{7.154}$$

The efficiency of the $D/M/1/B$ queue is given by

$$\eta = \frac{\text{Th}}{N_a(\text{in})} = \frac{1}{N} \times \sum_{j=0}^{B} s_{n-1,j} \tag{7.155}$$

Packets are lost in the $M/D/1/B$ queue when the queue is full and a packet arrives but does not leave. The average lost traffic is given by

$$N_a(\text{lost}) = a \sum_{i=0}^{n-2} s_{i,B} \tag{7.156}$$

The packet loss probability L is given by

$$L = \frac{N_a(\text{lost})}{N_a(\text{in})} = \frac{1}{N} \sum_{i=0}^{n-2} s_{i,B} \tag{7.157}$$

The average queue size is given by

$$Q_a = \sum_{i=0}^{n-1} \sum_{j=0}^{B} i \ s_{i,j} \tag{7.158}$$

7.11 Systems of Communicating Markov Chains

In the previous sections, we investigated the behavior of single Markov chains or single queues. More often than not, communication systems are composed of interconnected or interdependent Markov chains or queues. As an example, let us consider Figure 7.10. This system represents an entity that is connected to a bus or a communication channel in general. This could be the Ethernet network interface

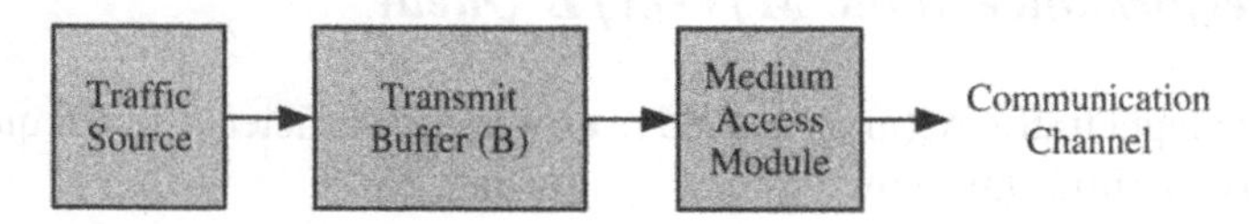

Fig. 7.10 A system of several Markov chains or queues

card (NIC) on a computer connected to a local-area network. The system to be analyzed consists of

1. Traffic source
2. Packet or transmit buffer
3. Medium access module

Each one of these components can be modeled individually using Markov chain analysis. For example, the traffic source can be modeled using the techniques in Chapter 11. The transmit buffer is of course modeled using any of the queuing models discussed in this chapter. The medium access module could be modeled using the techniques discussed in Chapter 10.

Let us assume that the Markov models for the three components are characterized by s_1 states for the traffic source, s_2 states for the queue, and s_3 states for the medium access module. We could develop a unified Markov model for our system and this would probably require $s_1 \times s_2 \times s_3$ states. This is the *state explosion problem* associated with most Markov chain models. If we, on the other hand, treat the problem as a system of dependent Markov chains, then we are in effect dealing with the individual components and the total number of states is $s_1 + s_2 + s_3$. Of course, we cannot solve each system separately since the state of each component depends on the other components communicating with it.

We can generalize the problem by considering a communicating Markov chain system composed of n modules. Module i is characterized by four quantities:

1. State transition matrix $\mathbf{P}_i$
2. Steady-state distribution vector $\mathbf{s}_i$
3. Probabilities x_i corresponding to the module inputs
4. Probabilities y_i corresponding to the module outputs

The parameters of each transition matrix $\mathbf{P}_i$ $(0 \leq i < n)$ depend, in general, on the values of $\mathbf{s}_j$, x_j, and y_j of all the other modules or systems. Furthermore, the module inputs and outputs x_i and y_i might also depend on the queue parameters such as $\mathbf{s}_j$. In this analysis, we assume that x_i and y_i are independent variables for simplicity. Figure 7.11 shows the system of n interdependent Markov chains.

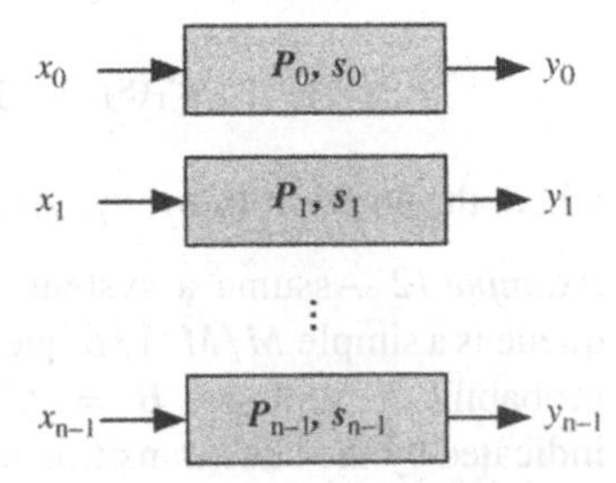

Fig. 7.11 Model of a network of interdependent Markov chains

If we attempt to model the system in Fig. 7.11 as a single Markov chain, the number of states we have to deal with would be given by

$$\text{Merged number of states} = \prod_{i=0}^{n-1} K_i$$

where K_i is the number of states of module i. If all the modules had the same number of states (K), then the total number of states would have been

$$\text{Merged number of states} = K^n$$

If we deal with the system as a system of communicating and dependent Markov chains, then the total number of states we have to deal with would be given by

$$\text{Number of states} = \sum_{i=0}^{n-1} K_i$$

where K_i is the number of states of module i. If all the modules had the same number of states (K), then the total number of states would have been

$$\text{Number of states} = nK$$

For typical systems, the quantity nK is much smaller than K^n. As an example, assume a modest system where each module has $K = 10$ states and we have n modules where $n = 5$. The number of states using the two approaches yields

$$\text{Number of states} = K^n = 10^5 = 100,000$$
$$\text{Number of states} = nK = 50$$

The situation is even worse and had $K = 100$ where our states would have been 10 billion using the former approach compared to 500 using the latter.

This is one reason that researchers and engineers alike attempt to use Petri nets to describe typical systems [6, 7]. However, Petri nets are basically graphical tools. Our approach in this section could be very loosely considered as approaching Petri nets through establishing the communicating channels and dependencies between the different modules of the system.

For each module in the system, we can write the steady-state equation

$$\mathbf{P}_i(\mathbf{s}_j, x_j, y_j) \times \mathbf{s}_i = \mathbf{s}_i \quad 0 \leq i, j < n \tag{7.159}$$

where the notation $\mathbf{P}_i(\mathbf{s}_j, x_j, y_j)$ indicates that $\mathbf{P}_i$ is a function of $\mathbf{s}_j$, x_j, and y_j.

Example 12 Assume a system of two interdependent queues such that the first queue is a simple $M/M/1/B$ queue with parameters: arrival probability a, departure probability c, and size $B = 2$. The second queue depends on the first queue as indicated by the state transition matrix

$$\mathbf{Y} = \begin{bmatrix} \alpha & 1 - x_0 \\ 1 - \alpha & x_0 \end{bmatrix}$$

where $\alpha < 0$ and x_0 is the probability that the first queue is empty. Explain how the steady-state distribution vectors $\mathbf{x}$ and $\mathbf{y}$ of the two queues could be obtained.

We have for the first queue

$$\mathbf{X} = \begin{bmatrix} 1 - ad & bc & 0 \\ ad & f & bc \\ 0 & ad & 1 - bc \end{bmatrix} \qquad \mathbf{x} = \begin{bmatrix} x_0 & x_1 & x_2 \end{bmatrix}^t$$

where $f = ac + bd$. Since the first queue is not dependent on the second queue, the steady-state distribution vector is found using Equation (7.17)

$$\mathbf{x} = \frac{1 - \rho}{1 - \rho^3} \times \begin{bmatrix} 1 & \rho & \rho^2 \end{bmatrix}^t$$

where $\rho = ad/bc$.

For the second queue, we have

$$\mathbf{Y} = \begin{bmatrix} \alpha & 1 - x_0 \\ 1 - \alpha & x_0 \end{bmatrix} \qquad \mathbf{y} = \begin{bmatrix} y_0 & y_1 \end{bmatrix}^t$$

Matrix $\mathbf{Y}$ is determined since we know the value of x_0. Thus, we can find the value of $\mathbf{y}$ using any of the techniques that we studied in Chapter 4. ■

7.11.1 A General Solution for Communicating Markov Chains

In the general case, a simple solution for the system is not possible due to the complexity of the transition matrices. The general steps we recommend to employ can be summarized as follows:

Step 0: Initialization of $\mathbf{s}_i$
Set initial nonzero values for all the distribution vectors $\mathbf{s}_i(0)$ for all $0 \le i < n$. Of course, each vector must be normalized in the sense that the sum of its components must always be 1. In general, at iteration k, the estimated value of $\mathbf{s}_i(k)$ is known. The notation $\mathbf{s}_i(k)$ indicates the value of $\mathbf{s}_i$ at iteration k.

Step 1: Estimating $\mathbf{P}_i$
The elements of the transition matrices $\mathbf{P}_i(k)$ at iteration k can now be estimated since $\mathbf{s}_i(k)$, x_i, and y_i are known. Initially, $k = 0$ to start our iterations and we are then able to calculate the initial $\mathbf{P}_i(0)$.

Step 2: Calculating $\mathbf{s}_i$
Knowing $\mathbf{P}_i(k)$ we can now calculate the values $\mathbf{s}_{i,c}(k)$, where the notation $\mathbf{s}_{i,c}(k)$ indicates the calculated value of $\mathbf{s}_i(k)$ which will most probably be different from the assumed $\mathbf{s}_i(k)$. We calculate $\mathbf{s}_{i,c}(k)$ using the equation

$$\mathbf{s}_{i,c}(k) = \mathbf{P}_i(k)\,\mathbf{s}_i(k) \tag{7.160}$$

Again, initially $k = 0$ at the start of iterations.
Step 3: Calculating the Estimation Error $\mathbf{e}_i(k)$
The calculated value $\mathbf{s}_{i,c}(k)$ will not be equal to the values $\mathbf{s}_i(k)$. The estimation error vector is calculated as

$$\mathbf{e}_i(k) = \mathbf{s}_{i,c}(k) - \mathbf{s}_i(k) \tag{7.161}$$

The magnitude of the estimation error for $\mathbf{s}_i(k)$ is given by

$$\epsilon_i(k) = \sqrt{\mathbf{e}_i^t(k)\,\mathbf{e}_i(k)} \tag{7.162}$$

Step 4: Updating Value of $\mathbf{s}_i(k)$
We are now able to update our guess of the state vectors and obtain new and better values $\mathbf{s}_i(k+1)$ according to the update equations

$$\mathbf{s}_i(k+1) = \mathbf{s}_i(k) + \alpha\mathbf{e}_i(k) \qquad 0 \le i < n \tag{7.163}$$

where $\alpha \ll 1$ is a small correction factor to ensure smooth convergence.

Step 5: Improving the Estimated Values of $\mathbf{s}_i(k)$
We repeat steps 1–4 with the new values $\mathbf{s}_i(k+1)$ until the total error measure ϵ_t is below an acceptable level

$$\epsilon_t = \sum_{i=0}^{n} \epsilon_i \le \gamma \tag{7.164}$$

where γ is the acceptable error threshold.

We are "confident" that convergence will take place since at each iteration the matrices $\mathbf{P}_i(k)$ are all column stochastic and their eigenvalues satisfy the inequality $\lambda \le 1$.

Problems

Throughput and Efficiency

7.1 Consider a switch that generates its own traffic, N_a(internal), in addition to the traffic arriving at its input, N_a(in), according to the discussion in Section 7.3. Define the throughput and the efficiency for this system in terms of N_a(in), N_a(internal), and N_a(out), such that η never exceeds unity.

*M/M/*1 *Queue*

7.2 Prove that (7.13) on page 229 is true when the $M/M/1$ queue is stable.

7.3 Consider an $M/M/1$ with a distribution vector ρ very close to unity such that

$$\rho = 1 - \epsilon$$

where $\epsilon \ll 1$. Find the equilibrium distribution vector.

7.4 Consider an $M/M/1$ queue with arrival probability $a = 0.5$ and departure probability $c = 0.6$.

(a) Construct the first six rows and columns of the transition matrix **P**.
(b) Find the values of the first ten components of the equilibrium distribution vector.
(c) Calculate the queue performance.

7.5 Repeat Problem 7.4 when the departure probability becomes almost equal to the arrival probability (e.g., $c = 0.55$).

7.6 Consider an $M/M/1$ queue with arrival probability $a = 0.1$ and departure probability $c = 0.5$.

(a) Construct the first six rows and columns of the transition matrix **P**.
(b) Find the values of the first ten components of the equilibrium distribution vector.
(c) Calculate the queue performance.

7.7 Repeat Problem 7.6 when the departure probability becomes almost equal to the arrival probability (e.g., $c = 0.11$).

7.8 In an $M/M/1$ queue, it was found out that the average queue size $Q_a = 5$ packets and the average waiting time is $W = 20$ time steps. Calculate the queue arrival and departure probabilities and find the first ten entries of the distribution vector.

7.9 In an $M/M/1$ queue, it was found out that the average queue size $Q_a = 2$ packets and the average waiting time is $W = 100$ time steps. Calculate the queue arrival and departure probabilities and find the first ten entries of the distribution vector.

7.10 Equation (7.7) describes the $M/M/1$ queue when a packet could be served in the same time step at which it arrives. Suppose that an arriving packet cannot be served until the next time step. What will be the expression for the state matrix? Compare your result to (7.7).

7.11 Derive the performance for the $M/M/1$ queue described in Problem 7.10.

7.12 In the $M/M/1$ queue in Problem 7.10, it was found that the average queue size is $Q_a = 2$ packets and the average waiting time is $W = 100$ time steps. Calculate the queue arrival and departure probabilities and find the first ten entries of the distribution vector.

$M/M/1/B$ *Queue*

7.13 Prove the average queue size formula for the $M/M/1/B$ queue given in (7.47) on page 236.

7.14 Consider an $M/M/1/B$ queue with arrival probability $a = 0.5$, departure probability $c = 0.6$, and the maximum queue size is $B = 4$.

(a) Construct the transition matrix **P**.
(b) Find the values of the components of the equilibrium distribution vector.
(c) Calculate the queue performance.
(d) Compare your results with those of the $M/M/1$ queue in Problem 7.4 having the same arrival and departure probabilities.

7.15 Repeat Problem 7.14 when the departure probability becomes almost equal to the arrival probability (e.g., $c = 0.55$). Compare your results with those of the $M/M/1$ queue in Problem 7.5 having the same arrival and departure probabilities.

7.16 Repeat Problem 7.14 when the departure probability actually exceeds the arrival probability (e.g., $c = 0.8$). Compare your results with those of the $M/M/1$ queue in Problem 7.5 having the same arrival and departure probabilities.

7.17 Consider an $M/M/1/B$ queue with arrival probability $a = 0.1$, departure probability $c = 0.5$, and maximum queue size $B = 5$.

(a) Construct the transition matrix **P**.
(b) Find the values of the equilibrium distribution vector.
(c) Calculate the queue performance.
(d) Compare your results with those of the $M/M/1$ queue in Problem 7.6 having the same arrival and departure probabilities.

7.18 Repeat Problem 7.17 when the departure probability becomes almost equal to the arrival probability (i.e., $c = 0.11$). Compare your results with those of the $M/M/1$ queue in Problem 7.7 having the same arrival and departure probabilities.

7.19 Equation (7.30), on page 234, describes the $M/M/1/B$ queue when a packet could be served in the same time step at which it arrives. Suppose that an arriving packet cannot be served until the next time step. What will be the expression for the state matrix? Compare your result to (7.30).

7.20 Derive the performance for the $M/M/1/B$ queue described in Problem 7.19.

7.21 Consider an $M/M/1/B$ queue with arrival probability $a = 0.4$, departure probability $c = 0.39$, and maximum queue size $B = 5$. The queue is not stable since the arrival probability is larger than the departure probability.

(a) Construct the transition matrix **P**.
(b) Find the values of the equilibrium distribution vector.
(c) Calculate the queue performance.

7.22 In the $M/M/1$ queue in Problem 7.20 it was found out that the average queue size $Q_a = 2$ packets and the average waiting time is $W = 100$ time steps. Calculate the queue arrival and departure probabilities and find the first ten entries of the distribution vector.

$M^m/M/1/B$ Queue

7.23 Write down the entries for the transition matrix of an $M^m/M/1/B$ queue in terms of the arrival and departure statistics a_k and c, respectively. Consider the case when $B = 4$.

7.24 Equation (7.63), on page 241, describes the $M^m/M/1/B$ queue when up to m packets could arrive at one time step. Prove that for the special case when $m = 1$, (7.63) becomes identical to (7.30).

7.25 Equation (7.63) describes the $M^m/M/1/B$ queue when a packet could be served in the same time step at which it arrives. Suppose that an arriving packet cannot be served until the next time step. What will be the expression for the state matrix? Compare your result to (7.63).

7.26 Consider an $M^m/M/1/B$ queue with arrival probability $a = 0.2$, $m = 2$, departure probability $c = 0.39$, and maximum queue size $B = 5$. The queue is not stable since the average number of arrivals is larger than the average number of departures.

(a) Construct the transition matrix **P**.
(b) Find the values of the equilibrium distribution vector.
(c) Calculate the queue performance.

7.27 In the analysis of the $M^m/M/1$ queue, it was assumed that when a packet arrives, it can be serviced at the same time step. Suppose the arriving packet is serviced the next time step. Draw the state transition diagram and at write down the corresponding transition matrix. Derive the main performance equations of such a queue.

$M/M^m/1/B$ Queue

7.28 Write down the entries for the transition matrix of an $M/M^m/1/B$ queue in terms of the arrival and departure statistics a and c_k, respectively. Consider the case when $B = 4$.

7.29 Equation (7.94), on page 248, describes the $M/M^m/1/B$ queue when up to m packets could leave at one time step. Prove that for the special case when $m = 1$, (7.94) becomes identical to (7.30).

7.30 Equation (7.94) describes the $M^m/M/1/B$ queue when a packet could be served in the same time step at which it arrives. Suppose that an arriving packet cannot be served until the next time step. What will be the expression for the state matrix? Compare your result to (7.94).

7.31 Consider an $M/M^m/1/B$ queue with arrival probability $a = 0.5$, departure probability $c = 0.2$, $m = 2$, and maximum queue size $B = 5$. The queue is not stable since the average number of arrivals is larger than the average number of departures.

(a) Construct the transition matrix **P**.
(b) Find the values of the equilibrium distribution vector.
(c) Calculate the queue performance.

7.32 In the analysis of the $M/M^m/1$ queue, it was assumed that when a packet arrives, it can be serviced at the same time step. Suppose the arriving packet is serviced the next time step. Draw the state transition diagram and at write down the corresponding transition matrix. Derive the main performance equations of such a queue.

References

1. F. Elguibaly, "Analysis and design of arbitration protocols", *IEEE Transactions on Computers*, vol. 38, No. 2, pp. 1168–1175, 1989.
2. D.G. Kendall, "Stochastic processes occurring in the theory of queues and their analysis by means of the embedded Markov chain", Annals of Mathematical Statistics, vol. 24, pp. 338–354, 1953.
3. M. E. Woodward, *Communication and Computer Networks*, IEEE Computer Society Press, Los Alamitos, CA, 1994.
4. I. Peterson, *Fatal Defect: Chasing Killer Computer Bugs*, Random House, New York, 1995.
5. W.J. Stewart, *Introduction to Numerical Solutions of Markov Chains*, Princeton University Press, Princeton, New Jersey, 1994.
6. J. Billington, M. Diaz and G. Rozenberg (eds.), *Application of Petri Nets to Communication Networks: Advances in Petri Nets*, Springer, New York, 1999.
7. K. Jensen, *Coloured Petri Nets: Basic Concepts, Analysis Methods and Practical Use*, Springer-Verlag, New York, 1997.

Chapter 8
Modeling Traffic Flow Control Protocols

8.1 Introduction

In this chapter, we illustrate how to develop queuing models for three protocols of traffic management:

1. The leaky bucket algorithm
2. The token bucket algorithm
3. Virtual scheduling algorithm (VS)

Modeling a protocol or a system is just like designing a digital system, or any system for that matter. There are many ways to model a protocol based on the assumptions that one makes. My motivation here is simplicity and not taking a guided tour through the maze of protocol modeling. My recommendation to the reader is to read the discussion on each protocol and then lay down the outline of a model that describes the protocol. The model or models developed here should then be compared with the one attempted by the reader.

8.2 The Leaky Bucket Algorithm

Computer traffic is seldom uniform and is characterized by periods of burstiness. Traffic bursts tax the network resources such as switch buffers and lead to network congestion and data loss. Because it is impossible for the network to accept only uniform traffic, mechanisms have been proposed to regulate or smooth out these bursts.

Thus traffic shaping, also known as traffic policing, aims at regulating the average rate of traffic flow even in the presence of occasional bursts [1]. This helps to manage the congestion problem at the switches.

When a user accesses the network, the important parameter to describe the traffic is the average data rate (λ_a). This is estimated by observing the number of packets (N) sent over a long time interval t and finding the average data rate as $\lambda_a = N/t$. This rate is compared to a maximum rate (λ_b) that is specified by the leaky bucket algorithm. As long as $\lambda_a < \lambda_b$, the user is classified as conforming, and data is

F. Gebali, *Analysis of Computer and Communication Networks*,
DOI: 10.1007/978-0-387-74437-7_8, © Springer Science+Business Media, LLC 2008

accepted. Users that obey this traffic contract are termed *conforming users*, while users that violate this contract are termed *violating* or *nonconforming users*. Traffic policing to ensure that each user is conforming is done at the points where users access the network (ingress points).

Leaky bucket is a *rate-based* algorithm for controlling the *maximum rate* of traffic arriving from a source. If the input data rate is less than the maximum rate specified by the algorithm, leaky bucket accepts the data. If the input data rate exceeds the maximum rate, leaky bucket passes the data at the maximum rate and excess data is buffered. If the buffer is full, then excess data is discarded. In our modeling of the leaky bucket algorithm, we are interested only in the state of the data buffer since the state of that buffer dictates the actions to be done on the incoming traffic. Our aim then is to simply model the buffer state so that we are able to predict the performance of the algorithm.

Figure 8.1(a) shows the leaky bucket buffer. The buffer accepts incoming data and releases the stored packets at a data rate that does not exceed λ_b. Depending on the data arrival rate λ_a and the state of occupancy of the packet buffer, the following scenarios could take place.

1. $\lambda_a < \lambda_b$: Data arrive at a rate (λ_a) lower than the maximum rate (λ_b) specified by the leaky bucket algorithm. In that case, data will be accepted as shown by the arrival of packets 1 and 2 in Fig. 8.1(b). The long time interarrival time between packets 1 and 2 indicate a low data arrival rate. We assumed in the figure that the maximum departure rate λ_b is equivalent to three time steps between packets. As soon as these packets arrive, they are passed through by the leaky bucket algorithm.

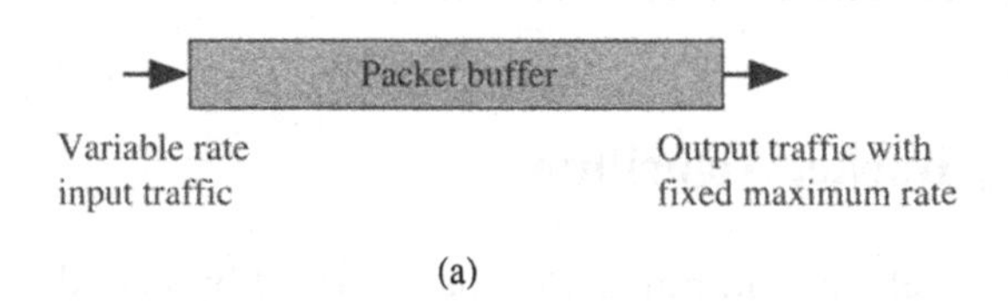

(a)

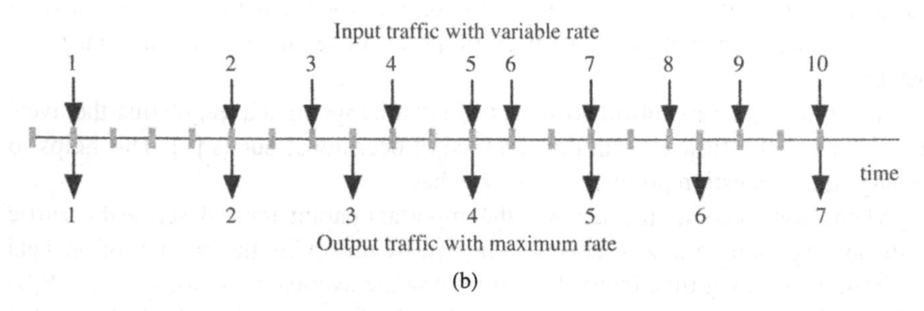

(b)

Fig. 8.1 The leaky bucket algorithm smoothes input variable rate traffic by buffering it and regulating the maximum buffer output traffic rate: (**a**) Packet buffer; (**b**) Packet arrival and departure

2. $\lambda_a > \lambda_b$: Data arrive at a rate higher than λ_b, and the data buffer is not full. In that case, data will be buffered so that it can be issued at the maximum rate λ_b. This is shown by the arrival of packets 3, 4, and 5 in Fig. 8.1(b). Note that packets 3, 4, and 5 arrive close together in time at the input indicating a high-input data rate. The interarrival time is 2 time steps, which indicates higher data rate than λ_b which is equivalent to two time steps. At the output, the spacing between these packets is equivalent to data transmitted at the rate λ_b.
3. $\lambda_a > \lambda_b$: Data arrive at a rate higher than λ_b, and the data buffer is full. In that case, data will be discarded or labeled as nonconforming. This is shown by, arrival of packets 6 and 7 in Fig. 8.1(b).

Figure 8.2 shows the variation of output data rate in relation to the input data rate. Grey areas indicate input data rate and solid black lines indicate maximum data rate λ_b. Output data rate does not exceed λ_b, and any excess input traffic is buffered or lost. The figure shows two occasions when the input data rate exceeds λ_b. When this situation happens, excess data are buffered and then released when the output data rate becomes low again.

8.2.1 Modeling the Leaky Bucket Algorithm

In this section, we perform Markov chain analysis of the leaky bucket algorithm. The states of the Markov chain represent the number of packets stored in the leaky bucket buffer.

Packets arrive at the input of the buffer at a rate λ_{in}, which varies with time because of the burstiness of the source. To model the source burstiness in a simple manner, we assume that the data source has the following parameters:

λ_a average data rate of source
σ source burst rate when it is nonconforming
λ_a and σ typically satisfy the relations

$$\lambda_a < \lambda_b \tag{8.1}$$

$$\sigma > \lambda_b \tag{8.2}$$

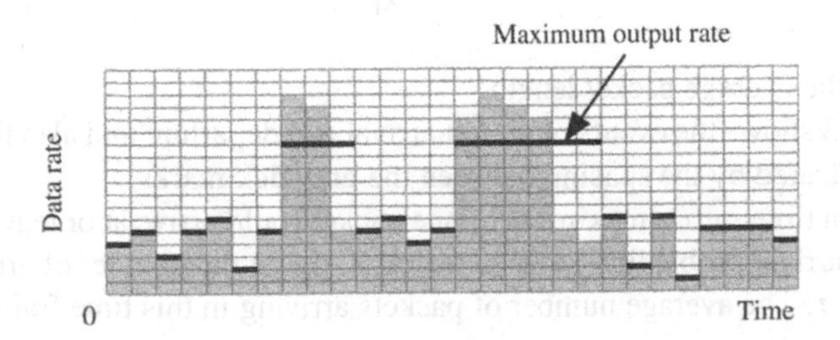

Fig. 8.2 Control of data rate by the leaky bucket algorithm. Data rate at the input is indicated by the grey areas, and data rate at the output is indicated by the *black lines*

where λ_b is the maximum data departure rate as determined by the leaky bucket algorithm.

On the other hand, packets leave the buffer with an output rate λ_{out} given by

$$\lambda_{out} = \begin{cases} \min(\lambda_{in}, \lambda_b) & \text{when packet buffer is empty} \\ \lambda_b & \text{when packet buffer is not empty} \end{cases} \tag{8.3}$$

Notice that the output data rate is governed by the state of the data buffer and not by the input data rate.

The leaky bucket algorithm can be modeled using two different types of queues depending on our choice of the time step. These two approaches are explained in the following two sections.

8.2.2 Single Arrival/Single Departure Model ($M/M/1/B$)

In this approach to modeling the leaky bucket algorithm, we take the time step equal to the inverse of the maximum data rate on the line.

$$T = \frac{1}{\lambda_l} \tag{8.4}$$

where the time step value is measured in units of seconds, and λ_l is the *maximum* input line rate (in units of packets/second) such that

$$\lambda_b < \lambda_l$$

The above inequality is true since the line is shared by many users. The time T is the time between packet arrivals at the maximum allowable rate on the input line. When λ_l is specified in units of bits per second (bps), T is obtained as

$$T = \frac{A}{\lambda_l} \tag{8.5}$$

where A is the average packet length.

Figure 8.3 shows the events of packet arrival and departure and also the time step value as indicated by the spacing between the grey tick marks.

At a given time step, a maximum of one packet could arrive at or leave the buffer. The packet arrival probability (a) is given by studying the number of arriving packets in a time t. The average number of packets arriving in this time frame is

$$N(\text{in}) = \lambda_a\, t \tag{8.6}$$

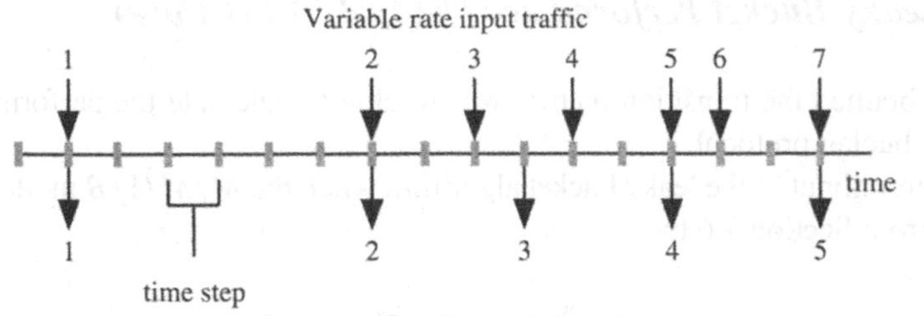

Fig. 8.3 The leaky bucket algorithm where the time step is chosen equal to the inverse of the maximum line rate

But during this time period, we have N time steps with $N = t/T$. The average number of arriving packets is estimated also using the binomial distribution as

$$N(\text{in}) = a\,N = a\,\frac{t}{T} \tag{8.7}$$

From the above two equations, we get

$$a = \lambda_a\,T = \frac{\lambda_a}{\lambda_l} \tag{8.8}$$

Of course, when the source is conforming, the departure probability is $c = 1$. Using a similar argument, the minimum packet departure probability (c) is give by

$$c = \frac{\lambda_{\text{out}}}{\lambda_l} \tag{8.9}$$

Therefore, we have a single-input, single-output data buffer whose size is assumed B. The queue, we are studying becomes $M/M/1/B$ queue, and the transition matrix is $(B+1) \times (B+1)$ and is given by

$$\mathbf{P} = \begin{bmatrix} f_0 & bc & 0 & 0 & \cdots & 0 & 0 & 0 & 0 \\ ad & f & bc & 0 & \cdots & 0 & 0 & 0 & 0 \\ 0 & ad & f & bc & \cdots & 0 & 0 & 0 & 0 \\ 0 & 0 & ad & f & \cdots & 0 & 0 & 0 & 0 \\ \vdots & \vdots & \vdots & \vdots & \ddots & \vdots & \vdots & \vdots & \vdots \\ 0 & 0 & 0 & 0 & \cdots & f & bc & 0 & 0 \\ 0 & 0 & 0 & 0 & \cdots & ad & f & bc & 0 \\ 0 & 0 & 0 & 0 & \cdots & 0 & ad & f & bc \\ 0 & 0 & 0 & 0 & \cdots & 0 & 0 & ad & 1-bc \end{bmatrix} \tag{8.10}$$

where $b = 1 - a$, $d = 1 - c$, $f_0 = 1 - ad$, and $f = 1 - ad - bc$.

8.2.3 Leaky Bucket Performance ($M/M/1/B$ Case)

Having obtained the transition matrix, we are able to calculate the performance of the leaky bucket protocol.

The throughput of the leaky bucket algorithm when the $M/M/1/B$ model is used is given from Section 7.6 by

$$\text{Th} = c\ (1 - b s_0) \tag{8.11}$$

The throughput is measured in units of packets/time step. The throughput in units of packets/second is expressed as

$$\text{Th}' = \frac{\text{Th}}{T} = \text{Th} \times \frac{\lambda_l}{A} \tag{8.12}$$

where we assumed λ_l was given in units of bits/second.

The average number of lost or tagged packets per time step is obtained using the results of the $M/M/1/B$ queue

$$N_a(\text{lost}) = s_B\ ad \tag{8.13}$$

The lost traffic is measured in units of packets/time step. And the number of packets lost per second is

$$\begin{aligned} N_a'(\text{lost}) &= \frac{N_a(\text{lost})}{T} \\ &= N_a(\text{lost}) \times \frac{\lambda_l}{A} \end{aligned} \tag{8.14}$$

where we assumed λ_l was given in units of bits/second. The packet loss probability is given by

$$L = \frac{N_a(\text{lost})}{N_a(\text{in})} = \frac{s_B\ ad}{\lambda_a} \tag{8.15}$$

The average queue size is given by

$$Q_a = \sum_{i=0}^{B} i\ s_i \tag{8.16}$$

where s_i is the probability that the data buffer has i packets.

Using Little's result, the average wait time in the buffer is

$$W = \frac{Q_a}{\text{Th}} \tag{8.17}$$

The wait time is measured in units of time steps. The wait time in units of seconds is given by

$$W' = \frac{Q_{\mathrm{a}}}{\mathrm{Th}'} \tag{8.18}$$

Example 1 A leaky bucket traffic shaper has the following parameters.

$$\begin{array}{ll} \lambda_{\mathrm{a}} = 1\,\mathrm{Mbps} & \sigma = 4\,\mathrm{Mbps} \\ \lambda_{\mathrm{b}} = 1.5\,\mathrm{Mbps} & \lambda_{\mathrm{l}} = 50\,\mathrm{Mbps} \\ A = 400\,\mathrm{bits} & B = 5\,\mathrm{packets} \end{array}$$

Derive the performance of this protocol using the $M/M/1/B$ modeling approach. The arrival probability is

$$a = \frac{\lambda_{\mathrm{a}}}{\lambda_{\mathrm{l}}} = 0.02$$

The minimum departure probability is

$$c = \frac{\lambda_{\mathrm{b}}}{\lambda_{\mathrm{l}}} = 0.03$$

We see that under the assumed traffic conditions the arrival probability is larger than the departure probability and we expect the packet buffer to be filled.

The transition matrix will be

$$\mathbf{P} = \begin{bmatrix} 0.9399 & 0.0281 & 0 & 0 & 0 & 0 \\ 0.0601 & 0.9117 & 0.0281 & 0 & 0 & 0 \\ 0 & 0.0601 & 0.9117 & 0.0281 & 0 & 0 \\ 0 & 0 & 0.0601 & 0.9117 & 0.0281 & 0 \\ 0 & 0 & 0 & 0.0601 & 0.9117 & 0.0281 \\ 0 & 0 & 0 & & 0.0601 & 0.9719 \end{bmatrix}$$

The equilibrium distribution vector is

$$\mathbf{s} = \begin{bmatrix} 0.0121 & 0.0258 & 0.0551 & 0.1177 & 0.2516 & 0.5377 \end{bmatrix}^t$$

Since $s_5 = 0.5377$, we conclude that 53.77% of the time the packet data buffer is full. The other performance parameters are

Th	=	0.0297	packets/time step
Th′	=	3.7076×10^3	packets/s
N_a(lost)	=	0.0323	packets/time step
N'_a(lost)	=	4.0424×10^3	packets/s
L	=	1.0432×10^{-8}	
Q_a	=	4.1843	packets
W	=	141.0713	times steps
W'	=	94.048	μs

We note that the leaky bucket is tagging or dropping 76.02% of the incoming packets. ∎

8.2.4 Multiple Arrival/Single Departure Model ($M^m/M/1/B$)

In this approach to modeling the leaky bucket algorithm, we take the time step equal to the inverse of the maximum data rate as dictated by the leaky bucket algorithm for that particular source.

$$T = \frac{1}{\lambda_b} \tag{8.19}$$

where the time step is measured in units of seconds and the leaky bucket rate λ_b is in units of packets/second. Usually, λ_b is specified in units of bits/second. In that case, T is obtained as

$$T = \frac{A}{\lambda_b} \tag{8.20}$$

where A is the average packet length.

Figure 8.4 shows the events of packet arrival and departure and also the time step value as indicated by the spacing between the successive output packets.

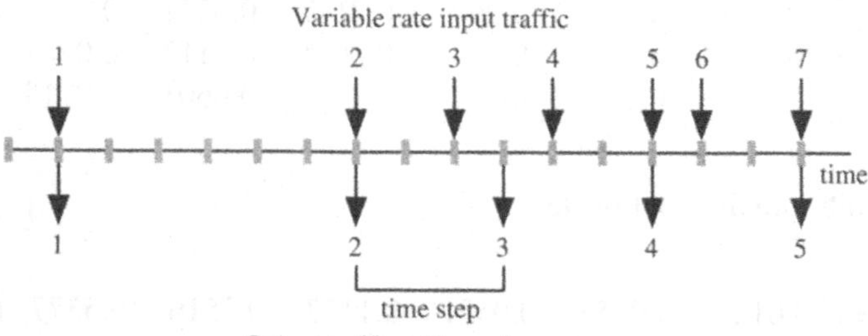

Fig. 8.4 Events of packet arrival and departure for the leaky bucket algorithm. The time step value is equal to the time between two adjacent output packets

At a given time step, one or more packets could arrive at the buffer and only one packet can leave if the buffer is not empty. Therefore, we have an $M^m/M/1/B$ queue to describe the state of the packet buffer.

The average data rate of the source as seen by the leaky bucket algorithm is given by λ_a. The maximum number of packets that could arrive at the queue input in one time step is determined by the maximum burst rate σ

$$N = \lceil \sigma \times T \rceil = \left\lceil \frac{\sigma}{\lambda_b} \right\rceil \tag{8.21}$$

where $\lceil x \rceil$ the is ceiling function which produces the smallest integer that is larger than or equal to x.

The probability of k packets arriving in one time step is given by

$$a_k = \binom{N}{k} a^k\, b^{N-k} \qquad k = 0, 1, 2, ..., N \tag{8.22}$$

where a is the probability that a packet arrives and $b = 1 - a$.

The average number of packets arriving in one time step is estimated as

$$N_a(\text{in}) = \lambda_a\, T \tag{8.23}$$

The average input packets is estimated also using the binomial distribution as

$$N_a(\text{in}) = a\, N = a\, \lceil \sigma\, T \rceil \tag{8.24}$$

From the above two equations, we get

$$a = \frac{\lambda_a\, T}{\lceil \sigma\, T \rceil} \leq \frac{\lambda_a}{\sigma} \tag{8.25}$$

Because of our choice for the time step size, the queue size can only decrease by one at most at any instant with probability $c = 1$. Assuming the packet buffer size is B, the transition matrix will be $(B + 1) \times (B + 1)$ and will be slightly modified from the form given by (7.63) on page 241:

$$\mathbf{P} = \begin{bmatrix} x & a_0 & 0 & 0 & 0 & \cdots & 0 \\ a_2 & a_1 & a_0 & 0 & 0 & \cdots & 0 \\ a_3 & a_2 & a_1 & a_0 & 0 & \cdots & 0 \\ a_4 & a_3 & a_2 & a_1 & a_0 & \cdots & 0 \\ \vdots & \vdots & \vdots & \vdots & \vdots & \ddots & 0 \\ a_B & a_{B-1} & a_{B-2} & a_{B-3} & a_{B-4} & \cdots & a_0 \\ z_B & z_{B-1} & z_{B-2} & z_{B-3} & z_{B-4} & \cdots & z_0 \end{bmatrix} \tag{8.26}$$

where $x = a_0 + a_1$ and

$$z_i = 1 - \sum_{j=0}^{i} a_j \tag{8.27}$$

8.2.5 Leaky Bucket Performance ($M^m/M/1/B$ Case)

Having obtained the transition matrix, we are able to calculate the performance of the leaky bucket protocol.

The throughput of the leaky bucket algorithm when the $M^m/M/1/B$ model is used is given from Section 7.7 on page 241 with a departure probability $c = 1$

$$\text{Th} = 1 - a_0\, s_0 \tag{8.28}$$

The throughput is measured in units of packets/time step and the throughput in units of packets/second is

$$\text{Th}' = \text{Th} \times \lambda_b \qquad \text{packets/s} \tag{8.29}$$

The lost or tagged packets are given by

$$\begin{aligned} N_a(\text{lost}) &= N_a(\text{in}) - N_a(\text{out}) \\ &= Na - (1 - a_0\, s_0) \end{aligned} \tag{8.30}$$

The lost traffic is measured in units of packets/time step. The average lost traffic per second is given by

$$\begin{aligned} N_a'(\text{lost}) &= \frac{N_a(\text{lost})}{T} \\ &= [Na - (1 - a_0\, s_0)]\, \lambda_b \end{aligned} \tag{8.31}$$

The packet loss probability l is the ratio of lost traffic relative to the input traffic

$$\begin{aligned} L &= \frac{N_a(\text{lost})}{N_a(\text{in})} \\ &= 1 - \frac{1 - a_0\, s_0}{Na} \end{aligned} \tag{8.32}$$

The average queue size is given by

$$Q_a = \sum_{i=0}^{B} i\, s_i \tag{8.33}$$

where s_i is the probability that the data buffer has i packets.

Using Little's result, the average wait time in the buffer is

$$W = \frac{Q_a}{\text{Th}} \tag{8.34}$$

The wait time is measured in units of time steps. The wait time in units of seconds is given by

$$W' = \frac{Q_a}{\text{Th}'} \tag{8.35}$$

Example 2 Repeat Example 1 using the $M^m/M/1/B$ modeling approach.

The average input data rate is

$$\lambda_a = 1 \times 0.3 + 4 \times 0.7 = 3.1 \qquad \text{Mbps}$$

N is found as

$$N = \left\lceil \frac{\sigma}{\lambda_b} \right\rceil = 3$$

The packet arrival probability is

$$a = \lambda_a \times T = 0.6889$$

The probability that k packets arrive in one time step is

$$a_k = \binom{3}{k} a^k b^{3-k} \qquad K = 0, 1, 2, 3$$

The transition matrix will be

$$\mathbf{P} = \begin{bmatrix} 0.2301 & 0.0301 & 0 & 0 & 0 & 0 \\ 0.4429 & 0.2000 & 0.0301 & 0 & 0 & 0 \\ 0.3269 & 0.4429 & 0.2000 & 0.0301 & 0 & 0 \\ 0 & 0.3269 & 0.4429 & 0.2000 & 0.0301 & 0 \\ 0 & 0 & 0.3269 & 0.4429 & 0.2000 & 0.0301 \\ 0 & 0 & 0 & 0.3269 & 0.7699 & 0.9699 \end{bmatrix}$$

The equilibrium distribution vector is

$$\mathbf{s} = \begin{bmatrix} 0 & 0 & 0.0001 & 0.0014 & 0.0370 & 0.9615 \end{bmatrix}^t$$

We see that 96% of the time the packet buffer is full. The other performance parameters are:

Th	= 1	packets/time step
Th′	$= 3.7500 \times 10^3$	packets/s
N_a(lost)	= 1.0667	packets/time step
L	= 0.5161	
Q_a	= 4.9600	packets
W	= 4.96	time steps
W'	= 1.3	ms

Comparing these performance figures with those obtained for the same system using the $M/M/1/B$ queue, we see that the packet throughput and delay in the system are approximately similar. One possible reason for the small variation is the use of the ceiling function in the $M^m/M/1/B$ analysis. ■

8.3 The Token Bucket Algorithm

Token bucket is a *credit-based* algorithm for controlling the *volume* of traffic arriving from a source. Tokens are issued at a constant rate and arriving packets can leave the system only if there are tokens available in the token buffer. Figure 8.5 shows the token bucket algorithm as it applies to a packet buffer. Figure 8.5(a) shows the input data buffer and token buffer. Input data arrive with a variable rate while the tokens arrive with a constant rate. There is mutual coupling or feedback between the two

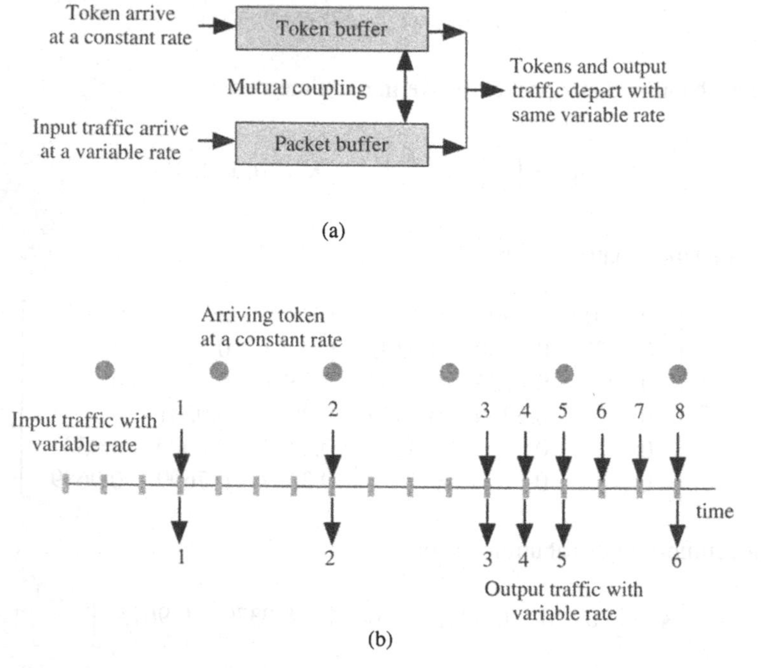

Fig. 8.5 The token bucket algorithm smoothes input variable rate traffic by buffering it and regulating the maximum buffer output traffic rate: (**a**) Packet buffer; (**b**) Packet arrival and departure

buffers such that both the tokens and the packets depart at the same rate which varies depending on the states of both queues.

Figure 8.5(b) shows the events of packet arrival and departure as well as token arrival (indicated by the grey circles). Depending on the data arrival rate and the state of occupancy of the token buffer, the following scenarios could take place.

1. No data arrive and tokens are stored in the buffer and count as credit for the source. This is similar to the situation before the arrival of packet 1 in Fig. 8.5(b).
2. Data arrive at a rate lower than the token issue rate. In that cases, data will be accepted and the excess tokens will be stored in the token buffer as credit since the source is conforming. This is similar to the arrival of packets 1, 2 and 3 in Fig. 8.5(b). More token arrive than data and token buffer starts filling up.
3. Data arrive at a rate higher than the token issue rate. In that cases, data will be accepted as long as there are tokens in the buffer. This is similar to the arrival of packets 4 and 5 in Fig. 8.5(b). Packets 4 and 5 go through although no tokens have arrived because the token buffer was not empty. When the token buffer becomes empty, data will be treated as in the following step.
4. Data arrive but no tokens are present. Only a portion of the data will be accepted at a rate equal to the token issue rate. The excess data can be discarded or tagged as nonconforming. This is similar to the arrival of packets 6, 7, and 8 in Fig. 8.5(b). Packets 6 and 7 arrive when the token buffer is empty and do not go through. The packets get stored in the packet buffer and when a token arrives packet 6 goes through.

Figure 8.6 shows the variation of output data rate in relation to the input data rate. Grey areas indicate input data rate and solid black lines indicate output data rate. The dashed line in the middle of the figure indicates the token arrival rate. We see that output data rate can temporarily exceed the token rate but only for a short time. The duration of this burst depends on the amount of tokens stored in the token buffer. Bigger token buffer size allows for longer bursts from the source. However, a bursty source will not allow the token buffer enough time to fill up.

On the other hand, the packet buffer allows for temporary storage of arriving packets when there are no tokens in the token buffer.

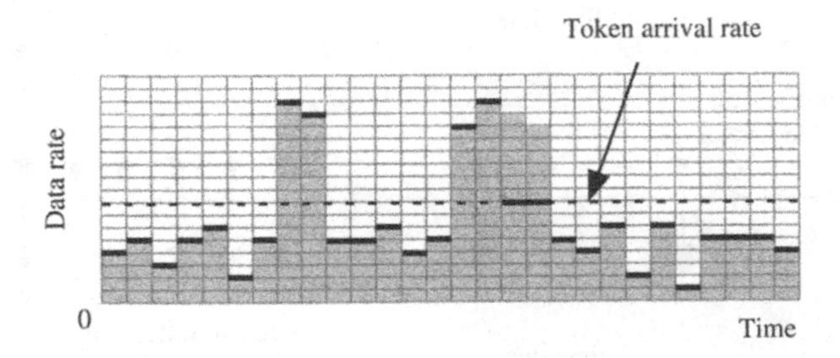

Fig. 8.6 Control of data rate by the token bucket algorithm. Data rate at the input is indicated by the grey areas, data rate at the output is indicated by the black lines and the dashed line is the token issue rate

8.3.1 Modeling the Token Bucket Algorithm

In this section, we perform Markov chain analysis of the token bucket algorithm. The states of the Markov chain represent the number of tokens stored in the token buffer or bucket and the number of packets stored in the packet buffer.

We cannot model the token buffer separately from the packet buffer since the departure from one buffer depends on the state of occupancy of the other buffer. In a sense, we have two mutually coupled queues as was shown in Fig. 8.5(a).

The token bucket algorithm can be modeled using two different types of queues depending on our choice of the time step. These two approaches are explained in the following two sections.

8.3.2 Single Arrival/Single Departures Model ($M/M/1/B$)

In this approach to modeling the token bucket algorithm, we take the time step equal to the inverse of the maximum data rate on the line

$$T = \frac{1}{\lambda_l} \tag{8.36}$$

where T is measured in units of seconds and λ_l is the *maximum* input line rate in units of packets/second. Usually, λ_l is specified in units of bits per second. In that case, T is obtained as

$$T = \frac{A}{\lambda_l} \tag{8.37}$$

where A is the average packet length.

Figure 8.7 shows the events of packet arrival and departure and also the time step value as indicated by the spacing between the grey tick marks.

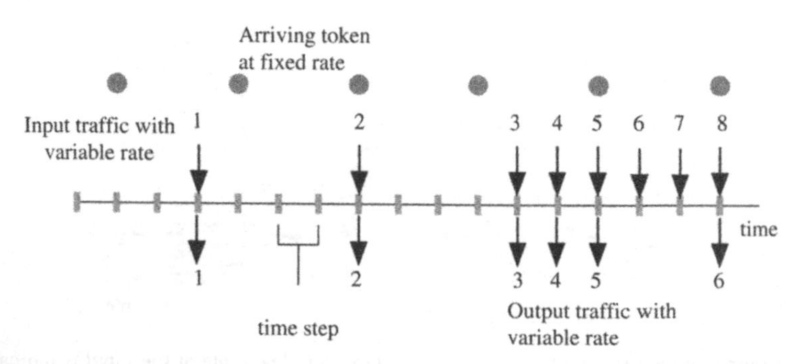

Fig. 8.7 The token bucket algorithm where the time step is chosen equal to the inverse of the maximum line rate

Packets arrive at the input of the buffer at a rate λ_a which varies with time because of the burstiness of the source. To model the source burstiness in a simple manner, we assume the data source has the following parameters:

λ_a Source average data rate.
σ Source burst rate.
λ_a and σ typically satisfy the relations

$$\lambda_a < \lambda_t \tag{8.38}$$

$$\sigma > \lambda_t \tag{8.39}$$

where λ_t is the token arrival rate.

The overall average data rate of the source as seen by the token bucket algorithm is given by λ_a. At a given time step a maximum of one token could arrive at or leave the token buffer. Thus the packet buffer is described by an $M/M/1/B$ queue. The token arrival probability (a) is given by

$$a = \frac{\lambda_t}{\lambda_l} \tag{8.40}$$

We also define $b = 1 - a$ as the probability that a token does not arrive.

At a given time step, a maximum of one packet could arrive or leave the packet buffer. Since a token leaves the token buffer each time a packet arrives, the token departure probability c is given by

$$c = \frac{\lambda_a}{\lambda_l} \tag{8.41}$$

We also define $d = 1 - c$ as the probability that a token does not leave the token buffer.

The state of occupancy of the *token buffer* depends on the statistics of token and packet arrivals as follows

1. The token buffer will stay at the same state with probability $ac + bd$, i.e., when a token arrives and a packet arrives or when no token arrives and no packet arrives too.
2. The token buffer will increase in size by one with probability ad; i.e., if a token arrives but no packets arrive.
3. The token buffer will decrease in size by one with probability bc; i.e., if no token arrives but one packet arrives.

The state of occupancy of the *packet buffer* depends on the statistics of the token and packet arrivals as follows

1. The packet buffer will stay at the same state with probability $ac+bd$, i.e., when a token arrives and a packet arrives or when no token arrives and no packet arrives too.
2. The packet buffer will increase in size by one with probability bc; i.e., if no token arrives but one packet arrives.
3. The packet buffer will decrease in size by one with probability ad; i.e., if a token arrives but no packets arrive.

Based on the above discussion, we have two single-input, single-output buffers. The size of the token buffer is assumed B_t and the size of the packet buffer is assumed B_p.

Figure 8.8 shows the Markov chain transition diagram for the system comprising the token and packet buffers. The upper row represents the states of the token buffer and the lower row represents the states of the packet buffer. The transition probabilities are dictated by the token and packet arrival probabilities and the states of occupancy of the token and packet buffers. The figure shows a token buffer whose size is $B_t = 4$ and a packet buffer whose size is $B_p = 3$.

The numbering of the states is completely arbitrary and does not necessarily represent the number of tokens or packets in a buffer. The meaning of each state is shown in Table 8.1.

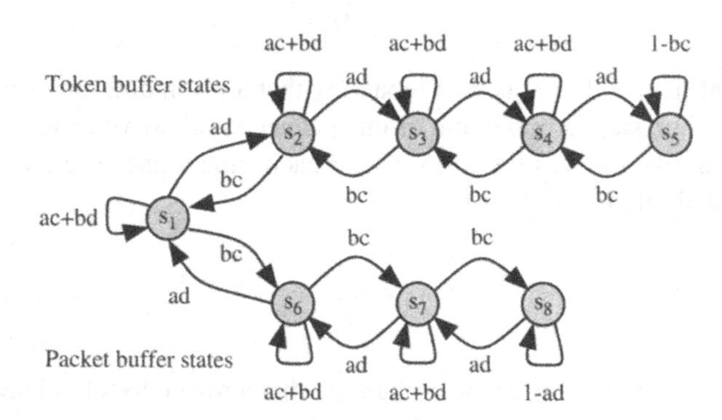

Fig. 8.8 Markov chain transition diagram for the system comprising the token and packet buffers. Token buffer size is $B_t = 4$ and packet buffer size is $B_p = 3$

Table 8.1 Defining the binomial coefficient

State	Token buffer occupancy	Packet buffer occupancy
s_1	0	0
s_2	1	0
s_3	2	0
s_4	3	0
s_5	4	0
s_6	0	1
s_7	0	2
s_8	0	3

The transition matrix of the composite system is a $\left(B_t + B_p + 1\right) \times \left(B_t + B_p + 1\right)$ tridiagonal matrix. For the case when $B_t = 4$ and $B_p = 3$, the matrix is given by

$$\mathbf{P} = \begin{bmatrix} f & bc & 0 & 0 & 0 & ad & 0 & 0 \\ ad & f & bc & 0 & 0 & 0 & 0 & 0 \\ 0 & ad & f & bc & 0 & 0 & 0 & 0 \\ 0 & 0 & ad & f & bc & 0 & 0 & 0 \\ 0 & 0 & 0 & ad & 1-bc & 0 & 0 & 0 \\ bc & 0 & 0 & 0 & 0 & f & ad & 0 \\ 0 & 0 & 0 & 0 & 0 & bc & f & ad \\ 0 & 0 & 0 & 0 & 0 & 0 & bc & 1-ad \end{bmatrix} \tag{8.42}$$

where $b = 1 - a$, $d = 1 - c$, and $f = ac + bd = 1 - ad - bc$.

8.3.3 Token Bucket Performance (M/M/1/B Case)

Having obtained the transition matrix, we are able to calculate the performance of the token bucket protocol.

The throughput of the token bucket algorithm is the average number of packets per time step that are produced without being tagged or lost. To find the throughput we must study all the states of the combined system. It is much easier to obtain the throughput using the traffic conservation principle after we find the lost traffic.

Packets are lost or tagged for future discard if they arrive when the packet buffer is full and no token arrives at that time step. The average number of lost or tagged packets per time step is

$$N_a(\text{lost}) = b\, c\; s_{B_t + B_p + 1} \tag{8.43}$$

The lost traffic is measured in units of packets/time step. And the number of packets lost in units of packets/second is

$$N_a'(\text{lost}) = \frac{N_a(\text{lost})}{T} = N_a(\text{lost})\, \lambda_l \tag{8.44}$$

The average number of packets arriving per time step is given by

$$N_a(\text{in}) = c \tag{8.45}$$

The packet loss probability is

$$\begin{aligned} L &= \frac{N_a(\text{lost})}{N_a(\text{in})} \\ &= \frac{b\, c\; s_{B_t + B_p + 1}}{c} \\ &= b\; s_{B_t + B_p + 1} \end{aligned} \tag{8.46}$$

The throughput of the token bucket algorithm is obtained using the traffic conservation principle

$$\begin{aligned} \mathrm{Th} &= N_\mathrm{a}(\mathrm{in}) - N_\mathrm{a}(\mathrm{lost}) \\ &= c - b\,c\ s_{B_\mathrm{t}+B_\mathrm{p}+1} \\ &= c\left(1 - b\, s_{B_\mathrm{t}+B_\mathrm{p}+1}\right) \end{aligned} \tag{8.47}$$

The throughput is measured in units of packets/time step. The throughput in units of packets/second is expressed as

$$\mathrm{Th}' = \frac{\mathrm{Th}}{T} = \mathrm{Th} \times \frac{\lambda_\mathrm{l}}{A} \tag{8.48}$$

where we assumed λ_l was given in units of bits/second.

The packet acceptance probability p_a or η of the token bucket algorithm is

$$\begin{aligned} p_\mathrm{a} &= \frac{\mathrm{Th}}{N_\mathrm{a}(\mathrm{in})} \\ &= 1 - L \\ &= 1 - b\ s_{B_\mathrm{t}+B_\mathrm{p}+1} \end{aligned} \tag{8.49}$$

We remind the reader that packet acceptance probability is just another name for the efficiency of the token bucket algorithm. It merely indicates the percentage of packets that make it through that traffic regulator without getting lost or tagged.

The average queue size for the tokens is given by

$$Q_\mathrm{t} = \sum_{i=1}^{B_\mathrm{t}} i\ s_{i+1} \tag{8.50}$$

Notice the range of values of the state index in the above equation.

Similarly, the average queue size for the packets is given by

$$Q_\mathrm{p} = \sum_{i=1}^{B_\mathrm{p}} i\ s_{i+B_\mathrm{t}+1} \tag{8.51}$$

Notice the range of values of the state index in the above equation.

Using Little's result, the average wait time or delay for the packets in the packet buffer is

$$W = \frac{Q_\mathrm{p}}{\mathrm{Th}} \tag{8.52}$$

The wait time is measured in units of time steps. The wait time in seconds is given by

$$W' = \frac{Q_p}{\mathrm{Th}'} \tag{8.53}$$

Example 3 Find the performance of a token bucket traffic shaper that has the following parameters, where A is the average packet length.

$$\begin{array}{ll} \lambda_a = 10\,\text{Mbps} & \sigma = 50\,\text{Mbps} \\ \lambda_t = 15\,\text{Mbps} & \lambda_l = 100\,\text{Mbps} \\ A = 400\,\text{bits} & B_t = 2\text{ packets} \\ B_p = 3\text{ packets} & \end{array}$$

The token arrival probability is

$$a = \frac{\lambda_t}{\lambda_l} = 0.15$$

The average data rate at the input is

$$\lambda_a = p\,\lambda_a + (1-p)\,\sigma = 26 \qquad \text{Mbps}$$

The token departure probability is

$$c = \frac{\lambda_a}{\lambda_l} = 0.26$$

The transition matrix will be

$$\mathbf{P} = \begin{bmatrix} 0.6680 & 0.2210 & 0 & 0.1110 & 0 & 0 \\ 0.1110 & 0.6680 & 0.2210 & 0 & 0 & 0 \\ 0 & 0.1110 & 0.7790 & 0 & 0 & 0 \\ 0.2210 & 0 & 0 & 0.6680 & 0.1110 & 0 \\ 0 & 0 & 0 & 0.2210 & 0.6680 & 0.1110 \\ 0 & 0 & 0 & 0 & 0.2210 & 0.8890 \end{bmatrix}$$

The equilibrium distribution vector is

$$\mathbf{s} = \begin{bmatrix} 0.0641 & 0.0322 & 0.0162 & 0.1276 & 0.2541 & 0.5059 \end{bmatrix}^t$$

We see that 50.59% of the time the packet buffer is full indicating that the source is misbehaving. The other performance parameters are

$$\begin{aligned} N_a(\text{lost}) &= 0.1118 && \text{packets/time step} \\ N_a'(\text{lost}) &= 2.7949 \times 10^4 && \text{packets/s} \end{aligned}$$

L	=	0.7453	
Th	=	0.1482	packets/time step
Th′	=	3.7051×10^4	packets/s
p_a	=	0.57	
Q_a	=	2.1533	packets
W	=	14.5294	times steps
W'	=	5.8118×10^{-5}	s

■

Example 4 Investigate the effect of doubling the token buffer or the packet buffer on the performance of the token bucket algorithm in the above example.

Doubling the token buffer to $B_t = 4$ or doubling the packet buffer result in the following parameters:

Parameter	$B_t = 2, B_p = 3$	$B_t = 4, B_p = 3$	$B_t = 2, B_p = 6$
N_a(lost)	0.1118	0.1104	0.1102
L	0.4300	0.4248	0.4239
Th	0.1482	0.1496	0.1498
p_a	0.57	0.5752	0.5761
Q_a	2.1533	2.1274	5.0173
W	14.5294	14.2250	33.4988

We note that increasing the buffer size improves the system performance. However, doubling the packet buffer size doubles the delay without too much improvement in throughput compared to doubling the token buffer size. ■

8.3.4 Multiple Arrivals/Single Departures Model ($M^m/M/1/B$)

In this approach to modeling the token bucket algorithm, we take the time step equal to the inverse of the fixed token arrival rate.

$$T = \frac{1}{\lambda_t} \tag{8.54}$$

where T is measured in units of seconds and λ_t is the token arrival rate in units of packets/s. Usually, λ_l is specified in units of bits per second. In that case, T is obtained as

$$T = \frac{A}{\lambda_t} \tag{8.55}$$

where A is the average packet length.

Figure 8.9 shows the events of packet arrival and departure and also the time step value as indicated by the spacing between the successive tokens.

Thus at a given time step only one token arrives at the buffer and one or more tokens can leave if the buffer is not empty and data arrives. The token arrival probability per time step is given by

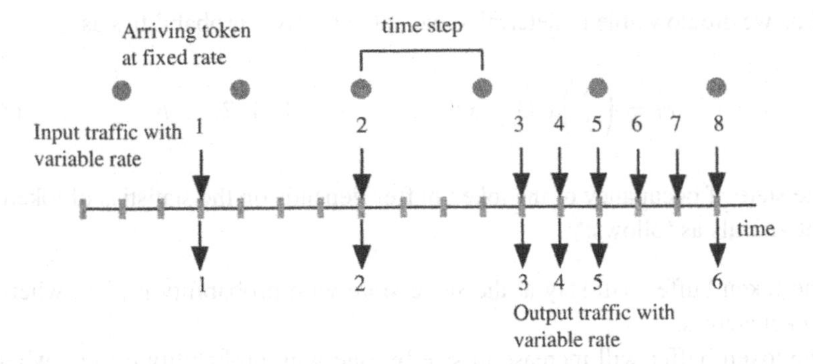

Fig. 8.9 Events of packet arrival and departure for the token bucket algorithm. The time step value is equal to the time between two adjacent arriving tokens

$$a = 1 \tag{8.56}$$

The average number of arriving packets per time step is given by

$$N_a(\text{in}) = \frac{\lambda_a}{\lambda_t} \tag{8.57}$$

where the average input data rate (λ_a) is given as before by

$$\lambda_a = p\lambda_a + (1 - p)\sigma \tag{8.58}$$

where p is the probability that the source is producing data at the rate λ_a and $1 - p$ is the probability that the source is producing data at the burst rate σ. λ_a could be smaller or larger than λ_t depending on the probability p.

The maximum number of packets (N) that could arrive at the queue input as determined by the burst rate σ

$$N = \left\lceil \frac{\sigma}{\lambda_t} \right\rceil \tag{8.59}$$

with $\lceil x \rceil$ is the smallest integer that is larger than or equal to x. The average number of packets that arrive per time step is given from (8.57) and (8.59) according to the binomial distribution as

$$N\, x = N_a(\text{in}) \tag{8.60}$$

where x is the probability of a packet arriving. The above equation gives x as

$$x = \frac{N_a(\text{in})}{N} \tag{8.61}$$

Thus we are now able to determine the packet arrival probabilities as

$$c_i = \binom{N}{i} x^i (1-x)^{L-i} \qquad i = 0, 1, 2, ..., N \tag{8.62}$$

The state of occupancy of the token buffer depends on the statistics of token and packet arrivals as follows:

1. The token buffer will stay at the same state with probability c_1, i.e., when one packet arrives.
2. The token buffer will increase in size by one with probability c_0; i.e., when no packets arrive.
3. The token buffer will decrease in size by one with probability c_2; i.e., when two packets arrive.
4. The token buffer will decrease in size by i with probability c_{i+1}; i.e., when $i+1$ packets arrive with $i < N$

The state of occupancy of the packet buffer depends on the statistics of token and packet arrivals as follows

1. The packet buffer will stay at the same state with probability c_1, i.e., when one packet arrives.
2. The packet buffer will decrease in size by one with probability c_0; i.e., when no packets arrive.
3. The packet buffer will increase in size by one with probability c_2; i.e., when two packets arrive.
4. The packet buffer will increase in size by i with probability c_{i+1}; i.e., when $i+1$ packets arrive with $i < N$

Based on the above discussion, we have two buffers to hold the tokens and the packets. The token buffer is single-input multiple output while the packet queue is multiple input, single-output. The size of the token buffer is assumed B_t and the size of the packet buffer is assumed B_p.

Figure 8.10 shows the Markov chain transition diagram for the system comprising the token and packet buffers. The upper row represents the states of the token buffer and the lower row represents the states of the packet buffer. The transition probabilities are dictated by the packet arrival probabilities and the states of occupancy of the token and packet buffers. The figure shows a token buffer whose size is $B_t = 4$ and a packet buffer whose size is $B_p = 3$. The maximum number of packets that could arrive in one time step is assumed $N = 3$. Figure 8.10 shows the c_0, c_1, and c_2 transitions. The c_3 transitions are only shown out of states s_4 and s_6 in order to reduce the clutter.

The meaning of each state is shown in Table 8.2.

The transition matrix of the composite system is a $(B_t + B_p + 1) \times (B_t + B_p + 1)$ tridiagonal matrix. For the case when $B_t = 4$, $B_p = 3$, and, $N = 3$, the matrix is given by

$$
\mathbf{P} = \begin{bmatrix}
c_1 & c_2 & c_3 & 0 & 0 & c_0 & 0 & 0 \\
c_0 & c_1 & c_2 & 0 & 0 & 0 & 0 & 0 \\
0 & c_0 & c_1 & c_2 & 0 & 0 & 0 & 0 \\
0 & 0 & c_0 & c_1 & c_2 & 0 & 0 & 0 \\
0 & 0 & 0 & c_0 & 1-c_2 & 0 & 0 & 0 \\
\\
c_2 & c_3 & 0 & 0 & 0 & c_1 & c_0 & 0 \\
c_3 & 0 & 0 & 0 & 0 & c_2 & c_1 & c_0 \\
0 & 0 & 0 & 0 & 0 & c_3 & c_2+c_3 & 1-c_0
\end{bmatrix} \tag{8.63}
$$

8.3.5 Token Bucket Performance (Multiple Arrival/Departure Case)

Having obtained the transition matrix, we are able to calculate the performance figures of the token bucket protocol.

The throughput of the token bucket algorithm is the average number of packets per time step that are produced without being tagged or lost. To find the throughput

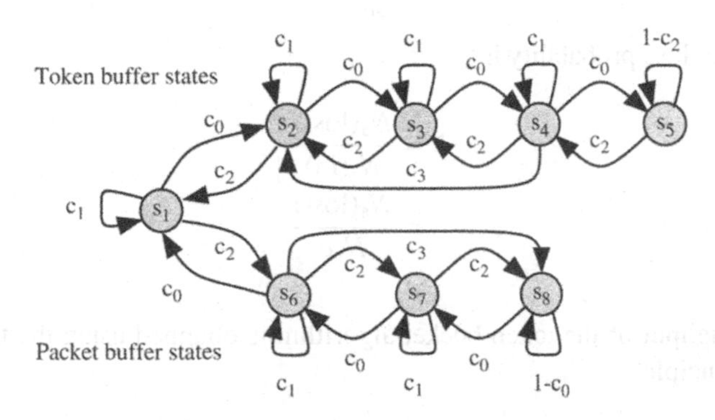

Fig. 8.10 Markov chain transition diagram for the system comprising the token and packet buffers. Token buffer size is $B_t = 4$, packet buffer size is $B_p = 3$, and $N = 3$. The c_3 transitions are only shown out of states s_4 and s_6 in order to reduce the clutter

Table 8.2 Defining the binomial coefficient

State	Token buffery occupancy	Packet buffer occupancy
s_1	0	0
s_2	1	0
s_3	2	0
s_4	3	0
s_5	4	0
s_6	0	1
s_7	0	2
s_8	0	3

we must study all the states of the combined system. It is much easier to obtain the throughput using the traffic conservation principle after we find the lost traffic.

Packets are lost or tagged for future discard if more than one packet arrives when the packet buffer cannot accommodate all of them.

When $N < B_p$, the average number of lost or tagged packets per time step is given by

$$N_a(\text{lost}) = \sum_{i=1}^{B_p-1} s_{B_t+B_p+2-i} \sum_{j=i+1}^{N} i\,(j-1)c_j \tag{8.64}$$

The lost traffic is measured in units of packets/time step. And the number of packets lost per second is

$$N'_a(\text{lost}) = \frac{N_a(\text{lost})}{T} = N_a(\text{lost})\,\lambda_l \tag{8.65}$$

The average number of packets arriving per time step is given by

$$N_a(\text{in}) = \sum_{i=0}^{N} i\,c_i = N\,x \tag{8.66}$$

The packet loss probability is

$$\begin{aligned} L &= \frac{N_a(\text{lost})}{N_a(\text{in})} \\ &= \frac{N_a(\text{lost})}{N\,c} \end{aligned} \tag{8.67}$$

The throughput of the token bucket algorithm is obtained using the traffic conservation principle

$$\begin{aligned} \text{Th} &= N_a(\text{in}) - N_a(\text{lost}) \\ &= N\,c - N_a(\text{lost}) \end{aligned} \tag{8.68}$$

The throughput is measured in units of packets/time step. The throughput in units of packets/second is expressed as

$$\text{Th}' = \frac{\text{Th}}{T} = \text{Th} \times \lambda_l \tag{8.69}$$

The packet acceptance probability of the token bucket algorithm is

$$\begin{aligned} p_a &= \frac{\text{Th}}{N_a(\text{in})} \\ &= 1 - \frac{N_a(\text{lost})}{N\,c} \end{aligned} \tag{8.70}$$

We remind the reader that packet acceptance probability is just another name for the efficiency of the token bucket algorithm. It merely indicates the percentage of packets that make it through that traffic regulator without getting lost or tagged.

The average queue size for the tokens is given by

$$Q_t = \sum_{i=1}^{B_t} i \; s_{i+1} \tag{8.71}$$

Notice the value of the state index in the above equation.

The average queue size for the packets is given by

$$Q_p = \sum_{i=1}^{B_p} i \; s_{i+B_t+1} \tag{8.72}$$

Notice the value of the state index in the above equation.

Using Little's result, the average wait time or delay for the packets in the packet buffer is

$$W = \frac{Q_p}{\text{Th}} \tag{8.73}$$

The wait time is measured in units of time steps. The wait time in seconds is given by

$$W' = \frac{Q_p}{\text{Th}'} \tag{8.74}$$

Example 5 Repeat Example 3 using the multiple arrival/departure modeling approach.

The maximum number of packets that could arrive in one time step m is found as

$$N = \left\lceil \frac{\sigma}{\lambda_{\text{out}}} \right\rceil = 4$$

The average input data rate is

$$\lambda_a = 10 \times 0.6 + 50 \times 0.4 = 26 \qquad \text{Mbps}$$

The packet arrival probability is

$$x = \frac{26}{100} = 0.4333$$

The probability that k packets arrive in one time step is

$$c_k = \binom{4}{k} a^k b^{4-k} \qquad k = 0, 1, 2$$

The transition matrix will be 6 × 6 and is given by

$$\mathbf{P} = \begin{bmatrix} 0.3154 & 0.3618 & 0.1844 & 0.1031 & 0 & 0 \\ 0.1031 & 0.3154 & 0.3618 & 0 & 0 & 0 \\ 0 & 0.1031 & 0.4184 & 0 & 0 & 0 \\ 0.3618 & 0.1844 & 0.0353 & 0.3154 & 0.1031 & 0 \\ 0.1844 & 0.0353 & 0 & 0.3618 & 0.3154 & 0.1031 \\ 0.0353 & 0 & 0 & 0.2197 & 0.5815 & 0.8965 \end{bmatrix}$$

The equilibrium distribution vector is

$$\mathbf{s} = \begin{bmatrix} 0.0039 & 0.0006 & 0.0001 & 0.0231 & 0.1388 & 0.8334 \end{bmatrix}^t$$

State s_2 indicates that there is a 0.0006 probability that there is one token in the token buffer. State s_6 indicates that 83.34% of the time the packet buffer is full indicating the source is misbehaving and the packet buffer is full.

The throughput of the queue is given by

$$\begin{aligned}
N_a(\text{lost}) &= 0.3279 && \text{packets/time step} \\
N'_a(\text{lost}) &= 1.2298 \times 10^4 && \text{packets/s} \\
L &= 0.1892 \\
\text{Th} &= 1.4054 && \text{packets/time step} \\
\text{Th}' &= 5.2702 \times 10^4 && \text{packets/s} \\
p_a &= 0.8108 \\
Q_a &= 2.8 && \text{packets} \\
W &= 1.9931 && \text{time steps} \\
W' &= 5.3149 \times 10^{-5} && \text{s}
\end{aligned}$$

■

Example 6 Investigate the effect of doubling the token buffer or the packet buffer on the performance of the token bucket algorithm in the above example.

Doubling the token buffer to $B_t = 4$ or doubling the packet buffer result in the following parameters:

Parameter	$B_t = 2, B_p = 3$	$B_t = 4, B_p = 3$	$B_t = 2, B_p = 6$
$N_a(\text{lost})$	0.3279	0.3279	0.3279
L	0.1892	0.1892	0.1892
Th	1.4054	1.4054	1.4054
p_a	0.8108	0.8108	0.8108
Q_a	2.8011	2.8011	5.8002
W	1.9931	1.9931	4.1271

Because the token buffer was nearly empty in the original system, doubling the size of the token buffer has no impact on the system performance as can be seen by comparing the second and third columns of the above table. Doubling the packet buffer size doubles the delay without noticeable improvement in throughput. ■

8.4 Virtual Scheduling (VS) Algorithm

The virtual scheduling (VS) algorithm manages the ATM network traffic by closely monitoring the cell arrival rate. When a cell arrives, the algorithm calculates the theoretical arrival time (TAT) of the next cell according to the formula

$$\mathrm{TAT} = \frac{1}{\lambda_a} \tag{8.75}$$

where λ_a is the expected average data rate (units of cells/second). TAT is measured by finding the difference between the arrival times of the headers of two consecutive ATM cells as explained in Fig. 8.11. This is *not* the time between the last bit of one cell and the first bit of the other.

If the cell arrival rate is in units of bits/second, then TAT is written as

$$\mathrm{TAT} = \frac{A}{\lambda_a} \tag{8.76}$$

where A is the size of an ATM cell which is 424 bits.

Assuming the time difference between the current cell and the next cell is t, then the cell is treated as conforming if t satisfies the following inequality

$$t \geq \mathrm{TAT} - \Delta \tag{8.77}$$

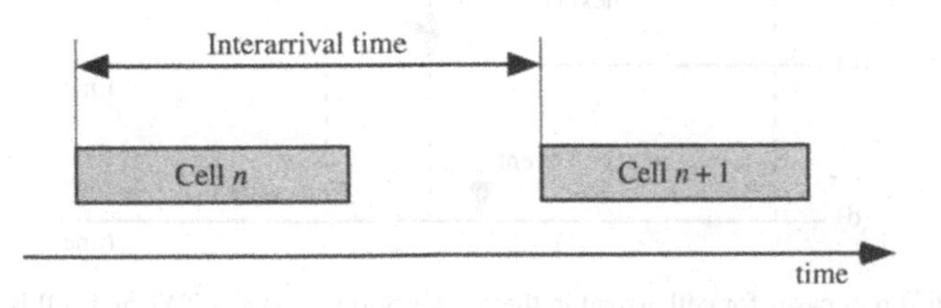

Fig. 8.11 Measuring the interarrival time between two consecutive ATM cells

where Δ is a small time value to allow for the slight variations in the data rate. The cell is treated as misbehaving, or nonconforming, when the cell arrival time satisfies the inequality

$$t < \text{TAT} - \Delta \tag{8.78}$$

The problem with the above two equations is that a source could keep sending data at a rate slightly higher than λ_a and still be conforming if every cell arrives within the bound of (8.77).

Figure 8.12 shows the different cases for cell arrival in VS. Figure 8.12(a) shows a conforming cell because the cell arrival time satisfies (8.77). Figure 8.12(b) shows another conforming cell because the arrival time still satisfies (8.77). Figure 8.12(c) also shows a conforming cell because the arrival time satisfies the equality part of (8.77). Figure 8.12(d) shows a nonconforming cell because the arrival time does not satisfy (8.77).

8.4.1 Modeling the VS Algorithm

In this section, we perform Markov chain analysis of the virtual scheduling algorithm. We make the following assumptions for our analysis of the virtual scheduling algorithm.

1. The states of the Markov chain represent how many times the arriving cells from a certain flow have been nonconforming. In other words, state s_i of the penalty queue indicates that the source has been nonconforming i times.

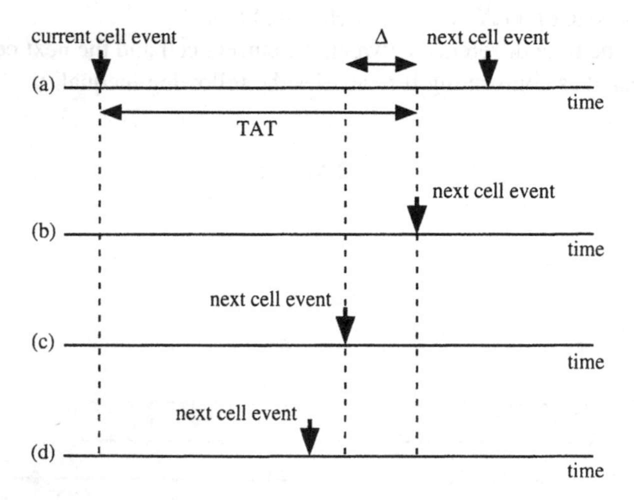

Fig. 8.12 Different cases for cell arrival in the VS algorithm. **(a)** $t > \text{TAT}$ and cell is conforming. **(b)** $t = \text{TAT}$ and cell is conforming. **(c)** $t = \text{TAT} - \Delta$ and cell is conforming. **(d)** $t < \text{TAT} - \Delta$ and cell is nonconforming

2. The number of states (B) of the queue will dictate the maximum burst size tolerated. Which is equal to the maximum number of penalties allowed for that source.
3. The queue changes states upon arrival of each cell.
4. Credit is given to the source each time it is conforming.
5. A penalty is given to the source each time it is nonconforming.
6. a is the probability that the arriving cell satisfies the inequality $t \geq \text{TAT}$. In that case, credit is issued to the source.
7. b is the probability that the arriving cell satisfies the following inequality

$$\text{TAT} - \Delta \leq t \leq \text{TAT}$$

   In that case, no credit or penalty is issued.
8. c is the probability that the arriving cell satisfies the inequality $t < \text{TAT}$. In that case, a penalty is issued to the source.

Of course, $c = 1 - a - b$ since the source cannot be in any other state.

Based on the above assumptions, we have a single arrival, single departure $M/M/1/B$ queue with $B + 1$ states. Figure 8.13 shows the state transitions for the VS queue.

It is interesting to note that the state transition diagram of the virtual scheduling algorithm in Fig. 8.13 is a special case of the state transition diagram for the token bucket algorithm in Fig. 8.8 when the token bucket buffer size is $B_t = 1$.

The corresponding transition matrix $\mathbf{P}$ will be $(B + 1) \times (B + 1)$ and will have the following entries for the case $B = 5$.

$$\mathbf{P} = \begin{bmatrix} 1-c & a & 0 & 0 & 0 & 0 \\ c & b & a & 0 & 0 & 0 \\ 0 & c & b & a & 0 & 0 \\ 0 & 0 & c & b & a & 0 \\ 0 & 0 & 0 & c & b & a \\ 0 & 0 & 0 & 0 & c & 1-a \end{bmatrix} \tag{8.79}$$

Notice that the transition matrix is tridiagonal because of the single arrival, single departure feature of the queue.

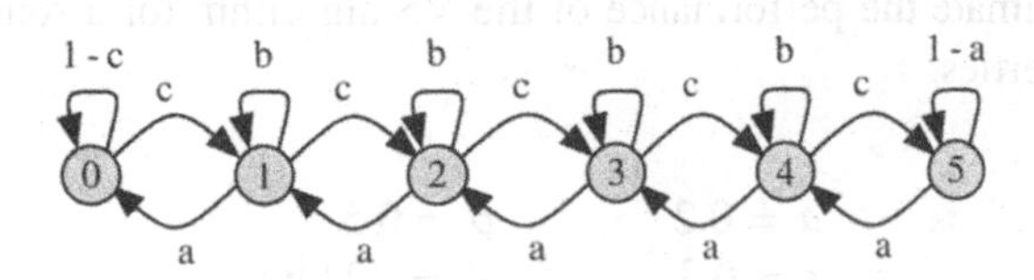

Fig. 8.13 State transition diagram for the VS queue for the case $B = 5$

8.4.2 VS Protocol Performance

Having obtained the transition matrix, we are able to calculate the performance of the VS protocol.

The probability that an arriving cell is marked for discard is considered equal to the cell loss probability. This happens when the source exceeds the maximum number of penalties allowed. Thus cell loss probability is given by

$$L = c\, s_B \tag{8.80}$$

where c is the probability that a cell arrived while the source is nonconforming and s_B is the probability that the penalty queue is full.

The average number of cells that are dropped per time step is given by

$$N_a(\text{lost}) = L\, \lambda_a \tag{8.81}$$

where λ_a is the average input data rate (units cells/time step).

The efficiency or access probability p_a is the probability that an arriving cell is not dropped or marked for future discard. p_a is given by

$$p_a = 1 - L = 1 - c\, s_B \tag{8.82}$$

The average number of packets that are accepted without being dropped per time step is the system throughput and is given by

$$\begin{aligned} \text{Th} &= p_a\, \lambda_a \\ &= (1 - c\, s_B)\, \lambda_a \end{aligned} \tag{8.83}$$

The maximum burst size allowed from the packet source is determined by the size of the queue.

$$\text{Max. burst size} = B \qquad \text{cells} \tag{8.84}$$

Example 7 Estimate the performance of the VS algorithm for a source having the following properties:

$$\begin{aligned} a &= 0.2 & b &= 0.5 \\ c &= 0.3 & A &= 424 \text{ bits} \\ \lambda &= 150\,\text{Mbps} & B &= 5 \end{aligned}$$

The transition matrix for the VS protocol is given by

$$\mathbf{P} = \begin{bmatrix} 0.7 & 0.2 & 0 & 0 & 0 & 0 \\ 0.3 & 0.5 & 0.2 & 0 & 0 & 0 \\ 0 & 0.3 & 0.5 & 0.2 & 0 & 0 \\ 0 & 0 & 0.3 & 0.5 & 0.2 & 0 \\ 0 & 0 & 0 & 0.3 & 0.5 & 0.2 \\ 0 & 0 & 0 & 0 & 0.3 & 0.8 \end{bmatrix}$$

The distribution vector for the system is

$$\mathbf{s} = \begin{bmatrix} 0.0481 & 0.0722 & 0.1083 & 0.1624 & 0.2436 & 0.3654 \end{bmatrix}^t$$

The performance figures of the protocol are as follows:

$$\begin{aligned} L &= 0.1096 \\ N_a(\text{lost}) &= 16.4436 \times 10^6 \quad \text{packets/s} \\ p_a &= 0.8904 \\ \text{Th} &= 133.5564 \times 10^6 \quad \text{packets/s} \end{aligned}$$

■

Example 8 What would happen in the above example if the VS algorithm uses a buffer whose size is $B = 2$?

The following table illustrates the effect of reducing the buffer size.

Parameter	$B = 8$	$B = 2$
L	0.1096	0.1421
N_a(lost) (M packets/s)	16.4436	21.3158
p_a	0.8904	0.8579
Th (M packets/s)	133.5564	128.6842

As expected, the reduced penalty buffer results in decreased performance such as higher cell loss probability and lower throughput. ■

Problems

Leaky Bucket Algorithm

8.1 In a leaky bucket traffic shaper, the packet arrival rate for a certain user is on the average 5 Mbps with a maximum burst rate of 30 Mbps. The output rate is dictated by the algorithm to be 10 Mbps. Derive the performance of this protocol using the $M/M/1/B$ modeling approach assuming packet buffer size to be $B = 5$ and the maximum line rate is 200 Mbps. Assume different values for the probability that the source is conforming.

8.2 Repeat problem 8.1 using the $M^m/M/1/B$ modeling approach.

8.3 An alternative to modeling the leaky bucket algorithm using the $M^m/M/1/B$ queue is to assume the Bernoulli probability of packet arrival to be given as

$$a = \frac{\lambda_a}{\lambda_l}$$

which is similar to the packet arrival statistics for the $M/M/1/B$ queue. Study this situation using the data given in Example 1 and comment on your results.

Token Bucket Algorithm

8.4 Consider the single arrival/departure model for the token bucket algorithm. Assume that arriving data are buffered in a packet buffer. Now we have two buffers to consider: the token buffer and the data buffer. Model the data buffer based on the results obtained for the token buffer

8.5 Estimate the maximum burst size in the token bucket protocol.

8.6 In a token bucket traffic shaper, the packet arrival rate for a certain user is on the average 15 Mbps with a maximum burst rate of 30 Mbps. Assume the source is conforming 30% of the time and the token arrival rate is dictated by the algorithm to be 20 Mbps. Study the state of the token buffer using the $M/M/1/B$ modeling approach assuming its size to be $B = 5$ and the maximum line rate is 100 Mbps.

8.7 Repeat Problem 8.6 using the multiple arrival/departure modeling approach.

8.8 Draw the Markov state transition diagram for the multiple arrival/departure model when $B_t = 4$, $B_p = 3$, and $m = 4$.

8.9 Write down the transition matrix for the above problem and compare with the same system that had $m = 3$ in (8.63) on page 291.

8.10 Write down the transition matrix for the multiple arrival/departure model when $B_t = 4$, $B_p = 6$, and $m = 8$.

Virtual Scheduling Algorithm

8.11 Analyze the virtual scheduling algorithm in which an arriving cell is conforming if $t \geq \text{TAT} - \Delta$ and is nonconforming if $t < \text{TAT} - \Delta$.

8.12 Compare the performance of the VS algorithm treated in the text to the VS algorithm analyzed in Problem 8.11.

8.13 Analyze the virtual scheduling algorithm in which an arriving cell is issued a credit or penalty according to the criteria:

$$\begin{aligned}
\text{credit}: &\quad t &&\geq \text{TAT} + \Delta \\
\text{no action}: &\quad \text{TAT} &&\leq t < \text{TAT} + \Delta \\
\text{no action}: &\quad \text{ATA} &&-\Delta \leq t < \text{TAT} + \Delta \\
\text{credit}: &\quad t &&\geq \text{TAT} - \Delta
\end{aligned}$$

References

1. A.S. Tanenbaum, *Computer Networks*, Prentice Hall PTR, Upper Saddle River, New Jersey, 1996.

Chapter 9
Modeling Error Control Protocols

9.1 Introduction

Modeling a protocol or a system is just like designing a digital system, or any system for that matter — There are many ways to model a protocol based on the assumptions that one makes. My motivation here is simplicity and not taking a guided tour through the maze of protocol modeling. My recommendation to the reader is to read the discussion on each protocol and then lay down the outline of a model that describes the protocol. The model or models developed here should then be compared with the one attempted by the reader.

Automatic-repeat-request (ARQ) techniques are used to control transmission errors caused by channel noise [1]. All ARQ techniques employ some kind of error coding of the transmitted data so that the receiver has the ability to detect the presence of errors. When an error is detected, the receiver requests a retransmission of the faulty data. ARQ techniques are simple to implement in hardware and they are especially effective when there is a reliable feedback channel connecting the receiver to the transmitter such that the round-trip delay is small.

There are three main types of ARQ techniques:

- Stop-and-wait ARQ (SW ARQ)
- Go-back-N ARQ (GBN ARQ)
- Selective-repeat ARQ (SR ARQ)

We discuss and model each of these techniques in the following sections.

9.2 Stop-and-Wait ARQ (SW ARQ) Protocol

Stop-and-wait ARQ (SW ARQ) protocol is a simple protocol for handling frame transmission errors when the round-trip time ($2\tau_p$) for frame propagation and reception of acknowledgment is smaller than frame transmission time (τ_t). The propagation delay τ_p is given by

$$\tau_p = \frac{d}{c}$$

F. Gebali, *Analysis of Computer and Communication Networks*,
DOI: 10.1007/978-0-387-74437-7_9,

where c is speed of light and d is the distance between transmitter and receiver. The transmission delay τ_t is given by

$$\tau_t = \frac{L}{\lambda}$$

where L is the number of bits in a frame and λ is the transmission rate in bits per second.

Thus, ARQ protocols are efficient and useful when we have

$$2\tau_p \ll \tau_t \tag{9.1}$$

If the above inequality is not true, then forward error correction (FEC) techniques should be used [1].

When the sender transmits a frame on the forward channel, the receiver checks it for errors. If there are no errors, the receiver acknowledges the correct transmission by sending an acknowledge (ACK) signal through the feedback channel. In that case, the transmitter proceeds to send the next frame. If there were errors in the received frame, the receiver sends a negative acknowledgment signal (NAK) and the sender sends the same frame again. If the receiver does not receive ACK or NAK signals due to some problem in the feedback channel, the receiver waits for a certain timeout period and sends the frame again.

Based on the above discussion, we conclude that the time between transmitted frames is equal to $2\tau_p$, where τ_p is the one-way propagation delay.

Figure 9.1 shows an example of transmitting several frames using stop-and-wait ARQ. Frame 1 was correctly received as indicated by the ACK signal and the sender starts sending frame 2.

Frame 2 was received in error as indicated by NAK and the grey line. The transmitter sends frame 2 one more time. For some reason, no acknowledgment signals were received (indicated by short grey line) and the sender sends frame 2 for the third time after waiting for the proper timeout period.

Frame 2 was received correctly as indicated by the ACK signal and the sender starts sending frame 3.

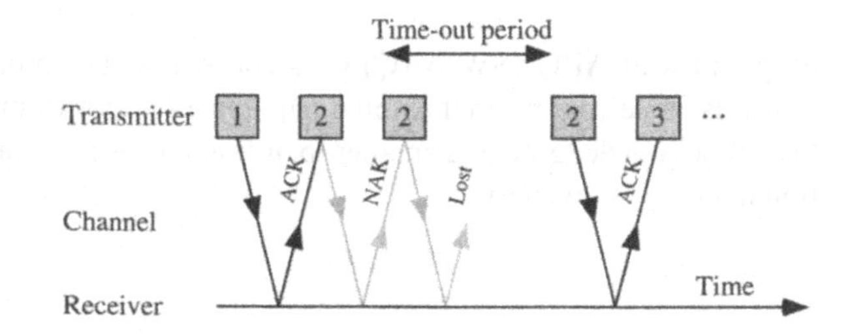

Fig. 9.1 Stop-and-wait ARQ protocol

9.2.1 Modeling Stop-and-Wait ARQ

In this section, we perform Markov chain analysis of the stop-and-wait algorithm. We make the following assumptions for our analysis of the stop-and-wait ARQ (SW ARQ):

1. The average length of a frame is n bits.
2. The forward channel has random noise and the probability that a bit will be received in error is ϵ. Another name for ϵ is bit error rate (BER).
3. The feedback channel is assumed noise-free so that acknowledgment signals from the receiving station will always be transmitted to the sending station.
4. The sender will keep sending a frame until it is correctly received. The effect of limiting the number of retransmissions is discussed in Problem 9.2.

The state of the sender while attempting to transmit a frame depends only on the outcome of the frame just sent. Hence we can represent the state of the sender as a Markov chain having the following properties:

1. State i of the Markov chain indicates that the sender is retransmitting the frame for the ith time. State 0 indicates error-free transmission.
2. The number of states is infinite since no upper bound is placed on the number of retransmissions.
3. The time step is taken equal to the sum of transmission delay and round-trip delay $T = \tau_t + 2\tau_p$.

The state transition diagram for the SW ARQ protocol is shown in Fig. 9.2. In the figure, e represents the probability that the transmitted frame contained an error. e is given by the expression

$$e = 1 - (1 - \epsilon)^n \tag{9.2}$$

For a noise-free channel $\epsilon = 0$ and so $e = 0$. When the average number of errors in a frame is very small (i.e., $\epsilon\, n \ll 1$), we can write

$$e \approx \epsilon\, n \tag{9.3}$$

The quantity $\epsilon\, n$ is an approximation of the average number of bits in error in a frame (see Problems 9.4 and 9.5). Naturally, we would like the number of errors to be small so as not to waste the bandwidth in retransmissions. Thus, we must have

$$e = \epsilon\, n \ll 1 \tag{9.4}$$

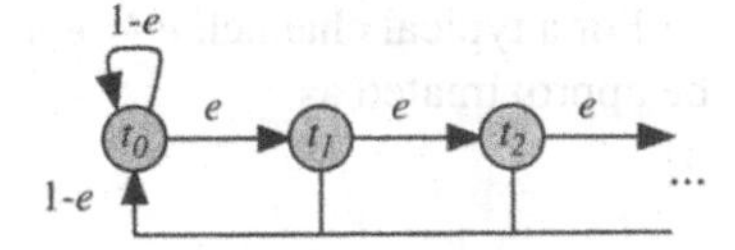

Fig. 9.2 State transition diagram of a sending station using the SW ARQ error control protocol

Equation (9.2) assumed no forward error correction (FEC) coding is implemented. Problem 9.4 requires you to derive an expression for e when FEC is employed.

We organize the state distribution vector as follows:

$$\mathbf{s} = \begin{bmatrix} s_0 & s_1 & s_2 & \cdots \end{bmatrix}^t \tag{9.5}$$

where s_i corresponds to retransmitting the frame for the ith time. s_0 corresponds to transmitting the frame once with zero retransmissions. This is the case when the frame was correctly received without having to retransmit it.

The corresponding transition matrix of the channel is given by

$$\mathbf{P} = \begin{bmatrix} 1-e & 1-e & 1-e & 1-e & \cdots \\ e & 0 & 0 & 0 & \cdots \\ 0 & e & 0 & 0 & \cdots \\ 0 & 0 & e & 0 & \cdots \\ \vdots & \vdots & \vdots & \vdots & \ddots \end{bmatrix} \tag{9.6}$$

At equilibrium, the distribution vector is obtained by solving the two equations

$$\mathbf{P}\,\mathbf{s} = \mathbf{s} \tag{9.7}$$

$$\sum \mathbf{s} = 1 \tag{9.8}$$

The solution to the above two equations is simple:

$$\mathbf{s} = (1-e) \times \begin{bmatrix} 1 \; e \; e^2 \cdots \end{bmatrix}^t \tag{9.9}$$

9.2.2 SW ARQ Performance

The average number of retransmissions for a frame is given by

$$\begin{aligned} N_t &= (1-e) \times (s_1 + 2s_2 + 3s_3 + \cdots) \\ &= (1-e)^2 \times \sum_{i=0}^{\infty} i\, e^i \\ &= e \qquad \text{transmissions/frame} \end{aligned} \tag{9.10}$$

For a noise-free channel, $e = 0$ and the average number of retransmissions is also 0. This indicates that a frame is sent once for a successful transmission.

For a typical channel, $e \approx \epsilon\, n \ll 1$ and the average number of transmissions can be approximated as

$$N_t \approx \epsilon\, n \tag{9.11}$$

We define the efficiency of the SW ARQ protocol as the inverse of the total number of transmissions which includes the first transmission plus the average number of retransmissions. In that case, η is given by

$$\eta = \frac{1}{1+N_t} = \frac{1}{1+e} \tag{9.12}$$

For an error-free channel, $N_t = 0$ and $\eta = 100\%$. For a typical channel, $e \approx \epsilon\, n \ll 1$ and the efficiency is given by

$$\eta \approx 1 - \epsilon\, n \tag{9.13}$$

This indicates that the efficiency decreases with increase in bit error rate or frame size. Thus, we see that the system performance will degrade gradually with any increase in the number of bits in the frame or any increase in the frame error probability.

The throughput of the transmitter can be expressed as

$$\mathrm{Th} = \eta = 1 - e \qquad \text{frames/time step} \tag{9.14}$$

Thus, for an error-free channel, $\eta = 1$ and arriving frames are guaranteed to be transmitted on the first try. We could have obtained the above expression for the throughput by estimating the number of frames that are successfully transmitted in each transmitter state:

$$\begin{aligned} \mathrm{Th} &= (1-e)\sum_{i=0}^{\infty} s_i \\ &= 1 - e \end{aligned} \tag{9.15}$$

When errors are present in the channel, then $\eta < 1$ and so is the system throughput.

Example 1 Assume an SW ARQ protocol in which the frame size is $n = 1000$ and the bit error rate is $\epsilon = 10^{-4}$. Find the performance of the SW ARQ protocol for this channel. Repeat the example when the bit error rate increases by a factor of 10.

According to (9.10), the average number of transmissions for a window is

$$N_t = 0.1052$$

and the efficiency is

$$\eta = 90.48\%$$

Notice that because the bit error rate is low, we need just about one transmission to correctly receive a frame.

Now we increase the bit error rate to $e = 10^{-3}$ and get the following results.

$$N_t = 1.7196$$

and the efficiency is

$$\eta = 36.77$$

Notice that when the bit error rate is increased by one order of magnitude, the average number of frame retransmission is increased by a factor of 16.35. ∎

9.3 Go-Back-N (GBN ARQ) Protocol

In the go-back-N protocol, the transmitter keeps sending frames but keeps a copy in a buffer, which is called the transmission window. The number of frames in the buffer, or the window, is N which equals the number of frames sent during one round-trip time.

When the sender transmits the frames on the forward channel, the receiver checks them for errors. If there are no errors, the receiver acknowledges each frame by sending acknowledge (ACK) signals through the feedback channel. Upon reception of an ACK for a certain frame, the receiver drops it from the head of its buffer. If a received frame is in error, the receiver sends a negative acknowledgment signal (NAK) for that particular frame. When the transmitter receives the NAK signal, it resends all N frames in its buffer starting with the frame in error.

Figure 9.3 shows an example of transmitting several frames using go-back-N where the buffer size is $N = 3$. Solid arrows indicate ACK signals and grey arrows indicate NAK signals. Frame 1 was correctly received while frame 2 was received in error. We see that the transmitter starts to send frame 2 and the three subsequent frames that were in its buffer.

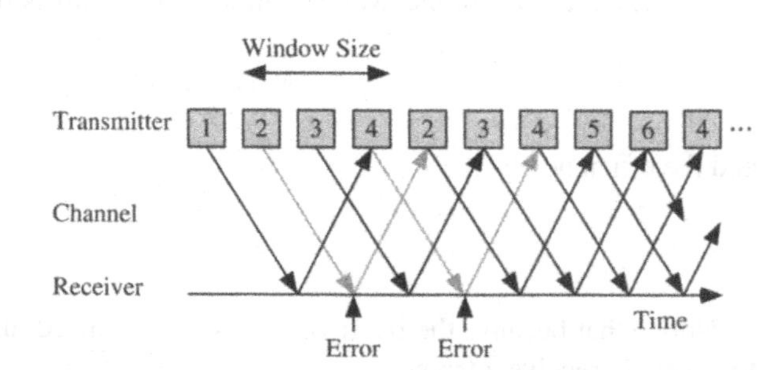

Fig. 9.3 Go-back-N ARQ protocol with buffer size $N = 3$. *Solid arrows* indicate ACK signals and *grey arrows* indicate NAK signals

9.3.1 Modeling the GBN ARQ Protocol

In this section, we perform Markov chain analysis of the GBN ARQ protocol. We make the following assumptions for our analysis of the go-back-N protocol.

1. Each window contains N frames.
2. The average length of a frame is n bits.
3. The forward channel has random noise and the probability that a bit will be received in error is ϵ. Another name for ϵ is bit error rate (BER).
4. The feedback channel is assumed noise-free so that acknowledgment signals from the receiving station will always be transmitted to the sending station.
5. The maximum number of retransmissions is k_m after which the sender will declare the channel to be not functioning.

The state of the sender while attempting to transmit a frame depends only on the outcome of the frame just sent. Hence we can represent the state of the sender as a Markov chain having the following properties:

1. The states of the Markov chain are grouped into the sets $\mathcal{T}$, $\mathcal{R}_1$, $\mathcal{R}_2$, etc. These sets are explained below.
2. The number of states is infinite since no upper bound is placed on the number of retransmissions.
3. The time step is taken equal to the sum of transmission delay of one frame $T = \tau_\text{t}$. Thus, a window that contains N frames will require N time steps to be transmitted.

The set $\mathcal{T}$ represents the states of the sender while it is transmitting a window for freshly arrived N frames:

$$\mathcal{T} = \left\{ t_1 \; t_2 \; \cdots \; t_N \right\} \tag{9.16}$$

The set $\mathcal{R}_1$ represents the states of the sender while it is retransmitting frames for the first time due to a damaged or lost frame. This set is the union of several subsets

$$\mathcal{R}_1 = R_{1,1} \cup R_{1,2} \cup \cdots \cup R_{1,N} \tag{9.17}$$

where subset $R_{1,i}$ is the subset of $\mathcal{R}_1$ that contains i states corresponding to retransmitting i frames for the first time. In other words, the first $N - i$ frames have been correctly received. With the help of (9.23) below, we can verify that all states in subset $R_{1,i}$ are equal so that we can write the i states associated with $R_{1,i}$ as

$$R_{1,i} = \left\{ r_{1,i} \; r_{1,i} \; \cdots \; r_{1,i} \right\} \tag{9.18}$$

Thus $\mathcal{R}_1$ has N *unique* subsets such that subset $R_{1,i}$ has i equal states $r_{1,i}$:
$r_{1,1}$ corresponding to last frame in error
$r_{1,2}$ corresponding to frame before last in error
$\vdots$
$r_{1,N}$ corresponding first frame in error
Similarly, the set $\mathcal{R}_2$ represents the states of the sender while it is retransmitting frames for the second time. This set is the union of several subsets

$$\mathcal{R}_2 = R_{2,1} \cup R_{2,2} \cup \cdots \cup R_{1,N} \tag{9.19}$$

where subset $R_{2,i}$ is the subset of $\mathcal{R}_2$ that contains i states corresponding to retransmitting i frames for the second time. With the help of (9.23), we can verify that all states in subset $R_{2,i}$ are equal so that we can write the i states associated with $R_{2,i}$ as

$$R_{2,i} = \left\{ r_{2,i}\ r_{2,i}\ \cdots\ r_{2,i} \right\} \tag{9.20}$$

Thus $\mathcal{R}_2$ has N *unique* subsets such that subset $R_{2,i}$ has i equal states $r_{2,i}$:

$r_{2,1}$ corresponding to last frame in error
$r_{2,2}$ corresponding to frame before last in error
$\vdots$
$r_{2,N}$ corresponding first frame in error

The state transition diagram for the GBN ARQ protocol is shown in Fig. 9.4. The sets of states $\mathcal{T}$, $\mathcal{R}_1$, and $\mathcal{R}_2$ are shown. To reduce clutter, only transitions in and out of $\mathcal{R}_1$ are indicated. The thick lines indicate multiple transitions lumped together. However, (9.23) shows all the transitions between states.

In the figure, all states in each column are equal due to the fact that the transition probabilities between them is 1.

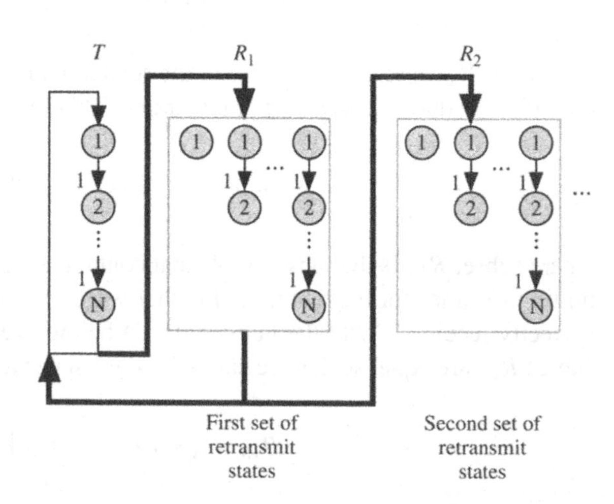

Fig. 9.4 State transition diagram of a sending station using the GBN ARQ error control protocol. Only transitions in and out of $\mathcal{R}_1$ are indicated

We organize the state distribution vector in the following order:

$$\mathbf{s} = \begin{bmatrix} T & \mathcal{R}_1 & \mathcal{R}_2 & \cdots \end{bmatrix}^t \tag{9.21}$$

For the case when $N = 2$, the distribution vector can be written as

$$\mathbf{s} = \begin{bmatrix} t & t & | & r_{1,1} & r_{1,2} & r_{1,2} & | & r_{2,1} & r_{2,2} & r_{2,2} & | & \cdots \end{bmatrix}^t \tag{9.22}$$

The corresponding transition matrix of the sender for the case when $N = 2$ is given by

$$\mathbf{P} = \begin{bmatrix} 0 & p_{2,0} & p_{1,0} & 0 & p_{2,0} & p_{1,0} & 0 & p_{2,0} & \cdots \\ 1 & 0 & 0 & 0 & 0 & 0 & 0 & 0 & \cdots \\ 0 & p_{2,1} & 0 & 0 & 0 & 0 & 0 & 0 & \cdots \\ 0 & p_{2,2} & 0 & 0 & 0 & 0 & 0 & 0 & \cdots \\ 0 & 0 & 0 & 1 & 0 & 0 & 0 & 0 & \cdots \\ 0 & 0 & p_{1,1} & 0 & p_{2,1} & 0 & 0 & 0 & \cdots \\ 0 & 0 & 0 & 0 & p_{2,2} & 0 & 0 & 0 & \cdots \\ 0 & 0 & 0 & 0 & 0 & 0 & 1 & 0 & \cdots \\ \vdots & \vdots & \vdots & \vdots & \vdots & \vdots & \vdots & \vdots & \ddots \end{bmatrix} \tag{9.23}$$

In the above matrix, the transition probability $p_{i,j}$ is the probability that the last j frames need to be retransmitted, given that a frame of i frames was sent. For instance, $p_{5,2}$ indicates the probability that the last two frames have to be retransmitted, given that five frames were sent and the first three frames were received without error. $p_{i,j}$ is given by the expression

$$p_{i,j} = (1 - e)^{i-j} e \tag{9.24}$$

where e is the probability that a frame contained one or more errors:

$$e = 1 - (1 - \epsilon)^n \tag{9.25}$$

At equilibrium, the distribution vector is obtained by solving the two equations

$$\mathbf{P}\,\mathbf{s} = \mathbf{s} \tag{9.26}$$

$$\sum \mathbf{s} = 1 \tag{9.27}$$

When k_{m} is the maximum number of retransmissions, the dimension of s would be given by

$$\dim(\mathbf{s}) = N + \frac{k_{\mathrm{m}} N(N+1)}{2} \tag{9.28}$$

As an example, for a window size $N = 32$ frames and the maximum number of retransmissions is $k_m = 16$, the size of **s** would be 33,344 and the state transition matrix would be of size $33,344 \times 33,344$. We can use MATLAB to find the distribution vector **s**.

9.3.2 Using Iterations to Find s

An alternative approach is to find **s** using iterations. From the structure of the matrix **P**, we can easily prove that all transmit states in the set $\mathcal{T}$ are equal:

$$t = t_1 = t_2 = \cdots = t_N \tag{9.29}$$

In fact, all states in any column of the matrix are equal. For example, we can write the N *unique states* of $\mathcal{R}_1$ as

$$r_{1,N} = t\ p_{N,N} \tag{9.30}$$
$$r_{1,N-1} = t\ p_{N,N-1} \tag{9.31}$$
$$r_{1,N-2} = t\ p_{N,N-2} \tag{9.32}$$
$$\vdots$$
$$r_{1,1} = t\ p_{N,1} \tag{9.33}$$

In general, we have

$$r_{1,j} = t\ p_{N,j} \qquad j = 1,\ 2, \ldots,\ N \tag{9.34}$$

In that case, state $r_{1,1}$ will be repeated once. State $r_{1,2}$ will be repeated twice, and finally $r_{1,N}$ will be repeated N times.

For the rest of the retransmission states, we use iterative expressions as follows. The N unique states of $\mathcal{R}_2$ are expressed in terms of the unique states of $\mathcal{R}_1$ as

$$r_{2,N} = r_{1,N}\ p_{N,N} \tag{9.35}$$
$$r_{2,N-1} = \sum_{i=N-1}^{N} r_{1,i}\ p_{i,N-1} \tag{9.36}$$
$$r_{2,N-2} = \sum_{i=N-2}^{N} r_{1,i}\ p_{i,N-2} \tag{9.37}$$
$$\vdots$$
$$r_{2,1} = \sum_{i=1}^{N} r_{1,i}\ p_{i,1} \tag{9.38}$$

In general, we have

$$r_{2,j} = \sum_{i=j}^{N} r_{1,i}\ p_{i,j} \qquad j = 1,\ 2,\ \ldots,\ N \tag{9.39}$$

The states associated with the kth retransmission $\mathcal{R}_k$ are given by the iterative expression

$$r_{k,j} = \sum_{i=j}^{N} r_{k-1,i}\ p_{i,j} \qquad \begin{array}{lcllll} k & = & 2, & 3, & \cdots & \\ j & = & 1, & 2, & \ldots, & N \end{array} \tag{9.40}$$

with the initial condition

$$r_{1,j} = t\ p_{N,j} \qquad j = 1,\ 2,\ \cdots\ N \tag{9.41}$$

9.3.3 Algorithm for Finding s *by Iterations*

The above iterations express all the retransmission states in terms of the transmit state t. In order to find the distribution vector **s**, we follow this algorithm:

1. Assign to each transmit state some value, for example,

$$t = 1 \tag{9.42}$$

2. Estimate the retransmit states for $\mathcal{R}_1$ using the iterative expression (9.34).
3. Estimate the values of the other retransmit states $\mathcal{R}_k$ using the iterative expression (9.40).
4. Find the sum of all states

$$\mathcal{S} = N\,t + \sum_{k=1}^{k_m} \sum_{j=1}^{N} j\ r_{k,j} \tag{9.43}$$

5. The normalized value of the distribution vector is given by the following equation:

$$\mathbf{s} = \frac{1}{\mathcal{S}} \left[\begin{array}{ccccccccccc} t & t & | & r_{1,1} & r_{1,2} & r_{1,2} & | & r_{2,1} & r_{2,2} & r_{2,2} & | & \cdots \end{array}\right]^t \tag{9.44}$$

9.3.4 GBN ARQ Performance

As long as the sender keeps retransmitting frames that were received in error, the next frame cannot be sent. Therefore, we are interested in estimating the average number of retransmissions for a given frame (R_a), the average number of frames sent in each retransmission attempt (N_a), and the average delay a frame takes to be transmitted when errors are present (T_a).

Estimating average number of retransmissions R_a

The probability that the source is in the kth retransmission state is given by

$$\alpha_k = \sum_{j=1}^{N} j \, r_{k,j} \tag{9.45}$$

α_k is also equal to the average number of frames sent at the kth retransmission (n_k).

The average number of retransmissions by the source for a given frame is given by

$$R_a = \sum_{k=1}^{k_m} k \, \alpha_k \tag{9.46}$$

Estimating average delay T_a

The delay associated with the kth retransmission is given by the accumulation of all the frames that were previously sent by earlier retransmissions. Thus, we can write

$$t_k = \sum_{j=1}^{k} \alpha_j \tag{9.47}$$

The average delay for transmitting a given frame is given by

$$T_a = \sum_{k=1}^{k_m} t_k \, \alpha_k \tag{9.48}$$

Estimating the average number of frames sent N_a

The average number of frames sent due to all retransmissions is given by

$$N_a = \sum_{k=1}^{k_m} n_k \, \alpha_k \tag{9.49}$$

GBN ARQ efficiency η *and throughput* (Th)
The efficiency of the GBN ARQ protocol is the ratio of frame size to the total number of frames transmitted:

$$\eta = \frac{N}{N + N_a} \tag{9.50}$$

When there are no errors, $N_a = 0$ and we get 100% efficiency.
The throughput of the transmitter can be expressed as

$$\text{Th} = \eta \qquad \text{frames/time step} \tag{9.51}$$

Thus for an error-free channel, $\eta = 1$ and arriving frames are guaranteed to be transmitted on the first try.
When errors are present in the channel, $\eta < 1$ and so is the system throughput.

Example 2 Assume a GBN ARQ protocol with the following parameters.

$$\begin{aligned} n &= 500 \quad \text{bits} \\ N &= 8 \quad \text{frames} \\ \epsilon &= 10^{-4} \\ k_m &= 16 \end{aligned}$$

Find the performance of the GBN ARQ protocol for this channel.
Using the technique in Section 9.3.2, the average number of retransmissions for a given frame is

$$R_a = 0.2195$$

the average delay for a given frame is

$$T_a = 0.0306$$

the average number of retransmitted frames for a given frame is

$$N_a = 0.026$$

and the efficiency is

$$\eta = 99.68\%$$

■

9.4 Selective-Repeat (SR ARQ) Protocol

The selective-repeat protocol is a general strategy for handling frame transmission errors when the round-trip time for frame transmission and reception of the acknowledgment is comparable to frame transmission time. SR ARQ is used by the TCP transport protocol. In this protocol, the transmitter groups the frames into windows so that each window contains N frames. When the sender sends frames within a window, the receiver stores the frames of the current window and checks for errors. After a complete window has been received, or after the proper timeout period, the receiver instructs the transmitter to resend only the frames that contained errors. That results in a more efficient protocol compared to GBN ARQ that resends frames in error as well as error-free frames.

Figure 9.5 shows an example of transmitting several frames using selective-repeat protocol where the buffer size is $N = 3$. Solid arrows indicate ACK signals and grey arrows indicate NAK signals. Frame 1 was correctly received while frame 2 was received in error. We see that the transmitter starts to send frame 2 as soon as the corresponding NAK is received. Frame 4 was also received in error, and we can see that it is retransmitted as soon as its NAK signal was received.

9.4.1 Modeling the SR ARQ Protocol

In this section, we perform Markov chain analysis of the SR ARQ protocol. We make the following assumptions for our analysis of the selective-repeat protocol.

1. Each window contains N frames.
2. The average length of a frame is n bits.
3. The forward channel has random noise and the probability that a bit will be received in error is ϵ. Another name for ϵ is bit error rate (BER).
4. The feedback channel is assumed noise-free so that acknowledgment signals from the receiving station will always be transmitted to the sending station.
5. The maximum number of retransmissions is k_{m} after which, the sender will declare the channel to be not functioning.

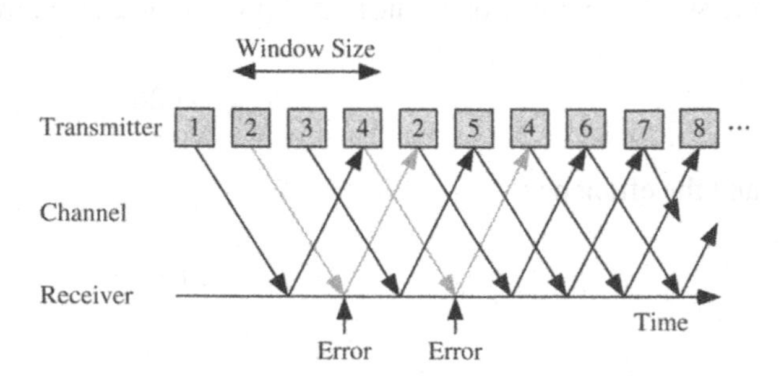

Fig. 9.5 Selective-repeat ARQ protocol with buffer size $N = 3$. *Solid arrows* indicate ACK signals and *grey arrows* indicate NAK signals

The state of the sender while attempting to transmit a frame depends only on the outcome of the frame just sent. Hence we can represent the state of the sender as a Markov chain having the following properties:

1. The states of the Markov chain are grouped into the sets $\mathcal{T}$, $\mathcal{R}_1$, $\mathcal{R}_2$, etc. These sets are explained below.
2. The number of states is infinite since no upper bound is placed on the number of retransmissions.
3. The time step is taken equal to the sum of transmission delay of one frame $T = \tau_t$. Thus, a window that contains N frames will require N time steps to be transmitted.

The set $\mathcal{T}$ represents the states of the sender while it is transmitting a window for freshly arrived N frames:

$$\mathcal{T} = \{ \ t_1 \quad t_2 \quad \cdots \quad t_N \ \} \tag{9.52}$$

The set $\mathcal{R}_1$ represents the states of the sender while it is retransmitting frames for the first time. This set is the union of several subsets

$$\mathcal{R}_1 = R_{1,1} \cup R_{1,2} \cup \cdots \cup R_{1,N} \tag{9.53}$$

where subset $R_{1,i}$ is the subset of $\mathcal{R}_1$ that contains i states corresponding to retransmitting i frames for the first time. With the help of (9.59), we can verify that all states in subset $R_{1,i}$ are equal so that we can write the i states associated with $R_{1,i}$ as

$$R_{1,i} = \{ \ r_{1,i} \quad r_{1,i} \quad \cdots \quad r_{1,i} \ \} \tag{9.54}$$

Thus $\mathcal{R}_1$ has N *unique* states:

$r_{1,1}$ repeated once,
$r_{1,2}$ repeated twice,

$\vdots$

$r_{1,N}$ repeated N times.

Similarly, the set $\mathcal{R}_2$ represents the states of the sender while it is retransmitting frames for the second time. This set is the union of several subsets

$$\mathcal{R}_2 = R_{2,1} \cup R_{2,2} \cup \cdots \cup R_{1,N} \tag{9.55}$$

where subset $R_{2,i}$ is the subset of $\mathcal{R}_2$ that contains i states corresponding to retransmitting i frames for the second time. With the help of (9.59), we can verify that all states in subset $R_{2,i}$ are equal so that we can write the i states associated with $R_{2,i}$ as

$$R_{2,i} = \{ \ r_{2,i} \quad r_{2,i} \quad \cdots \quad r_{2,i} \ \} \tag{9.56}$$

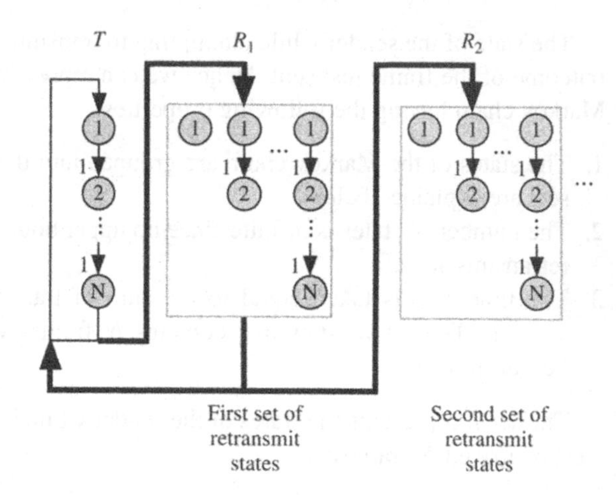

Fig. 9.6 State transition diagram of a sending station using the SR ARQ error control protocol. Only transitions in and out of $\mathcal{R}_1$ are indicated

Thus $\mathcal{R}_2$ has N *unique* states:

$r_{2,1}$ repeated once,
$r_{2,2}$ repeated twice,
⋮
$r_{2,N}$ repeated N times.

The state transition diagram for the SR ARQ protocol is shown in Fig. 9.6. The sets of states $\mathcal{T}$, $\mathcal{R}_1$, and $\mathcal{R}_2$ are shown. To reduce clutter, only transitions in and out of $\mathcal{R}_1$ are indicated. The thick lines indicate multiple transitions lumped together. However, (9.59) shows all the transitions between states.

In the figure, all states in each column are equal due to the fact that the transition probabilities between them is 1.

We organize the state distribution vector in the following order:

$$\mathbf{s} = \begin{bmatrix} \mathcal{T} & \mathcal{R}_1 & \mathcal{R}_2 & \cdots \end{bmatrix}^t \tag{9.57}$$

For the case when $N = 2$, the distribution vector can be written as

$$\mathbf{s} = \begin{bmatrix} t & t & | & r_{1,1} & r_{1,2} & r_{1,2} & | & r_{2,1} & r_{2,2} & r_{2,2} & | & \cdots \end{bmatrix}^t \tag{9.58}$$

The corresponding transition matrix of the sender for the case when $N = 2$ is given by

$$
\mathbf{P} = \begin{bmatrix}
0 & p_{2,0} & p_{1,0} & 0 & p_{2,0} & p_{1,0} & 0 & p_{2,0} & \cdots \\
1 & 0 & 0 & 0 & 0 & 0 & 0 & 0 & \cdots \\
 & & & & & & & & \\
0 & p_{2,1} & 0 & 0 & 0 & 0 & 0 & 0 & \cdots \\
 & & & & & & & & \\
0 & p_{2,2} & 0 & 0 & 0 & 0 & 0 & 0 & \cdots \\
0 & 0 & 0 & 0 & 0 & 0 & 0 & 0 & \cdots \\
 & & & & & & & & \\
0 & 0 & p_{1,1} & 0 & p_{2,1} & 0 & 0 & 0 & \cdots \\
 & & & & & & & & \\
0 & 0 & 0 & 0 & p_{2,2} & 0 & 0 & 0 & \cdots \\
0 & 0 & 0 & 0 & 0 & 0 & 1 & 0 & \cdots \\
\vdots & \vdots & \vdots & \vdots & \vdots & \vdots & \vdots & \vdots & \ddots
\end{bmatrix} \tag{9.59}
$$

In the above matrix, the transition probability $p_{i,j}$ is the probability that j frames need to be retransmitted, given that a frame of i frames was sent. For instance, $p_{5,2}$ indicates the probability that two frames have to be retransmitted, given that five frames were sent. $p_{i,j}$ is given by the expression

$$
p_{i,j} = \binom{i}{j} (1 - e)^{i-j} \, e^{j} \tag{9.60}
$$

where e is the probability that a frame contained one or more errors:

$$
e = 1 - (1 - \epsilon)^{n} \tag{9.61}
$$

At equilibrium, the distribution vector is obtained by solving the two equations

$$
\mathbf{P}\,\mathbf{s} = \mathbf{s} \tag{9.62}
$$

$$
\sum \mathbf{s} = 1 \tag{9.63}
$$

When k_{m} is the maximum number of retransmissions, the dimension of $\mathbf{s}$ would be given by

$$
\dim(\mathbf{s}) = N + \frac{k_{\mathrm{m}} N (N + 1)}{2} \tag{9.64}
$$

As an example, for a window size $N = 32$ frames and the maximum number of retransmissions is $k_{\mathrm{m}} = 16$, the size of $\mathbf{s}$ would be 33,344 and the state transition matrix would be of size $33,344 \times 33,344$. We can use MATLAB to find the distribution vector $\mathbf{s}$.

9.4.2 SR ARQ Performance

As long as the sender keeps retransmitting frames that were received in error, the next frame cannot be sent. Therefore, we are interested in estimating the average number of retransmissions for a given frame (R_a), the average number of frames sent in each retransmission attempt (N_a), and the average delay a frame takes to be transmitted when errors are present (T_a).

Estimating average number of retransmissions R_a
The probability that the source is in the kth retransmission state is given by

$$\alpha_k = \sum_{j=1}^{N} r_{k,j} \tag{9.65}$$

The average number of retransmissions by the source for a given frame is given by

$$R_a = \sum_{k=1}^{k_m} k\, \alpha_k \tag{9.66}$$

Estimating average delay T_a
The average number of frames sent at the kth retransmissions is given by

$$n_k = \sum_{j=1}^{N} j\, r_{k,j} \tag{9.67}$$

The delay associated with the kth retransmission is given by the accumulation of all the frames that were previously sent by earlier retransmissions. Thus, we can write

$$t_k = \sum_{j=1}^{k} n_j \tag{9.68}$$

The average delay for transmitting a given frame is given by

$$T_a = \sum_{k=1}^{k_m} t_k\, \alpha_k \tag{9.69}$$

Estimating the average number of frames sent N_a
The average number of frames sent due to all retransmissions is given by

$$N_a = \sum_{k=1}^{k_m} n_k\, \alpha_k \tag{9.70}$$

SR ARQ efficiency η and throughput (Th)
The efficiency of the SR ARQ protocol is the ratio of frame size to the total number of frames transmitted:

$$\eta = \frac{N}{N + N_a} \tag{9.71}$$

When there are no errors, $N_a = 0$ and we get 100% efficiency.
The throughput of the transmitter can be expressed as

$$\text{Th} = \eta \qquad \text{frames/time step} \tag{9.72}$$

Thus, for an error-free channel, $\eta = 1$ and arriving frames are guaranteed to be transmitted on the first try.
When errors are present in the channel, $\eta < 1$ and so is the system throughput.

Example 3 Assume an SR ARQ protocol with the following parameters.

$$\begin{aligned} n &= 500 \quad \text{bits} \\ N &= 8 \quad \text{frames} \\ e &= 10^{-4} \\ k_m &= 16 \end{aligned}$$

Find the performance of the SR ARQ protocol for this channel.
Using the technique in Section 9.3.2, the average number of retransmissions for a given frame is

$$R_a = 0.37$$

the average delay for a given frame is

$$T_t = 0.14$$

the average number of retransmitted frames for a given frame is

$$N_a = 0.13$$

and the efficiency is

$$\eta = 98.41\%$$

Comparing these results with those of GBN ARQ, we note that SR ARQ performs better for the same parameters. ■

Problems

Stop-and-Wait Automatic-Repeat-Request (SW ARQ) Protocol

9.1 Prove that the probability e that a frame in the SW ARQ protocol is in error is $e \approx \epsilon\, n$ when $\epsilon \ll 1$.

9.2 One of the assumptions in Section 9.2.1 of the SW ARQ protocol was that the sender will keep retransmitting the frame until it is correctly received. Assume the maximum number of retransmissions is limited to k_m. How will this impact the state transition diagram, state transition matrix, the distribution vector, and the system performance?

9.3 Assume a SW ARQ protocol in which the frame size is 100 bits and the probability that a received frame is in error is $\epsilon = 10^{-5}$. Find the performance of the protocol.

9.4 Assume a forward error control (FEC) coding is used such that the receiving station can correctly decode a received frame if the number of errors does not exceed k errors. Obtain an expression for the frame error probability e under this scheme and compare the expression you get to Equation (9.2).

9.5 Assume a SW ARQ protocol in which the frame size is n bits but forward error correction (FEC) is used to improve the performance. The FEC code employed can correct only up to $k = 3$ bits in error.

1. Draw the transition diagram for such a protocol and compare with the standard SW ARQ protocol discussed in the text.
2. Derive the transition matrix and compare with the standard SW ARQ protocol.
3. Estimate the performance of this protocol and compare with the standard SW ARQ protocol.

9.6 Equations (9.11) and (9.13) indicate that SW ARQ performance will not change if we scale n to αn, where $\alpha > 1$, and decrease ϵ by the same scale factor ϵ/α when $\epsilon \ll 1$. Verify these assertions using SW ARQ parameters of $n = 100$, $\epsilon = 10^{-4}$.

9.7 Assume SW ARQ where the sender has a transmit buffer of size B. Study the sender transmit buffer.

GBN ARQ Protocol

9.8 Obtain the transition matrix for the GBN ARQ protocol having the following parameters (chosen to make the problem manageable):

$$N = 3 \quad k_m = 1$$

Assume two cases of the channel: a very noisy channel ($\epsilon = 0.01$) and for a less noisy channel ($\epsilon = 10^{-5}$). Compare the two matrices and state your conclusions.

9.9 Given a GBN ARQ protocol in which the window size is $N = 20$ frames and the probability that a received frame is in error is $e = 10^{-4}$, find the performance of such a protocol for this channel assuming $k_m = 8$.

9.10 Given a GBN ARQ protocol in which the window size is $N = 20$ frames and the probability that a received frame is in error is $e = 5 \times 10^{-4}$, find the performance of such a protocol for this channel.

9.11 Assume the GBN ARQ protocol is now modified such that if the received window contained one or more frames in error, then the whole window is discarded and a request is issued to retransmit the entire window again. This is repeated for a maximum of k_m times until an error-free window is received.

1. Identify the states of this system.
2. Write down the transition matrix.
3. The transition matrix that results will be reducible. Derive the steady-state distribution vector.

Selective-Repeat (SR ARQ) Protocol

9.12 Obtain the transition matrices for the SR ARQ protocol having the following parameters (chosen to make the problem manageable):

$$N = 3 \quad k_m = 1$$

Assume two cases of the channel: a very noisy channel ($e = 0.01$) and for a less noisy channel ($e = 10^{-5}$). Compare the two matrices and state your conclusions.

9.13 Given a SR ARQ protocol in which the window size is $N = 20$ frames and the probability that a received frame is in error is $e = 10^{-4}$, find the performance of such a protocol for this channel assuming $k_m = 8$.

9.14 Given a SR ARQ protocol in which the window size is $N = 20$ frames and the probability that a received frame is in error is $e = 5 \times 10^{-4}$, find the performance of such a protocol for this channel.

9.15 Consider a SR ARQ protocol where forward error correction (FEC) coding is employed. In that scheme, the sender adds extra correction bits to each frame or frame. The receiver is thus able to correct frames that have up to two errors per window. Draw the transition diagram for this system and compare with the SR ARQ protocol discussed in this chapter. Derive the relevant performance for this modified protocol.

9.16 In the SR ARQ protocol discussed in the test, when a window is received with i errors, the i frames are retransmitted until all of them are received without any errors. Now consider the case when the receiver only requests to retransmit j frames out of the i frames that contained i errors originally. Do you expect this protocol to perform better than the SR ARQ protocol discussed in text?

References

1. S. Lin, D.J. Costello, Jr., and M.J. Miller, "Automatic-repeat-request error-control schemes", *IEEE Communications Magazine*, vol. 22, no. 12, pp. 5–17, 1984.

Chapter 10
Modeling Medium Access Control Protocols

10.1 Introduction

In this chapter, we illustrate how to develop models for several medium access control (MAC) protocols that are commonly used in computer communications. We will model the following medium access protocols:

1. IEEE Standard 802.1p: The static priority scheduling algorithm
2. Pure ALOHA
3. Slotted ALOHA
4. IEEE Standard 802.3: carrier sense multiple access with collision detection (CSMA/CD)
5. Carrier sense multiple access with collision avoidance (CSMA/CA)
6. IEEE Standard 802.11: for ad hoc wireless LANs using the distributed coordination function (DCF)
7. IEEE Standard 802.11: for infrastructure wireless LANs using the point coordination function (PCF) and a p-persistent backoff strategy
8. IEEE Standard 802.11: for infrastructure wireless LANs using the point coordination function (PCF) and a 1-persistent backoff strategy

The static, or fixed, priority scheduling algorithm is lumped with media access algorithms since static priority is also used as a medium access protocol. Because MAC belongs to the data link layer, our unit of data transfer is the frame since packets are the business of the higher layer like the network layer and above.

Modeling a protocol is just like designing a digital system, or any system for that mater—There are many ways to model a protocol based on the assumptions that one makes. Our motivation here is simplicity and not taking a guided tour through the maze of protocol modeling. Our recommendation to the reader is to read the discussion on each protocol and then lay down the outline of a model that describes the protocol. The model or models developed here should then be compared with the one attempted by the reader.

F. Gebali, *Analysis of Computer and Communication Networks*,
DOI: 10.1007/978-0-387-74437-7_10, © Springer Science+Business Media, LLC 2008

10.2 IEEE Standard 802.1p: Static Priority Protocol

The IEEE 802.1p is based on the static priority scheduling algorithm. IEEE 802.1p standard has a priority scheme such that frames queued in a lower priority queue are not sent if there are frames queued in higher priority queues. The frames are sent only when all higher priority queues are empty. In that sense, the static priority protocol is a scheduling protocol to provide access to an outgoing link among several competing queues.

The IEEE 802.1p enables creating priority classes for network traffic. This enables quality of service (QoS) support. The analysis given here provides insight into how to use Markov chains to derive important performance figures.

10.2.1 Modeling the IEEE 802.1p: Static Priority Protocol

In this section, we assume there are N priority classes and each class has its own queue to store incoming traffic for that class. The state of each queue depends only on its immediate past history and we can model the queues using Markov chain analysis. To start our analysis, we employ the following assumptions:

1. The states of the Markov chain represent the occupancy of the priority queues.
2. The time step is taken equal to the transmission delay of a frame.
3. There are N priority classes, with class 1 having the highest priority and so on till class N which has the lowest priority.
4. The size of the queue in priority class i is equal to B_i.
5. a_i is the frame arrival probability for queue i.
6. c_i is the frame departure probability for queue i.
7. Arrivals are processed at the same time step.
8. All frames have equal lengths.

Figure 10.1 illustrates the flows into and out of each queue. The downward arrows represent lost flows. Notice that some data are lost by each queue when the arrival rate exceeds the departure rate. The highest priority queue does not suffer any data loss since its data are guaranteed service. The highest priority queue does not need a buffer to store incoming data when preemptive static priority is employed. If a nonpreemptive scheme is employed, then the highest priority queue will require a buffer of size one to store incoming data until the frame being sent is finished.

From the above assumptions, we can write for queue 1 the following frame arrival and departure probabilities:

$$u_1(\text{arrival}) = a_1 \tag{10.1}$$

$$u_1(\text{departure}) = c_1 = 1 \tag{10.2}$$

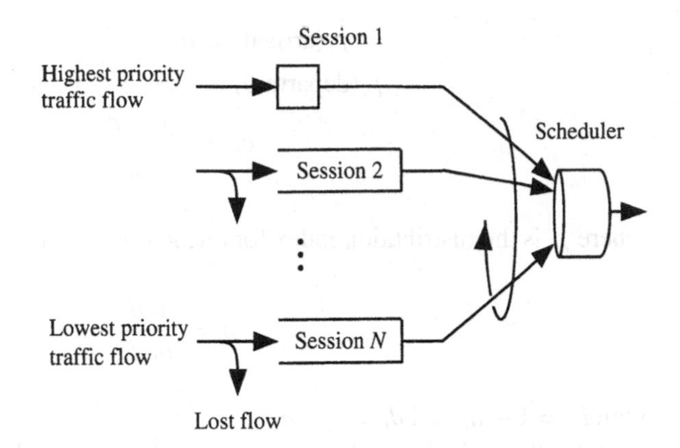

Fig. 10.1 The flows into and out of each priority queue in the static priority scheduling protocol. The *downward arrows* represent lost flows

For queue 2, we can write the following frame arrival and departure probabilities:

$$u_2(\text{arrival}) = a_2 \tag{10.3}$$

$$u_2(\text{departure}) = c_2 = e_1 \tag{10.4}$$

where $e_1 = b_1 = 1 - a_1$ is the probability that queue 1 is empty since no frames arrive. Thus queue 2 can access the output channel. The probability that queue 2 is empty is given by the expression for state s_0 of the $M/M/1/B$ queue:

$$e_2 = \frac{1 - \rho_2}{1 - \rho_2^{B_2+1}} \tag{10.5}$$

where B_2 is the size of the queue 2 buffer and ρ_2 is the distribution index for queue 2:

$$\rho_2 = \frac{a_2\, d_2}{b_2\, c_2} \tag{10.6}$$

with $b_2 = 1 - a_2$ and $d_2 = 1 - c_2$.

For queue 3, we can write the following frame arrival and departure probabilities:

$$p_3(\text{arrival}) = a_3 \tag{10.7}$$

$$p_3(\text{departure}) = c_3 = e_2 \times e_1 \tag{10.8}$$

$$e_3 = \frac{1 - \rho_3}{1 - \rho_3^{B_3+1}} \tag{10.9}$$

In general, we can write the following iterative expression for queue i, where $1 < i \leq N$:

$$p_i(\text{arrival}) = a_i \tag{10.10}$$

$$p_i(\text{departure}) = c_i = e_{i-1}\, e_{i-2} \cdots e_1 \tag{10.11}$$

$$e_i = \frac{1-\rho_i}{1-\rho_i^{B_i+1}} \tag{10.12}$$

where ρ_i is the distribution index for queue i and is given by

$$\rho_i = \frac{a_i\, d_i}{b_i\, c_i} \tag{10.13}$$

with $b_i = 1 - a_i$ and $d_i = 1 - c_i$.

After the arrival and departure probabilities of each queue are found, we can estimate the queue parameters according to the $M/M/1/B$ analysis in Section 7.6.

The performance parameters for queue 1 are a bit unique due to its high priority:

$$
\begin{array}{lll}
\text{Th}_1 & = 1 & \text{frames/time step} \\
p_{a,1} & = 1 & \\
N_{a,1}(\text{lost}) & = 0 & \text{frames/time step} \\
L_1 & = 0 & \\
Q_{a,1} & = 0 & \text{frames} \\
W_1 & = 0 & \text{time steps}
\end{array}
$$

For queue i with $1 < i \leq N$, we can write

$$
\begin{array}{lll}
\text{Th}_i & = c_i\ (1 - b_i\ e_i) & \text{frames/time step} \\
p_{a,i} & = \dfrac{c_i\,(1 - b_i\ e_i)}{a_i} & \\
N_{a,i}(\text{lost}) & = e_i\ a_i\ d_i & \text{frames/time step} \\
L_i & = e_i\ d_i & \\
Q_{a,i} & = \dfrac{\left[\rho_i - (B_i+1)\rho_i^{B_i} + B_i \rho_i^{B_i+1}\right]}{(1-\rho_i)\left(1-\rho_i^{B_i+1}\right)} & \text{frames} \\
W_i & = \dfrac{Q_{a,i}}{\text{Th}_i} & \text{time steps}
\end{array}
$$

Example 1 Consider a static priority protocol serving four users where the frame arrival probabilities for all users are equal (i.e., $a_i = a$ for all $1 \leq i \leq 4$) and all users have the same buffer size (i.e., $B_i = B$ for all $1 \leq i \leq 4$). Estimate the performance of each user.

The following table shows the performance parameters for each user starting with the highest priority user:

	1	2	3	4
Th	1	0.3965	0.3167	0.0376
p_a	1	0.9912	0.7916	0.0940
N_a(lost)	0	0.0035	0.0833	0.3624
L	0	0.0088	0.2084	0.9060
Q_a	0	0.6941	3.8790	4.9377
W	0	1.7507	12.2502	131.2926

As expected, the least priority queue has the worst performance. ■

10.3 ALOHA

ALOHA was developed by N. Abramson in the 1970s at the University of Hawaii to allow several computers spread over a wide geographical area to communicate using a broadcast wireless channel. The technique chosen was simple and applies to any system where several users attempt to access a shared resource without the help of a central controller. ALOHA did not require global time synchronization and this impacted its performance.

In ALOHA network, a user that has a frame to transmit does so without waiting to see if the channel is busy or not. When two users transmit at the same time both colliding frames will be received in error due to interference. The collision phenomenon that occurs in a shared medium is also known as *contention*. The sender knows that contention has occurred by *listening* to the channel to check if the frame it just sent is in error or not. If errors are detected, the sender retransmits the frame after waiting for a random amount of time. Another way for the sender to sense collision is to wait for an acknowledgment from the receiver.

Figure 10.2 illustrates the ALOHA contention problem. The figure assumes all transmitted frames have equal lengths and each frame has a duration T. The time T is equal to the maximum propagation delay between any pair of stations in the network. Because of the "free for all" situation, anything can happen. Let us see how a conflict might arise while attempting to transmit one frame, shown as the shaded block in Fig. 10.2. Table 10.1 summarizes potential conflict situations.

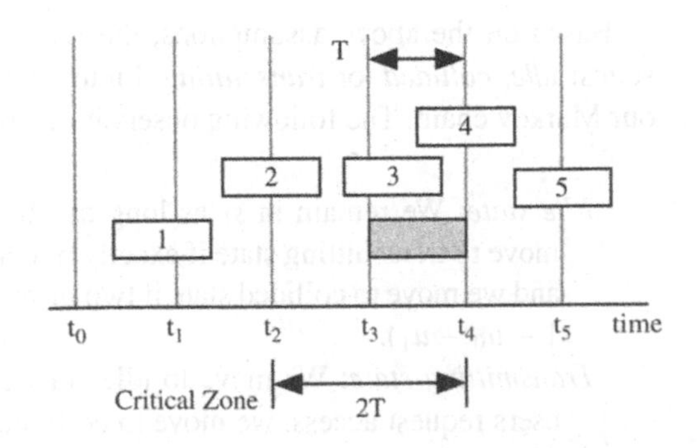

Fig. 10.2 ALOHA contention problem. Illustrating all possible conflict situations encountered by the transmitted frame (*shaded rectangle*)

Table 10.1 ALOHA contention problem: illustrating all possible conflict situations encountered by the transmitted frame (shaded rectangle in Fig. 10.2)

Frame	Start time t	Contention with shaded frame
1	$t_0 < t < t_1$	No
2	$t_1 < t < t_2$	No
3	$t_2 < t < t_3$	Yes
4	$t_3 < t < t_4$	Yes
5	$t_4 < t < t_5$	No

We make the following conclusions based on Fig. 10.2 and Table 10.1. Any frame transmitted in the period T before our shaded frame will cause contention (e.g., frame 3). Any frame transmitted during the period T when our shaded frame is being sent will cause contention (e.g., frame 4). The critical zone during transmission of the shaded frame is shown at the bottom of Fig. 10.2. Thus for a successful transmission at a given time step, the channel must be quiet and all users must be idle for previous time step.

10.3.1 Modeling the ALOHA Network

In this section, we perform Markov chain analysis of the ALOHA network. We make the following assumptions for our analysis of ALOHA:

1. The states of the Markov chain represent the status of the wireless channel: *idle*, *transmitting*, and *collided*.
2. The propagation delay between any pair of users is less than the frame time T.
3. The time step value T is taken equal to the frame transmission delay.
4. There are N users in the system.
5. Users can transmit any time they want.
6. The probability that a user transmits a frame in one time step is a.
7. All frames have equal lengths and the duration of each frame is T.
8. Contention occurs if a frame is sent at time t and there are transmissions during the time period $T - t$ to $t + T$.
9. A user retransmits a corrupted frame after waiting a random amount of time.

Based on the above assumptions, the wireless channel can be in one of three states: *idle, collided*, or *transmitting*. Figure 10.3 shows the transition diagram for our Markov chain. The following observations help explain the figure:

Idle state: We remain in s_1 as long as all users are idle (probability u_0). We move to transmitting state if exactly one user requests access (probability u_1) and we move to collided state if two or more users request access (probability $1 - u_0 - u_1$).

Transmitting state: We move to idle state if all users are idle. If one or more users request access, we move to collided state since there will be no period of calm before the next transmission.

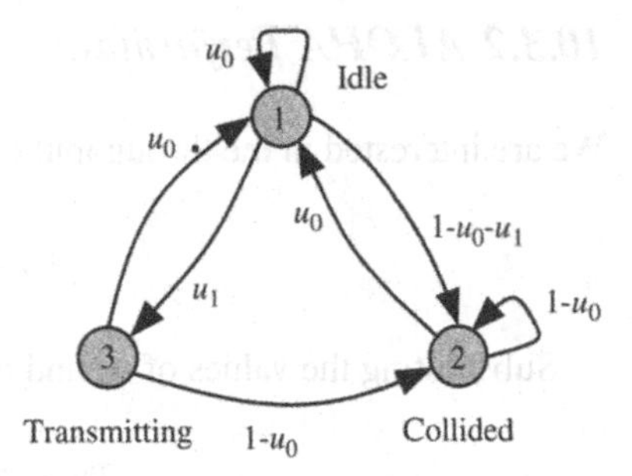

Fig. 10.3 State transition diagram for the ALOHA channel

Collided state: We move to idle state if all users are idle. Any arriving requests keep the system in collided state.

Note in the figure that the channel can move to the transmitting state in the next time step only if it is presently idle. If the channel is presently in the collided state, it cannot move to the transmitting state in the next time step since this would violate condition 8 above. The channel must first become idle and quiet since any user that attempts to transmit will only succeed in jamming the channel again. When the channel is transmitting, a period of calm has to be maintained for one time step and the channel then moves to the transmitting state. Otherwise, the channel will become collided again.

The probability that k users request access during a time step is given by binomial statistics:

$$p_k = \binom{N}{k} a^k (1-a)^{N-k} \tag{10.14}$$

The transition matrix of the system is

$$\mathbf{P} = \begin{bmatrix} u_0 & u_0 & u_0 \\ 1-u_0-u_1 & 1-u_0 & 1-u_0 \\ u_1 & 0 & 0 \end{bmatrix} \tag{10.15}$$

At equilibrium, the distribution vector is obtained by solving the two equations

$$\mathbf{P}\,\mathbf{s} = \mathbf{s} \tag{10.16}$$

$$s_1 + s_2 + s_3 = 1 \tag{10.17}$$

The solution to the above two equations is simple:

$$\begin{aligned} s_1 &= u_0 \\ s_2 &= 1 - u_0 - u_0\,u_1 \\ s_3 &= u_0\,u_1 \end{aligned} \tag{10.18}$$

10.3.2 ALOHA Performance

We are interested in the throughput of ALOHA, which is given by

$$\text{Th} = s_3 = u_0\, u_1 \tag{10.19}$$

Substituting the values of u_0 and u_1 from (10.14), we get

$$\text{Th} = N\, a\, (1 - a)^{2N-1} \tag{10.20}$$

For large N, the throughput can be expressed in terms of the input traffic as

$$\text{Th} = Nae^{-2Na} \tag{10.21}$$

The throughput is measured in units of frames/time step. The throughput in units of frames/s is

$$\text{Th}' = \frac{\text{Th}}{T} \tag{10.22}$$

The dotted line in Fig. 10.4 shows the variation of the throughput against the average number of users transmitting frames per time step for the case when $N = 10$ users. The solid line is for the slotted ALOHA network, which is discussed in the next section.

The input traffic is found from the binomial distribution

$$N_a(\text{in}) = N\, a \tag{10.23}$$

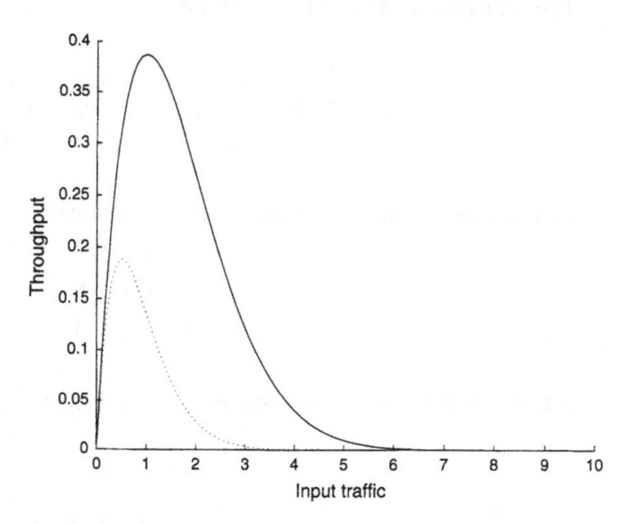

Fig. 10.4 The throughput for ALOHA versus the average number of requests per time step for the case $N = 10$. *Dotted line* is for ALOHA and *solid line* is for slotted ALOHA

We define the *acceptance probability* or *efficiency* p_a as the probability of a successful transmission for a given user. p_a is given by

$$p_a = \frac{\text{Th}}{N_a(\text{in})}$$
$$= (1-a)^{2N-1} \quad (10.24)$$
$$\approx \mathrm{e}^{-2aN} \quad (10.25)$$

The last equation relies on the approximation

$$\lim_{n\to\infty}\left(1-\frac{x}{n}\right)^n = \mathrm{e}^n$$

The average energy required to transmit a frame is estimated as

$$E = E_0 \sum_{i=0}^{\infty}(i+1)(1-p_a)^i\, p_a$$
$$= \frac{E_0}{p_a} \quad (10.26)$$

where E_0 is the energy required to send the one frame. In dB, the above equation can be written as

$$E = -10\log_{10} p_a \quad \text{dB} \quad (10.27)$$
$$\approx 20aN\log_{10}\mathrm{e} = 8.7\,aN \quad \text{dB} \quad (10.28)$$

We see that the average energy to transmit a packet increases exponentially with increasing traffic.

The average number of attempts before a successful transmission is

$$n_a = \sum_{n=0}^{\infty} n\;(1-p_a)^n\;p_a \quad (10.29)$$

This evaluates to

$$n_a = \frac{1-p_a}{p_a} \quad (10.30)$$

If the average duration of the random wait period is τ (in seconds), then the average wait for a frame before successful transmission is

$$W = n_a \times \tau = \frac{\tau\,(1-p_a)}{p_a} \quad \text{seconds} \quad (10.31)$$

Assume that the number of stations is fixed. In that case, we can vary the arrival probability a and investigate how the throughput varies with input traffic. The maximum throughput occurs when

$$\frac{\mathrm{d\,Th}}{\mathrm{d}\,a} = 0 \tag{10.32}$$

Differentiating (10.20) indicates that the maximum throughput for ALOHA occurs when

$$a_0 = \frac{1}{2\,N} \tag{10.33}$$

Thus, the maximum throughput is theoretically equal to

$$\mathrm{Th(max)} = \frac{1}{2}\left(1 - \frac{1}{2N}\right)^{2N-1} \tag{10.34}$$

The above equation gives the maximum throughput for any number of users N. A simple expression is obtained when the user population is very large, $N \rightarrow \infty$:

$$\begin{aligned} \mathrm{Th(max)} &= \frac{1}{2\,\mathrm{e}} \\ &= 18.394\% \end{aligned} \tag{10.35}$$

In summary, maximum throughput occurs when $N \rightarrow \infty$ and we would have

$$\mathrm{Th} = 0.18394 \quad \text{frames/time step} \tag{10.36}$$

$$N_a(\mathrm{in}) = 0.5 \quad \text{frames/time step} \tag{10.37}$$

$$a_0 = 1/(2N) \tag{10.38}$$

$$p_a = \mathrm{Th}/N_a(\mathrm{in}) = 0.367 \tag{10.39}$$

This discrete-time, and very general, result compares extremely well with the throughput estimate obtained using fluid flow analysis [1] for a continuous-time system with Poisson traffic. Thus when the number of users increases, the transmission request probability must decrease in proportion. For example, when the system has $N = 50$ users, the maximum throughput is approximately 18.487%, using Equation (10.34), and the transmission request probability for maximum throughput must be $a \approx 0.01$. The problems at the end of the chapter confirm our predictions.

Example 2 Assume an ALOHA network supporting $N = 20$ users and each user issues a request per time step with probability $a = 0.01$. Find

(a) The throughput
(b) The average number of time steps before successful transmission
(c) The maximum throughput
(d) The value of a for maximum throughput

The performance figures are as follows:

Th	= 0.1351	frames/time step
n_a	= 0.4799	attempts
Th(max)	= 0.1839	frames/time step
a_0	= 0.025	for maximum throughput

■

10.4 Slotted ALOHA

Slotted ALOHA was proposed to improve the throughput of the original ALOHA [2]. As the name implies, time in slotted ALOHA is divided into slots and the value of one time step equals the frame time T. Users know about the start of a new slot through a synchronizing signal that is transmitted by a source. A user with data to send is permitted to transmit only at the start of a time step. This removes the chaos of pure ALOHA and improves the efficiency as we shall see below.

10.4.1 Modeling the Slotted ALOHA Network

In this section, we perform Markov chain analysis of the slotted ALOHA network. We employ the same assumptions that we used to model ALOHA in Section 10.3.1 with the only exception that users are allowed to transmit only at the start of a time step and not at any time as before.

Based on our assumptions, the wireless channel can be in one of three states: *idle, collided*, or *transmitting*. Figure 10.5 shows the transition diagram for our Markov chain. Further, the time step value is naturally chosen equal to the slot time value. Now is a good time for the reader to compare that figure with the pure ALOHA transition diagram of Fig. 10.3. What is the major difference?

The major difference is that the channel can move from collided state to transmitting state in one time step. This basically creates more chances for the channel to move to the transmitting state and this enhances the throughput and performance in general.

Idle state: We remain in s_1 as long as all users are idle (probability u_0). We move to transmitting state if exactly one user requests access (probability u_1) and we move to collided state if two or more users request access (probability $1 - u_0 - u_1$).

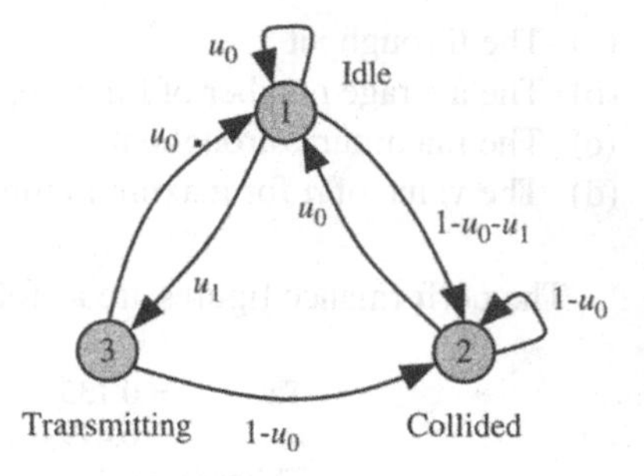

Fig. 10.5 State transition diagram for the slotted ALOHA channel

Transmitting state: We move to idle state if all users are idle. We move to transmitting state if exactly one user requests access (probability u_1) and we move to collided state if two or more users request access (probability $1 - u_0 - u_1$).

Collided state: We move to idle state if all users are idle. We move to transmitting state if exactly one user requests access (probability u_1) and we stay in collided state if two or more users request access (probability $1 - u_0 - u_1$).

The probability that k users request access during a time step is given as before by binomial statistics:

$$p_k = \binom{N}{k} a^k (1-a)^{N-k} \tag{10.40}$$

The transition matrix of the system is

$$\mathbf{P} = \begin{bmatrix} u_0 & u_0 & u_0 \\ 1-u_0-u_1 & 1-u_0-u_1 & 1-u_0-u_1 \\ u_1 & u_1 & u_1 \end{bmatrix} \tag{10.41}$$

At equilibrium, the distribution vector is obtained by solving the two equations

$$\mathbf{P}\,\mathbf{s} = \mathbf{s} \tag{10.42}$$

$$s_1 + s_2 + s_3 = 1 \tag{10.43}$$

The solution to the above two equations is simple:

$$\begin{aligned} s_1 &= u_0 \\ s_2 &= 1 - u_0 - u_1 \\ s_3 &= u_1 \end{aligned} \tag{10.44}$$

Now is a good time for the reader to compare the above equation with the corresponding equation for the equilibrium distribution vector of pure ALOHA in (10.18). Several observations can be made even before we get any numerical values:

1. Pure ALOHA and slotted ALOHA channels have the same probability of being in state s_1 (the idle state).
2. Slotted ALOHA has a *lower* probability of being in state s_2 (the collided state) compared to pure ALOHA.
3. Slotted ALOHA has a *higher* probability of being in state s_3 (the transmitting state) compared to pure ALOHA.

All these factors contribute to enhance the performance of slotted ALOHA.

10.4.2 Slotted ALOHA Performance

We are interested in the throughput of slotted ALOHA, which is given by

$$\mathrm{Th} = s_3 = u_1 = N\, a\, (1-a)^{N-1} \tag{10.45}$$

A frame will be transmitted if only one user requests to access the channel irrespective of the previous state of the channel. For large N, the throughput can be expressed in terms of the input traffic as

$$\mathrm{Th} = Nae^{-Na} \tag{10.46}$$

Slotted ALOHA has higher throughput compared to ALOHA by a factor of $u_0^{-1} = (1-a)^{-N}$. We expect the two networks to perform almost the same for very low traffic conditions ($a \ll 1$). Slotted ALOHA will perform better than ALOHA by *orders of magnitude* for high values of traffic ($a \approx 1$), especially for systems with many users ($N \gg 1$).

The throughput in units of frames/s is

$$\mathrm{Th}' = \frac{\mathrm{Th}}{T} \tag{10.47}$$

The solid line in Fig. 10.4 shows the variation of the throughput against the average number of users transmitting frames per time step for the case when $N = 10$ users. The average number of users is given by the binomial distribution

$$N_a(\mathrm{in}) = N\, a \tag{10.48}$$

We define the *acceptance probability* or *efficiency* p_a as the probability of a successful transmission for a given user. p_a is given by

$$\begin{aligned} p_a &= \frac{\mathrm{Th}}{N_a(\mathrm{in})} \\ &= (1-a)^{N-1} \qquad (10.49) \\ &\approx \mathrm{e}^{-aN} \qquad (10.50) \end{aligned}$$

The average energy required to transmit a frame is estimated as

$$E = E_0 \sum_{i=0}^{\infty} (i+1)(1-p_a)^i \; p_a = \frac{E_0}{p_a} \tag{10.51}$$

where E_0 is the energy required to send one frame. In dB, the above equation can be written as

$$E = -10 \log_{10} p_a \quad \text{dB} \tag{10.52}$$

$$\approx 10aN \log_{10} \mathrm{e} = 4.3\, aN \quad \text{dB} \tag{10.53}$$

We see that average energy to transmit a packet increases exponentially with increasing traffic but at a rate half that of pure ALOHA.

The average number of attempts before a successful transmission is

$$n_a = \sum_{n=0}^{\infty} n \; (1-p_a)^n \; p_a \tag{10.54}$$

This evaluates to

$$n_a = \frac{1-p_a}{p_a} \tag{10.55}$$

If the average duration of the random wait period is τ (in seconds), then the average wait for a frame before successful transmission is

$$W = n_a \tau = \frac{\tau\,(1-p_a)}{p_a} \qquad \text{seconds} \tag{10.56}$$

Assume that the number of stations is fixed. In that case, we can vary the arrival probability a and investigate how the throughput varies. The maximum throughput occurs when

$$\frac{\mathrm{d\,Th}}{\mathrm{d}\,a} = 0 \tag{10.57}$$

Differentiating (10.20) indicates that the maximum throughput for ALOHA occurs when

$$a_0 = \frac{1}{N} \tag{10.58}$$

Thus, the maximum throughput is theoretically equal to

$$\text{Th(max)} = \left(1 - \frac{1}{N}\right)^{N-1} \tag{10.59}$$

The above equation gives the maximum throughput for any number of users N. A simple expression is obtained when the user population is very large, $N \to \infty$:

$$\text{Th(max)} = \frac{1}{\text{e}}$$

$$= 36.788\% \tag{10.60}$$

In summary, maximum throughput occurs when $N \to \infty$ and we have

$$\text{Th} = 36.788\% \quad \text{frames/time step} \tag{10.61}$$

$$N_a(\text{in}) = 1 \quad \text{frames/time step} \tag{10.62}$$

$$a_0 = 1/N \tag{10.63}$$

$$p_a = \text{Th}/N_a(\text{in}) = 0.367 \tag{10.64}$$

This compares very well with the throughput estimate obtained using fluid flow analysis [1] for a continuous-time system with Poisson traffic. Thus when the number of users increases, the transmission request probability must decrease in proportion. For example, when the system has $N = 100$ users, the maximum throughput is approximately 36.788% and the transmission request probability must be $a \approx 0.01$.

Comparing these predictions with pure ALOHA, we find that slotted ALOHA could support double the number of users and achieve double the throughput.

We note that at maximum throughput, pure ALOHA and slotted ALOHA have the same efficiency. This is a bit surprising since it was not previously reported in the literature. In fact, the efficiency for pure ALOHA is given by the expressions

$$p_a = (1 - a_0)^{2N-1}$$
$$a_0 = 1/2N$$

And the efficiency for slotted ALOHA is given by the expressions

$$p_a = (1 - a_0)^{N-1}$$
$$a_0 = 1/N$$

Both of these expressions evaluate to e^{-1} as $N \to \infty$ and the arrival probability at maximum throughput is taken as $a_0 = 1/2N$ (for pure ALOHA) and $a_0 = 1/N$ (for slotted ALOHA).

Both systems show maximum efficiency at very low traffic. As traffic increases, the efficiencies of both systems start to decrease. However, the efficiency of slotted ALOHA decreases at a slower rate compared to pure ALOHA.

Example 3 Repeat Example 2 assuming a slotted ALOHA network.

The performance figures are as follows:

Th	= 0.1652	frames/time step
n_a	= 0.2104	attempts
Th (max)	= 0.3679	frames/time step
a_0	= 0.05	for maximum throughput

We note that the throughput of slotted ALOHA is slightly higher than pure ALOHA but it is not its double since we are far from the optimum traffic arrival probability for either systems.

Similarly, the average number of tries is less for slotted ALOHA. ■

10.5 IEEE Standard 802.3 (CSMA/CD)

The IEEE 802.3 standard is used for *wired LANs* where the time required for *one bit* to travel between the two farthest stations (propagation time) is much smaller than the time required for *one frame* to be sent by the sender (transmission delay). The IEEE Standard 802.3 specifies a carrier sense multiple access with collision detection (CSMA/CD).

Signals on the channel travel very close to the speed of light and it takes a finite amount of time before all stations become aware that a channel is starting to access the medium. Therefore, a collision is said to take place when two or more stations start transmitting within the frame propagation delay. Because during this time, transmitting stations think that the medium is idle. When that happens, the two colliding stations *stop transmitting* and wait for a random amount of time before attempting to transmit again. This reduces the chance that the stations will once again transmit simultaneously. The maximum distance limitation for CSMA/CD is about 2500 m (1.5 miles). At this value, the ratio of propagation delay to transmission delay is less than 0.1 [3].

To summarize, in CSMA/CD protocol, all stations monitor the channel to determine when it is free. This is done by special *carrier sensing* circuits in each station. If the channel is busy, a station backs off and starts sensing the channel with probability p. This is called p-persistent CSMA/CD. The station refrains from transmitting on an idle channel with probability $1 - p$. This reduces the probability of collisions. If the channel is sensed free, the station starts to transmit. Transmitting stations monitor the signal on the channel and compare it to the transmitted signal to decide if a collision is taking place or not. This is done by special *collision detection* circuits. When the LAN contains N stations, p is chosen such that $Np < 1$ [3].

The IEEE 802.3 standard describes a 1-persistent CSMA/CD with exponential backoff strategy which is more commonly known as Ethernet.

At this point, it is worthwhile mentioning three different CSMA access strategies:

- 1-persistent CSMA
- Nonpersistent CSMA
- p-persistent CSMA

In a 1-persistent CSMA, a station with frame to send senses the channel. If the channel is sensed free, the frame is sent. If the channel is busy, the station *continuously* monitors the channel and sends the frame when the channel is sensed idle.

In a nonpersistent CSMA, a station with frame to send senses the channel. If the channel is sensed free, the frame is sent. If the channel is busy, the station *waits* for a random amount of time before monitoring the channel and sends the frame when the channel is sensed idle. If the channel is sensed busy, the station repeats the random wait.

In a p-persistent CSMA, a station with frame to send senses the channel. If the channel is sensed free, the frame is sent with probability p. The transmission is deferred for the next time slot with probability $1 - p$. This process is repeated until the frame is sent.

10.5.1 IEEE 802.3 (CSMA/CD) Model Assumptions

A simple analysis of IEEE 802.3 was given in [4] and [5]. A more complicated analysis of IEEE 802.3 can be found [6] which is by no means more accurate since it makes drastic assumptions about the channel states and the probability of a successful transmission. We follow here the middle ground and provide a tractable analysis making reasonable approximations.

Let us define τ_c to be the delay time required for a user to detect that a collision has taken place and τ_t to be the transmission delay for one frame. Typically, $\tau_c \ll \tau_t$ since collision detection is done using fast electronic circuits. Therefore, periods of transmission are separated by one or more *contention minislots* [7]. Similar to ALOHA, a user could determine if there is contention or not during a time period equal to the propagation delay, i.e., τ_p.

To start our analysis, we employ a set of assumptions for IEEE 802.3 model as follows:

1. Since the current state of the channel depends only on its immediate past history, we can model the channel using Markov chain analysis.
2. The states of the Markov chain represent the states of the channel: *idle*, *transmitting*, and *collided*.
3. The channel is shared among N stations.
4. There is a single station class (equal priority).
5. The frame arrival probability per time step is a.
6. All transmitted frames have equal lengths.
7. A frame duration is equal to the transmission delay of a frame τ_t.
8. The time step of the Markov chain is chosen equal to the collision detection delay, i.e., $T = \tau_c$.

9. We define n as the ratio of transmission delay to propagation delay, i.e., $n = \tau_t/\tau_c$. We assume that $n > 1$, which is true for small LANs carrying long frames or operating at low transmission speeds.
10. The time required to detect a collision is less than or equal to the time step value T.
11. Probability that an idle station receives a frame for transmission during a frame transmission time (or frame duration) is a.
12. A p-persistent CSMA/CD is assumed. This is equivalent to a 1-persistent CSMA/CD with a backoff strategy.
13. An *adaptive backoff strategy* is assumed where a collided user starts to sense the channel with probability α, which is taken equal to the frame arrival probability (i.e., $\alpha = a$). In that sense, each collided user adapts its request probability α to be proportional to its incoming traffic probability a.
14. A station can have at most one message waiting for transmission.

The alert reader will note that the above assumptions lead to significant simplification. We were able to justify that collided users behave the same as uncollided users through the adoption of 1-persistence and the adaptive backoff strategy.

10.5.2 IEEE 802.3 (CSMA/CD) State Transition Diagram

Based on the above assumptions, the wireless channel can be in one of three states: *idle, collided*, or *transmitting*. Figure 10.6 shows the state diagram of CSMA/CD protocol. The channel has several transmitting states because the time required for transmitting one frame (τ_t) is bigger than the propagation delay τ_p.

The probability that i users are active at a given time step is given by

$$u_i = \binom{N}{i} a^i (1-a)^{N-i} \tag{10.65}$$

where a is the probability that a station requests a transmission during a time step.

We organize the distribution vector at equilibrium as follows:

$$\mathbf{s} = \begin{bmatrix} s_i & s_{t_1} & s_{t_2} & \cdots & s_{t_n} & s_c \end{bmatrix}^t \tag{10.66}$$

The corresponding transition matrix of the channel is given by

$$\mathbf{P} = \begin{bmatrix} u_0 & 0 & 0 & 0 & \cdots & 0 & 1 & 1 \\ u_1 & 0 & 0 & 0 & \cdots & 0 & 0 & 0 \\ 0 & 1 & 0 & 0 & \cdots & 0 & 0 & 0 \\ 0 & 0 & 1 & 0 & \cdots & 0 & 0 & 0 \\ \vdots & \vdots & \vdots & \vdots & \ddots & \vdots & \vdots & \vdots \\ 0 & 0 & 0 & 0 & \cdots & 0 & 0 & 0 \\ 0 & 0 & 0 & 0 & \cdots & 1 & 0 & 0 \\ 1-u_0-u_1 & 0 & 0 & 0 & \cdots & 0 & 0 & 0 \end{bmatrix} \tag{10.67}$$

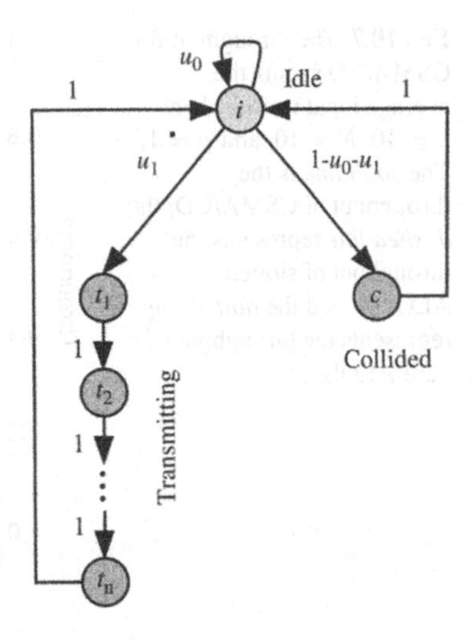

Fig. 10.6 State transition diagram for the IEEE 802.3 CSMA/CD channel

At equilibrium, the distribution vector is obtained by solving the two equations

$$\mathbf{P}\,\mathbf{s} = \mathbf{s} \tag{10.68}$$

$$\sum \mathbf{s} = 1 \tag{10.69}$$

The solution to the above two equations is simple:

$$\mathbf{s} = \frac{1}{2 + u_1(n-1) - u_0} \begin{bmatrix} 1 \\ u_1 \\ u_1 \\ \vdots \\ u_1 \\ 1 - u_0 - u_1 \end{bmatrix} \tag{10.70}$$

10.5.3 IEEE 802.3 (CSMA/CD) Protocol Performance

The throughput is given by the equation

$$\begin{aligned} \text{Th} &= \sum_{i=1}^{n} s_{t_i} \\ &= \frac{n u_1}{2 + u_1(n-1) - u_0} \end{aligned} \tag{10.71}$$

For large values of n, the throughput approaches 100% which is expected since little time is wasted during the collision.

Figure 10.7 shows the throughput of the IEEE 802.3 protocol when $n = 10$, $N = 10$, and $p = 1$. The solid line is the throughput of CSMA/CD, the dashed

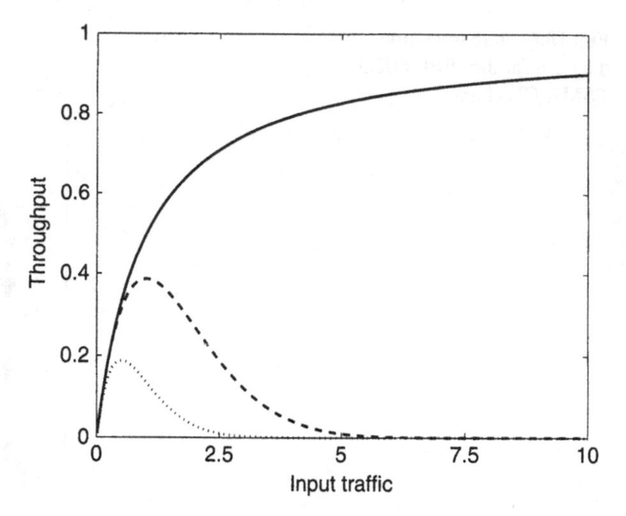

Fig. 10.7 The throughput for CSMA/CD versus the average input traffic when $n = 10$, $N = 10$, and $p = 1$. The *solid line* is the throughput of CSMA/CD, the *dashed line* represents the throughput of slotted ALOHA, and the *dotted line* represents the throughput of pure ALOHA

line represents the throughput of slotted ALOHA, and the dotted line represents the throughput of pure ALOHA.

The access probability for the user in the CSMA/CD protocol is given by

$$p_a = \frac{\text{Th}}{Na} \tag{10.72}$$

Figure 10.8 shows the access probability of CSMA/CD protocol. The solid line is the access probability of CSMA/CD, the dashed line represents the access probability of slotted ALOHA, and the dotted line represents the access probability of pure ALOHA.

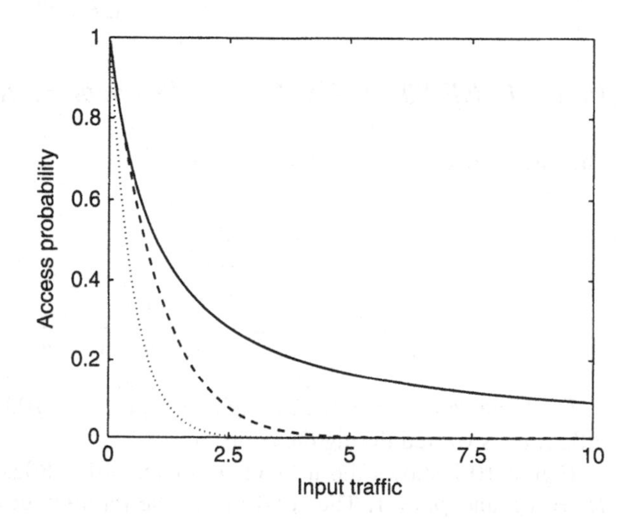

Fig. 10.8 Access probability of the IEEE 802.3 CSMA/CD protocol versus the average input traffic per time step for the case $n = 10$, $N = 10$, and $p = 1$. The *solid line* is the access probability of CSMA/CD, the *dashed line* represents the access probability of slotted ALOHA, and the *dotted line* represents the access probability of pure ALOHA

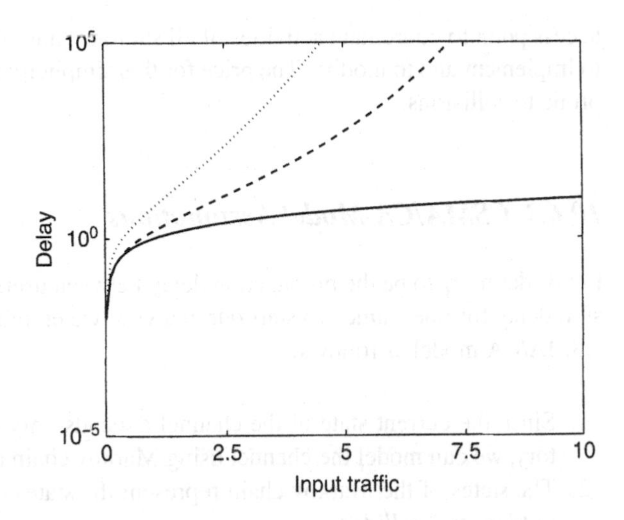

Fig. 10.9 Frame delay for the IEEE 802.3 CSMA/CD protocol versus the average input traffic frame for the case $n = 10$, $N = 10$, and $p = 1$. The *solid line* is the delay of CSMA/CD, the *dashed line* represents the delay of slotted ALOHA, and the *dotted line* represents the delay of pure ALOHA

The average number of attempts for a successful transmission is

$$n_a = \sum_{i=0}^{\infty} i\,(1 - p_a)^i\; p_a = \frac{1 - p_a}{p_a} \tag{10.73}$$

Figure 10.9 shows the delay of the IEEE 802.3 CSMA/CD protocol when $n = 10$, $N = 10$, and $p = 1$. The solid line is the delay of CSMA/CD, the dashed line represents the delay of slotted ALOHA, and the dotted line represents the delay of pure ALOHA.

10.6 Carrier Sense Multiple Access-Collision Avoidance (CSMA/CA)

Carrier sense multiple access with collision avoidance (CSMA/CA) is used in *wireless LANs* where a transmitting station is unable to determine if a collision occurred while transmitting or not. Collision detection, as is employed in Ethernet, cannot be used for the radio frequency transmissions. The reason is that when a node is transmitting it cannot hear any other node in the system which may be transmitting, since its own signal will drown out other signals arriving at the node. A station will ultimately know when a collision has taken place by reception of negative acknowledgment or by timeout mechanisms.

An ad hoc network is a collection of communicating nodes that do not have established infrastructure or centralized administration [8]. CSMA/CA protocol is useful in ad hoc networks where access to the network is decentralized since each station coordinates its own decisions for accessing the medium. There is no central access point to coordinate activities of all station. Thus ad hoc networks are simpler to implement and to modify. The price for this simplicity is that ad hoc networks are prone to collisions.

10.6.1 CSMA/CA Model Assumptions

Let us define τ_p to be the propagation delay between users and τ_t to be the transmission delay for one frame. To start our analysis, we employ a set of assumptions for CSMA/CA model as follows:

1. Since the current state of the channel depends only on its immediate past history, we can model the channel using Markov chain analysis.
2. The states of the Markov chain represent the states of the channel: *idle*, *transmitting*, and *collided.*
3. The channel is shared among N stations.
4. There is a single station class (equal priority).
5. The frame arrival probability per time step is a.
6. All transmitted frames have equal lengths.
7. A frame duration is equal to the transmission delay of a frame τ_t.
8. The time step of the Markov chain is chosen equal to the propagation delay, i.e., $T = \tau_p$.
9. We define n as the ratio of transmission delay to propagation delay, i.e., $n = \tau_t/\tau_p$. We assume that $n > 1$, which is true for small LANs carrying long frames or operating at low transmission speeds.
10. A 1-persistent CSMA/CA is assumed.
11. A station can have at most one message waiting for transmission.

Based on the above assumptions, the wireless channel can be in one of three states: *idle, collided*, or *transmitting*. Figure 10.10 shows the state diagram of CSMA/CA protocol. The channel has several transmitting states because the time required for transmitting one frame (τ_t) is bigger than the propagation delay τ_p.

The probability that i users are active at a given time step is given by

$$u_i = \binom{N}{i} a^i (1-a)^{N-i} \tag{10.74}$$

We organize the distribution vector at equilibrium as follows.

$$\mathbf{s} = \begin{bmatrix} s_i & s_{t_1} & s_{t_2} & \cdots & s_{t_n} & s_{c_1} & s_{c_2} & \cdots & s_{c_n} \end{bmatrix}^t \tag{10.75}$$

The corresponding transition matrix of the channel for the case when $n = 3$ is given by

$$\mathbf{P} = \begin{bmatrix} u_0 & 0 & 0 & 1 & 0 & 0 & 1 \\ u_1 & 0 & 0 & 0 & 0 & 0 & 0 \\ 0 & 1 & 0 & 0 & 0 & 0 & 0 \\ 0 & 0 & 1 & 0 & 0 & 0 & 0 \\ 1-u_0-u_1 & 0 & 0 & 0 & 0 & 0 & 0 \\ 0 & 0 & 0 & 0 & 1 & 0 & 0 \\ 0 & 0 & 0 & 0 & 0 & 1 & 0 \end{bmatrix} \tag{10.76}$$

At equilibrium, the distribution vector is obtained by solving the two equations

$$\mathbf{P}\,\mathbf{s} = \mathbf{s} \tag{10.77}$$

$$\sum \mathbf{s} = 1 \tag{10.78}$$

The solution to the above two equations is simple:

$$\mathbf{s} = \frac{1}{n(1-u_0)+1} \begin{bmatrix} 1 \\ u_1 \\ u_1 \\ \vdots \\ u_1 \\ 1-u_0-u_1 \\ 1-u_0-u_1 \\ \vdots \\ 1-u_0-u_1 \end{bmatrix} \tag{10.79}$$

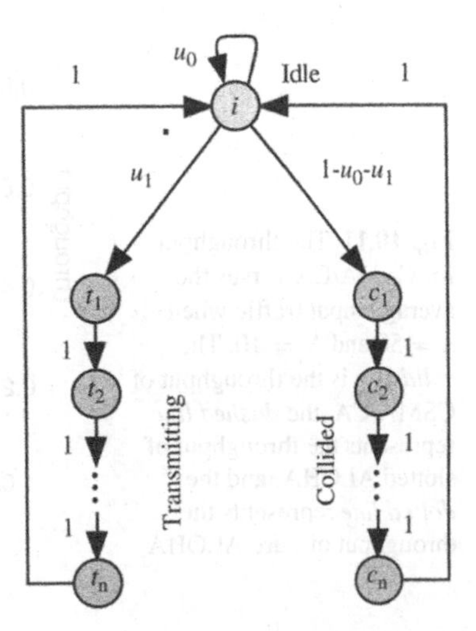

Fig. 10.10 State transition diagram for the CSMA/CA channel

10.6.2 CSMA/CA Protocol Performance

The throughput is given by the equation

$$\begin{aligned} \text{Th} &= \sum_{i=1}^{n} s_{t_i} \\ &= \frac{nu_1}{n(1-u_0)+1} \end{aligned} \tag{10.80}$$

For large values of N, we can write the throughput as

$$\text{Th} = \frac{n\mathrm{e}^{-\lambda}}{n\left(1-\mathrm{e}^{-\lambda}\right)+1} \tag{10.81}$$

For large values of n, the throughput approaches the expression

$$\text{Th} \rightarrow u_1/(1-u_0) < 100\% \tag{10.82}$$

$$\approx \frac{\lambda \mathrm{e}^{-\lambda}}{1-\mathrm{e}^{-\lambda}} = \frac{\lambda}{\mathrm{e}^{\lambda}-1} < 100\% \tag{10.83}$$

which is expected since little time is wasted during collisions.

Figure 10.11 shows the throughput of CSMA/CA when $n = 50$ and $N = 10$. The solid line is the throughput of CSMA/CA, the dashed line represents the throughput of slotted ALOHA, and the dotted line represents the throughput of pure ALOHA.

It is interesting to compare Fig. 10.7 for CSMA/CD and Fig. 10.11 for CSMA/CA. The latter protocol shows lower throughput for the same set of parameters as the

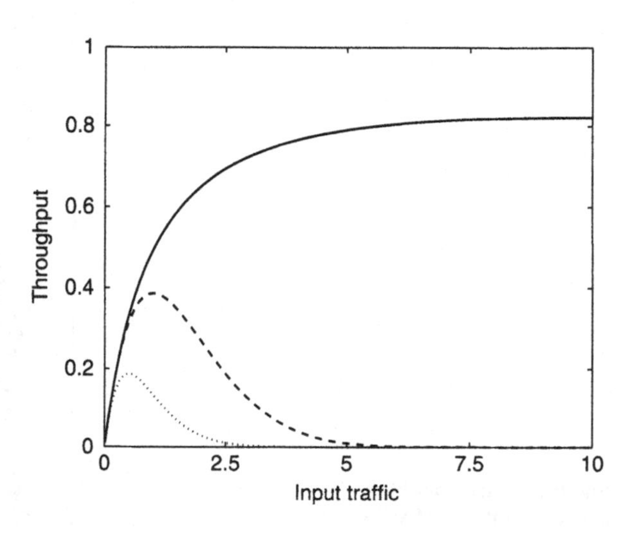

Fig. 10.11 The throughput for CSMA/CA versus the average input traffic when $n = 50$ and $N = 10$. The *solid line* is the throughput of CSMA/CA, the *dashed line* represents the throughput of slotted ALOHA, and the *dotted line* represents the throughput of pure ALOHA

former. This is due to the fact that CSMA/CA keeps transmitting frames even when collisions have taken place. Therefore, precious bandwidth and time are wasted transmitting frames while CSMA/CD stops the transmission promptly.

The access probability for a user in the CSMA/CA protocol is given by

$$p_a = \frac{\text{Th}}{Na} \tag{10.84}$$

Figure 10.12 shows the access probability of CSMA/CA when $n = 50$ and $N = 10$. The solid line is the access probability of CSMA/CA, the dashed line represents the access probability of slotted ALOHA, and the dotted line represents the access probability of pure ALOHA.

The average energy required to transmit a frame is estimated as

$$\begin{aligned} E &= E_0 \sum_{i=0}^{\infty} (i+1)(1-p_a)^i \, p_a \\ &= \frac{E_0}{p_a} \end{aligned} \tag{10.85}$$

where E_0 is the energy required to send one frame. In dB, the above equation can be written as

$$E = -10 \log_{10} p_a \quad \text{dB} \tag{10.86}$$

Figure 10.13 shows the average energy required to transmit a frame for the CSMA/CA when $n = 50$ and $N = 10$. The solid line is the average energy of

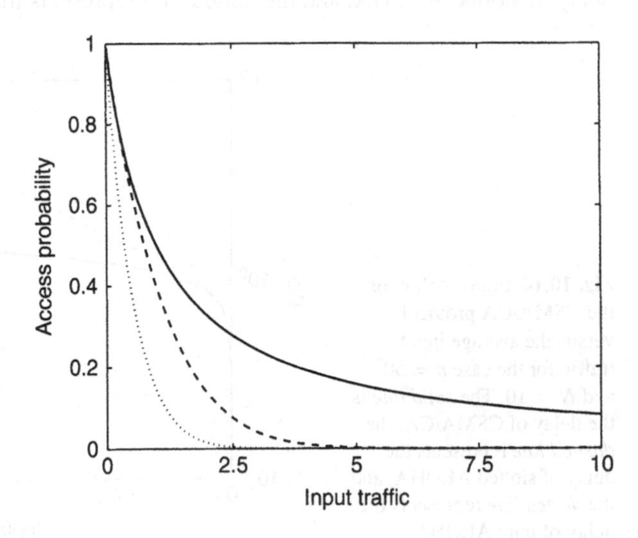

Fig. 10.12 Access probability of CSMA/CA protocol versus the average input traffic per time step for the case $n = 50$ and $N = 10$. The *solid line* is the access probability of CSMA/CA, the *dashed line* represents the access probability of slotted ALOHA, and the *dotted line* represents the access probability of pure ALOHA

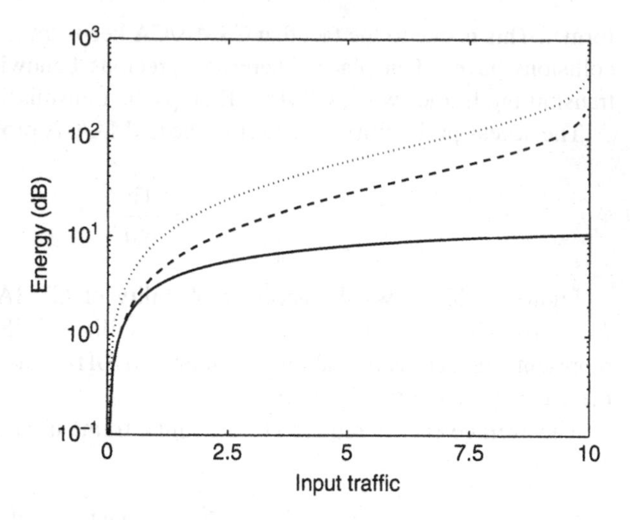

Fig. 10.13 Average energy to transmit a frame for the CSMA/CA protocol versus the average input traffic per time step for the case $n = 50$ and $N = 10$. The *solid line* is the energy for CSMA/CA, the *dashed line* represents the energy for slotted ALOHA, and the *dotted line* represents the energy for pure ALOHA

CSMA/CA, the dashed line represents the average energy of slotted ALOHA, and the dotted line represents the average energy of pure ALOHA.

The average number of attempts for a successful transmission is

$$\begin{aligned} n_a &= \sum_{i=0}^{\infty} i\,(1-p_a)^i\,p_a \\ &= \frac{1-p_a}{p_a} \end{aligned} \tag{10.87}$$

Figure 10.14 shows the delay of the CSMA/CA protocol when $n = 50$, $N = 10$, and $p = 1$. The solid line is the delay of CSMA/CA, the dashed line represents the delay of slotted ALOHA, and the dotted line represents the delay of pure ALOHA.

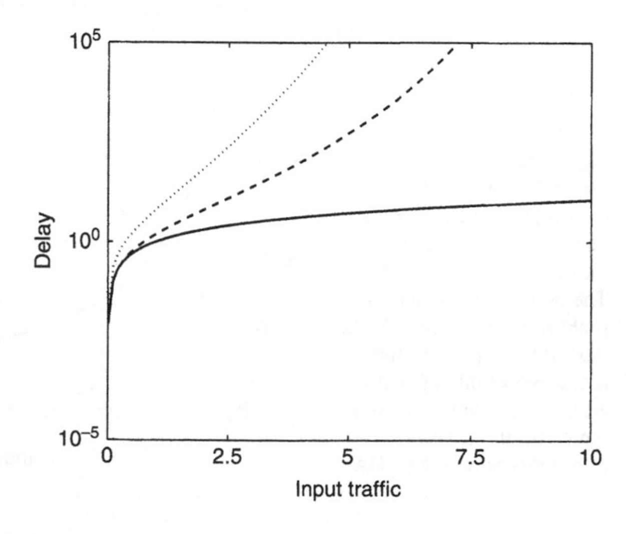

Fig. 10.14 Frame delay for the CSMA/CA protocol versus the average input traffic for the case $n = 50$ and $N = 10$. The *solid line* is the delay of CSMA/CA, the *dashed line* represents the delay of slotted ALOHA, and the *dotted line* represents the delay of pure ALOHA

10.7 IEEE 802.11: DCF Function for Ad Hoc Wireless LANs

A major advance in wireless communications is introduction of the IEEE 802.11 medium access control (MAC) and physical layer (PHY) standard. The specification includes three sublayers of the physical (PHY) layer: a frequency-hopping spread-spectrum (FHSS) physical layer, a direct-sequence spread-spectrum (DSSS) link layer, and a diffused infrared layer [9, 10]. IEEE 802.11 wireless LAN standard is used for infrastructure as well as ad hoc networks.

Infrastructure wireless networks have a central controller called *access point* (AP) that coordinates medium access among the users. This part of the protocol is referred to as point coordination function (PCF) and it occupies a short time period at the start of each transmitted frame as shown in Fig. 10.15. The frame starts with a point coordination function (PCF) period, which is a time period to enable prioritized access for control messages and time-critical traffic [11–13]. This form of *centralized medium access scheme* enables the IEEE 802.11 to offer some quality of service (QoS) guarantees through implementation of a *scheduling algorithm* at the AP.

AD hoc wireless networks do not have a central controller. Instead, each node or user attempts to access the shared medium on its own. This part of the protocol is referred to as distributed coordination function (DCF) and it follows the PCF period of each transmitted frame as shown in Fig. 10.15. This is a form of *distributed reservation scheme* that could provide statistical QoS guarantees.

10.7.1 IEEE 802.11: DCF Medium Access Control

The DCF MAC part of IEEE 802.11 standard is based on CSMA-CA (listen-before-talk) with rotating backoff window [11]. When a node receives a frame to be transmitted, it chooses a random backoff value, which determines the amount of time the node must wait until it is allowed to transmit its frame. A node stores this backoff value in a *backoff counter*. During periods in which the channel is clear, the node decrements its backoff counter. (When the channel is busy it does not decrement its backoff counter.) When the backoff counter reaches zero, the node transmits the frame. Since the probability that two nodes will choose the same backoff factor is small, collisions between frames are minimized. Collision detection, as is employed in Ethernet, cannot be used for the radio frequency transmissions of IEEE 802.11. The reason is that when a node is transmitting it cannot hear any other node in the system which may be transmitting, since its own signal will drown out any others arriving at the node.

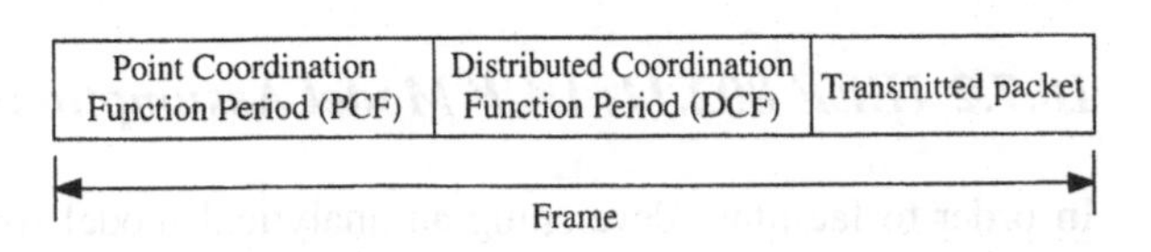

Fig. 10.15 Location of PCF and DCF periods in the frame of IEEE 802.11 protocol

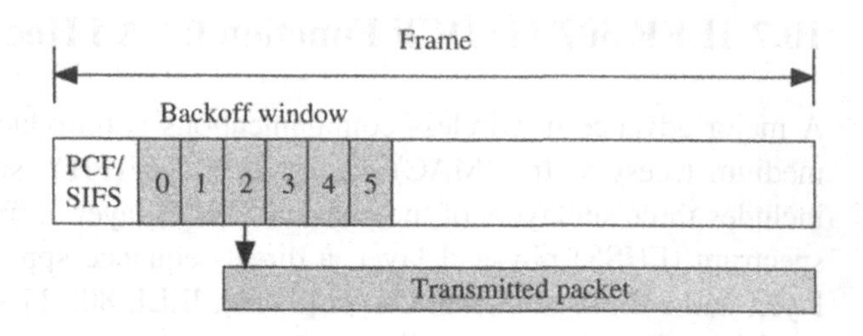

Fig. 10.16 DCF part of the IEEE 802.11 frame

In that sense, 802.11 could be classified as CSMA/CA but with provisions for reducing the chance of collisions through adoption of the reservation slots using the backoff counters. The slots have the effect of ensuring that a reduced number of users compete for access to the channel during any given reservation slot.

Figure 10.16 shows the DCF part of the IEEE 802.11 frame. After the PCF period (i.e., SIFS), there is the DCF period (i.e., DIFS) which is a contention window that is divided into reservation slots. The figure shows six such slots. The duration of a reservation slot depends on the propagation delay between stations. The rest of the frame is dedicated to transmitting the frames.

A station that intends to transmit senses if the channel is busy. It will then wait for the end of the current transmission and the PCF delay. It then randomly selects a reservation slot within the backoff window. The figure shows that a station in time reservation slot 2 starts transmitting a frame since the channel was not used during reservation slots 0 and 1.

Collisions occur when two or more stations select the same reservation slot. If another station started transmission at an earlier reservation slot, the station freezes its backoff counter and waits for the remaining content of this counter after the current transmission ends. We consider the behavior of one user, which we term the tagged user. Figure 10.17 shows the IEEE 802.11 MAC scheme as viewed by a certain user (called the *tagged user*). Figure 10.17 indicates that the tagged user, as indicated by the black circle, randomly selected reservation slot 7 to start transmission. So its backoff counter contains the value 7 now. However, another user starts transmission at reservation slot 2 as indicated by the grey box. Since the channel was quiet for two reservation slots (slot 0 and 1), the backoff counter of the tagged user will contain the value 5 at the end of the current frame.

Figure 10.17(b) shows the next frame. However, another user at reservation slot 1 started transmission. Since the channel was quiet for one reservation slots (slot 0), the backoff counter of the tagged user will contain the value 4 at the end of the current frame.

Figure 10.17(c) shows the next frame. The tagged user is successful in starting transmission since the channel was quite for four reservation slots (0, 1, 2, and 3).

10.7.2 IEEE 802.11: DCF Model Assumptions

In order to facilitate developing an analytical model for the 802.11 protocol, we model the backoff counters in each station in terms of allocation to a reservation

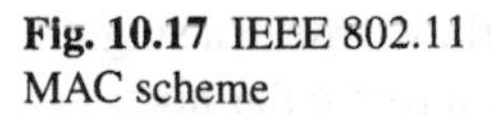

Fig. 10.17 IEEE 802.11 MAC scheme

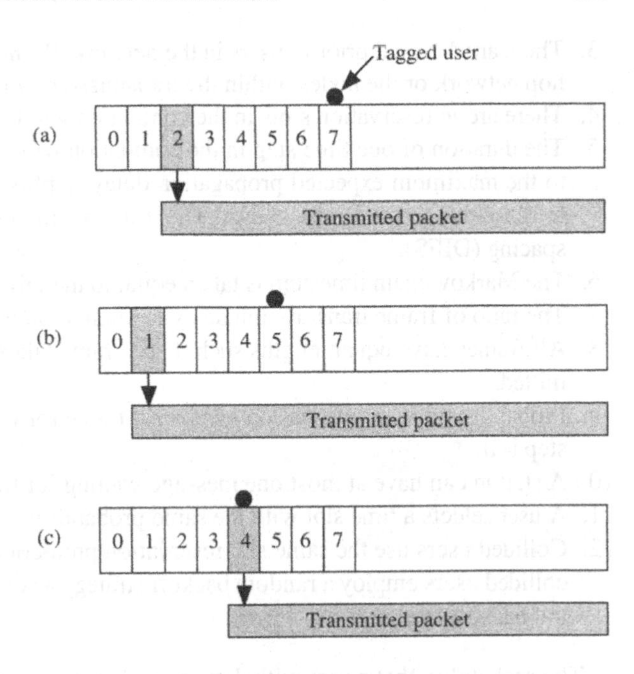

slot. Figure 10.18 shows the backoff window divided into w reservation slots. The value of w equals the maximum value that can be stored in the backoff counter of any user.

At the start of the frame, users in slot 0 sense that the channel is idle and are able to start transmission. When that happens, all users in the other slots will sense that the channel is busy and will refrain from transmitting their data. A collision could happen if two or more users in slot 0 begin to transmit simultaneously. However, the likelihood of this event happening is small since the number of users in slot 0 is lesser than the total number of users.

If all users in slot 0 remain idle, then users in slot 1 can transmit if they have frames to send. The same argument can be applied to the rest of the reservation slots.

We employ the following simplifying assumptions:

1. Since the current state of the channel depends only on its immediate past history, we can model the channel using Markov chain analysis.
2. The states of the Markov chain represent the states of the channel: idle, transmitting, and collided.

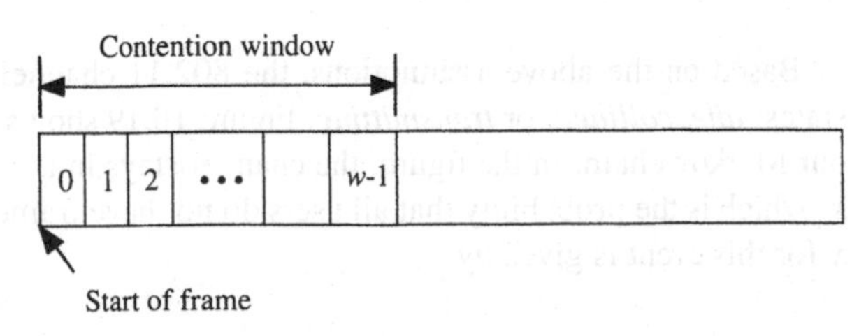

Fig. 10.18 Modeling the backoff counters in IEEE 802.11 MAC scheme in terms of w reservation slots

3. There are N equal priority users in the network. By network, we mean a single-hop network or the nodes within the transmission range of a particular node.
4. There are w reservation slots in the contention window.
5. The duration of one time step in the contention window is roughly taken equal to the maximum expected propagation delay τ_{p} plus the time it takes a station to sense the presence of a carrier. This time is called the distributed interframe spacing (DIFS).
6. The Markov chain time step is taken equal to the DIFS period.
7. The ratio of frame transmission delay to contention window delay is $n > 1$.
8. All frames have equal lengths such that a frame takes n time steps to be transmitted.
9. Probability that an idle station receives a frame for transmission during a time step is a.
10. A station can have at most one message waiting for transmission.
11. A user selects a time slot with the same probability.
12. Collided users use the same slot reservation protocol as uncollided users—i.e., collided users employ a random backoff strategy as opposed to binary exponential backoff strategy.

The probability that a user with data to send reserves a particular reservation slot is given by

$$\alpha = \frac{1}{w} \tag{10.88}$$

Slot reservation requests are representative of incoming traffic. Therefore, the user population seen by each slot in the contention window is given by

$$N' = \alpha N = \frac{N}{w} \tag{10.89}$$

The above equation indicates that the number of users that compete to access the medium is reduced from N to N'. We expect therefore that the chance of collisions is reduced compared to the case when there were no reservation slots.

The probability that k users attempt a transmission during a given reservation slot is given by

$$u_k = \binom{N'}{k} a^k (1-a)^{N'-k}; \qquad 0 \le k \le N' \tag{10.90}$$

Based on the above assumptions, the 802.11 channel could be in one of three states: *idle, collided*, or *transmitting*. Figure 10.19 shows the transition diagram for our Markov chain. In the figure, the channel stays in the idle state with probability x, which is the probability that all users do not have frames to send. The probability x for this event is given by

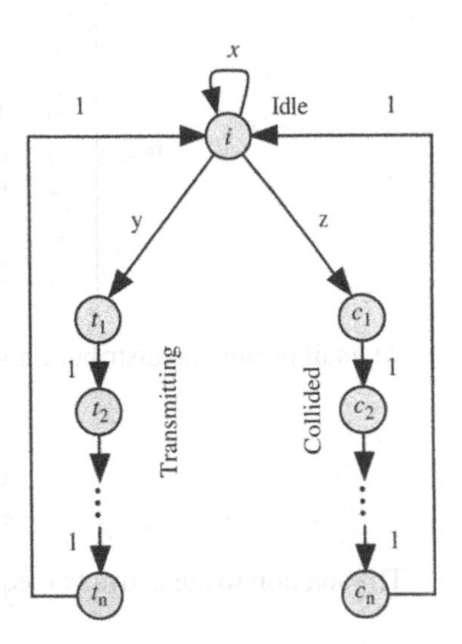

Fig. 10.19 State transition diagram for the 802.11 channel

$$x = u_0^w = (1 - a)^N \tag{10.91}$$

The channel moves from idle to transmitting state with probability y, which is the probability that only one user in any of the reservation slots requests a transmission and all users in the previous slots did not request to access the channel. The probability y for this event is given by

$$y = u_1 + u_0 u_1 + u_0^2 u_1 + \cdots + u_0^{w-1} u_1 = \frac{u_1 \left(1 - u_0^w\right)}{1 - u_0} \tag{10.92}$$

The channel moves from the idle to the collided state with probability z, which is the probability that more than one user in any of the reservation slots requests a transmission and all other users in the previous slots are idle. The probability z for this event is simply given by

$$z = 1 - x - y \tag{10.93}$$

We organize the distribution vector at equilibrium as follows:

$$\mathbf{s} = \left[\begin{array}{cccccccc} s_i & s_{t_1} & s_{t_2} & \cdots & s_{t_n} & s_{c_1} & s_{c_2} & \cdots & s_{c_n} \end{array}\right]^t \tag{10.94}$$

The corresponding state transition matrix of the channel for the case when $n = 3$ is given by

$$
\mathbf{P} = \begin{bmatrix} x & 0 & 0 & 1 & 0 & 0 & 1 \\ y & 0 & 0 & 0 & 0 & 0 & 0 \\ 0 & 1 & 0 & 0 & 0 & 0 & 0 \\ 0 & 0 & 1 & 0 & 0 & 0 & 0 \\ z & 0 & 0 & 0 & 0 & 0 & 0 \\ 0 & 0 & 0 & 0 & 1 & 0 & 0 \\ 0 & 0 & 0 & 0 & 0 & 1 & 0 \end{bmatrix} \tag{10.95}
$$

At equilibrium, the distribution vector is obtained by solving the two equations

$$
\mathbf{P}\,\mathbf{s} = \mathbf{s} \tag{10.96}
$$

$$
\sum \mathbf{s} = 1 \tag{10.97}
$$

The solution to the above two equations is simple:

$$
\mathbf{s} = \frac{1}{n(1-x)+1} \begin{bmatrix} 1 \\ y \\ y \\ \vdots \\ y \\ z \\ z \\ \vdots \\ z \end{bmatrix} \tag{10.98}
$$

When $w = 1$, the equilibrium distribution vector for IEEE 802.11 above becomes identical to the equilibrium distribution vector for CSMA/CA in Eq. (10.79).

10.7.3 IEEE 802.11: DCF Protocol Performance

The throughput is given by the equation

$$
\begin{aligned} \mathrm{Th} &= \sum_{i=1}^{n} s_{t_i} \\ &= \frac{ny}{n(1-x)+1} \end{aligned} \tag{10.99}
$$

Figure 10.20 shows the throughput of IEEE 802.11 when $w = 8$, $n = 10$, and $N = 32$. The upper solid line is the throughput of IEEE 802.11/DCF, the lower solid line is the throughput of CSMA/CA, the dashed line represents the throughput of slotted ALOHA, and the dotted line represents the throughput of pure ALOHA.

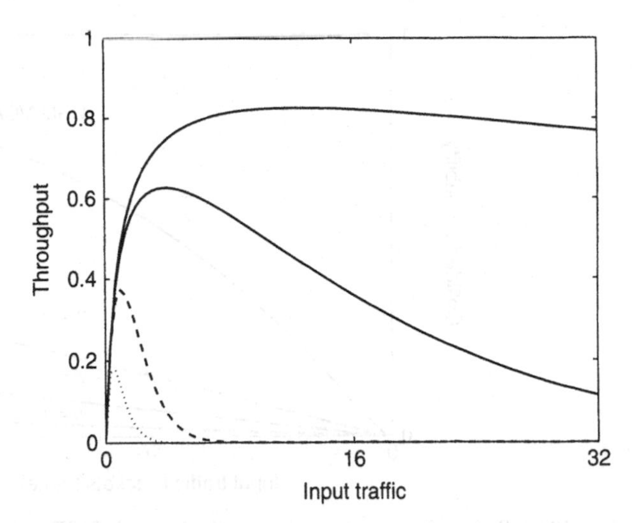

Fig. 10.20 Throughput for the IEEE 802.11/DCF protocol versus the average input traffic when $w = 8$, $n = 10$, and $N = 32$. The *upper solid line* is the throughput of IEEE 802.11/DCF, the *lower solid line* is the throughput of CSMA/CA, the *dashed line* represents the throughput of slotted ALOHA, and the *dotted line* represents the throughput of pure ALOHA

The throughput is high due to several factors such as distribution of users among reservation channels and assuming the wireless medium is error free.

Changing the number of reservation slots has a direct effect on the throughput. To test this, we selected values of w as 4, 8, and 16. Figure 10.21 shows the throughput for the IEEE 802.11/DCF protocol versus the average input traffic when $n = 10$ and $N = 32$ and $w = 4$, 8, and 16. Dashed line represents the throughput of slotted ALOHA, and the dotted line represents the throughput of pure ALOHA. We note two things from this figure. First, reducing the value of n leads to lesser throughput since more significant amount of time is spent listening to the carrier.

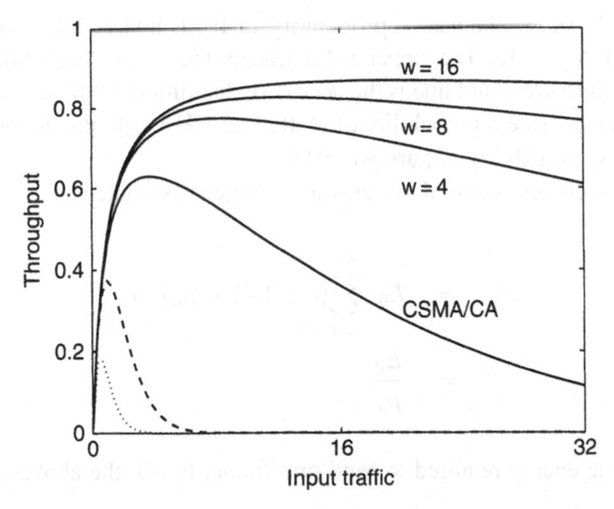

Fig. 10.21 Throughput for the IEEE 802.11/DCF protocol versus the average input traffic when $n = 10$ and $N = 32$ and $w = 4$, 8, and 16. *Dashed line* represents the throughput of slotted ALOHA and the *dotted line* represents the throughput of pure ALOHA

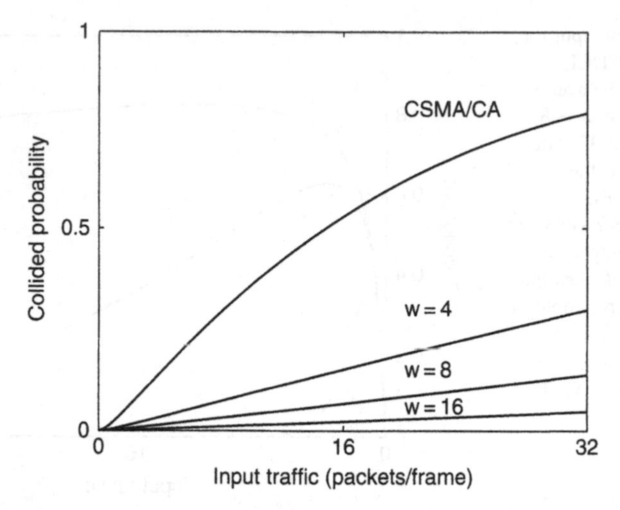

Fig. 10.22 Probability of collision for the IEEE 802.11/DCF protocol versus the average input traffic when $n = 10$ and $N = 32$ and $w = 4$, 8, and 16

Second, adoption of reservation slots distributes the user requests among the slots. This leads to less chance of collisions and more throughput results. To prove this last point, the collision probability is shown in Fig. 10.22.

The user access probability is given by

$$p_a = \frac{\text{Th}}{Na} \tag{10.100}$$

Figure 10.23 shows the access probability of IEEE 802.11/DCF when $w = 8$, $n = 10$, and $N = 32$. The upper solid line is the access probability of IEEE 802.11/DCF, the lower solid line is the access probability of CSMA/CA, the dashed line represents the access probability of slotted ALOHA, and the dotted line represents the access probability of pure ALOHA.

The average energy required to transmit a frame is estimated as

$$\begin{aligned} E &= E_0 \sum_{i=0}^{\infty} (i+1)(1-p_a)^i \, p_a \\ &= \frac{E_0}{p_a} \end{aligned} \tag{10.101}$$

where E_0 is the energy required to send one frame. In dB, the above equation can be written as

$$E = -10 \log_{10} p_a \quad \text{dB} \tag{10.102}$$

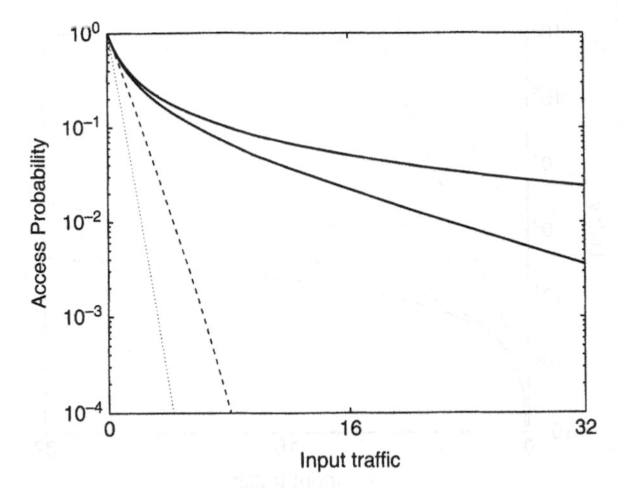

Fig. 10.23 Access probability for the IEEE 802.11 protocol versus the average input traffic when $w = 8$, $n = 10$, and $N = 32$. The *upper solid line* is the access probability of IEEE 802.11/DCF, the *lower solid line* is the access probability of CSMA/CA, the *dashed line* represents the access probability of slotted ALOHA, and the *dotted line* represents the access probability of pure ALOHA

Figure 10.24 shows the energy of IEEE 802.11/DCF when $w = 8$, $n = 10$, and $N = 32$. The lower solid line is the energy of IEEE 802.11/DCF, the upper solid line is the energy of CSMA/CA, the dashed line represents the energy of slotted ALOHA, and the dotted line represents the energy of pure ALOHA.

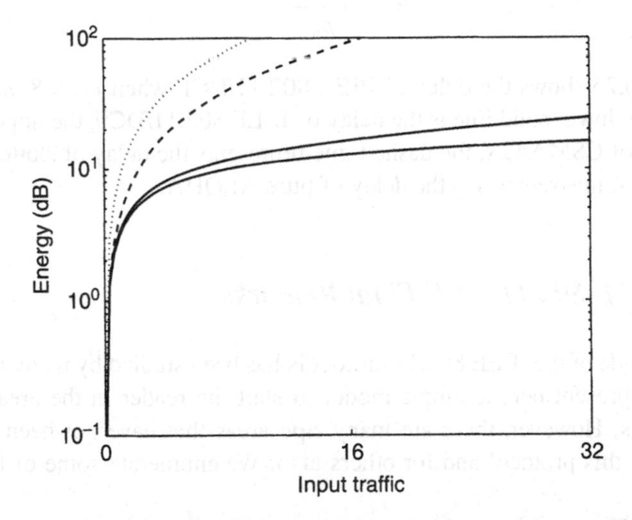

Fig. 10.24 Energy for the IEEE 802.11/DCF protocol versus the average input traffic when $w = 8$, $n = 10$, and $N = 32$. The *lower solid line* is the energy of IEEE 802.11/DCF, the *upper solid line* is the energy of CSMA/CA, the *dashed line* represents the energy of slotted ALOHA, and the *dotted line* represents the energy of pure ALOHA

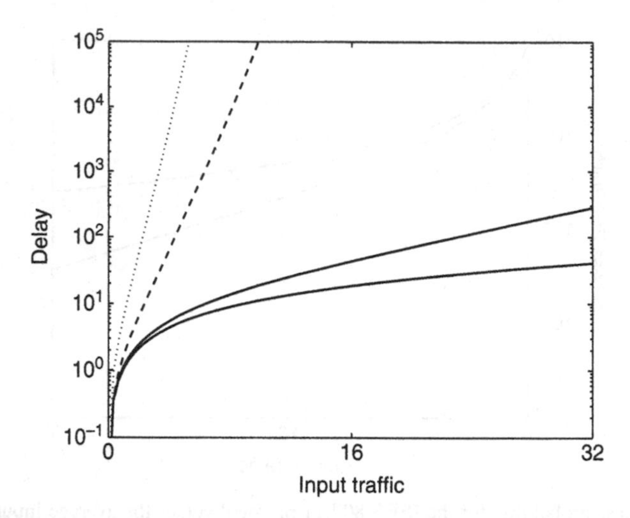

Fig. 10.25 Delay for the IEEE 802.11/DCF protocol versus the average input traffic when $w = 8$, $n = 10$, and $N = 32$. The *lower solid line* is the delay of IEEE 802.11/DCF, the *upper solid line* is the delay of CSMA/CA, the *dashed line* represents the delay of slotted ALOHA, and the *dotted line* represents the delay of pure ALOHA

The average number of attempts for a successful transmission is

$$\begin{aligned} n_a &= \sum_{i=0}^{\infty} i\,(1-p_a)^i\,p_a \\ &= \frac{1-p_a}{p_a} \end{aligned} \tag{10.103}$$

Figure 10.25 shows the delay of IEEE 802.11/DCF when $w = 8$, $n = 10$, and $N = 32$. The lower solid line is the delay of IEEE 802.11/DCF, the upper solid line is the delay of CSMA/CA, the dashed line represents the delay of slotted ALOHA, and the dotted line represents the delay of pure ALOHA.

10.7.4 IEEE 802.11/DCF Final Remarks

The DCF mode of the IEEE 802.11 protocols has been studied by many researchers. We tried to present here a simple model to start the reader in the area of modeling protocols. However, there are many ripe areas that have not been adequately explored for this protocol and for others also. We enumerate some of these directions:

1. Channel errors have not been considered. This is a physical layer problem but could also be considered in a cross-layer modeling. What matters here is to obtain the probability that a frame or packet is in error.

2. Simple backoff strategies were used here in order to obtain simple expressions. Adopting binary exponential backoff (BEB) strategy would result in a model that requires an iterative solution.
3. Perhaps the most serious deficiency in the models in this chapter is the implicit assumption that no new traffic arrives while the user is attempting to access the channel. This amounts to assuming the *transmit* buffer has a single storage location only. Using a more realistic buffer would require solving two queuing systems.

10.8 IEEE 802.11: PCF Function for Infrastructure Wireless LANs

In the IEEE 802.11, there is a feature to support quality of service (QoS) in infrastructure LANs through implementation of the point coordination function (PCF). A central controller polls the users to determine which user can access the medium in the next frame. Such medium access scheme is used in *infrastructure* wireless LANs where a central controller is provided to coordinate the activities of the users or mobile terminals. The central controller in wireless LANs such as HiperLAN/2 is called the access point (AP).

The central controller prevents collisions from taking place since it coordinates access to the channel through implementation of some kind of reservation or scheduling algorithm. Thus we can classify the IEEE 802.11/PCF as a reservation-based scheme that is collision-free.

In IEEE 802.11, using the PCF function, time is divided into time steps as shown in Fig. 10.26. There are two directions of communication: the uplink from the users to the central controller and downlink from the central controller to the users. The uplink frames consist of a request phase and a data phase. In the request phase, users issue their requests and in the data phase, users send their data when they are told to do so by the central controller or scheduler.

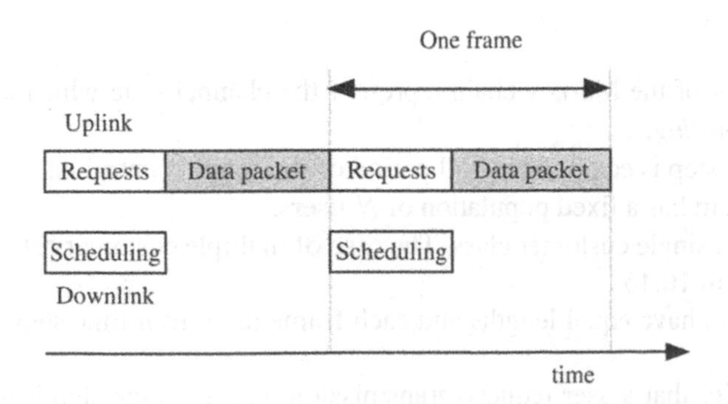

Fig. 10.26 Frame structure for IEEE 802.11 protocol using the PCF function

The data phase transmits one frame in the uplink direction from one user based on some scheduling policy. The receiver is told by the central controller to expect an arriving frame. Alternatively, all users receive the transmitted frame and decide whether it was destined to them or not.

IEEE 802.11/PCF is efficient when the time taken by the request phase is much smaller than the data phase if both the uplink and downlink use the same channel. If there are two separate channels, then the protocol is efficient as long as the delay of the request phase is smaller than the duration of the data phase. If the central controller requires a time that is comparable to the data phase, then the channel is not well used and other MAC techniques should be used.

10.8.1 IEEE 802.11: PCF Medium Access Control

The central controller communicates with the users through the downlink. The controller polls all users in the system and schedules only one user to access the medium based on some scheduling strategy. In addition, the controller informs the receivers to be ready for the transmitted frames. In a nonpersistent PCF protocol, users that were not granted access to the channel issue another request to transmit with probability p when polled by the access point on the next poll period. A large value for the persistence probability p would ensure that the user gets to compete in the next poll period. On the other hand, a small p reduces the number of contending users. We propose here to adopt a *variable* or *adaptive* persistence strategy. A user with heavy traffic would use a large value for p, while a user with low traffic would use a small value for p.

10.8.2 IEEE 802.11: Nonpersistent PCF Model Assumptions

In this section, we perform Markov chain analysis of the IEEE 802.11 (PCF) protocol. We make the following assumptions for our analysis of the simple protocol:

1. The states of the Markov chain represent the channel state which could be *idle* or *transmitting*.
2. The time step is equal to the poll period of the central controller.
3. The system has a fixed population of N users.
4. There is a single customer class. The case of multiple customer classes is treated in Problem 10.46.
5. All frames have equal lengths and each frame requires n time steps to be transmitted.
6. Probability that a user requests transmission during a time step is a and probability that the user is idle during a time step is $b = 1 - a$.

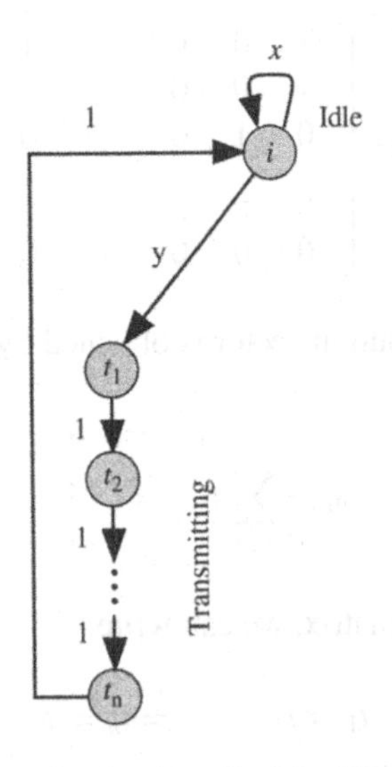

Fig. 10.27 State transition diagram for the channel of the IEEE 802.11 protocol employing the PCF function

7. If a user is refused access to the channel, it will issue a request at the next poll period with persistence probability *equal* to the frame arrival probability (i.e., $p = a$).
8. The channel is assumed to be always available for transmission in every time step.

The state diagram for the channel is shown in Fig. 10.27. In the figure, x is the probability that all users are idle and do not issue a request to access the channel during the current time step and $y = 1 - x$. No collision takes place when more than one request is present due to polling. The controller will then decide which station is granted access and inform the others of that decision through ACK/NAK mechanism. We can write

$$x = b^N \tag{10.104}$$

We organize the distribution vector at equilibrium as follows:

$$\mathbf{s} = \begin{bmatrix} s_i & s_{t_1} & s_{t_2} & \cdots & s_{t_n} \end{bmatrix}^t \tag{10.105}$$

where s_i is the idle state and s_{t_j} is the state when the channel is transmitting the jth part of the frame. The corresponding transition matrix of the channel is given by

$$\mathbf{P} = \begin{bmatrix} x & 0 & 0 & \cdots & 1 \\ y & 0 & 0 & \cdots & 0 \\ 0 & 1 & 0 & \cdots & 0 \\ \vdots & \vdots & \vdots & \ddots & \vdots \\ 0 & 0 & 0 & \cdots & 0 \end{bmatrix} \tag{10.106}$$

At equilibrium, the distribution vector is obtained by solving the two equations

$$\mathbf{P}\,\mathbf{s} = \mathbf{s} \tag{10.107}$$

$$s_i + \sum_{i=1} t_i = 1 \tag{10.108}$$

From the structure of the matrix, we can write

$$t_1 = t_2 = \cdots = t_n = t \tag{10.109}$$

We also have

$$t = y\, s_i \tag{10.110}$$

Using simple algebra, we can finally obtain the distribution vector as

$$\mathbf{s} = \frac{1}{ny+1} \begin{bmatrix} 1 & y & \cdots & y \end{bmatrix}^t \tag{10.111}$$

10.8.3 IEEE 802.11: Nonpersistent PCF Protocol Performance

Having obtained the transition matrix, we are able to find the performance of the nonpersistent IEEE 802.11/PCF channel.

The input traffic is defined here as the average number of requests per time step and is given by

$$\begin{aligned} N_a(\text{in}) &= \sum_{i=0}^{N} i \binom{N}{i} a^i \,(1-a)^{N-i} \\ &= N\,a \end{aligned} \tag{10.112}$$

The throughput of the system is given by

$$\text{Th} = \sum_{i=1} n s_{t_i} = \frac{ny}{ny+1} \tag{10.113}$$

Notice that when $n \rightarrow \infty$, the maximum throughput approaches 1. This is because comparatively little time is wasted polling the users. On the other hand,

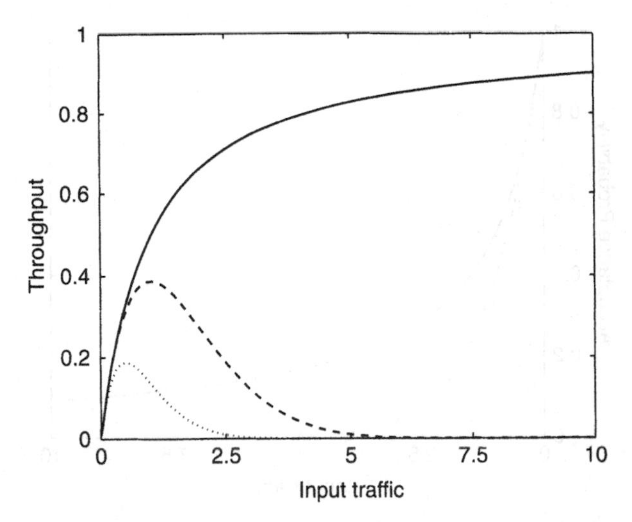

Fig. 10.28 The throughput for nonpersistent IEEE 802.11/PCF versus the average input traffic when $n = 50$ and $N = 10$. The *solid line* is the throughput of nonpersistent IEEE 802.11/PCF, the *dashed line* represents the throughput of slotted ALOHA, and the *dotted line* represents the throughput of pure ALOHA

when $n \approx 1$, the maximum throughput approaches 50%. This is because approximately 50% of the time the access point is polling the users and no frames are being transmitted during the period.

Figure 10.28 shows the throughput of the nonpersistent IEEE 802.11/PCF protocol when $n = 50$ and $N = 10$. The solid line is the throughput of nonpersistent IEEE 802.11/PCF, the dashed line represents the throughput of slotted ALOHA, and the dotted line represents the throughput of pure ALOHA.

Because all users are the same, the throughput seen by one user is given by

$$\text{Th(user)} = \frac{\text{Th}}{N} = \frac{ny}{N(ny+1)} \tag{10.114}$$

When n is very large, the maximum throughput for each user will approach $1/N$. This is to be expected due to the fair sharing of the medium among the users.

Define p_a as the access probability for one user which is given by

$$p_a = \frac{\text{Th(user)}}{a} = \frac{\text{Th}}{N_a(\text{in})} \tag{10.115}$$

Figure 10.29 shows the access probability of nonpersistent IEEE 802.11/PCF when $n = 50$ and $N = 10$. The solid line is the access probability of nonpersistent IEEE 802.11/PCF, the dashed line represents the access probability of slotted ALOHA, and the dotted line represents the access probability of pure ALOHA.

Having found the acceptance probability, we are able to determine the average number of attempts before a user gets access to the medium:

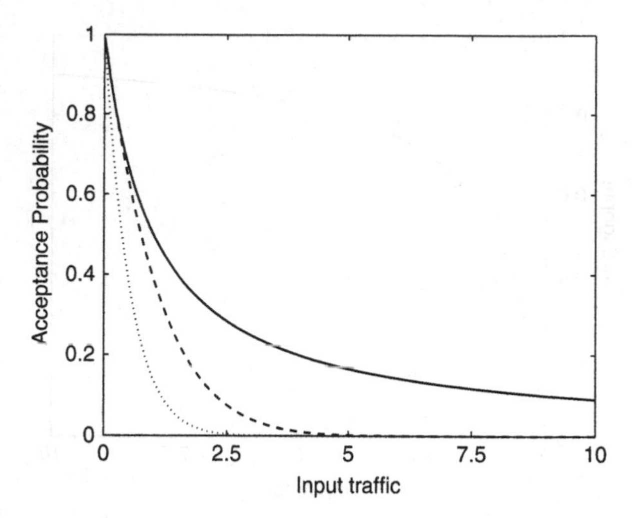

Fig. 10.29 Delay for the nonpersistent IEEE 802.11 protocol versus the average input traffic per time step when $n = 50$ and $N = 10$. The *solid line* is the access probability of IEEE 802.11/PCF, the *dashed line* represents the access probability of slotted ALOHA, and the *dotted line* represents the access probability of pure ALOHA

$$n_a = \sum_{k=0}^{\infty} k\,(1 - p_a)^k \; p_a \tag{10.116}$$

$$= \frac{1 - p_a}{p_a} \tag{10.117}$$

Figure 10.30 shows the delay of nonpersistent IEEE 802.11/PCF when $n = 50$ and $N = 10$. The solid line is the delay of nonpersistent IEEE 802.11, the dashed line represents the delay of slotted ALOHA, and the dotted line represents the delay of pure ALOHA.

10.8.4 IEEE 802.11: 1-Persistent PCF

This section deals with modeling the point coordination function (PCF) of the 1-persistent IEEE 802.11 protocol. We make the following assumptions for our analysis of the simple protocol.

1. The states of the Markov chain represent the number of queued users requesting access to the channel at the start of any time step.
2. Time is divided into time steps such that each time step starts with a control phase and then a data phase. In the control phase, all arriving requests are processed and one access is granted. In the data phase, the user that was granted access to the medium transfers its data.
3. The system has a fixed population of N users.

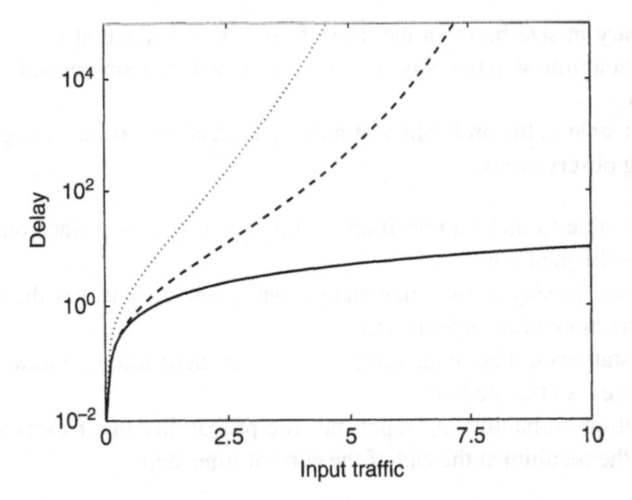

Fig. 10.30 Delay for the nonpersistent IEEE 802.11/PCF protocol versus the average input traffic per time step when $n = 50$ and $N = 10$. The *solid line* is the delay of nonpersistent IEEE 802.11/PCF, the *dashed line* represents the delay of slotted ALOHA, and the *dotted line* represents the delay of pure ALOHA

4. There is a single customer class. The case of multiple customer classes is treated in Problem 10.46.
5. A user can have, at most, one message waiting for transmission. At the end of certain time step, users that have requests pending cannot issue more requests at the next time step.
6. A user that is transmitting at a certain time step can issue a new request at the next time step.
7. Users that were denied access will attempt a retransmission at the next time step.
8. Probability that a user requests transmission is a and probability that the user is idle is $b = 1 - a$.
9. The channel is assumed to be always available for transmission in every time step.

The state diagram for the system is shown in Fig. 10.31 where the user population is N. Only transitions out of state 2 are shown for simplicity. In the figure, state i indicates that there are i users requesting transmission. Based on the above assumptions, it is impossible for the system to have N queued requests since one request is always guaranteed to be processed in the same time step. The request

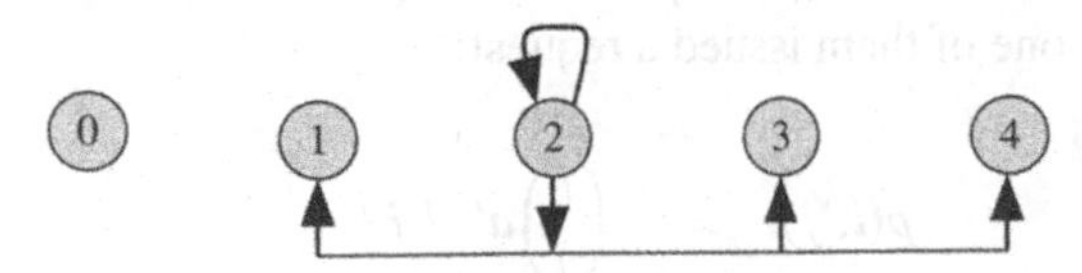

Fig. 10.31 State transition diagram for the single access reservation-based MAC protocol when the system has $N = 5$ users. Only transitions out of state 2 are shown for simplicity

queue will vary in size between the limits 0 and $N - 1$ since at worst N requests could arrive at a time step but only $N - 1$ requests will remain queued at the end of the time step.

Starting at state j, the probability of making a transition to state i is governed by the following observations:

- It is impossible to make a transition to state i if $i < j - 1$ since only one user can access the medium.
- It is possible to stay in the same state j when one user leaves the system and exactly one more user requests access.
- The next state would be in the range $j \le i < N$ depending on how many users request access to the medium.
- The transition probability p_{ij} represents the probability that i users still request access to the medium at the end of the current time step.

According to the assumptions we employed, the resulting transition matrix is lower $N \times N$ Hessenberg matrix in which all the element $p_{ij} = 0$ for $j > i + 1$. Suppose there are j queued requests at the end of a time step. In the next time step, we are sure that j requests will be issued plus a possible 0 to $N - j$ additional requests coming from the other users.

We organize the distribution vector at equilibrium as follows:

$$\mathbf{s} = \begin{bmatrix} s_0 & s_1 & s_2 & \cdots & s_{N-1} \end{bmatrix}^t \tag{10.118}$$

where s_i corresponds to the system state when there are i users with unsatisfied (queued) requests. The corresponding transition matrix of the channel is given by

$$\mathbf{P} = \begin{bmatrix} y & p(N-1,0) & \cdots & 0 & 0 \\ p(N,2) & p(N-1,1) & \cdots & 0 & 0 \\ p(N,3) & p(N-1,2) & \cdots & 0 & 0 \\ p(3,4) & p(N-1,3) & \cdots & 0 & 0 \\ \vdots & \vdots & \ddots & \vdots & \vdots \\ p(N,N-2) & p(N-1,N-3) & \cdots & p(2,0) & 0 \\ p(N,N-1) & p(N-1,N-2) & \cdots & p(2,1) & p(1,0) \\ p(N,N) & p(N-1,N-1) & \cdots & p(2,2) & p(1,1) \end{bmatrix} \tag{10.119}$$

where $p(i, j)$ is the probability that there were i idle users and j of them issued requests to access the medium. y represents the probability that there were N idle users and at most one of them issued a request:

$$p(i,j) = \binom{i}{j} a^{i-j}\, b^j \tag{10.120}$$

$$y = p(N,0) + p(N,1) \tag{10.121}$$

For the case $N = 5$, the transition matrix is 5×5 and is given by

$$\mathbf{P} = \begin{bmatrix} y & p(4,0) & 0 & 0 & 0 \\ p(5,2) & p(4,1) & p(3,0) & 0 & 0 \\ p(5,3) & p(4,2) & p(3,1) & p(2,0) & 0 \\ p(5,4) & p(4,3) & p(3,2) & p(3,1) & p(1,0) \\ p(5,5) & p(4,4) & p(3,3) & p(2,2) & p(1,1) \end{bmatrix} \qquad (10.122)$$

10.8.5 1-Persistent IEEE 802.11/PCF Performance

Having obtained the transition matrix, we are able to find the performance of the 1-persistent IEEE 802.11/PCF channel.

The throughput of the system is given by

$$\text{Th} = s_0 \left[1 - p(N,0)\right] + (1 - s_0) \qquad (10.123)$$

$$= 1 - s_0 p(N,0) \qquad (10.124)$$

The first term corresponds to the probability that the queue is empty and one or more requests arrive and the second term corresponds to the probability that the queue is not empty. Figure 10.32 shows the throughput of the 1-persistent IEEE 802.11 protocol when $N = 10$. The solid line is the throughput of 1-persistent IEEE 802.11/PCF, the dashed line represents the throughput of slotted ALOHA, and the dotted line represents the throughput of pure ALOHA.

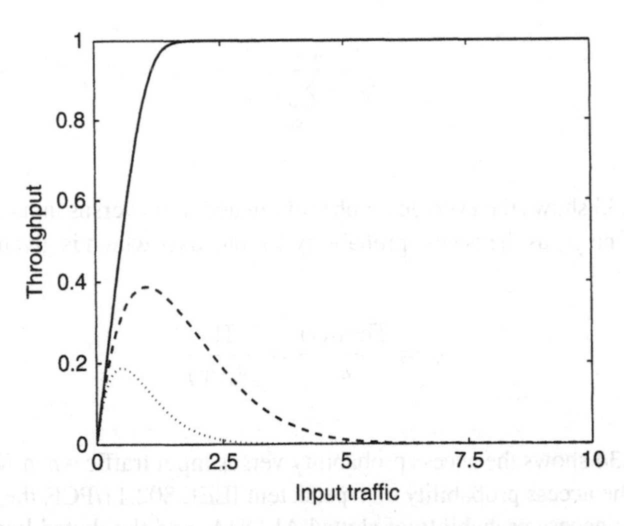

Fig. 10.32 The throughput for 1-persistent IEEE 802.11/PCF versus the average input traffic when $N = 10$. The *solid line* is the throughput of 1-persistent IEEE 802.11/PCF, the *dashed line* represents the throughput of slotted ALOHA, and the *dotted line* represents the throughput of pure ALOHA

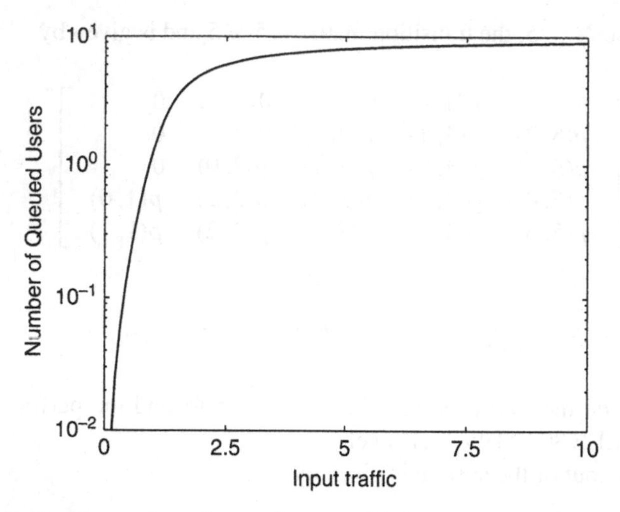

Fig. 10.33 The average number of queued users with frames to send in 1-persistent IEEE 802.11/PCF versus the average input traffic when $N = 10$

Notice that the throughput reaches 100% as soon as the input traffic approaches 1 since at this level there is more than one user with frames to send and there are no collisions. Further, the use of transmit buffers ensures that arriving requests are not dropped as long as the buffer is not full.

The average number of queued users with frames to send is given by

$$Q_a = \sum_{i=0}^{N-1} i \ s_i \tag{10.125}$$

Figure 10.33 shows the average number of queued users versus input traffic when $N = 10$. Define p_a as the access probability for one user which is given by

$$p_a = \frac{\text{Th(user)}}{a} = \frac{\text{Th}}{N_a(\text{in})} \tag{10.126}$$

Figure 10.34 shows the access probability versus input traffic when $N = 10$. The solid line is the access probability of 1-persistent IEEE 802.11/PCF, the dashed line represents the access probability of slotted ALOHA, and the dotted line represents the access probability of pure ALOHA.

Having found the acceptance probability, we are able to determine the average number of attempts before a user gets access to the medium:

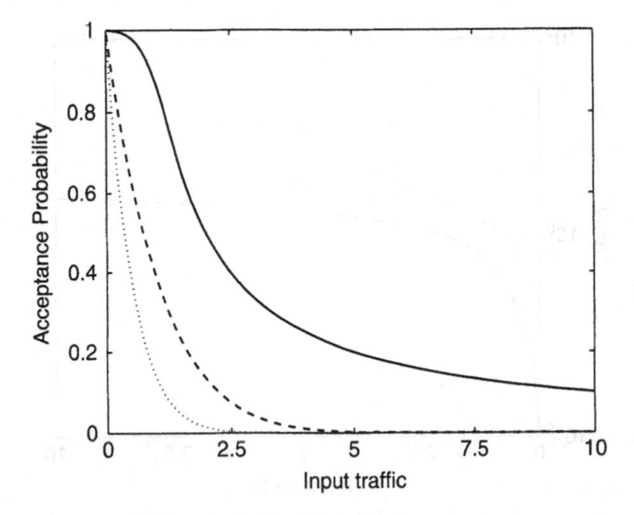

Fig. 10.34 The access probability in IEEE 802.11/PCF versus the average input traffic when $N = 10$. The *solid line* is the access probability of 1-persistent IEEE 802.11/PCF, the *dashed line* represents the access probability of slotted ALOHA, and the *dotted line* represents the access probability of pure ALOHA

$$n_a = \sum_{k=0}^{\infty} k\,(1-p_a)^k\ p_a \qquad (10.127)$$

$$= \frac{1-p_a}{p_a} \qquad (10.128)$$

Figure 10.35 shows the delay of IEEE 802.11 when $N = 10$. The solid line is the delay of 1-persistent IEEE 802.11, the dashed line represents the delay of slotted ALOHA, and the dotted line represents the delay of pure ALOHA.

10.8.6 1-Persistent IEEE 802.11/PCF User Performance

The previous section modeled the states of all the N users of the 1-persistent IEEE 802.11/PCF. In this section, we study an individual user, usually called the tagged user. Specifically, we would like to study the access probability p_a which represents the probability that a request for transmitting data will be granted.

Because all users are the same, the throughput seen by the tagged user is given by

$$\text{Th(user)} = \frac{\text{Th}}{N} \qquad (10.129)$$

Define p_a as the access probability for our tagged user which is given by

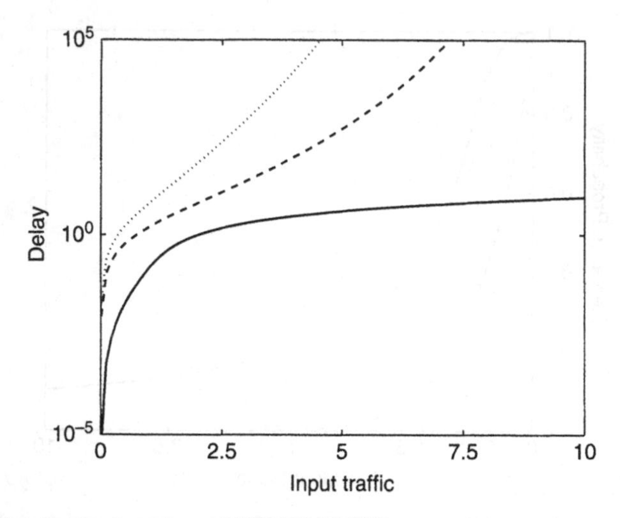

Fig. 10.35 Delay for the 1-persistent IEEE 802.11/PCF protocol versus the average input traffic when $N = 10$. The *solid line* is the delay of 1-persistent IEEE 802.11/PCF, the *dashed line* represents the delay of slotted ALOHA, and the *dotted line* represents the delay of pure ALOHA

$$p_a = \frac{\text{Th(user)}}{a} = \frac{\text{Th}}{N_a(\text{in})} \tag{10.130}$$

Thus we obtain

$$p_a = \frac{1 - s_0 p(N, 0)}{aN} \tag{10.131}$$

At light loading condition such that $aN \ll 1$, the acceptance probability becomes $p_a = 1$, which makes sense since most of the time only one user has a request and is immediately granted access to the medium. For heavy traffic $a \approx 1$, the acceptance probability $p_a \approx 1/N$, which also makes sense since most of the time all users have requests and only one is granted access with probability $1/N$.

Having found the acceptance probability, we are able to determine the average number of attempts before a user gets access to the medium:

$$n_a = \sum_{k=0}^{\infty} k\,(1 - p_a)^k \; p_a \tag{10.132}$$

$$= \frac{1 - p_a}{p_a} \tag{10.133}$$

Example 4 A 1-persistent IEEE 802.11 employing the PCF function serves 8 customers and the channel bit rate is 1,920 kbps, which is used for a wireless microcellular system. Each customer is assumed to issue 20 requests each second and the average length of a frame is 5.12 kb. Obtain the performance of this system.

First, we must find the time step value knowing the duration of an average frame length:

$$\text{time step value} = 5.12/1,920 = 2.7 \text{ ms}$$

To find the frame arrival probability a per time step, we need to estimate the number of requests issued by the user over some observation time and we also need to find the number of time steps during this same observation period. Take the observation period to be 1 s. Thus the total user traffic during this period is

$$N_a(\text{in}) = \text{user request rate} \times 1 = 20 \text{ requests}$$

The number of time steps during this period n is given by

$$n = \left\lceil \frac{1 \text{ s}}{T} \right\rceil = 375$$

The number of requests from the binomial distribution is given from the relation

$$a \times n = N_a(\text{in})$$

Thus we have the user request probability as

$$a = 0.0533$$

The transition matrix will be

$$\mathbf{P} = \begin{bmatrix} 0.9357 & 0.6814 & 0 & 0 & 0 & 0 & 0 & 0 \\ 0.0573 & 0.2687 & 0.7198 & 0 & 0 & 0 & 0 & 0 \\ 0.0065 & 0.0454 & 0.2433 & 0.7603 & 0 & 0 & 0 & 0 \\ 0.0005 & 0.0043 & 0.0343 & 0.2433 & 0.8031 & 0 & 0 & 0 \\ 0 & 0.0002 & 0.0026 & 0.0343 & 0.1810 & 0.8484 & 0 & 0 \\ 0 & 0 & 0.0001 & 0.0026 & 0.0153 & 0.1434 & 0.8962 & 0 \\ 0 & 0 & 0 & 0.0001 & 0.0006 & 0.0081 & 0.1010 & 0.9467 \\ 0 & 0 & 0 & 0 & 0 & 0.0002 & 0.0028 & 0.0533 \end{bmatrix}$$

The distribution vector for the request queue states is

$$\mathbf{s} = \begin{bmatrix} 0.8988 & 0.0848 & 0.0145 & 0.0018 & 0.0002 & 0 & 0 & 0 \end{bmatrix}^t$$

We see that approximately 90% of the time the queue is empty, which is the value of the first element of the vector. The performance figures are

$$\begin{aligned} \text{Th} &= 0.4203 \quad \text{frames/time step} \\ Q_a &= 0.1198 \quad \text{requests} \\ p_a &= 0.832 \\ n_a &= 0.202 \quad \text{time steps} \end{aligned}$$

■

10.9 IEEE 802.11e: Quality of Service Support

The IEEE 802.11e provided enhancements over the legacy IEEE 802.11 to be able to provide quality of service support such as voice over wireless and streaming multimedia. True to our strategy throughout this book, we strive to provide simplified analyses. The reader can then carry this approach further and develop more sophisticated models.

Let us start by briefly explaining how quality of service is supported in this protocol. The distributed coordination function (DCF) of legacy IEEE 802.11 is now replaced with enhanced distributed channel access (EDCA). High-priority traffic is given better chance of accessing the channel compared to low-priority traffic. On average, high-priority traffic waits less and experiences higher throughputs. We can extend the analysis in Section 10.7 to describe the IEEE 802.11e using the following assumptions:

1. There are two user priority classes: class 1 is the high-priority users and class 2 is the lower-priority users.
2. There are two contention windows: w_1 for class 1 and $w_2 < w_1$ for class 2.
3. The probability that a user has a frame to send is a.
4. $\gamma < 1$ is the probability that a transmitted frame belongs to class 1, and $1 - \gamma$ the probability that the frame belongs to class 2 traffic.

The probability that a class 1 user with data to send reserves a particular reservation slot is given by

$$\alpha_1 = \frac{1}{w_1} \tag{10.134}$$

Similarly, $\alpha_2 = 1/w_2$ is defined for class 2 users. Slot reservation requests are representative of incoming traffic. The user population seen by each slot in the contention windows of the two service classes is given by

$$N_1 = \frac{N}{w_1} \tag{10.135}$$

$$N_2 = \frac{N}{w_2} \tag{10.136}$$

The probability that k class 1 users attempt a transmission during a given reservation slot is given by

$$u_k = \binom{N_1}{k} a_1^k (1 - a_1)^{N_1 - k}; \qquad 0 \le k \le N_1 \tag{10.137}$$

Likewise, the probability that k class 2 users attempt a transmission during a given reservation slot is given by

$$v_k = \binom{N_2}{k} a_2^k (1 - a_2)^{N_2 - k}; \qquad 0 \le k \le N_2 \tag{10.138}$$

Figure 10.36 shows the IEEE 802.11e channel state transition diagram. We notice that as far as the channel is concerned, it could be in one of four states:

- *idle* when no frames are being transmitted
- *collided* when two or more frames are being transmitted simultaneously
- *transmitting class 1 traffic* when exactly one class 1 frame is being transmitted
- *transmitting class 2 traffic* when exactly one class 2 frame is being transmitted

The channel stays in the idle state with probability x given by

$$x = u_0\, v_0 = (1 - a_1)^N\, (1 - a_2)^N \tag{10.139}$$

The channel moves from idle to class 1 transmitting state with probability y_1, which is the probability that only one class 1 user in any of the w_1 reservation slots requests a transmission and all users in the previous slots did not request to access the channel. The probability y_1 for this event is given by

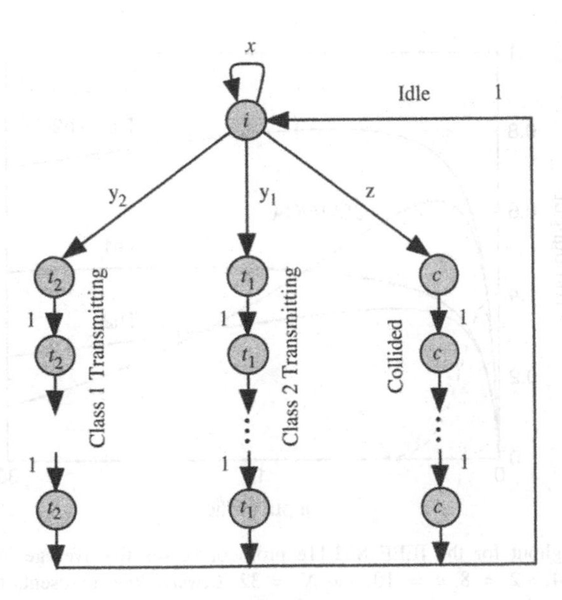

Fig. 10.36 IEEE 802.11e channel state transition diagram

$$\begin{aligned} y_1 &= u_1 v_0 + u_0 v_0^2 u_1 + u_0^2 v_0^3 u_1 + \cdots + u_0^{w_1 - 1} v_0^{w_1} u_1 \\ &= u_1 v_0 \times \frac{1 - (u_0 v_0)^{w_1}}{1 - u_0} \end{aligned} \tag{10.140}$$

In a similar fashion, the probability y_2 is given by

$$y_2 = u_0 v_1 \times \frac{1 - (u_0 v_0)^{w_1}}{1 - u_0 v_0} + v_1 (u_0 v_0)^{w_1} \times \frac{1 - v_0^{w_2 - w_1}}{1 - v_0} \tag{10.141}$$

We can now proceed to find the distribution vector at equilibrium using Fig. 10.36. We can write the following equations:

$$t_1 = y_1 \, i \tag{10.142}$$

$$t_2 = y_2 \, i \tag{10.143}$$

$$c = z \, i \tag{10.144}$$

It is easy to prove that the state probabilities are given by

$$i = 1/D \tag{10.145}$$

$$t_1 = y_1/D \tag{10.146}$$

$$t_1 = y_1/D \tag{10.147}$$

$$c = z/D \tag{10.148}$$

$$D = n(1 - x) + 1 \tag{10.149}$$

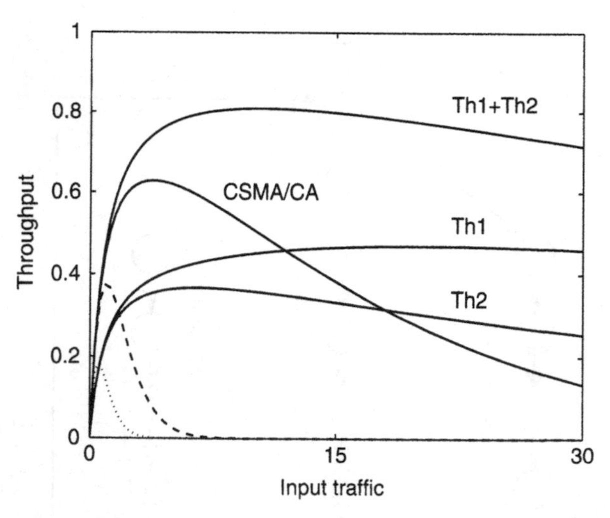

Fig. 10.37 Throughput for the IEEE 802.11e protocol versus the average input traffic when $\gamma = 0.5$, $w1 = 4$, $w2 = 8$, $n = 10$, and $N = 32$. *Dashed line* represents the throughput of slotted ALOHA and the *dotted line* represents the throughput of pure ALOHA

The throughputs for both priority classes are given by

$$\mathrm{Th}_1 = n y_1 / D \quad (10.150)$$

$$\mathrm{Th}_2 = n y_2 / D \quad (10.151)$$

Figure 10.37 shows the throughput of IEEE 802.11e when $\gamma = 0.5$, $w1 = 4$, $w2 = 8$, $n = 10$, and $N = 32$. The average input traffic $N_a(\text{in})$ is defined as

$$N_a(\text{in}) = a\ N \quad (10.152)$$

We notice, as expected, that class 1 throughput is higher than class 2 throughput. What is really important for quality of service support is not the absolute value of each throughput component since incoming traffic belonging to a certain service class could be low to start with. In that case, the resulting throughput would be low but this does not indicate low performance by any means. What is important here is the packet acceptance probability p_a for each class. We define the packet acceptance probabilities for the two service classes as

$$p_{a1} = \frac{\mathrm{Th}_1}{\gamma N_a(\text{in})} \quad (10.153)$$

$$p_{a2} = \frac{\mathrm{Th}_2}{(1-\gamma) N_a(\text{in})} \quad (10.154)$$

We would like to be assured that $p_{a1} > p_{a2}$ for all levels of incoming traffic. Figure 10.38 shows the frame access probability for the IEEE 802.11e protocol

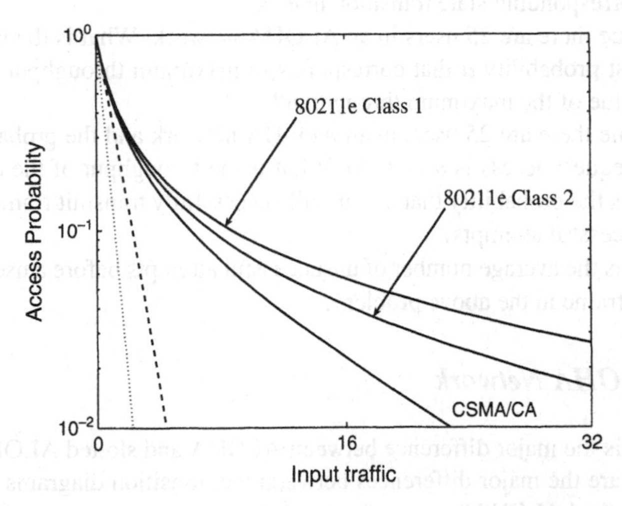

Fig. 10.38 Frame access probability for the IEEE 802.11e protocol versus the average input traffic when $\gamma = 0.5$, $w1 = 4$, $w2 = 8$, $n = 10$, and $N = 32$. *Dashed line* represents the access probability of slotted ALOHA and the *dotted line* represents the probability of pure ALOHA

versus the average input traffic when $\gamma = 0.5$, $w1 = 4$, $w2 = 8$, $n = 10$, and $N = 32$. Dashed line represents the throughput of slotted ALOHA and the dotted line represents the throughput of pure ALOHA.

We note from the figure that the access probability for the class 1 traffic is higher than that of class 2. This will lead to lesser delay for class 1 traffic, which is the desired performance.

Problems

ALOHA Network

10.1 Use Equation (10.20) to find the maximum value for the throughput of an ALOHA network and the value of a at the maximum.

10.2 Using the results of Section A.6 in Appendix A on page 599, prove that the maximum throughput of the ALOHA network approaches the value $1/2e$ as $N \rightarrow \infty$.

10.3 Assume an ALOHA network that is operating at its optimum conditions. What is the average number of attempts for a user to be able to transmit a frame under these conditions?

10.4 Assume an ALOHA network where the frame length is a multiple of some unit of length with an upper limit on the maximum frame size. Draw a possible transition diagram for such system and write down the corresponding state transition matrix.

10.5 Assume an ALOHA network where the propagation delay is bigger than the frame time T. What would be a good choice for the time step of the Markov chain? Draw a possible transition diagram for such system and write down the corresponding state transition matrix.

10.6 Assume there are 25 users in an ALOHA network. What is the transmission request probability a that corresponds to maximum throughput and what is the value of the maximum throughput?

10.7 Assume there are 25 users in an ALOHA network and the probability that a user request access is $a = 0.06$. What is the throughput of the channel and what is the probability that a user will successfully transmit frame after three unsuccessful attempts?

10.8 What is the average number of unsuccessful attempts before a user can transmit a frame in the above problem?

Slotted ALOHA Network

10.9 What is the major difference between ALOHA and slotted ALOHA?

10.10 What are the major differences between the transition diagrams of ALOHA and slotted ALOHA?

10.11 Write down the expressions for the steady-state distribution vectors for ALOHA and slotted ALOHA and comment on their similarities and differences. Explain why slotted ALOHA is expected to perform better than ALOHA.

10.12 Write down the ratio of throughput for slotted ALOHA compared to ALOHA. Approximate the expression for the limits when $a \ll 1$ and $a \approx 1$.

10.13 Use Equation (10.45) to find the maximum value for the throughput of a slotted ALOHA network and the value of a at the maximum.

10.14 Using the results of Section A.6 in Appendix A on page 599, prove that the maximum throughput of the slotted ALOHA network approaches the value $1/e$ as $N \rightarrow \infty$.

10.15 Assume a slotted ALOHA network that is operating at its optimum conditions. What is the average number of attempts for a frame to be transmitted under these conditions?

10.16 Assume a slotted ALOHA network where the frame length is a multiple of some unit of length with an upper limit on the maximum frame size. Draw a possible transition diagram for such system and write down the corresponding transition matrix.

10.17 Assume a slotted ALOHA network where the propagation delay is bigger than the frame time T. Draw a possible transition diagram for such system and write down the corresponding transition matrix.

10.18 Assume there are 25 users in a slotted ALOHA network. What is the transmission request probability a that corresponds to maximum throughput and what is the value of the maximum throughput?

10.19 Assume there are 25 users in a slotted ALOHA network and the probability that a user request access is $a = 0.06$. What is the throughput of the channel and what is the probability that a user will successfully transmit frame after three unsuccessful attempts?

10.20 What is the average number of unsuccessful attempts before a user can transmit a frame in the above problem?

10.21 Compare the maximum throughput values for ALOHA and slotted ALOHA and the level of activity for the sources under these conditions assuming the same number of users in both systems.

10.22 Assume an ALOHA network and a slotted ALOHA network. Both systems support N users and each user is active with probability $a = 0.02$. What is the optimum value of N for maximum throughput for both systems?

IEEE 802.3 (CSMA/CD, Ethernet)

10.23 Prove that the throughput for the IEEE 802.3 protocol in Section 10.5 approaches $n/(n+1)$ when under heavy traffic.

10.24 Assume an IEEE 802.3 network where the frame length is not constant. In that case, the number of time slots required by the transmitted frames could take between $n_{\min}$ and $n_{\max}$ slots.

(a) Draw the resulting state transition diagram.

(b) Indicate on the diagram the transition probabilities.
(c) Write down the state transition matrix.

10.25 Assume the backoff parameter $f = 1$. What will be the expressions for the IEEE 802.3 protocol

(a) Throughput
(b) Input traffic at maximum throughput
(c) Maximum throughput value

10.26 Assume an IEEE 802.3 network in which a transmitting user could move on to transmit another frame without turning to the idle state. The probability of this event happening is c.

(a) Identify the possible states of this Markov chain.
(b) Study the state transition probabilities and write down the state transition matrix.

10.27 Analyze the IEEE 802.3 for p-persistent CSMA/CD where all collided users attempt to access the channel with probability p.

10.28 Assume in the 802.3 that a collided station attempts a retransmission for a limited number of times (assume two only). If it fails after two attempts, it returns back to being idle. Analyze this situation.

10.29 In our analysis of CSMA/CD, we did not implement the exponential backoff strategy for collided stations. How could the model we developed here be modified to include this? Is there an alternative way to analyze this situation?

IEEE 802.11

10.30 Explain what is meant by the following PHY layer terms: frequency-hopping spread-spectrum (FHSS), direct-sequence spread-spectrum (DSSS) link layer, orthogonal frequency division multiple access (OFDM).

10.31 Explain what is meant by infrastructure wireless networks.

10.32 Explain what is meant by ad hoc wireless networks.

10.33 Explain what is meant by wireless sensor networks. How do these differ from ad hoc networks?

10.34 Explain the operation of the distributed coordination function (DCF) and indicate the type of LAN that uses it (infrastructure or ad hoc).

10.35 Explain how DCF reduces the probability of collisions.

10.36 Develop a discrete-time Markov chain model for the IEEE 802.11 user under the DCF function. Choose a time step value equal to the propagation delay.

10.37 Simulate the performance of the IEEE 802.11 channel for different values of the number of reservation slots w.

10.38 The analysis of the IEEE 802.11/DCF protocol assumed equally likely assignment of users to the w reservation slots. Develop a model of the channel when the assignment of active users to reservation slots follows a distribution different from the uniform distribution.

10.39 The analysis of the IEEE 802.11/DCF protocol assumed that the backoff counters of active users decrement by one when the channel is free. Develop a model of channel when the backoff counters assume a new random value each time the channel is busy. Only users that can transmit a frame are the ones that happen to have a backoff counter value of 0.

10.40 Explain the operation of the point coordination function (PCF) and indicate the type of LAN that uses it (infrastructure or ad hoc).

10.41 Explain why PCF eliminates collisions.

10.42 Investigate the types of scheduling protocols that could be used in the PCF portion of the IEEE 802.11 channel.

10.43 Develop a discrete-time Markov chain model for the IEEE 802.11 user under the PCF function. Choose a time step value equal to the poll period.

10.44 Repeat Example 4 for the case when the number of requests per user increases to 100 each second. Comment on your results.

10.45 Assume the IEEE 802.11/PCF protocol that serves 10 customers and the channel speed is 1 Mbps. Each customer is assumed to issue requests at a rate of 100 requests/s and the average length of a frame is 5.12 kb. Obtain the performance of this system.

10.46 Analyze the IEEE 802.11/PCF protocol in which there are two customer classes. Class 1 has N_1 customers and class 2 has N_2 customers. Users in class 1 can access the channel when they issue a request, while users in class 2 can only access the channel when none of the users of class 1 has a request. The system can be modeled as two separate queues for each class but the transition probabilities in class 2 queue depends on the state of requests of class 1.

10.47 In the analysis of the IEEE 802.11/PCF protocol, where users had transmit buffers, it was assumed that requests arriving at a given time step are processed in that time step. Develop a new analysis that processes these requests at the next time step.

References

1. A.S. Tanenbaum, *Computer Networks*, Prentice Hall PTR, Upper Saddle River, New Jersey, 1996.
2. L. Roberts, "Extension of frame communication technology to a hand held personal terminal", *NCC, AFIPS*, pp. 711–716, 1973.
3. S. Keshav, *An Engineering Approach to Computer Networks*, Addison-Wesley, Reading, Massachusetts, 1997.
4. R.M. Metcalfe and D.R. Boggs, "Ethernet: Distributed frame switching for local computer networks", *Communications of the ACM*, vol. 19, pp. 395–404, 1976.
5. F.A. Tobagi, "Analysis of a two-hop centralized frame radio network: Part I — Slotted Aloha", *IEEE Transactions on Communications*, vol. COM-28, pp. 196–207, 1980.

6. M.E. Woodward, *Communication and Computer Networks*, IEEE Computer Society Press, Los Alamitos, California, 1994.
7. A. Leon-Garcia and I. Widjaja, *Communication Networks*, McGraw-Hill, New York, 2000.
8. D.B. Jhonson and D.A. Maltz, "Dynamic source routing in Ad Hoc wireless networks", in *Mobile Computing*, T. Imielinski and H. Korth, Eds., Kluwer Academic Publishers, 1996.
9. P. Karn, "MACA - A new channel access method for frame radio", *ARRL/CRRL Amateur Radio 9th Computer Networking Conference*, September,1990.
10. S. Kapp, "802.11a. More bandwidth without the wires", *IEEE Internet Computing*, vol. 6, no. 4, pp. 75–79, 2002.
11. J. Weinmiller, H. Woesner, and A. Wolisz, "Analyzing and improving the IEEE 802.11-MAC protocol for wireless LANs", *Proceedings of the Fourth International Workshop on Modeling, Analysis, and Simulation of Computer and Telecommunication Systems (MASCOTS'96)*, pp. 200–206, February, 1996.
12. S. Khurana, A. Kahol, S.K.S. Gupta, and P.K. Srimani, "Performance evaluation of distributed coordination function for IEEE802.11 wireless LAN protocol in presence of mobile and hidden terminals", *Proceedings of 7th International Symposium on Modeling, Analysis and Simulation of Computer and Telecommunication Systems*, pp. 40–47, Oct. 1999.
13. J. Liu, D.M. Nicol, L.F. Perrone, and M. Liljenstam, "Towards high performance modeling of the 802.11 wireless protocol", *Proceedings of the Winter Simulation Conference*, vol. 9, pp. 1315–1320, December 2001.

Chapter 11
Modeling Network Traffic

11.1 Introduction

Models that describe and generate telecommunication traffic are important for several reasons [1]:

Traffic description: Network users might be required to give a traffic description to the service provider. Based on that, the service provider decides whether the new connection can be admitted with a guaranteed quality of service and without violating the quality of service for established connections.

System simulation: Future networks and new equipment could be designed and the expected network performance checked.

Different models are used to describe different types of traffic. For example, voice traffic is commonly described using the on–off source or the Markov modulated Poisson process. Studies suggest that traffic sources such as variable-bit-rate video and Ethernet traffic are better represented by self-similar traffic models [2–7]. The important characteristics of a traffic source are its *average data rate*, *burstiness*, and *correlation*. The average data rate gives an indication of the expected traffic volume for a given period of time. Burstiness describes the tendency of traffic to occur in clusters. A traffic burst affects buffer occupancy and leads to network congestion and data loss. Data burstiness is manifested by the *autocorrelation function* which describes the relation between packet arrivals at different times. It was recently discovered that network traffic exhibits long-range dependence, i.e., the autocorrelation function approaches zero very slowly in comparison with the exponential decay characterizing short-range-dependent traffic [2–7]. Long-range-dependent traffic produces a wide range in traffic volume away from the average rate. This great variation in traffic flow also affects buffer occupancy and network congestion. In summary, high burstiness or long-term correlation leads to buffer overflow and network congestion. We begin by discussing the different models describing traffic time arrival statistics.

Simple traffic models are sometimes called *point processes* since they are basically *counting processes* that count the number of packets that arrive in a time

F. Gebali, *Analysis of Computer and Communication Networks*,
DOI: 10.1007/978-0-387-74437-7_11, © Springer Science+Business Media, LLC 2008

interval. These point processes sometime give the random sequence representing the time separations between packets. Several random processes are grouped together to give more complex traffic patterns. The Internet traffic archive (http://ita.ee.lbl.gov/index.html) provides data sets for network traffic and some useful software.

The other extreme for traffic modeling is to use fluid flow models. Fluid flow modeling groups the traffic into flows that are characterized by average and burst data rates. The object in these models is to investigate traffic at the *aggregate level* such as Ethernet traffic or traffic arriving at ingress and egress points of some Internet service provider (ISP). Fluid flow models do not concern themselves with the details of individual packet arrivals or departures.

The difference between point processes and fluid flow models is similar to the difference in modeling an electric current in terms of the individual electrons or in terms of the current equations.

11.2 Flow Traffic Models

Flow traffic or fluid traffic models hide the details of the different traffics flowing in the network and replace them with flows that have a small set of characterizing parameters. The resulting models are easily generated, measured, or monitored.

For an end-to-end application, a *flow* has constant addressing and service requirements [8]. These requirements define a flow specification or *flowspec*, which is used for bandwidth planning and service planning. *Individual flows*, belonging to single sessions or applications, are combined into *composite flows* that share the same path, link, or service requirements. Composite flows, in turn, are combined into *backbone flows* when the network achieves a certain level of hierarchy. Describing flows in this fashion makes it easier to combine flow characteristics and to work with a smaller set of data. For example, a core router might separate incoming data into individual flows, composite flows, and backbone flows depending on the quality of service (QoS) required by the users. This results in smaller number of service queues and simpler implementation of the scheduling algorithm implemented in the router. In most networks, the majority of the flows are low-performance backbone flows; there will also be some composite flows, and there will be few high-performance individual flows. The high-performance flows will influence the design of the scheduling algorithm in the switch, size, and number of the queues required since they usually have demanding delay and/or bandwidth requirements. The backbone flows will influence the buffer size required since they will usually constitute the bulk of the traffic and most of the storage within the switch.

11.2.1 Modulated Poisson Processes

In a Markov modulated traffic model, states are introduced where the source changes its characteristic based on the state it is in. The state of the source could represent its data rate, its packet length, etc. When the Markov process represents data rate, the source can be in any of several active states and generates traffic with a rate that is determined by the state. This is commonly called Markov modulated Poisson process (MMPP). The simplest model is the on–off model, and more complex models are described in the next section for video traffic.

11.2.2 On–Off Model

A popular model for bursty sources is the on–off model, where the source switches between an active state, producing packets, and a silent state, where no packets are produced. In that sense, the on–off model is a two-state MMPP. Traffic from this type of source is characterized by many variable length bursts of activity, interspersed with variable length periods of inactivity. This model is commonly used to describe constant bit rate (CBR) traffic in ATM [9, 10]. Figure 11.1 shows the two-state model for the on–off source.

The source stays in the active state with probability a and stays in the silent state with probability s. When the source is in the active state (called the soujorn time), the source generates data at a rate λ in units of bits per second or packets per second. The traffic pattern generated by this source is shown in Fig. 11.2.

The probability that the length of the active period is n time steps is given by the geometric distribution

$$A(n) = a^n(1-a) \qquad n \geq 1 \tag{11.1}$$

The average duration of the active period is given by

$$T_{\mathrm{a}} = \frac{a}{1-a} \text{ time steps} \tag{11.2}$$

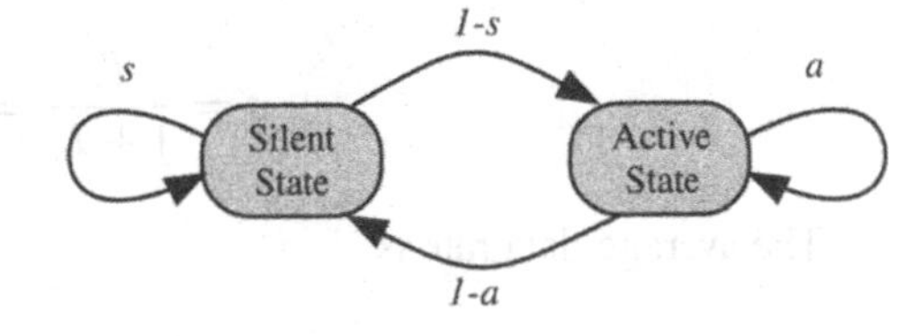

Fig. 11.1 An on–off source model

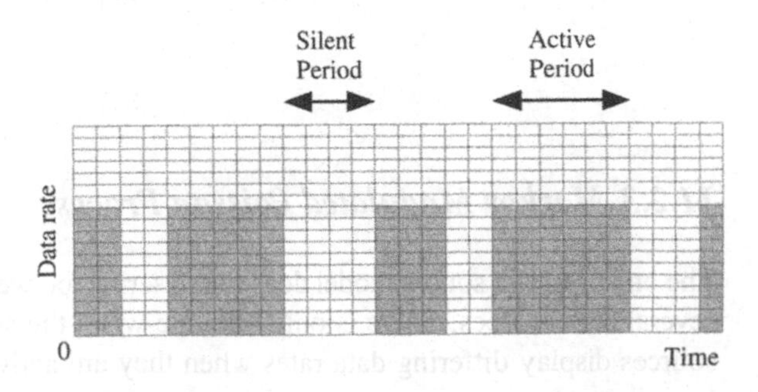

Fig. 11.2 Packet pattern for an on–off source model

Similarly, the probability that the length of the silent period is n time steps is given by the geometric distribution

$$S(n) = s^n(1-s) \qquad i \geq 1 \tag{11.3}$$

The average duration of the silent period is given by

$$T_s = \frac{s}{1-s} \text{ time steps} \tag{11.4}$$

Assume that λ is the data rate when the source is in the active state. In that case, the average data rate is obtained as

$$\begin{aligned} \lambda_a &= \frac{\lambda \times T_a}{T_a + T_s} \\ &= \frac{\lambda}{1 + T_s/T_a} \leq \lambda \end{aligned} \tag{11.5}$$

Example 1 Assume a 64 kbit/s voice source which is modeled as an on–off source with an average duration of the active period of $T_a = 0.45$ s and average duration of the silent period is $T_s = 1.5$ s. Estimate the source parameters and the average data rate.

From (11.2), the probability that the source remains in active state is

$$a = \frac{T_a}{1 + T_a} = 0.3103$$

From (11.4) the probability that the source remains silent is

$$s = \frac{T_s}{1 + T_s} = 0.6$$

The average data rate is

$$\lambda_a = 14.77 \text{ kbps}$$

■

11.2.3 Markov Modulated Poisson Process

The on–off traffic source model does not describe too well the effect of multiplexing several data sources. There is only one rate when the source is active while actual sources display differing data rates when they are active. To handle this situation,

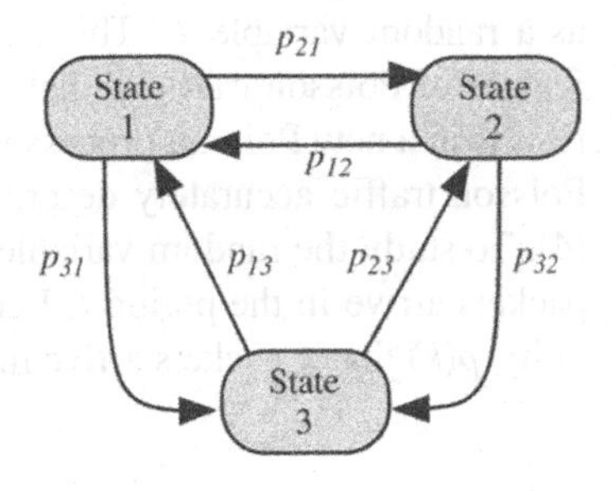

Fig. 11.3 A three-state MMPP source model

more states are added to the MMPP. Figure 11.3 shows a three-state configuration which is naturally called three-state MMPP.

For an MMPP with N states, we construct an $N \times N$ state transition matrix **P** whose element p_{ij} represents the probability of making a transition from state j to state i. This choice is consistent with our definition for the transition matrix of a Markov chain. Needless to say, this matrix is a column stochastic matrix.

When the source is in state i, $1 \leq i \leq N$, the packets are produced at a rate λ_i. When the number of states is $N = 2$, we have a *switched Poisson process* (SPP) [10]. When $N = 2$ and $\lambda_1 = 0$, we have an *interrupted Poisson process* (IPP), which is also the on–off model that was discussed above.

11.2.4 Autoregressive Models

Autoregressive models produce traffic with short-range dependence where the autocorrelation function decays exponentially. An autoregressive model of order N, denoted AR(N), is described by

$$X\,(n) = \sum_{k=1}^{N} a_i\; X\,(n-k) + \epsilon\,(n) \tag{11.6}$$

where $X\,(n)$ is a random variable indicating the traffic rate at that time; $\epsilon\,(n)$ is a random variable having a small range to fit experimental data. The above formula gives a simple method for generating the next random number, given the previous set of N random numbers which is computationally appealing.

Alternative forms of the above expression using moving average (MA) and autoregressive moving average (ARMA) expressions were also proposed [11].

11.3 Continuous-Time Modeling: Poisson Traffic Description

Poisson traffic description is a model often used by many researchers due to the simplicity of the model. A characteristic of traffic is the lull period in which no packets arrive. We can think of the interarrival time between two successive packets as a random variable T. This r.v. is continuous for Poisson traffic. The attractive feature of Poisson traffic is that the sum of several independent Poisson processes results in a new Poisson process whose rate is the sum of the component rates [12]. Poisson traffic accurately describes user-initiated TELNET and FTP connections [4]. To study the random variable T, we need to study the probability $p(0)$ that no packets arrive in the period t. Let us start by assuming Poisson traffic with probability $p(k)$ that k packets arrive in a time period t which is given by

$$p(k) = \frac{(\lambda_a t)^k}{k!} e^{-\lambda_a t} \tag{11.7}$$

where λ_a (packets/s) is the average packet arrival rate. Note that this expression for probability is valid for all values of $0 \leq k < \infty$. Of course, we do not expect an infinite number of packets to arrive in a time interval t, but this is the expression and that is what it predicts. Note that Poisson distribution really talks about numbers. It specifies the probability of getting a number of packets k in a given time period t.

For the interarrival time, we ask a different sort of question: what is the probability that the time separation between adjacent packets is t? To derive an expression for the pdf distribution for the interarrival time, we need to find the probability that no packets arrive in period t. The probability that no packets arrive in a time period t is obtained by substituting $k = 0$ in (11.7) to obtain

$$p(0) = e^{-\lambda_a t} \tag{11.8}$$

This probability is equivalent to the event $A : T > t$, and we can write

$$\begin{aligned} p(A : T > t) &= p(0) \\ &= e^{-\lambda_a t} \end{aligned} \tag{11.9}$$

The event A is basically the event that no packets arrived for a time period t. What happens after this time period is not specified. A packet might arrive or no packets arrive.

In order to find the pdf associated with the interarrival time, we need to define event B which is complementary to A as follows:

$$B : T \leq t$$

and the probability associated with event B is

$$\begin{aligned} p(B : T \leq t) &= 1 - p(A) \\ &= 1 - e^{-\lambda_a t} \end{aligned} \tag{11.10}$$

The cdf for the random variable T is given from (11.10) by

$$F_T(t) = p(T \leq t) = 1 - e^{-\lambda_a t} \tag{11.11}$$

The pdf for this random variable is obtained by differentiating the above equation

$$f_T(t) = \lambda_a e^{-\lambda_a t} \tag{11.12}$$

Thus the pdf for the interarrival time of Poisson traffic follows the *exponential distribution* that was discussed in Chapter 1.

Example 2 Find the average value for the exponentially distributed interarrival time having the distribution in (11.12)

The average time between arriving packets T_a is given by

$$\begin{aligned} T_a &= \int_{t=0}^{\infty} t\, \lambda_a e^{-\lambda_a t}\, dt \\ &= \frac{1}{\lambda_a} \qquad \text{s} \end{aligned}$$

We see that as the rate of packet arrival decreases ($\lambda_a \ll 1$), the average time between packets increases as expected. ■

Example 3 Consider an ATM channel where a source transmits data with an average data rate of 500 kbps. Derive the corresponding Poisson distribution and find the probability that 10 cells arrive in a period of 1 ms.

Since we are talking about cells, we have to convert all the data rates from bits per second quantities into cells/second using the information we have about average packet length A. We start by calculating the average arrival rate which is easily done since we know the size of an ATM cell.

$$\lambda_a = \frac{500 \times 10^3}{8 \times 53} = 1.179\,2 \times 10^3 \qquad \text{cells/s}$$

The probability of 10 cells arriving in the time period t according to the Poisson distribution is found using (11.7):

$$p(10) = \frac{(\lambda_a\, t)^{10}}{(10)!} e^{-\lambda_a t} = 4.407 \times 10^{-7}$$

■

11.3.1 Memoryless Property of Poisson Traffic

The memoryless property of Poisson traffic is defined using the following conditional probability expression related to the interarrival time:

$$p(T > t + \epsilon | T > t) = p(T > \epsilon) \quad \text{for all} \quad t, \ \epsilon > 0 \tag{11.13}$$

Basically, this equation states that the probability that no packets arrive for a time $t + \epsilon$, given that no packets arrived up to time t, does not depend on the value of t. It depends only on ϵ. So, in effect, the expression states that we know that we waited for t seconds and no packets arrived. Now we reset our clock and we ask the question: What is the probability that a packet arrives if we wait for a period ϵ seconds? The probability of this event only depends on our choice of ϵ value and will not use our prior knowledge of the period t.

Let us state this property using two examples of systems having the memoryless property. Assume that we are studying the interarrival times of buses instead of packets. Assume also that the time between bus arrivals is a random variable with memoryless property. We arrive at the bus stop at 9:00 a.m. and wait for 1 h yet no buses show up. Now we know that no buses showed up for the past hour, and we naturally ask the question: What are the odds that a bus will show up if we wait for five more minutes. The probability that no buses will come in the next five minutes will depend only on the wait period (5 min) and not on how long we have been waiting at the bus stop.

Another example of memoryless property is the case of an appliance (a television set for example). If the time between failures is a random variable with memoryless property, then the probability that the TV will fail after 1 h of use is the same at any time independent of when we bought the TV or how long the TV has been used.

Obviously, the time between failures in cars and airplanes has a memory property. That is why an older car breaks down more often when compared to a new car or when compared to an older car that is only driven on weekends in the summer months only.

Let us turn back to our interarrival time statistics. From (11.9), we could write

$$p(T > t) = e^{-\lambda_a t} \tag{11.14}$$

Changing the time value from t to $t + \epsilon$, we get

$$p(T > t + \epsilon) = e^{-\lambda_a (t+\epsilon)} \tag{11.15}$$

Equation (11.13) is a conditional probability, and we can write it as

$$p(A|B) = \frac{p\left(A \bigcap B\right)}{p(B)} \tag{11.16}$$

where the events A and B are defined as

$$A : T > t + \epsilon \tag{11.17}$$

$$B : T > t \tag{11.18}$$

But $A \bigcap B = A$ since $\epsilon > 0$ implies that if event A took place, then event B has taken place also. Thus we have

$$p(T > t + \epsilon | T > t) = \frac{p(A)}{p(B)} \tag{11.19}$$

$$= \frac{e^{-\lambda_a (t+\epsilon)}}{e^{-\lambda_a t}} \tag{11.20}$$

$$= e^{-\lambda_a \epsilon} \tag{11.21}$$

Thus we have proved that the interarrival time for the exponential distribution is memoryless.

11.3.2 Realistic Models for Poisson Traffic

The Poisson distribution and the interarrival time considered in Section 11.3 do not offer much freedom in describing realistic traffic sources since they contain one parameter only: λ (packets/s) that reflected the average data arrival rate. The minimum value for the interarrival time is zero. This implies that the time interval between two packet headers could be zero. An interarrival time value of zero implies two things: that our packets have zero length and that the data rate could be infinity. Both of these conclusions are not realistic.

A realistic bursty source could be described using the parameters.

λ_a	the average data rate
σ	the maximum data rate expected from the source

Since we are talking about rates in terms of packets/second, we need to make sure that the rates are in terms of packet/second. The source parameters above could be elaborated upon further, depending on our need. Section 11.2.1 discusses source with multiple data rates.

Now we ask the question: How can we write down an expression for a Poisson distribution that takes into account all of the source parameters? We have two options:

1. *Flow description*: This option allows us to specify the randomness of the instantaneous data rate produced by the source.
2. *Interarrival time description*: This option allows us to specify the randomness of the periods between adjacent packets produced by the source.

11.3.3 Flow Description

We start by writing the pdf for the instantaneous data rate in the form

$$f_\Lambda(\lambda) = b\,\mathrm{e}^{-b\lambda} \tag{11.22}$$

where λ is the data rate, and the parameter b is the *shape* parameter that determines the steepness of the exponential curve.

Figure 11.4 shows the distribution; λ_a in the figure indicates the average data rate. The distribution in (11.22) is a valid pdf since its integral equals unity.

To find the parameter b, we need to estimate the average data rate λ_a. The average data rate for the distribution given in (11.22) is

$$\lambda_\mathrm{a} = \int_0^\infty \lambda\, b\,\mathrm{e}^{-b\,\lambda}\, \mathrm{d}\lambda = \frac{1}{b} \tag{11.23}$$

Based on this equation, we can determine the pdf for the data rate produced by the source, given its average rate. If λ_a is the average data rate of a source, then the pdf for its rate is given by

$$f_\Lambda(\lambda) = \frac{1}{\lambda_\mathrm{a}}\mathrm{e}^{-\lambda/\lambda_\mathrm{a}} \tag{11.24}$$

Thus to describe the data rate of a source that follows the Poisson distribution, we need to specify its average data rate λ_a only.

11.3.4 Interarrival Time Description

The pdf description of the interarrival time for a Poisson source follows the exponential distribution which we repeat here for convenience.

$$f_T(t) = \lambda_\mathrm{a}\mathrm{e}^{-\lambda_\mathrm{a} t} \tag{11.25}$$

We mentioned that this equation is not sufficient to describe real traffic since it contains only one parameter, λ_a, which describes only the average data rate. We can

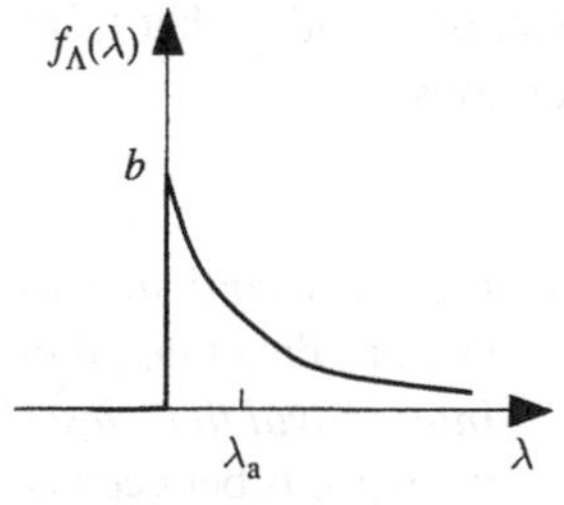

Fig. 11.4 Exponential distribution describing instantaneous rate of a Poisson source

modify the interarrival time distribution and obtain the *biased exponential distribution* as follows.

$$f_T(t) = \begin{cases} 0 & t < a \\ b \exp -b\,(t-a) & t \geq a \end{cases} \tag{11.26}$$

where $a \geq 0$ is the *position parameter* (units s) and $b > 0$ is the *shape parameter* (units s^{-1}). Basically, a represents the minimum time between adjacent packets, and b determines how fast the exponential function decays with time. Both a and b will determine the average packet rate as will be explained in Example 4.

Figure 11.5 shows the distribution given by the expression in (11.26). The figure details that a places the pdf at the desired position on the time axis and b determines how fast the exponential function decays with time. The distribution in (11.26) is a valid pdf since its integral equals unity. The next section explains how to obtain the correct values for a and b for a typical source.

Example 4 Find the average value for the exponentially distributed interarrival time with pdf given by (11.26)

The average time separation between arriving packets T_a is given by

$$\begin{aligned} T_a &= \int_{t=a}^{\infty} t b \exp(-b(t-a))\, dt \\ &= a + \frac{1}{b} \quad s \end{aligned}$$

The average interarrival time T_a depends on both a and b parameters. We see that as the shape parameter decreases ($b \ll 1$), the average time between packets increases.

On the other hand, when b is large ($b \gg 1$), the exponential function will approach a delta function, and the interarrival time will have its *minimum value* $T_a \approx a$.

The variance of the interarrival time for the shifted exponential distribution is given by

$$\sigma^2 = \frac{1}{b^2}$$

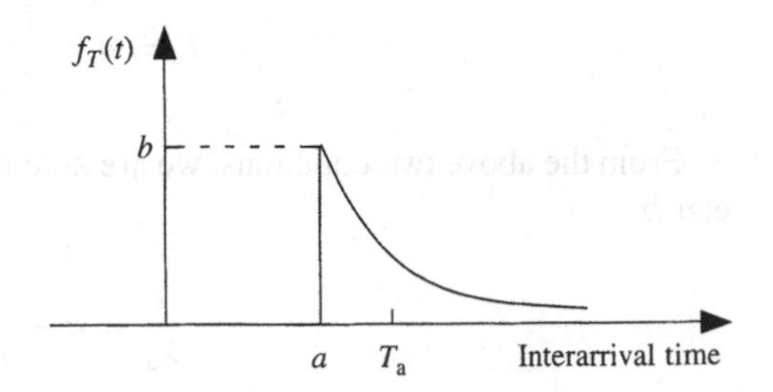

Fig. 11.5 A biased exponential distribution with two design parameters: position parameter a and shape parameter b

which depends only on the shape parameter. So large values for b will result in traffic with low burstiness approaching CBR (constant bit rate). Lower values for b result in more bursty traffic.

11.3.5 Extracting Poisson Traffic Parameters

In this section, we show how to find the values of the position parameter a and shape parameter b for a source whose average rate λ_a and burst rate σ are known.

The position parameter a is equivalent to the minimum time between two adjacent packets. In a time period t, the *maximum number* of packets that could be produced by the source is given by

$$N_m = \sigma\, t \tag{11.27}$$

where it was assumed that σ was given in units of packets/second. The minimum time between two adjacent packets was defined as a and is given by

$$\alpha = \frac{t}{N_m} = \frac{1}{\sigma} \quad \text{s} \tag{11.28}$$

Problem 11.5 discusses about obtaining the parameter a when σ is expressed in units of bits/second.

In a time period t, the *average number* of packets that could be produced by the source is given by

$$N_a = \lambda_a t \tag{11.29}$$

The average time between two adjacent packets is given by

$$T_a = \frac{t}{N_a} = \frac{1}{\lambda_a} \quad \text{s} \tag{11.30}$$

But from Example 4, we obtained an expression for the average interarrival time as

$$T_a = a + \frac{1}{b} \quad \text{s} \tag{11.31}$$

From the above two equations, we are able to obtain a value for the shape parameter b

$$\frac{1}{\lambda_a} = a + \frac{1}{b} \tag{11.32}$$

Therefore, we have

$$b = \frac{\sigma \ \lambda_a}{\sigma - \lambda_a} \quad \mathrm{s}^{-1} \tag{11.33}$$

Problem 11.6 discusses obtaining the parameter a when σ and λ_a are expressed in units of bits/second.

Example 5 A data source follows the Poisson distribution and has an average data rate $\lambda_a = 10^3$ packets/s and maximum burst rate of $\sigma = 3 \times 10^3$ packets/s. Estimate the exponential distribution parameters that best describe that source.

The position parameter is given from (11.28) by

$$a = \frac{1}{3 \times 10^6} = 3.33 \times 10^{-4} \quad \mathrm{s}$$

The shape parameter b is given from (11.33) by

$$b = 1500 \quad \mathrm{s}^{-1}$$

The pdf for the interarrival time is given by

$$f_T(t) = 1500 \exp -1000 \left(t - 3.3 \times 10^{-4}\right)$$

■

11.3.6 Poisson Traffic and Queuing Analysis

The previous subsection discussed how the biased exponential distribution parameters can be extracted given the system parameters:

λ_a the average data rate
σ the maximum data rate expected from the source

A Poisson source matching these given parameters has position parameter given by (11.28) and shape parameter given by (11.33). In this section, we ask the question: Given a Poisson source with known parameters that feed into a queue, what is the packet arrival probability for the queue? Remember that in queuing theory, the two most important descriptors are the arrival statistics and the departure statistics.

There are two cases that must be studied separately based on the values of the step size T and the position parameter a.

Case When $T \leq a$

The case $T \leq a$ implies that we are sampling our queue at a very high rate that is greater than the burst rate of the source. Therefore, when $T \leq a$ at most, one packet could arrive in one time step with a probability x that we have to determine. We use the symbol x for arrival probability since the symbol α is used here to describe the position parameter.

The number of time steps over a time period t is estimated as

$$n = \frac{t}{T} \tag{11.34}$$

The average number of packets arriving over a period t is given by

$$\begin{aligned} N_a &= \lambda_a t \\ &= \lambda_a \, n \, T \end{aligned} \tag{11.35}$$

where λ_a was assumed to be given in units of packets/second.

From the binomial distribution, the average number of packets in one step time is

$$N_a = x \, n \tag{11.36}$$

where x is the packet arrival probability in one time step.

From the above two equations, the packet arrival probability per time step is given by

$$x = \lambda_a \, T \tag{11.37}$$

We see in the above equation that as T gets smaller or as the source activity is reduced (small λ_a), the arrival probability is decreased, which makes sense.

Example 6 Estimate the packet arrival probability for a source with the following parameters. $\lambda_a = 50$ packets/s and $\sigma = 150$ packets/s. Assume that the time step value is $T = 1$ ms.

The position parameter is

$$\alpha = \frac{1}{\sigma} = 6.7 \quad \text{ms}$$

The shape parameter is

$$\beta = \frac{\lambda_a \sigma}{\sigma - \lambda_a} = 75 \quad \text{s}^{-1}$$

The packet arrival probability is

$$x = 0.05$$

■

Case When $T > a$

The case $T > a$ implies that we are sampling our queue at a rate that is slower than the burst rate of the source. Therefore, when $T > a$, more than one packet could arrive in one time step and we have to find the packet arrival statistics that describe this situation.

We start our estimation of the arrival probability x by determining the maximum number of packets that could arrive in one step time

$$N_m = \lceil \sigma \, T \rceil \tag{11.38}$$

The *ceiling* function was used here, after assuming that the receiver considers packets that partly arrive during one time step. If the receiver does not wait for partially arrived packets, then the *floor* function should be used.

The average number of packets arriving in the time period T is

$$N_a = \lambda_a \, T \tag{11.39}$$

From the binomial distribution, the average number of packets in one step time is

$$N_a = x \, N_m \tag{11.40}$$

From the above two equations, the packet arrival probability per time step is

$$x = \frac{\lambda_a \, T}{N_m} \leq \frac{\lambda_a}{\sigma} \tag{11.41}$$

The probability that k packets arrive at one time step T is given by the binomial distribution

$$p(k) = \binom{N_m}{k} x^k \, (1 - x)^{N_m - k} \tag{11.42}$$

Example 7 A data source follows the Poisson distribution and has an average data rate of $\lambda_a = 10^3$ packets/s and maximum burst rate of $\sigma = 5 \times 10^3$ packets/s. Find the biased Poisson parameters that describe this source and find the packet arrival probabilities if the time step is chosen equal to $T = 1$ ms.

The biased Poisson parameters are

$$
\begin{aligned}
a &= 2 \times 10^{-4} \quad \text{ms} \\
b &= 1.250 \times 10^{3} \quad \text{s}^{-1}
\end{aligned}
$$

The maximum number of packets that could arrive in one time step is

$$N_{\text{m}} = 5$$

The packet arrival probability per time step is

$$x = 0.2$$

The probability that k packets arrive per time step is

$$
\begin{aligned}
p(0) &= 3.2768 \times 10^{-4} \\
p(1) &= 4.0960 \times 10^{-1} \\
p(2) &= 2.0480 \times 10^{-1} \\
p(3) &= 5.1200 \times 10^{-2} \\
p(4) &= 6.4000 \times 10^{-3} \\
p(5) &= 3.2000 \times 10^{-4}
\end{aligned}
$$

■

11.4 Discrete-Time Modeling: Interarrival Time for Bernoulli Traffic

Poisson traffic description applies when time is treated as continuous. Bernoulli traffic is analogous to Poisson traffic when time is discrete. Discrete-time traffic is typically described by *Bernoulli trials* that give rise to a *binomial process* in which the probability that a packet arrives at a given time step is x and the probability that a packet does not arrive is $y = 1 - x$. We use the symbol x for arrival probability since the symbol a is used here to describe the position parameter.

We can think of Bernoulli traffic with binomial packet arrival distribution as the discrete version of Poisson traffic with exponential packet interarrival time distribution. The latter distribution could be termed a fluid flow traffic model since it deals with *flow rate* as opposed to counting the number of packets that arrive in a certain time period.

The probability that k packets arrive in n time steps is given by the binomial distribution

$$p(k) = \binom{n}{k} x^k \, y^{n-k} \tag{11.43}$$

We can think of the interarrival time n between two successive packets as a random variable N. This r.v. is discrete for Bernoulli traffic.

The probability $p(0)$ that no packets arrive for *at least n* consecutive time steps is given by

$$p(0) = y^n$$

The above equation simply states that we did not get any packets in n time steps. What happens during time step $n + 1$ is not specified in the above equation. We might get a packet or we might not. Notice that for large n, the probability that no packets arrive diminishes which makes sense.

The above probability is equivalent to the event $A : N > n$, and we can write

$$\begin{aligned} p(A : N > n) &= p(0) \\ &= (1 - x)^n \end{aligned} \tag{11.44}$$

This equation will help us in the next section to prove the memoryless property of Bernoulli traffic. However, we proceed here to find the pmf associated with the interarrival time.

In order to find the pmf associated with the interarrival time, we need to find the probability that the interarrival time *exactly equals n* time steps. In other words, our event now specifies that no packets arrived for n time steps followed by a packet arrival event at the next step when the time index is $n + 1$. The desired probability is given by

$$p = x \, (1 - x)^n \tag{11.45}$$

This probability is equal to the pmf of the interarrival time and we can write the pmf as

$$p(N = n) = x \, (1 - x)^n \tag{11.46}$$

Figure 11.6 illustrates the pmf for the interarrival time of Bernoulli traffic that has geometric distribution. A simple geometric distribution with one parameter x, the probability that exactly one packet arrives at the end of time interval n, is shown in Fig. 11.6.

Example 8 Find the average value for the interarrival time of Bernoulli traffic which is described by the pmf distribution of (11.46).

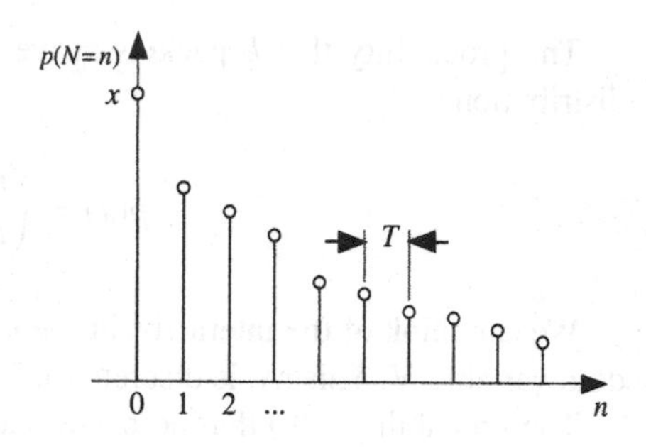

Fig. 11.6 A simple geometric distribution with one parameter x, the probability that exactly one packet arrives at the end of time interval n

The average number of time steps between packets n_a is given by

$$n_a = \sum_{n=0}^{\infty} n\, x\, y^n$$
$$= \frac{y}{1-y} = \frac{y}{x}$$

We see that as the probability for the packet arrival decreases ($x \ll 1$), the average number of steps between the packets increases as expected. On the other hand, when x approaches unity, the interarrival time becomes $n_a \approx 0$ as expected. This indicates that arriving packets have no empty time slots in between them. ■

Example 9 Consider communication channel where packets arrive with an average data rate of $\lambda_a = 2.3 \times 10^4$ packets/s and a burst rate limited only by the line rate, $\sigma = \lambda_l = 155.52$ Mbps. Derive the equivalent binomial distribution parameters and find the probability that 10 packets arrive in one sample time period of $T = 1$ ms.

The average number of packets received in one time step period is given by

$$N_a = 2.3 \times 10^4 \times 10^{-3} = 23$$

which represents the average traffic produced by the source.

The maximum number of packets that could be received in this time is found by estimating the duration of one packet as determined by the line rate.

$$\Delta = \frac{8 \times 53}{155.52 \times 10^6} = 2.7263 \quad \mu\text{s}$$

The maximum number of packets that could be received in 1 ms period is given by

$$N_m = \left\lceil \frac{T}{\Delta} \right\rceil = 367$$

The packet arrival probability per time step is given the equation

$$x\, N_m = N_a$$

which gives

$$x = \frac{23}{367} = 0.0627$$

and of course $y = 1 - x = 0.9373$.

The probability of 10 cells arriving out of potential N_m cells is

$$p(10) = \binom{N_m}{10} x^{10} \, y^{N_m - 10} = 9.3192 \times 10^{-4}$$

We note that $p(10)$ as obtained here is almost equal to $p(10)$ as obtained using the Poisson distribution in Example 4. ■

11.4.1 Realistic Models for Bernoulli Traffic

The Bernoulli distribution and the interarrival time considered in Section 11.4 do not offer much freedom in describing realistic traffic sources since they contain only one parameter: x represented the probability that a packet arrived at a given time step. The minimum value for the interarrival time is zero (i.e., $n = 0$). This implies that the time interval between two packet headers could be exactly one time step.

Equation (11.46) can be modified as follows

$$p(N = n) = x\,(1 - x)^{(n - n_0)} \qquad (11.47)$$

Now the traffic distribution can be described by two parameters x and n_0. The parameter n_0 represents the minimum number of time steps between adjacent packets. Figure 11.7 illustrates the pmf for the interarrival time of Bernoulli traffic that has geometric distribution. A simple shifted, or biased, geometric distribution with two parameters x and n_0 is shown in Fig. 11.7.

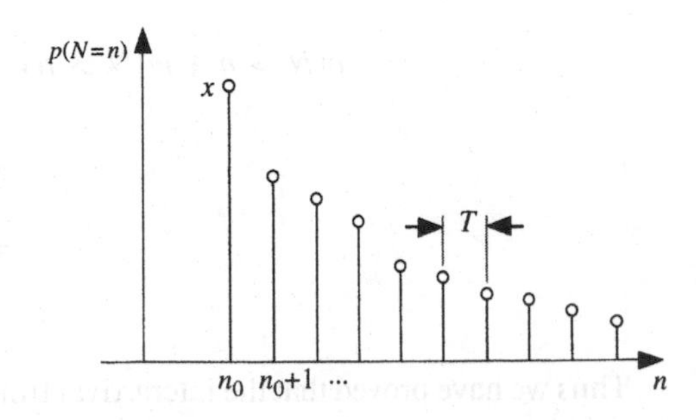

Fig. 11.7 A realistic model for the geometric distribution of Bernoulli traffic interarrival time. The distribution is described in terms of two parameters x and n_0

11.4.2 Memoryless Property of Bernoulli Traffic

Assume that n is the number of time steps between arriving packets. It is obvious that the value of n shows random variations from one packet to another. Define N as the random variable associated with n. The memoryless property of the interarrival time for Bernoulli traffic is defined using the following conditional probability expression for discrete random variables

$$p(N > n + m|N > n) = p(N > m) \quad \text{for all} \quad n, m > 0 \tag{11.48}$$

Basically, this equation states that the probability that a packet arrives after a time $n + m$, given that no packets arrived up to time n, does not depend on the value of n. It depends only on m.

From (11.44), we could write

$$p(n) = (1 - x)^n \tag{11.49}$$

where x is the packet arrival probability during one time step. Changing the time value from n to $n + m$, we get

$$p(n + m) = (1 - x)^{n+m} \tag{11.50}$$

Equation (11.48) is a conditional probability, and we can write it as

$$p(A|B) = \frac{p\left(A \bigcap B\right)}{p(B)} \tag{11.51}$$

where the events A and B are defined as

$$A : N > n + m \tag{11.52}$$

$$B : N > n \tag{11.53}$$

But $A \bigcap B = A$ since $m > 0$, which implies that if event A took place, then event B has also taken place. Thus we have

$$p(N > n + m|N > n) = \frac{p(A)}{p(B)} \tag{11.54}$$

$$= \frac{(1 - x)^{n+m}}{(1 - x)^n} \tag{11.55}$$

$$= (1 - x)^m \tag{11.56}$$

$$= p(N > m) \tag{11.57}$$

Thus we have proved that the interarrival time for Bernoulli traffic is memoryless.

11.4.3 Realistic Model for Bernoulli Traffic

The interarrival time considered in Section 11.4 does not offer much freedom in describing realistic traffic sources since it contains only one parameter: x the probability that a packet arrives at a certain time step. As mentioned, a realistic source is typically described using more parameters than just the average data rate.

λ_a the average data rate
σ the maximum data rate expected from the source

Then the question arises: How can we write down an expression similar to the one given in (11.46) that takes into account all of these parameters? We can follow a similar approach as we did for Poisson traffic in Section 11.3.2 when we modified the exponential distribution by shifting it by the position parameter a. We modify (11.46) to obtain the *biased geometric distribution* as follows.

$$p(N = n) = \begin{cases} 0 & n < \alpha \\ x\,(1 - x)^{n-\alpha} & n \geq \alpha \end{cases} \tag{11.58}$$

where n is the number of time steps between packet arrivals, $\alpha \geq 0$ is called the *position parameter*, in units of time steps, and x is the probability that a packet arrived during one time step. Basically, a represents the minimum number of time steps between adjacent packets.

Figure 11.8 shows the discrete exponential distribution. We see from the figure that a places the pmf at the desired position on the time axis, and x determines how fast the exponential function decays with time. The distribution in (11.58) is a valid pmf since its sum equals unity. The values of α and x will be derived in the next section.

Example 10 Find the average value for the geometrically distributed interarrival time given by (11.58).

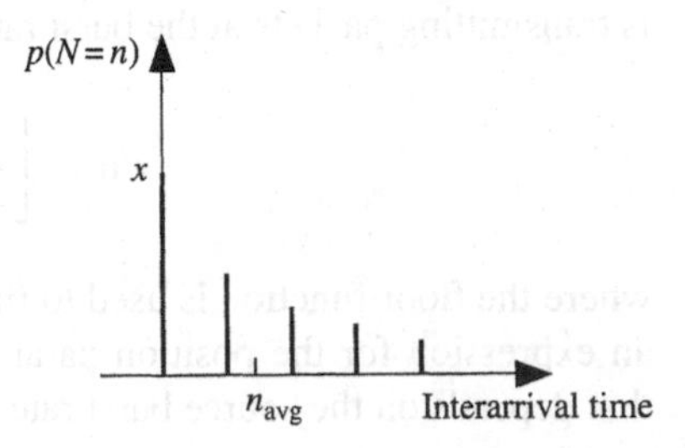

Fig. 11.8 A biased geometric distribution with two design parameters: position parameter α and packet arrival probability x

The average number of time steps between packets n_a is given by

$$n_a = \sum_{n=\alpha}^{\infty} n\, x\, (1-x)^{n-\alpha} = \frac{y}{x} + \alpha$$

We see that as the probability for packet arrival decreases ($x \ll 1$), the average number of steps between packets increases as expected. A high value for α implies slow traffic since the minimum separation between packets is large. This results in increased values for n_a.

The expression we obtained for the average interarrival time reduces to the expression obtained in Example 8 for the simple exponential distribution when the position parameter $\alpha \to 0$. ■

11.4.4 Extracting Bernoulli Traffic Parameters

In this section, we show how to find the values of the position parameter a and packet arrival probability x for a source whose average rate λ_a and burst rate σ are known.

The position parameter α in (11.58) is found by studying packet arrival during a period t. The time step T associated with this discrete arrival process is arbitrary and depends on the specifics of the system being studied.

If we study our system for a time period t, then the number of time steps n spanned by this time period is given by

$$n = \frac{t}{T} \tag{11.59}$$

The maximum number of packets N_m that could arrive during this time period depends on the burst rate σ

$$N_m = \sigma \times t \tag{11.60}$$

where σ was assumed to be given in units of packets/second.

The minimum number of time steps between adjacent packets when the source is transmitting packets at the burst rate is given by

$$\alpha = \left\lfloor \frac{n}{N_m} \right\rfloor = \left\lfloor \frac{1}{\sigma\, T} \right\rfloor \tag{11.61}$$

where the floor function is used to find a conservative estimate of α. Now we have an expression for the position parameter a for the biased exponential distribution that depends on the source burst rate and the time step value.

Now we turn our attention to estimating the arrival probability x. Consider a time period t again, which corresponds to n time steps. Because we have Bernoulli traffic, we can estimate x from the average number of packets received in that time period t which corresponds to n time steps:

$$N_a = x \times n \tag{11.62}$$

The average number of packets can also be estimated from the average data rate λ_a.

$$N_a = \lambda_a \times t \tag{11.63}$$

We can find a value of x from the above two equations as

$$x = \frac{\lambda_a \times t}{n} = \lambda_a\, T \tag{11.64}$$

Example 11 An ATM data source follows the binomial distribution and has an average data rate of $\lambda_a = 400$ packets/s and maximum burst rate of $\sigma = 10^3$ packets/s. Estimate the geometric distribution parameters that best describe that source if the time step value T chosen is equal to 0.1 ms.

The position parameter is given by

$$\alpha = 10 \qquad \text{time steps}$$

and the packet arrival probability per time step is given by

$$x = 0.04$$

Thus the pmf describing the interarrival time is given by

$$p(N = n) = 0.04 \times 0.96^{n-10}$$

■

11.4.5 Bernoulli Traffic and Queuing Analysis

The previous subsection discussed how the biased geometric distribution parameters can be extracted given the system parameters:

λ_a the average data rate
σ the maximum data rate expected from the source

A Bernoulli source matching these given parameters has position and shape parameters:

$$a = 1/(\sigma\ T) \tag{11.65}$$
$$x = \lambda_a\ T \tag{11.66}$$

In this section, we ask the question: Given a Bernoulli source with known parameters that feeds into a queue, what is the packet arrival probability for the queue? Remember that in queuing theory the two most important descriptors are the arrival statistics and the departure statistics. The analysis we undertake is very similar to the one done for Poisson traffic in Section 11.4.

There are two cases that must be studied separately based on the values of the step size T and the position parameter a.

Case When $T \leq a$

The case $T \leq a$ implies that we are sampling our queue at a very high rate that is greater than the burst rate of the source. Therefore, when $T \leq a$ at most, one packet could arrive in one time step with probability x that we have to determine.

The number of time steps over a time period t is estimated as

$$n = \frac{t}{T} \tag{11.67}$$

The average number of packets arriving over a period t is given by

$$\begin{aligned} N_a &= \lambda_a\ t \\ &= \lambda_a\ n\ T \end{aligned} \tag{11.68}$$

From the binomial distribution, the average number of packets in one step time is

$$N_a = x\ n \tag{11.69}$$

where x is the packet arrival probability in one time step.

From the above two equations, the packet arrival probability per time step is given by

$$x = \lambda_a\ T \tag{11.70}$$

We see in the above equation that as T gets smaller or as the source activity is reduced (small λ_a), the arrival probability is decreased, which makes sense.

Case When $T > a$

The case $T > a$ implies that we are sampling our queue at a rate that is slower than the burst rate of the source. Therefore, when $T > a$, more than one packet could arrive in one time step and we have to find the packet arrival statistics that describe this situation.

We start our estimation of the arrival probability x by determine the maximum number of packets that could arrive in one step time

$$N_m = \lceil \sigma \, T \rceil \tag{11.71}$$

The *ceiling* function was used here after assuming that the receiver will consider packets that partly arrived during one time step. If the receiver does not wait for partially arrived packets, then the *floor* function should be used.

The average number of packets arriving in the time period T is

$$N_a(\text{in}) = \lambda_a \, T \tag{11.72}$$

From the binomial distribution, the average number of packets in one step time is

$$N_a(\text{in}) = x \; N_m \tag{11.73}$$

From the above two equations, the packet arrival probability per time step is

$$x = \frac{\lambda_a \, T}{N_m} \tag{11.74}$$

The probability that k packets arrive at one time step T is given by the binomial distribution

$$p(k) = \binom{N_m}{k} x^k \, (1-x)^{N_m-k} \tag{11.75}$$

11.5 Self-Similar Traffic

We are familiar with the concept of periodic waveforms. A periodic signal repeats itself with *additive* translations of time. For example, the sine wave $\sin \omega t$ will have the same value if we add an integer, multiple of the period $T = 2\pi/\omega$ since

$$\sin \omega t = \sin \omega (t + \text{i} \; T)$$

On the other hand, a self-similar signal repeats itself with *multiplicative* changes in the time scale [13, 14]. Thus a self-similar waveform will have the same shape if we scale the time axis up or down. In other words, imagine we observe a certain waveform on a scope when the scope is set at 1 ms/division. We increase the resolution and set the scale to 1 μs/division. If the incoming signal is self-similar, the scope would display the same waveform we saw earlier at a coarser scale.

Self-similar traffic describes traffic on Ethernet LANs and variable-bit-rate video services [2–7]. These results were based on analysis of millions of observed packets over an Ethernet LAN and an analysis of millions of observed frame data generated by variable-bit-rate (VBR) video. The main characteristic of self-similar traffic is the presence of "similarly-looking" bursts at every time scale (seconds, minutes, hours) [12].

The effect of self-similarity is to introduce long-range (large-lag) autocorrelation into the traffic stream, which is observed in practice. This phenomenon leads to periods of high traffic volumes even when the average traffic intensity is low. A switch or router accepting self-similar traffic will find that its buffers will be overwhelmed at certain times even if the expected traffic rate is low. Thus switches with buffer sizes, selected based on simulations using Poisson, traffic, will encounter unexpected buffer overflow and packet loss. Poisson traffic models predict exponential decrease in data loss as the buffer size increases since the probability of finding the queue in a high-occupancy state decreases exponentially. Self-similar models, on the other hand, predict stretched exponential loss curves. This is why increasing link capacity is much more effective in improving performance than increasing buffer size. The rational being that it is better to move the data along than to attempt to store them since any buffer size selected might not be enough when self-similar traffic is encountered.

11.6 Self-Similarity and Random Processes

Assume that we have a discrete-time random process $X(n)$ that produces the set of random variables $\{X_0, X_1, \cdots\}$. We define the aggregated random process X^m as a random process whose data samples are calculated as

$$
\begin{aligned}
X_0^{(m)} &= \frac{1}{m}\left[X_0 + X_1 + \cdots + X_{m-1}\right] \\
X_1^{(m)} &= \frac{1}{m}\left[X_m + X_{m+1} + \cdots + X_{2m-1}\right] \\
X_2^{(m)} &= \frac{1}{m}\left[X_{2m} + X_{2m+1} + \cdots + X_{3m-1}\right] \\
&\vdots
\end{aligned}
$$

The random process is self-similar if it satisfies the following properties.

1. The processes X and $X^{(m)}$ are related by the equation

$$X^{(m)} = \frac{1}{m^{(1-H)}} X \tag{11.76}$$

   where H is the *Hurst parameter* $(0.5 < H < 1)$.
2. The means of X and $X^{(m)}$ are equal

$$E[X] = E[X^{(m)}] = \mu \tag{11.77}$$

3. The autocovariance functions of X and and $X^{(m)}$ are equal

$$E\left[(X(n+k)-\mu)(X(n)-\mu)\right] = E\left[\left(X(n+k)^{(m)}-\mu\right)\left(X(n)^{(m)}-\mu\right)\right] \tag{11.78}$$

A self-similar random process exhibits long-range dependence where the autocorrelation function $r_{XX}(n)$ or the autocovariance function $c_{XX}(n)$ do not vanish for large values of n. Distributions that have long-range dependence are sometimes called *heavy-tailed distributions*. A random process that displays no long-range dependence will have the autocorrelation, and autocovariance functions vanish for low values of n. A typical random process that has no long-range dependence is the Brownian motion.

Typically, self-similar phenomena are described using the Hurst parameter H whose value lies in the range

$$0.5 < H < 1 \tag{11.79}$$

The case $H = 0.5$ describes random walk problems or Brownian motion which exhibit no self-similarity. As $H \to 1$, the degree of self-similarity increases as well as the long-range dependence.

One way to model self-similar traffic is to use pdf distributions for the interarrival time that exhibit heavy-tailed distribution, as explained in the following section.

11.7 Heavy-Tailed Distributions

A heavy-tailed distribution gives rise to traffic that shows long-range dependence like in compressed video traffic. A distribution is heavy-tailed if it exhibits the following characteristics

1. Its variance is high or infinite.
2. Its cdf has the property

$$1 - F(x) = p(X > x) \sim \frac{1}{x^\alpha} \qquad x \to \infty \tag{11.80}$$

where $0 < \alpha < 2$ is the shape parameter and X is a random variable.

11.8 Pareto Traffic Distribution

The Pareto distribution that we studied in Section 1.20 on page 19 is used here to describe realistic traffic sources that have bursty behavior. The Pareto distribution is described by the pdf

$$f(x) = \frac{b\, a^b}{x^{b+1}} \tag{11.81}$$

where a is the position parameter, b is the shape parameter, and the random variable X has values limited in the range $a \leq x < \infty$. The Pareto distribution cdf is given by

$$F(x) = 1 - \left(\frac{a}{x}\right)^b \tag{11.82}$$

Notice that the Pareto distribution satisfies condition 2 of heavy-tailed distributions defined in Section 11.7.

The mean and variance for X are

$$\mu = \frac{b\,a}{b-1} \tag{11.83}$$

$$\sigma^2 = \frac{b\,a^2}{(b-1)^2\,(b-2)} \tag{11.84}$$

The mean is always positive as long as $b > 1$. The variance is meaningful only when $b > 2$. The variance of the Pareto distribution could be made high by properly choosing the shape parameter b to be close to 1 as the above equation indicates.

The Hurst parameter corresponding to the Pareto distribution is given by the equation

$$H = \frac{3-b}{2} \tag{11.85}$$

Table 11.1 shows the relation between the source burstiness and the two parameters H and Pareto distribution shape parameter b.

From the table, we conclude that in order to describe self-similar traffic using the Pareto distribution, we must have the shape parameter b close to one—typically H is chosen within the range 0.7–0.8 which would correspond to b values in the range 1.4–1.6. By proper choice of b, we can satisfy all the conditions defining heavy-tailed distributions defined in Section 11.7.

From 11.82, we can write

$$P(X > x) = 1 - F(x) = \left(\frac{a}{x}\right)^b \tag{11.86}$$

which means that the probability that the random variable has a value greater than x decreases at a rate that depends on the shape parameter b. If $b \approx 1$, the distribution has very large mean and variance [15].

Table 11.1 Relation between the source burstiness and the two parameters H and Pareto distribution shape parameter b.

Traffic statistics	H value	b value
Long-range dependent	$H \rightarrow 1$	$b \rightarrow 1$
Short-range dependent	$H \rightarrow 0.5$	$b \rightarrow 2$

A realistic bursty source is typically described using some or all of these parameter:

λ_{min}	the minimum data rate
λ_{a}	the average data rate
σ	the maximum data rate expected from the source

Since we are talking about rates in terms of packets/second, we need to convert these specifications into proper packet rates. The question becomes, how can we write down an expression for a Pareto distribution that takes into account all of these parameters?

We have two options:

1. *Flow description*: This option allows us to specify the randomness of the instantaneous data rate produced by the source.
2. *Interarrival time description*: This option allows us to specify the randomness of the periods between adjacent packets produced by the source.

11.8.1 Flow Description

We start by writing the pdf for the instantaneous data rate in the form

$$f_\Lambda(\lambda) = \begin{cases} 0 & \text{when} \quad \lambda < \lambda_{\text{min}} \\ b\, a^b/\lambda^{b+1} & \text{when} \quad \lambda \geq \lambda_{\text{min}} \end{cases} \tag{11.87}$$

where λ_{min} is the minimum data rate, which could be zero, a is the *position parameter* and b is the *shape parameter* that determines the steepness of the curve.

The values of the two parameters a and b can be found for a source with traffic descriptors $(\lambda_{\text{min}}, \lambda_{\text{a}}, \sigma)$ as follows.

$$a = \lambda_{\text{min}} \tag{11.88}$$

$$b = \frac{\lambda_{\text{a}}}{\lambda_{\text{a}} - \lambda_{\text{min}}} \tag{11.89}$$

To produce a bursty source, the value of b could be chosen close to 1, according to the data in Table 11.1.

11.8.2 Interarrival Time Description

The interarrival time following the Pareto distribution has a pdf that is given by

$$f_T(t) = \frac{b\, a^b}{t^{b+1}} \qquad \text{with } a \leq t < \infty \tag{11.90}$$

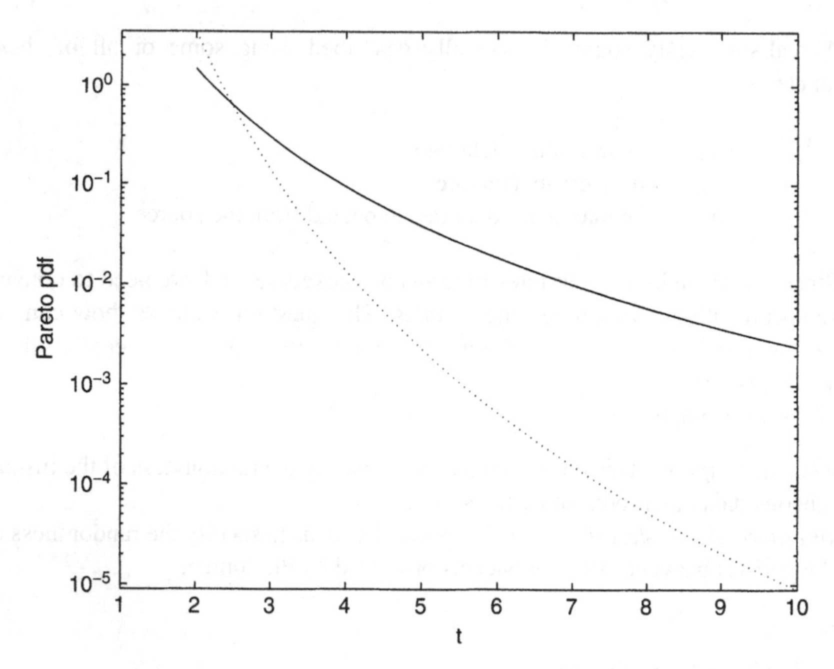

Fig. 11.9 The pdf distribution for the case when $a = 2$ and $b = 3$ (*solid line*) and $b = 7$ (*dashed line*)

where a (units s) is the position parameter and $b \geq 1$ is the shape parameter. Figure 11.9 shows the pdf distribution for the case when $a = 2$ and $b = 3$ (solid line), and $b = 7$ (dashed line). For the smaller value of shape parameter b, the pdf becomes flatter and has higher values at larger values of t. This results in larger variance in the interarrival time distribution.

11.8.3 Extracting Pareto Interarrival Time Statistics

A realistic source is typically described using some or all of these parameter:

λ_a the average data rate
σ the maximum data rate expected from the source

The question we pose here is, how to find a Pareto distribution that best matches the given source parameters? In a time period t, the maximum number of packets that could be produced by the source is given by

$$N_m = \sigma\, t \qquad (11.91)$$

We use this estimate to calculate the minimum time between two adjacent packets as follows.

$$a = \frac{t}{N_{\mathrm{m}}} = \frac{1}{\sigma} \quad \mathrm{s} \tag{11.92}$$

The position parameter depends only on the average packet size and burst rate.

In the time period t, the average number of packets that could be produced by the source is given by

$$N_{\mathrm{a}} = \lambda_{\mathrm{a}} t \tag{11.93}$$

The average time between two adjacent packets is given by

$$T_{\mathrm{a}} = \frac{t}{N_{\mathrm{a}}} = \frac{1}{\lambda_{\mathrm{a}}} \quad \mathrm{s} \tag{11.94}$$

But from the Pareto pdf distribution, the average interarrival time is given by

$$T_{\mathrm{a}} = \int_{t=a}^{\infty} t \, \frac{b \, a^b}{t^{b+1}} \, \mathrm{d}t = \frac{b \, a}{b-1} \quad \mathrm{s} \tag{11.95}$$

From the above two equations, we are able to obtain a value for the shape parameter b

$$\frac{1}{\lambda_{\mathrm{a}}} = \frac{b \, a}{b-1} \tag{11.96}$$

Therefore, we have

$$b = \frac{\sigma}{\sigma - \lambda_{\mathrm{a}}} \tag{11.97}$$

The shape parameter depends only on the average rate λ_a and burst rate σ. Furthermore, the shape parameter lies between the following extreme values

$$\begin{aligned} b &= 1 \qquad \text{when } \sigma \gg \lambda_{\mathrm{a}} \\ b &\to \infty \quad \text{when } \lambda_{\mathrm{a}} \to \sigma \end{aligned}$$

The first expression applies to a fairly bursty source, and the second expression applies to a constant-bit-rate source, where the average data rate equals the burst rate. Thus the range of the shape parameter b can be expressed as

$$1 \le b < \infty \tag{11.98}$$

Lower values of b imply bursty sources and higher values of b imply sources with little variations in the interarrival times since we would have a constant rate source.

Example 12 A bursty source produces data at an average rate of 5 Mbps, and its maximum burst rate is 20 Mbps. Estimate the Pareto parameters that best describe that source, assuming that the average packet size is 400 bits.

The position parameter is

$$a = \frac{A}{\sigma} = 20 \quad \mu\text{s}$$

The average data rate is used to determine the shape parameter b

$$b = \frac{\sigma}{\sigma - \lambda_a} = 1.333$$

■

11.8.4 Pareto Distribution and Queuing Analysis

The previous subsection discussed how the Pareto distribution parameters can be extracted, given the system parameters:

λ_a the source data rate
σ the maximum data rate expected from the source

A Pareto distribution matching these given source parameters has position and shape parameters:

$$a = 1/\sigma \quad \text{s} \tag{11.99}$$

$$b = \sigma/(\sigma - \lambda_a) \tag{11.100}$$

where we assumed the rates to be given in terms of packets/second.

In this section, we ask the question: Given a Pareto source with known parameters that feeds into a queue, what is the packet arrival probability? There are two cases that must be studied separately based on the values of the step size T and the position parameter a.

Case When $T \leq a$

When $T \leq a$ at most, one packet could arrive in one time step with probability x that we have to determine.

The number of time steps over a time period t is estimated as

$$n = \frac{t}{T} \tag{11.101}$$

The average number of packets in time period t is given by

$$\begin{aligned} N_a &= \lambda_a t \\ &= \lambda_a n T \end{aligned} \tag{11.102}$$

From the binomial distribution, the average number of packets that arrive during time t is given by

$$N_a = x\, n \tag{11.103}$$

where x is the packet arrival probability in one time step.

From the above two equations, we get

$$x = \lambda_a T \tag{11.104}$$

We see in the above equation that as T gets smaller or as the source activity is reduced (small λ_a), the arrival probability is decreased, which makes sense. The arrival probability for a Pareto distribution when $T \leq a$ is identical to the arrival probability for the Poisson distribution.

Example 13 Estimate the Pareto parameters and the packet arrival probability for a source with the following parameters. $\lambda_a = 10^3$ packets/s and $\sigma = 1.5 \times 10^4$ packets/s, Assume that the time step value is $T = 0.1$ ms.

The position parameter is

$$a = 6.6667 \times 10^{-4} \quad \text{s}$$

The shape parameter is

$$b = 1.0714$$

The arrival probability is

$$x = 0.1$$

■

Case When $T > a$

When $T > a$, more than one packet could arrive in one time step, and we have to find the binomial distribution parameters that describe this situation.

We start our estimation of the arrival probability x by finding the maximum number of packets that could arrive in one time step

$$N_m = \lceil \sigma\, T \rceil \tag{11.105}$$

The ceiling function was used here after assuming that the receiver will consider packets that partly arrived during one time step. If the receiver does not wait for partially arrived packets, then the floor function should be used. The average number of packets arriving in the time period T is

$$N_a = \lceil \lambda_a\, T \rceil \tag{11.106}$$

From the binomial distribution, the average number of packets in one time step is

$$N_a = x\, N_m \tag{11.107}$$

From the above two equations, the packet arrival probability per time step is

$$x = \frac{N_a}{N_m} \leq \frac{\lambda_a}{\sigma} \tag{11.108}$$

The probability that k packets arrive at one time step T is given by the binomial distribution

$$p(k) = \binom{N_m}{k} x^k\, (1 - x)^{N_m - k} \tag{11.109}$$

Example 14 A data source follows the Pareto distribution and has an average data rate $\lambda_a = 2 \times 10^3$ packets/s and maximum burst rate of $\sigma = 5 \times 10^3$ packets/s. Find the Pareto pdf parameters that describe this source and find the packet arrival probabilities if the time step chosen is equal to $T = 2$ ms.

The Pareto parameters are

$$a = 2 \times 10^{-4} \qquad \text{ms} < T$$

The maximum number of packets that could arrive in one time step is

$$N_m = 10$$

The average number of packets that could arrive in one time step is

$$N_{\mathrm{a}} = 4$$

The packet arrival probability per time step is

$$x = 0.4$$

■

11.9 Traffic Data Rate Modeling with Arbitrary Source Distribution

In this section, we attempt to model a traffic source that follows a general or arbitrary user-defined data rate traffic pattern. Assume that the probability mass function (pmf) of the source data rate is shown in Fig. 11.10. The number of pmf points is assumed K and the time resolution is T. The traffic model for this source is defined by two K-component vectors:

$$\mathbf{v}_p = \left[p_0 \; p_1 \; \cdots \; p_{K-1} \right]^t \tag{11.110}$$

$$\mathbf{v}_\lambda = \left[\lambda_0 \; \lambda_1 \; \cdots \; \lambda_{K-1} \right]^t \tag{11.111}$$

where the vector $\mathbf{v}_p$ contains the pmf probabilities and the vector $\mathbf{v}_\lambda$ contains the corresponding data rate values. The peak and average data rates (packets/s) are given by

$$\sigma = \lambda_{K-1} \tag{11.112}$$

$$\lambda_{\mathrm{a}} = \sum_{i=0}^{K-1} p_i \, \lambda_i \tag{11.113}$$

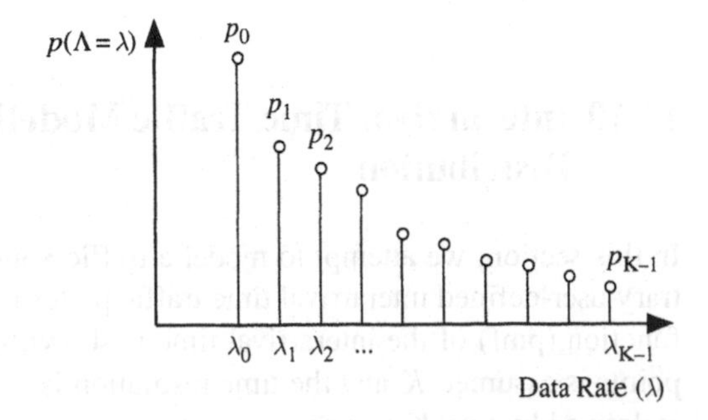

Fig. 11.10 pmf distribution for a source with arbitrary user-specified data rate statistics

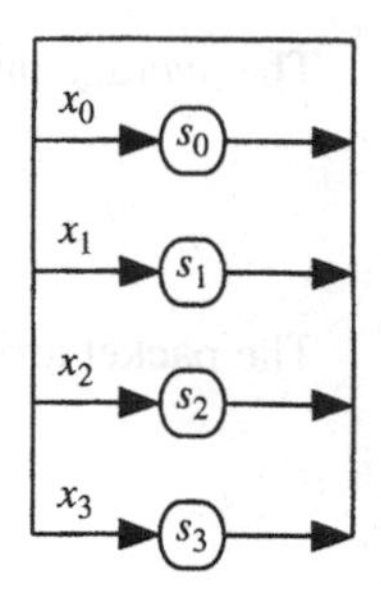

Fig. 11.11 The state transition diagram for a traffic source that follows a particular pmf distribution

To generate traffic that obeys that general distribution, we construct the source state transition diagram as shown in Fig. 11.11. State s_i of the source states in Fig. 11.11 corresponds to data rate λ_i.

We need to calculate the state transition probabilities x_i in the figure and see how they are related to the source probabilities p_i. We cannot just assume that the probabilities x_i are equal to p_i without some proof. From pmf definition and the figure, we can write the probability p_i as

$$p_i \equiv s_i \tag{11.114}$$

The RHS of the above equation indicates that the probability that the source data rate is λ_i given by the probability that the source state is in state s_i. At steady state, we can write

$$s_i = x_i \sum_{i=0}^{K-1} s_i = x_i \tag{11.115}$$

And from the above two equations, we determine the state transition probabilities x_i as

$$x_i = p_i \tag{11.116}$$

Although x_i was proved to be equal to p_i, this situation will not hold true for the interarrival traffic model in Section 11.10.

11.10 Interarrival Time Traffic Modeling with Arbitrary Source Distribution

In this section, we attempt to model a traffic source that follows a general or arbitrary user-defined interarrival time traffic pattern. Assume that the probability mass function (pmf) of the interarrival time is shown in Fig. 11.12. The number of pmf points is assumed K and the time resolution is T. The traffic model for this source is defined by two K-component vectors:

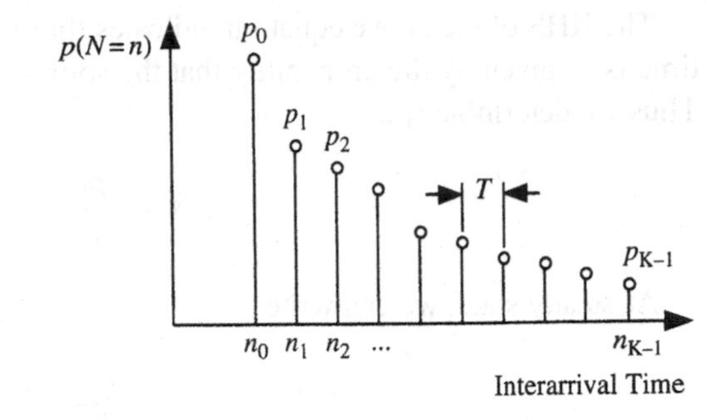

Fig. 11.12 pmf distribution for a source with arbitrary user-specified interarrival time statistics

$$\mathbf{v}_p = \left[p_0 \ p_1 \ \cdots \ p_{K-1} \right]^t \tag{11.117}$$

$$\mathbf{v}_n = \left[n_0 \ n_1 \ \cdots \ n_{K-1} \right]^t \tag{11.118}$$

where the vector $\mathbf{v}_p$ contains the pmf probabilities and the vector $\mathbf{v}_n$ contains the corresponding interarrival time values. The peak and average data rates (packets/s) are given by

$$\sigma = \frac{1}{n_0 T} \tag{11.119}$$

$$\lambda_a = \frac{1}{\sum_{i=0}^{K-1} p_i \ n_i} \tag{11.120}$$

To generate traffic that obeys that general distribution, we construct the source state transition diagram as shown in Fig. 11.13. We take the time step value T in the Markov chain equal to the time resolution value in Fig. 11.12. Row i of the source states in Fig. 11.13 corresponds to data arrival with an interarrival time value of t_i. The right-most state in each row is the state where data is actually generated.

We need to calculate the state transition probabilities x_i in the figure and see how they are related to the source probabilities p_i. From pmf definition and the figure, we can write the probability p_i as

$$p_i = n_i \ s_i \tag{11.121}$$

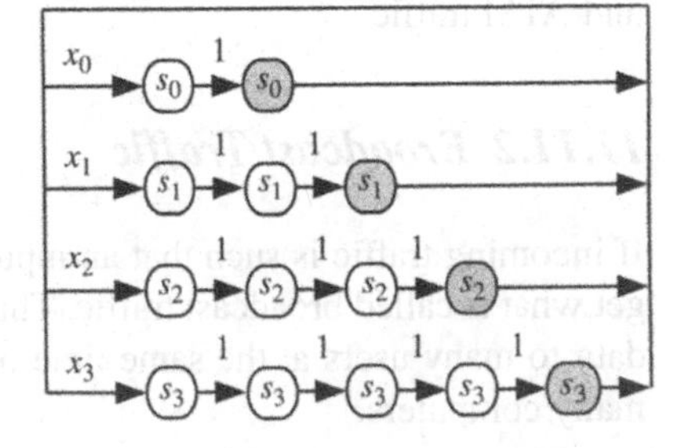

Fig. 11.13 The state transition diagram for a traffic source that follows a particular pmf distribution

The RHS of the above equation indicates that the probability that the interarrival time is n_i given by the probability that the source state is any of the states in row i. Thus we determine s_i as

$$s_i = \frac{p_i}{n_i} \tag{11.122}$$

At steady state, we can write

$$s_i = x_i \sum_{i=0}^{K-1} s_i \tag{11.123}$$

And from the above two equations, we determine the state transition probabilities x_i as

$$x_i = \frac{s_i}{\sum_{i=0}^{K-1} s_i} \tag{11.124}$$

11.11 Destination Statistics

Data or traffic arriving at the inputs of a switch or a router need to be routed to the desired output ports based on the information provided in the packet header and the routing table of the switch. The distribution of packets among the output ports is random, and we identify three types of destination statistics as discussed in the following sections.

11.11.1 Uniform Traffic

For a switch with N inputs and N outputs, uniform traffic implies that an incoming packet chooses a particular output port with probability $1/N$. This is true for any packet arriving at any input port. This model is referred to as the independent uniform traffic model [16]. Most studies assume uniform traffic to simplify the analysis. This assumption is true for traffic at routers or switching nodes in the network since the queuing and gradual release of the packets leads to randomization of the addressing [17]. These results apply to Ethernet LAN traffic as well as to WAN IP and ATM traffic.

11.11.2 Broadcast Traffic

If incoming traffic is such that an input port requests to access all output ports, we get what is called broadcast traffic. This type of traffic occurs when a site is sending data to many users at the same time or when a computer updates the databases of many computers.

Assume an $N \times N$ switch with N input ports and N output ports. We assume for simplicity that *each input port* carries two types of traffic flows: uniformly distributed traffic flow whose rate is λ_u and broadcast traffic flow whose rate is λ_b.

The traffic flows at each input and output port are given by

$$\lambda_{in} = \lambda_u + \lambda_b \quad (11.125)$$

$$\lambda_{out} = \lambda_u + N\,\lambda_b \quad (11.126)$$

Note that each output port now has to carry more output traffic than what came in on the average because of the amplification effect of data broadcast.

The *total* traffic flows at the input and output of the switch are given by

$$f(\text{in}) = N\,\lambda_u + N\,\lambda_b \quad (11.127)$$

$$f(\text{out}) = N\,\lambda_u + N^2\,\lambda_b \quad (11.128)$$

The amount of traffic through the network increases due to data broadcast.

11.11.3 Hot-Spot Traffic

If incoming traffic is such that many input ports request one particular output port, we get what is called hot-spot traffic. This type of traffic occurs when a popular web site is being browsed by many users at the same time or when many computers request access to the same server. Reference [18] models hot spot traffic as a fixed fraction of the arrival rate or arrival probability.

Assume an $N \times N$ switch with N input ports and N output ports. We assume for simplicity that *each input port* carries two types of traffic flows: uniformly distributed traffic flow whose rate is λ_u and hot-spot traffic flow whose rate is λ_h.

The traffic flow at each output port that is *not the destination* of the hot-spot traffic is given by

$$\lambda_r(\text{out}) = \lambda_u \quad (11.129)$$

The data rate at the output port that is *the destination* of the hot-spot traffic is given by

$$\lambda_h(\text{out}) = \lambda_u + N\lambda_h \quad (11.130)$$

Note that hot-spot traffic effectively increases the traffic at the hot-spot port.

The overall traffic flow at the input of the switch is given by

$$f(\text{in}) = N\,\lambda_u + N\,\lambda_h \quad (11.131)$$

The overall traffic flow at the output of the switch is given by

$$f(\text{out}) = N\,\lambda_\text{u} + N\,\lambda_\text{h} \tag{11.132}$$

The amount of traffic through the network does not increase due to hot-spot traffic.

11.12 Packet Length Statistics

Unlike ATM, many protocols produce packets that have variable lengths. Examples of protocols that have variable length packets are IP, Frame Relay- and Ethernet. Knowledge of the packet size is essential if one wants to estimate the buffer space required to store data having a certain arrival distribution statistics.

Poisson distribution could be used to provide a simple model for packet length statistics. The probability of receiving a packet of length A is given by

$$p(A) = \frac{\mu^A}{A!}\mathrm{e}^{-\mu} \tag{11.133}$$

where $A = 1, 2, \cdots$ units of length, and μ is the average packet length.

An exponential distribution could be used also in the form

$$f(A) = \frac{1}{\mu}\mathrm{e}^{-A/\mu} \tag{11.134}$$

where μ is the average packet length. Alternatively, we could use the binomial distribution to describe the probability of receiving a packet of length A

$$p(A) = \binom{N}{A} x^A (1-x)^{N-A} \tag{11.135}$$

where N is the maximum packet length and x is the probability that one byte is received. We could find the value of x if the average packet length μ is known:

$$x = \frac{\mu}{N} \tag{11.136}$$

If the packet length is highly irregular, then a Pareto distribution might be used in the form

$$f(A) = \frac{b\,a^b}{A^{b+1}} \tag{11.137}$$

where a is minimum packet length and $b > 1$ is a shape parameter. The average packet length for this distribution is

$$\mu = \frac{b\,a}{b-1} \tag{11.138}$$

We could also use the MMPP models to describe packet length statistics as was discussed in Section 11.2.3. In that model, we assume a Markov chain with N states such that in state s_i, the source produces a packet with length A_i. The probability of making transitions from one state to another is assumed based on experimental observations or based on model assumptions.

11.13 Packet Transmission Error Description

The previous sections dealt with issues related to network traffic such as data rate variation, packet length variation, and packet destination. When the packets are in transit, they are corrupted due to channel impairment or they could be totally lost due to congestion or address errors. We would like to model channel errors since these will affect the performance of the overall data transmission.

Figure 11.14 shows a model for adding errors to traffic during transmission. We have a data source that randomly generate frames as time progresses, such that the interarrival time between the generated frames follows one of the distributions discussed earlier.

We could even add another degree of freedom by randomly assigning different frame lengths. The number of packets per frame follows some distribution like Poisson, Bernoulli, or Pareto. The number of packets per frame is indicated in the figure by the numbers n_1, n_2, etc.

An error source also randomly generates errors with time. The number of errors per frame also follows some distribution like Poisson, Bernoulli, or Pareto. For example, a bursty error source could follow the Pareto distribution to generate lengthy error bursts. The number of packets in error per frame is indicated in the figure by the numbers e_1, e_2, etc. When the number of errors is either 0 or 1, we have a binary error source. When a Pareto distribution is used to generate the random numbers, we get bursts of errors with high probability.

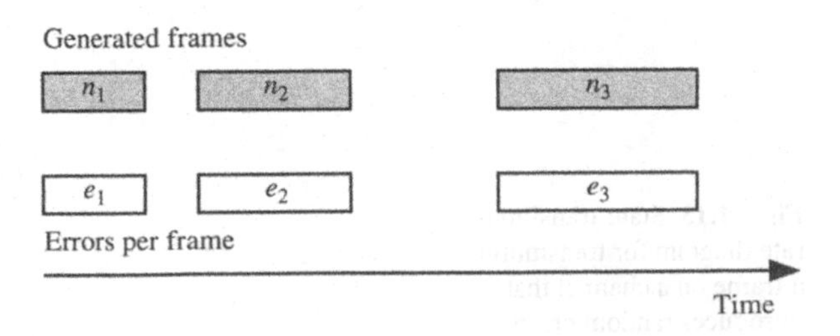

Fig. 11.14 Time series sequence of generated data and channel errors. A received frame is in error if it is generated at the same time that an error is generated

Example 15 Assume an on–off data source that generates equal length frames with probability a per time step. Assume for simplicity that each frame contains only one packet. The channel introduces errors in the transmitted frames such that the probability of a packet is received in error is e. Perform a Markov chain analysis of the system and derive its performance parameters.

The Markov chain model we use has four states:

State	Significance
1	Source is idle
2	Source is retransmitting a frame that was in error
3	Frame is transmitted with no errors
4	Frame is transmitted with an error

Figure 11.15 shows the Markov chain transition diagram, and the associated transition matrix for the system is given by

$$\mathbf{P} = \begin{bmatrix} 1-a & 0 & 1 & 0 \\ 0 & 0 & 0 & 1 \\ a(1-e) & 1-e & 0 & 0 \\ a\,e & e & 0 & 0 \end{bmatrix}$$

The system throughput is given by

$$\text{Th} = s_3$$

The average number of lost packets per time step is given by

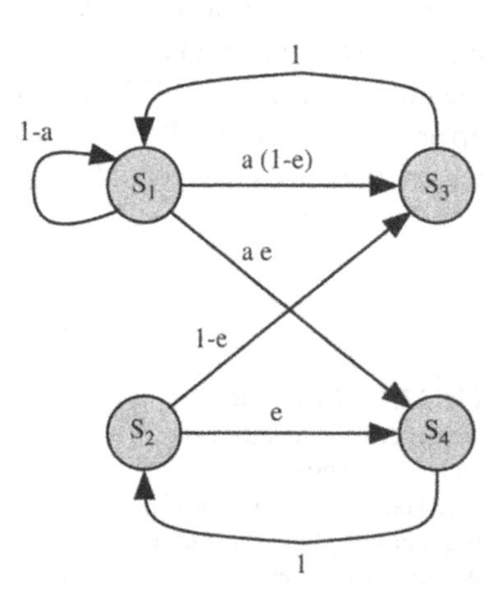

Fig. 11.15 State-transition-rate diagram for transmitting a frame on a channel that introduces random errors

$$\begin{aligned} N_a(\text{lost}) &= N_a(\text{in}) - N_a(\text{out}) \\ &= a - \text{Th} \\ &= a - s_3 \end{aligned}$$

The probability that the packet will be transmitted is

$$p_a = \frac{\text{Th}}{a} = \frac{s_3}{a}$$

The packet loss probability is

$$L = \frac{N_a(\text{lost})}{a} = 1 - \frac{s_3}{a}$$

The average number of retransmissions is given by

$$W = \frac{1 - p_a}{p_a} = \frac{a}{s_3} - 1$$

Example 16 Assume in the above example that $a = 0.7$ and $e = 0.1$. Calculate the performance of the system.

Figure 11.16 shows the variation of throughput (Th), delay (W), access probability (p_a), and loss probability (L) versus the input traffic (a). Two values of error probability are used $e1 = 0.1$ (solid line) and $e2 = 0.6$ (dotted line). We note that there is a maximum value on the throughput of Th(max) $= 0.5$ and that the system performance deteriorates rapidly when the error probability increases.

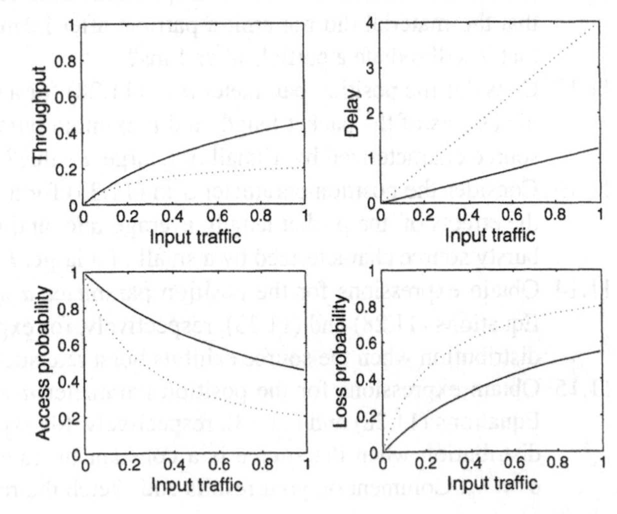

Fig. 11.16 Throughput (Th), delay (W), access probability (p_a), and loss probability (L) versus the input traffic (a). Two values of error probability are used $e1 = 0.1$ (*solid line*) and $e2 = 0.6$ (*dotted line*)

Problems

Traffic Modeling

11.1 Why do we need to develop models for network traffic?
11.2 What is meant when we say that a source is bursty?
11.3 What is meant by a point process in traffic modeling?
11.4 What is meant by a fluid flow model to describe network traffic?

Exponential Interarrival Time

11.5 Obtain (11.28) on page 394 when the source maximum data rate σ is given in units of bits/second.

11.6 Obtain (11.33) on page 395 when the source maximum data rate σ and average data rate λ_a are given in units of bits/second.

11.7 A Poisson packet source produces packets at the rate of 1500 packets/s. Find the probability that a packet will be received within a 0.5 ms interval. What is the average number of packets received within this same time interval?

11.8 A data source has the following parameters: $\lambda_{\mathrm{a}} = 200$ kbps and $\sigma = 300$ kbps. Find the pdf that describes the data rate distribution using the results of Example 5 on page 395.

11.9 Obtain the mean and variance for the random variable T for the exponential interarrival time whose pdf distribution is given in (11.12)

11.10 A radioactive material has a half-life of 1 ms. Find the average time interval between the emitted particles assuming a Poisson process. Write down an expression for the probability of detecting 5 radiated particles in a period of 0.5 ms.

11.11 A radioactive material has an average decay time of 10 ms. An observer finds that the material did not emit a particle after 20 ms, what is the probability that it will radiate a particle after 1 ms?

11.12 Consider the position parameter a in (11.28) for a Poisson source. What are the effects of the packet length and maximum burst rate? Is a high data rate source characterized by a small or a large a value?

11.13 Consider the position parameter a in (11.33) for a Poisson source. What are the effects of, the packet length, average rate, and maximum burst rate? Is a bursty source characterized by a small or a large, b value?

11.14 Obtain expressions for the position parameter a and shape parameter b in Equations (11.28) and (11.33), respectively, for exponential interarrival time distribution when the source exhibits burst rate such that $\sigma \gg \lambda_{\mathrm{a}}$.

11.15 Obtain expressions for the position parameter a and shape parameter b in Equations (11.28) and (11.33), respectively, for exponential interarrival time distribution when the source is a constant bit rate source (CBR) such that $\sigma = \lambda_{\mathrm{a}}$. Comment on your results and sketch the resulting pdf distributions.

11.16 A Poisson source has an interarrival time pdf distribution with the following two parameter values $a = 1$ ms and $b = 5000$ packets/s. Write down an expression for the probability that the source produces 20 packets in a period of 1 ms.

11.17 A data source follows the Poisson distribution and has an average data rate of 100 kbps and maximum burst rate of 500 kbps. Estimate the exponential distribution parameters that best describe that source assuming that the average packet size is 0.5 kB.

Discrete Exponential Interarrival Time

11.18 Because of the snow, the bus arrival times become messed up. Assume that buses should arrive on the average each 10 min. However, because of the snow, the probability that a bus arrives at this time on time is 30%. A passenger just missed the bus, what is the probability that he/she will have to wait for 1 h before the next bus arrives?

11.19 A data source has a exponential interarrival time pdf distribution with the following two parameter values $x = 0.1$ and $b = 5000$. Write down an expression for the probability that the source produces 20 packets in a period of 20 time steps.

11.20 A data source follows the binomial distribution and has an average data rate of 10 kbps and maximum burst rate of 100 kbps. Estimate the discrete exponential distribution parameters that best describe that source if packets are being transmitted on an ATM network operating at OC-3 (155.52 Mbps).

11.21 A data source follows the binomial distribution and has an average data rate of 10 packets/s and maximum burst rate of 100 packets/s. Estimate the discrete exponential distribution parameters that best describe that source if packets are being transmitted on an Ethernet network operating at 10 Mbps where the average packet length is 1024 bytes.

Pareto Interarrival Time

11.22 Prove that the Pareto distribution is not memoryless. This implies that if a burst is received, it is likely that the burst will continue.

11.23 A bursty source produces data at an average rate of 5 Mbps and its maximum burst rate is 20 Mbps. Estimate the Pareto parameters that best describe that source assuming the average packet size is 400 bits.

11.24 Find the average interarrival time for the source in the previous problem.

Packet Transmission Error Description

11.25 Use the results of Example 15 on page 424 to study the effect of channel error on data transmission. Pick some value for $a = 0.5$, say, and vary the error probability between $0.001 < e < 0.9$. Plot the system throughput and comment on your results.

11.26 Consider Example 15 and suppose that there is an upper limit on the number of retransmissions before the frame is considered lost. Obtain the resulting Markov transition diagram and the associated transition matrix.

11.27 Consider Example 15 again and suppose that no transmissions are allowed. This could be the case for real-time data or best-effort traffic. Obtain the resulting Markov transition diagram and the associated transition matrix.

11.28 Consider Example 15 again, but this time assume that number of errors per frame varies between 0 and 5. A forward error correction (FEC) scheme is used and frame is considered to be error free if it contains up to two packets in error. Obtain the resulting Markov transition diagram and the associated transition matrix.

11.29 Assume an adaptive forward error correction (FEC) scheme where three levels of error correction are employed:
FEC level 1: can correct one error in received frame only.
FEC level 2: can correct up to three errors.
FEC level 3: can correct up to five errors.
When the errors in the received frame can be corrected, the next frame is transmitted using the next lower FEC level. When the errors in the received frame cannot be corrected, the frame is retransmitted using the next higher FEC level. Assume each frame to contain no more than five errors in it due to packet size limitations. Derive the Markov chain transition diagram and the associated transition matrix.

References

1. D. Denteneer and V. Pronk, "Traffic models for telecommunication", *13th Symposium on Computational Statistics*, Bristol, Great Britain, pp. 269–274, 1998.
2. W. Leland, M. Taqqu, W. Willinger, and D. Wilson, "On the self-similar nature of Ethernet traffic", *IEEE/ACM Trans. on Networking*, vol. 2, pp. 1–15, 1994.
3. J. Beran, R. Sherman, M.S. Taqqu and W. Willinger, "Variable-bit-rate video traffic and long-range dependence", *IEEE/ACM Transactions on Networking*, vol. 2, pp. 1–15, 1994.
4. V. Paxson and S. Floyd, "Wide-area traffic: The failure of Poisson modeling", *Proceedings of the conference on communications architectures protocols and applications, ACM Sigcomm*'94, pp. 257–268, 1994.
5. S. Molnar and A. Vidas, "On modeling and shaping self-similar ATM traffic, teletraffic contributions for the information age", Proc. 15th International Teletraffic Congress, Washington D. C., June 1997, pp. 1409–1420, 1997.
6. Z. Sahinoglu and S. Tekinay, "On multimedia networks: Self-similar traffic and network performance", *IEEE Communications Magazine*, Jan. 1999.
7. M. Corvella and A. Bestavros, "Self-similarity in World Wide Web traffic: evidence and possible causes", *IEEE/ACM Transactions on Networking*, vol. 5, pp. 835–846, 1996.
8. J.D. McCabe, *Practical Computer Network Analysis and Design*, Morgan Kaufmann, San Francisco, 1998.
9. J.M. Pitts and J.A. Schormans, *Introduction to ATM Design and Performance*, John Wiley, New York, 1996.

10. H. Saito, *Teletraffic Technologies in ATM Networks*, Artech House, Boston, 1994.
11. G.E.P. Box and G.M. Jenkins, *Time Series Analysis: Forecasting and Control*, Prentice-Hall, New Jersey, 1976.
12. V.S. Frost and B. Melamed, "Traffic modeling for telecommunications networks", *IEEE Communications Magazine*, vol. 32, March 1994.
13. M.R. Schroeder, *Number Theory in Science and Communications*, Springer-Verlag, New York, 1986.
14. L.M. Sander, "Fractal growth", *Scientific American*, vol. 256, pp. 94–100, 1987.
15. T. Hettmansperger and M. Keenan,"Tailweight, statistical interference and families of distributions - A brief survey" in *Statistical Distributions in Scientific Work*, vol. 1, G.P. Patil et. al. ed., Kluwer, Boston, 1980.
16. F.A. Tobagi, T. Kwok, and F.M. Chiussi, "Architecture, performance, and implementation of the tandem-banyan fast packet switch", *Proceedings of the IEEE*, vol. 78, pp. 133–166, 1990.
17. V.J. Friesen and J.W. Wong, "The effect of multiplexing, switching and other factors on the performance of broadband networks", *Proceedings of the IEEE INFOCOM'93*, pp. 1194–1203, 1993.
18. G.F. Pfister and V.A. Norton, "Hot-spot contention and combining in multistage interconnection networks", *IEEE Transactions Computers*, vol. 34, pp. 943–948, 1993.
19. M. Guizani and A. Rayes, *Designing ATM Switching Networks*, McGraw-Hill, New York, 1999.
20. A. Erramilli, G. Gordon, and W. Willinger, "Applications of fractals in engineering for realistic traffic processes", *International Teletraffic Conference*, 14:35–44, 1994.
21. W. Stallings, *Data and Computer Communications*, Fifth edition, Prentice Hall, New Jersey, 1998.
22. N. Likhanov, B. Tsybakov, and N.D. Georganas, "Analysis of an ATM buffer with self-similar ("fractal") input traffic", *INFOCOM*, vol. 3, pp. 8b.2.1–8b.2.7, 1995.
23. I. Norros, "A storage model with self-similar input", *Queuing Systems*, vol. 16, 1994.
24. J. Gordon, "Pareto process as a model of self-similar packet traffic", *IEEE Global Telecommunications Conference* (Globecom 95), Singapore, 14 November, pp. 2232–2235, 1995.
25. F. Gebali, A queuing model for self-similar traffic, *The 1st IEEE International Symposium on Signal Processing and Information Technology*, December 28–30, 2001, Cairo, Egypt.
26. L. Kleinrock, *Queuing Systems*, volume I, John Wiley, New York, 1975.
27. P. Sen, B. Maglaris, N.-E. Rikli, and D. Anastassiou, "Models for packet switching of variable-bit-rate video sources", *IEEE Journal on Selected Areas in Communications*, vol. 7, pp. 865–869, 1989.
28. B. Maglaris, D. Anastassiou, P. Sen, G. Karlsson, and J. Robbins, "Performance models of statistical multiplexing in packet video communications", *IEEE Transactions on Communications*, vol. 36, pp. 834–844, 1988.
29. P. Sen, B. Maglaris, N. Rikli, D. Anastassiou, "Models for packet switching of variable-bit-rate video sources", *IEEE Journal on Selected Areas in Communications*, vol. 7, pp. 865–869, 1989.
30. M. Nomura, T. Fujii, and N. Ohta, "Basic characteristics of variable rate video coding in ATM environment", *IEEE Journal on Selected Areas in Communications*, vol. 7, pp. 752–760, 1989.
31. R. Grunenfelder, J.P. Cosmos, S. Manthorpe, and A. Odinma-Okafor, "Characterization of video codecs as autoregressive moving average process and related queuing system performance", *IEEE Journal on Selected Areas in Communications*, vol. 9, pp. 284–293, 1991.
32. P. Sen, "Models for packet switching of variable-bit-rate video sources", *IEEE Journal on Selected Areas in Communications*, vol. 7, pp. 865–869, 1989.
33. H.E. Jurst, "Long-term storage capacity of reservoirs", *Transactions of the American Society of Civil Engineers*, vol. 116, pp. 770–799, 1951.
34. B.B. Mandelbrot and J.W. van Ness, "Fractional Brownian motions, fractional noises and applications", *SIAM Review*, vol. 10, pp. 422–437, 1968.
35. R.Y. Awdeh and H.T. Muftah, "Survey of ATM switch architectures", *Computer Networks and ISDN Systems*, vol. 27, pp. 1567–1613, 1995.

Chapter 12
Scheduling Algorithms

12.1 Introduction

A scheduling algorithm must be implemented at each network router or switch to enable the sharing of the switch-limited resources among the packets traveling through it. The resources being shared include available link bandwidth and buffer space. The main functions of a scheduler in the network are to (1) provide required QoS for the different users, by making proper choices for *selecting* next packet for forwarding to the next node; (2) select next packet for *dropping* during periods of congestion when the switch buffer space is starting to get full; and (3) provide *fair sharing* of network resources among the different users.

12.1.1 Packet Selection Policy

In a typical switch or router, several packets will request to access a certain output port. Because only one packet can leave, the rest must be stored in an intermediate buffer. Somehow we must find a way to decide which stored packet must be sent next. Different selection policies could be implemented for different types of queues depending on the service classes of the queues. For example, some applications have rigid real-time constraints on delay and jitter, while other adaptive applications agree to modify their behavior based on the network status. At the time of writing, most Internet applications are handled using *best-effort* packet transfer policy with no bandwidth or delay guarantees [1].

12.1.2 Packet Dropping Policy

The fact that packets have to be stored in each switching node of the network implies that buffer storage in a switch is a resource that must be shared among the different users or sessions. During periods of congestion, the switch buffers become full and the scheduler must also decide which packets to drop.

F. Gebali, *Analysis of Computer and Communication Networks*,
DOI: 10.1007/978-0-387-74437-7_12, © Springer Science+Business Media, LLC 2008

12.1.3 Fair Sharing Policy

The switch resources such as available output link bandwidth and local buffer space must be shared among the switch users. Because of the different classes of service supported by the switch, an equal sharing of the resources is not the best option. Rather, the scheduler must allocate these finite resources in a *fair* manner so that each user can get its share based on its class of service. Another issue related to fair sharing is *isolation* or *protection*. This is required because not all users abide by their agreed-upon data rate. When this happens, the misbehaving (nonconforming) user starts to hog the resources at the expense of other well-behaving (conforming) users.

It is obvious from the previous discussion that data scheduling is required at each node in the network for three reasons: (a) *selection* of a packet for transmission from the population of queued packets destined to a certain switch output; (b) *provide QoS* for the different types of flows going through the switch; and (c) *drop* packets when the buffer space becomes full.

12.2 Scheduling as an Optimization Problem

From the above discussion, it is clear that the scheduling problem is an optimization problem since the scheduler distributes the system-limited resources among the user traffic which impacts the offered quality of service (QoS). The scheduling algorithms to be discussed in the following sections are all heuristic and have no solid proof that the proposed scheduling policy is optimal in any mathematical sense. Further, we will note that the scheduler discussed here reduces packet delay by allocating large bandwidth to the delay-sensitive traffic. This might be self-contradictory for some types of traffic, which is delay sensitive but does not require high bandwidth.

The author developed a hierarchical scheduler [2] that is based on the transportation problem optimization technique [3, 4]. The transportation problem technique allows optimizing different QoS types such as delay-sensitive traffic and bandwidth-sensitive traffic through the same switch.

Finding the optimum solution to a transportation problem is not simple and might consume a certain amount of time. However, the author developed a simple greedy algorithm for solving the transportation problem based on computational geometric concepts. The algorithm uses only simple add/subtract operations and hence should be fast to compute [2].

12.3 Scheduler Location in Switches

The scheduler will be located in the switch where packets are buffered and where packets must share a resource like the switch fabric or the output links.

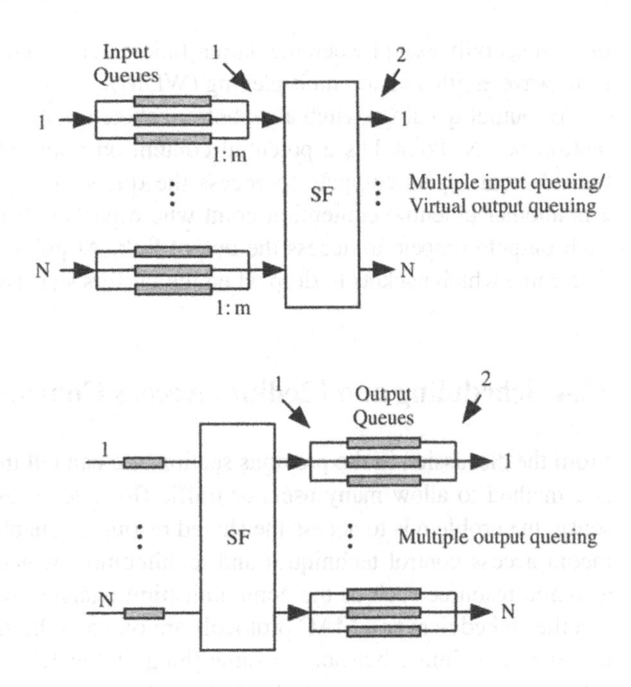

Fig. 12.1 Virtual output queuing (VOQ) and output queuing switches with different queues for each service class. The points at which packet contention occurs are shown

When a switch is capable of supporting different QoS classes, each service class will have its own dedicated buffer. At the extreme, each session or user channel will have its own dedicated buffer in each switch it encounters. Figure 12.1 shows two buffer location options in switches. The rectangles labeled "SF" indicated the switching fabric of the switch whose function is to provide connectivity between each input port and each output port. The figure on the top shows an input queuing switch where the buffers are located at each input port. The figure on the bottom shows an output queuing switch where the buffers are located at each output port. Multiple buffers are shown at each input or output port when multiple service classes are supported.

Irrespective of the buffering strategy employed, packets will contend for the shared resources and some form of scheduling algorithm will be required. The figure shows that there are three types of shared resources: (1) the buffer storage space; (2) the switch fabric (SF); and (3) the output links (i.e., available bandwidth)

For the input queuing switch, top sketch in Fig. 12.1, we can get two potential contention points. Point 1 is a potential contention point where all the packets from the different queues at an input port compete to access the switch fabric. Of course, a remedy for this problem would be to modify the switch fabric to allow more than one packet to access the fabric simultaneously. At point 1, the scheduler must also determine which packets to drop when the buffers start being full. Point 2 is another potential contention point where packets from the different inputs compete to access a certain output port. At this point, usually the output link is only able to transmit one packet only except when the output link is composed of multiple channels such as in wavelength division multiplexing (WDM).

For output queuing switch also, bottom sketch in Fig. 12.1, we can get two contention points. Point 1 is a potential contention point where all the packets from the different inputs compete to access the queues of a certain output port. Point 2 is another potential contention point where packets from the different queues in each output compete to access the output link. At point 2 the scheduler must also determine which packets to drop when the buffers start being full.

12.4 Scheduling and Medium Access Control

From the discussion in the previous section, we can tell that a scheduling algorithm is a method to allow many users or traffic flows to access the output link. In that sense, the problem is to access the shared resource. Chapters 10, 13, and 14 discuss media access control techniques and architectures which allow packets to access a shared resource such as the communication channel. A question naturally arises whether schedulers and MAC protocols are one and the same. The quick answer is that they are similar but not the same thing. Table 12.1 compares scheduler algorithms and media access control (MAC) techniques.

12.5 Scheduler Design Issues

There are several design issues related to schedulers. These issues are reviewed in the following subsections. In the discussion to follow we shall use the terms "user", "session", "connection", or "flow" to describe the traffic carried by the switch.

Table 12.1 Comparison between scheduling algorithms and media access control (MAC) techniques

Scheduling algorithms	MAC techniques
Designed to provide QoS guarantees	Designed to provide resource access only
Determine the user's share of bandwidth, buffer space allocation, and packet discard decisions	Does not deal with these issues
Shared resource is outgoing link of a switch/router and buffer space	Shared resource is a bus, a wireless channel, or a shared memory
Operate at switch or router output ports	Operate at output ports or each device connected to the medium
Used with switches or routers	Used with buses, wireless channels, etc
Physically exists inside a switch or router	Distributed among all users and MAC controllers
Implemented in software (at least for now)	Implemented in software and hardware
Operate at layers above the physical layer	Operate at the physical layer

12.5.1 Priority

Ideality dictates that equality is a good thing. However, real life tells us that special people are "more equal"! Alas, equality is not a good thing in computers or networks. The scheduler is only able to do a decision to select a certain user only when this user has higher priority compared to all other users. Priority assignment could be static or dynamic. A static priority assignment implies that users belonging to a certain class of service have a certain level of priority that does not change with time or with the traffic load. On the other hand, dynamic priority allocation changes the priority of a certain class of service with time or based on the traffic volume.

In addition to priority assignment, there is an *arbitration rule* that is invoked to resolve the conflicts between users with the same priority.

12.5.2 Degree of Aggregation

In order to provide guaranteed QoS, the scheduler has to keep *state information* about the performance of each user or connection. The problems with such schedulers are limitations on scaling, deployment difficulties, and the requirement of mapping between application and network service parameters [5]. The large number of states that must be maintained and references slow down the scheduler and limit the number of users that can be accepted. Other schedulers aggregate several connections into classes to reduce the amount of state information and workload. This approach is more promising since it deals with large aggregates of network traffic and its per-hop behavior is configurable [6]. The differentiated services scheduler provides constant *ratios* of the quality of service ratios between the service classes even when the quality level is varying. The price to be paid by aggregating traffic is loss of deterministic QoS guarantees since the state of each connection is lost. QoS guarantees for a high-level aggregation server are provided on a probabilistic basis. In other words, the scheduler loses specific information about the status of each user since it only keeps track of groups of users. This reduces the number of states that must be checked and updated. Guarantees can be provided on the QoS for groups of users, but each user cannot be guaranteed specific level of service.

12.5.3 Work-Conserving Versus Nonwork-conserving

A work-conserving algorithm is idle when all of the priority queues are empty. A nonwork-conserving algorithm might not transmit a packet even when the priority queues are occupied. This might happen to reduce the delay jitter for example [7]. This is nice in theory but is not implemented in practice. In general, it was found that work-conserving algorithms provide lower average delay than nonwork-conserving algorithm.

Examples of work-conserving algorithms include generalized processor sharing [8], weighted fair queuing [9], virtual clock [10], weighted round robin [11], delay-earliest-due date (Delay-EDD) [12], and deficit round robin [13].

Examples of nonwork-conserving algorithms are stop-and-go, jitter earliest-due date (jitter-EDD) [14], and rate-controlled static priorities (RCSP).

12.5.4 Packet Drop Policy

Schedulers not only select which packet to serve next, but also have to select which packet to drop when the system resources become overloaded. There are several options for dropping packets such as dropping packets that arrive after the buffer reaches a certain level of occupancy (this is known as tail dropping). Another option is to drop packets from any place in the buffer depending on their priority. As we shall see later, a third approach is to randomly select packets for dropping once the system resources become congested.

12.6 Rate-Based Versus Credit-Based Scheduling

Scheduling methods could be classified broadly as rate-based or time-based. A rate-based scheduler selects packets based on the data rate allocated for the service class. Rate-based scheduling methods include fair queuing (FQ) [9, 15], which is equivalent to virtual clock (VC) [10] weighted fair queuing (WFQ) [9], hierarchical round robin (HRR) [16], deficit round robin (DRR) [13], stop-and-go (S&G) [17, 18, 19, 20], and rate-controlled static priority (RCSP) [21]. Rate-based methods allow a variety of techniques for arriving at a service policy [22]. Some of the methods are fairly complex to implement since they require complex mathematical operations and require knowledge of the states of the different flows. However, they can only provide guarantees based on the maximum delay or delay jitter bounds since they translate these requirements into an equivalent bandwidth. This proves to be the weak point on these algorithms since providing bandwidth guarantees to effect delay guarantees ignores the actual bandwidth requirements of the actual traffic stream. Needless to say, this would lead to unfair bandwidth allocation to the other services that have real QoS bandwidth requirements.

A time-based scheduler selects packets based on their time of arrival. Time-based methods include earliest-due date for delay (EDD-D) [23], earliest-due date for jitter (EDD-J) [24], and smallest response time (SRT) [25]. Scheduler-based methods require keeping track of the arrival times of the different packets and calculate the packet priority based on the packet arrival time and the deadline imposed on it. To provide end-to-end guarantees on the delay and delay jitter, the scheduler must be implemented at all the switching nodes and the incoming traffic must conform very closely with the assumed model [22].

12.7 Scheduler Performance Measures

A good scheduling algorithm must satisfy several of the following performance measures [26]:

QoS: The scheduler must be able to support different types of QoS classes with varying requirements such as bandwidth (throughput), delay, delay jitter, and loss probability [27].

Fairness: The main goal of fairness is to serve sessions in proportion to some specified value [28]. The simplest fair allocation of resources is to equally divide the available bandwidth and buffer space among all the users and to drop excess packets equally from the different queues. However, if a user does not require all of its allocated share, then the excess share should be divided equally among the other users.

Isolation or protection: Isolation means that a misbehaving user should not adversely impact other users. The user becomes misbehaving when its packet arrival rate exceeds what is expected. Isolating the effects of misbehaving users is equivalent to *protecting* conforming users.

Simplicity: The scheduler must be easy to implement in hardware especially at high network speeds. This requires that computations to be done by the scheduler to be small in number and simple to calculate.

Scaling: The scheduler must perform well even when the number of sessions increases or when the link speed is increased. Some scheduling algorithm must keep state information about every user and must update this information very frequently. This places limitations on how many users can be supported by the scheduler and places limitations on the scheduler delay.

12.8 Analysis of Common Scheduling Algorithms

The remainder of this chapter discusses several scheduling algorithms that vary in performance from support of best-effort traffic with no performance guarantees to schedulers that support guaranteed services with bounds on bandwidth and delay. Models that describe the performance of each algorithm are also developed.

12.9 First-In/First-Out (FIFO)

First-in/first-out (FIFO) is also known as first-come/first-served (FCFS). The FIFO method sorts users according to their arrival time. The users that arrive early are served first [29]. In that sense, all users have the same priority and the *arbitration rule* is based on the time of arrival of the packets. This method is used naturally to store incoming packets in queues that could be associated with the input or output ports.

FIFO is simple to implement and packet insertion and deletion are particularly simple and do not require any state to be maintained for each session. However, this proves to be a disadvantage since the server has no way of distinguishing between the packets belonging to different users. Thus some users might be misbehaving and fill a large portion of the queue which increases the chance of dropping the packets of other users.

When all incoming flows are queued in a single queue, greedy users exhibiting long periods of activity (bursty behavior) will take the lion's share of the queue capacity at the expense of other flows. When the queue is filled, packets at the tail are dropped. This is the reason why this method is known as *FIFO with tail drop*. Most routers adopt this method because of its simplicity.

FIFO does not provide per-connection delay or rate guarantees since priority is based solely on arrival time. One way to provide delay bounds is to limit the size of the queue so that the maximum delay equals the time to serve a full queue. Of course, once the queue size is limited, there will be a probability that an arriving packet will be discarded if it arrives when the queue is full. To reduce the packet discard probability, the number of sessions should be limited.

12.9.1 Queuing Analysis of FIFO/FCFS

Let us perform a simple queuing analysis for a FIFO buffer. We make the following simplifying assumptions:

1. The size of the buffer is B.
2. The maximum number of customers that could arrive at the input of the buffer at a certain time step is m.
3. The average length of packets from any user is A.
4. The arrival probability for any user is a.
5. The departure probability for the buffer is c.

Figure 12.2 shows the FIFO buffer where several sessions converge at the input and only one packet can leave which corresponds to an $M^m/M/1/B$ queue. The transition matrix for such queue was derived in Section 7.7 on page 241. Based on that, we can derive expressions for the scheduling delay which corresponds to the queuing delay in that situation.

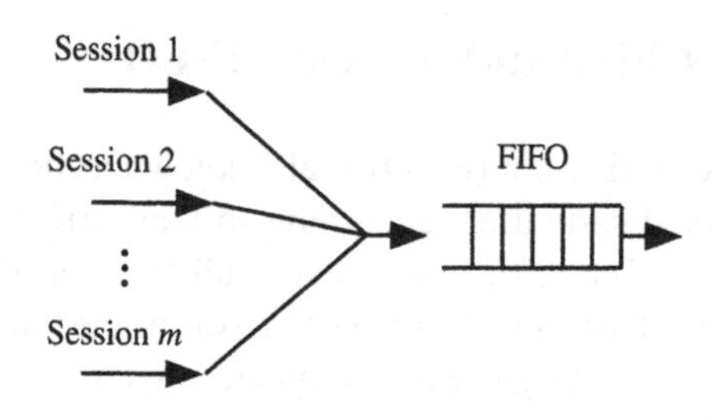

Fig. 12.2 A FIFO buffer with m flows at its input

The throughput of the FCFS queue was given in Section 7.7, which we repeat here for convenience.

The average throughput is estimated as

$$\text{Th} = c \ (1 - a_0 \ s_0) \tag{12.1}$$

where a_0 is the probability that no packets arrive during a time step and s_0 is the probability that the queue is empty.

The average lost traffic $N_a(\text{lost})$ is given by

$$\begin{aligned} N_a(\text{lost}) &= N_a(\text{in}) - N_a(\text{out}) \\ &= N_a(\text{in}) - \text{Th} \\ &= m \ a - c \ (1 - a_0 \ s_0) \end{aligned} \tag{12.2}$$

We refer the reader to Sections 7.7.1 and 7.7.2 for a more detailed discussion of the performance figures for the $M^m/M/1/B$ queue.

12.10 Static Priority (SP) Scheduler

In a static priority scheduler, separate queues are assigned different priorities. Incoming data are routed to the different queues depending on their priority. The scheduler serves packets in a lower priority queue only if all the higher priority queues are empty [30, 31, 29, 32, 33].

The static priority scheduler is also known as the IEEE 802.1p, which was discussed in Chapter 10. A queuing analysis of the static priority scheduler was performed in Section 10.2 on page 326.

12.11 Round Robin Scheduler (RR)

The round robin scheduler serves backlogged sessions one after another in a fixed order. When the scheduler finishes transmitting a packet from session 1, say, it moves on to session 2 and checks if there are any packets waiting for transmission. After the scheduler has finished going through all sessions, it comes back to the first session, hence the name round robin. Figure 12.3 shows a round robin scheduler serving four sessions.

The main features of this scheduler are ease of implementation in hardware and protection of all sessions even best-effort sessions. A greedy or misbehaving user will not be able to transmit more than one packet per round. Hence all other users will not be penalized. The misbehaving user will only succeed in filling its own buffer and data will start getting lost. However, the long packets will be transmitted since the scheduler serves whole packets. This could affect the delay experienced by other sessions that might have short packets to transmit.

Fig. 12.3 Round robin scheduler serving four sessions

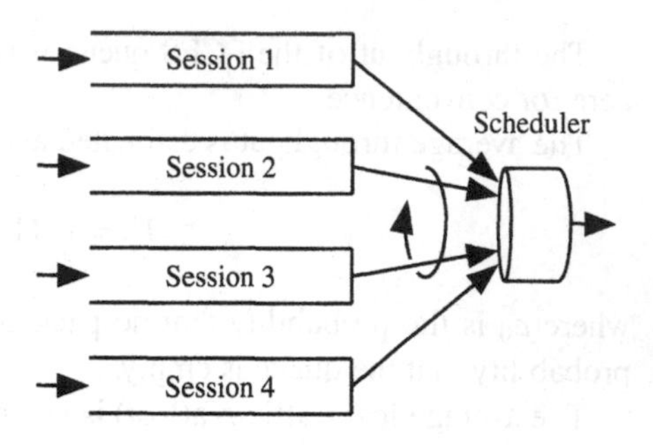

Assuming m sessions, the maximum bound on delay for a queue is given by

$$W(\max) = \sum_{i=1}^{m} \frac{A_i}{C} \quad \text{s} \tag{12.3}$$

where A_i is the head of the line packet length in bits for session i and C (bps) is the output link rate.

The ratio of service provided for session i relative to the total service provided to all sessions is equivalent to finding the ratio of number of bits moved from session i relative to the total number of bits moved in one round. We define f_i as that ratio which is written as

$$f_i = \frac{A_i}{\sum_{j=1}^{m} A_j} \tag{12.4}$$

Thus the bandwidth C_i given to session i relative to the total output bandwidth is simply given by

$$C_i = C\ f_i \tag{12.5}$$

In an ideal round robin algorithm, all packets have equal lengths and each session share would be

$$f_i = \frac{1}{m} \tag{12.6}$$

$$C_i = \frac{C}{m} \tag{12.7}$$

The above two equations assume that all sessions are backlogged and have packets with equal lengths. The algorithm is not fair when some sessions have variable length packets since the server will allocate more time for these sessions.

12.11.1 Queuing Analysis for RR

We can study the occupancy of each queue in a round robin scheduler using the following assumptions:

1. The outgoing link capacity is C.
2. The number of queues or sessions is m.
3. The size of queue i is B_i.
4. Time step equals T, the duration of one round of the scheduler.
5. The input data rate for session i is λ_i.
6. The maximum burst rate for queue i is σ_i.
7. The head of line packet length for queue i is A_i.
8. The arrival probability for queue i is a_i.
9. A maximum of N_i packets could arrive during one round into queue i.
10. The size of queue i is B_i.
11. The probability of departure from queue i is $c_i = 1$.

Based on the above assumptions, we find that we have an $M^m/M/1/B$ queue. The transition matrix for such queue was derived in Section 7.7 on page 241. Based on that, we can derive expressions for the scheduling delay, which corresponds to the queuing delay in that situation.

The arrival probability a_i can be found as follows. The duration of one round T is given by

$$T = \sum_{i=1}^{m} \frac{A_i}{C} \qquad \text{s} \tag{12.8}$$

and when all sessions have packets with equal length A, the above expression simplifies to

$$T = \frac{m\,A}{C} \qquad \text{s} \tag{12.9}$$

The average interarrival time for session i is given by

$$T_i = \frac{A_i}{\lambda_i} \qquad \text{s} \tag{12.10}$$

The probability of k arrivals in one time step is given by

$$p_{i,k} = \binom{N_i}{k} a_i^k \, b_i^{N_i-k} \qquad k = 0, 1, 2, \ldots, N_i \tag{12.11}$$

where a_i is the Bernoulli probability of packet arrival, $b_i = 1 - a_i$, and N_i is the maximum number of packets that could arrive at the queue input as determined by the maximum burst rate σ_i

$$N_i = \left\lceil \frac{\sigma_i \times T}{A_i} \right\rceil \tag{12.12}$$

with $\lceil x \rceil$ is the smallest integer that is larger than or equal to x. Assuming binomial distribution, we can estimate a_i from the average number of packets received on one time step:

$$a_i \times N_i = \frac{\lambda_i \times T}{A_i} \tag{12.13}$$

which gives

$$a_i = \frac{\lambda_i \times T}{N_i \; A_i} \tag{12.14}$$

Because of our choice for the step size, the queue size can only decrease by one at most at any instant with probability $c_i = 1$.

Assuming the packet buffer size B_i, the transition matrix for queue i will be $(B_i + 1) \times (B_i + 1)$ and is given by

$$\mathbf{P} = \begin{bmatrix} q & p_0 & 0 & 0 & 0 & \cdots & 0 \\ p_2 & p_1 & p_0 & 0 & 0 & \cdots & 0 \\ p_3 & p_2 & p_1 & p_0 & 0 & \cdots & 0 \\ p_4 & p_3 & p_2 & p_1 & p_0 & \cdots & 0 \\ \vdots & \vdots & \vdots & \vdots & \vdots & \ddots & 0 \\ p_{B_i} & p_{B_i-1} & p_{B_i-2} & p_{B_i-3} & p_{B_i-4} & \cdots & p_0 \\ f_1 & f_2 & f_3 & f_4 & f_5 & \cdots & f_{B_i+1} \end{bmatrix} \tag{12.15}$$

where $q = p_0 + p_1$ and

$$f_j = 1 - \sum_{k=0}^{B_i - j + 1} p_k \tag{12.16}$$

Of course, if $B_i > N_i$, then the terms $p_j = 0$ whenever $N_i < j \le p_{B_i}$.

The transition matrix helps us find performance parameters for queue i. For that, we need to determine the equilibrium distribution vector **s**. Repeating the same procedure for all other queues, we would then be able to find the performance of the round robin scheduler.

For queue i, the throughput is given by

$$\text{Th} = 1 - s_0 \qquad \text{packets/timestep} \tag{12.17}$$

where s_0 is the probability that the queue is empty. The throughput in units of packets/s is given by

$$\text{Th} = \frac{1 - s_0}{T} \qquad \text{packets/s} \tag{12.18}$$

And the throughput in units of bps is given by

$$\text{Th} = \frac{(1 - s_0)\, A_i}{T} \qquad \text{bps} \tag{12.19}$$

The average queue length is given by

$$Q_a = \sum_{j=0}^{B_i} i\, s_j \tag{12.20}$$

We can invoke Little's result to estimate the *wait time*, which is the average number of time steps a packet spends in the queue before it is served, as

$$Q_a = W \times \text{Th} \tag{12.21}$$

where W is the average number of time steps that a packet spends in the queue:

$$W = \frac{Q_a}{1 - s_0} \qquad \text{timesteps} \tag{12.22}$$

The wait time in units of seconds is given by

$$W = \frac{Q_a\, T}{(1 - s_0)\, A_i} \qquad \text{s} \tag{12.23}$$

Example 1 Assume a round robin scheduler in which all sessions are identical with the following parameters:

$$\begin{aligned} N &= 4 \text{ sessions} & C &= 1\,\text{Mbps} \\ B &= 5 \text{ packets} & L &= 1024\,\text{bits} \\ \lambda &= 50\,\text{kbps} & \sigma &= 500\,\text{kbps} \end{aligned}$$

Determine the transition matrix and determine the system performance parameters.

The duration of one round is

$$T = 4 \times \frac{1024}{5 \times 10^6} = 4.1 \qquad \text{ms}$$

Based on the data burst rate, a maximum of two packets could arrive during a time period T. The Bernoulli arrival probability at each queue during this time period is given by

$$a = 0.1$$

The transition matrix is 6×6:

$$\mathbf{P} = \begin{bmatrix} 0.99 & 0.81 & 0 & 0 & 0 & 0 \\ 0.01 & 0.18 & 0.81 & 0 & 0 & 0 \\ 0 & 0.01 & 0.18 & 0.81 & 0 & 0 \\ 0 & 0 & 0.01 & 0.18 & 0.81 & 0 \\ 0 & 0 & 0 & 0.01 & 0.18 & 0.81 \\ 0 & 0 & 0 & 0 & 0.01 & 0.19 \end{bmatrix}$$

The equilibrium distribution vector is

$$\mathbf{s} = \begin{bmatrix} 0.9877 \ 0.0122 \ 0.0002 \ 0 \ 0 \ 0 \end{bmatrix}^t$$

The throughput of the queue is

$$\begin{aligned} \text{Th} = 1 - s_0 &= 0.0124 \qquad \text{packets/timestep} \\ &= 3.0864 \qquad \text{kbps} \end{aligned}$$

The average queue length is

$$Q_{\text{a}} = 0.0125$$

The average wait time for a queue is

$$W = 4.05 \qquad \mu\text{s}$$

■

12.12 Weighted Round Robin Scheduler (WRR)

The round robin scheduler discussed in the previous section treats all connections equally. When each connection has a different weight, we get the weighted round robin (WRR) scheduler.

Assume session i has an integer weight w_i associated with it. Then in each round of service, the WRR scheduler transmits w_i packets for session i, and so on.

The fraction of the bandwidth for session i in this algorithm can be measured as the service associated with session i relative to the total service in one round. The service received by the head of the line packet in queue i is simply the number of bits transmitted:

$$f_i = \frac{w_i \, A_i}{\sum_{j=1}^{m} w_j \, A_j} \tag{12.24}$$

In an ideal WRR algorithm, all packets have equal lengths and the share would be

$$f_i = \frac{w_i}{\sum_{j=1}^{m} w_j} \tag{12.25}$$

The algorithm is not fair when some sessions have variable length packets since the server will allocate more time for these sessions.

12.12.1 Queuing Analysis for WRR

We can study the occupancy of each queue in a WRR scheduler using the following assumptions:

1. The outgoing link capacity is C.
2. The number of queues or sessions is m.
3. The size of queue i is B_i.
4. The weight associated with queue i is w_i.
5. Time step equals T, the duration of one round of the scheduler.
6. The input data rate for session i is λ_i.
7. The maximum burst rate for queue i is σ_i.
8. The head of line packet length for queue i is A_i.
9. The arrival probability for queue i is a_i.
10. A maximum of N_i packets could arrive during one round into queue i.
11. A maximum of w_i packets could leave during one round from queue i.
12. The probability of departure from queue i is $c_i = 1$.

Based on the above assumptions, we find that we have an $M^m/M^m/1/B$ queue.

The arrival probability a_i can be found by first selecting a time step value. We choose the time step to be equal to the duration of one round T

$$T = \sum_{i=1}^{m} \frac{w_i \, A_i}{C} \tag{12.26}$$

Next, we have to estimate how many packets arrive in one time step from session i. The average interarrival time for session i is given by

$$T_i = \frac{A_i}{\lambda_i} \tag{12.27}$$

The probability of k arrivals in one time step is given by

$$p_{i,k} = \binom{N_i}{k} a_i^k \, b_i^{N_i-k} \qquad k = 0, 1, 2, \ldots, N_i \tag{12.28}$$

where a_i is the Bernoulli probability of packet arrival, $b_i = 1 - a_i$, and N_i is the maximum number of packets that could arrive in one time step. N_i is determined by the maximum burst rate σ_i as follows

$$N_i = \lceil \sigma_i \times T \rceil \tag{12.29}$$

where the ceiling function $\lceil x \rceil$ is the smallest integer that is larger than or equal to x. Assuming binomial distribution, we can estimate a_i from the average number of packets received in one time step:

$$a_i \times N_i = \lambda_i \times T \tag{12.30}$$

which gives

$$a_i = \frac{\lambda_i \times T}{N_i} \tag{12.31}$$

Because of our choice for the step size, the queue size can decrease by w_i packets at most at any instant with probability $c_i = 1$.

From the above calculations, we are able to construct a transition matrix for the queue of session i. Having done that, we are able to obtain expressions for the queue parameters such as throughput, delay, and average queue length.

12.13 Max–Min Fairness Scheduling

Scheduling deals with the sharing of a resource among several users. However, not all users have the same demands on the resource. In many cases, not all users require an equal share of that bandwidth. How should we divide the bandwidth among all users in a fair manner? One way to do that is to use the max–min policy.

Max–min sharing policy is an iterative technique for fair sharing of the resource. For example, assume the outgoing link capacity is C and there are m users. Assume user i requires a bandwidth λ_i and the users are sorted such that

$$\lambda_1 \leq \lambda_2 \leq \cdots \leq \lambda_m \tag{12.32}$$

The allocation of the bandwidth proceeds as follows:

1. Allocate the bandwidth equally among all users C/m.
2. If $\lambda_1 < C/m$, then give user 1 only λ_1.

3. Allocate the remaining bandwidth $C - \lambda_1$ equally among the remaining users $(C - \lambda_1)/(m - 1)$.
4. If $\lambda_2 < (C - \lambda_1)/(m - 1)$, then give user 2 only λ_2.
5. Repeat the procedure until all users have been considered.

Example 2 Assume an outgoing link is being shared among five channels. The system parameters (in units of Mbps) are as follows:

$$\begin{aligned} C &= 155 \\ \lambda_1 &= 10 \\ \lambda_2 &= 20 \\ \lambda_3 &= 60 \\ \lambda_4 &= 80 \\ \lambda_5 &= 80 \end{aligned}$$

Find the rates assigned to each flow according to the max–min algorithm.

The sum of the flows due to all users is

$$10 + 20 + 60 + 80 + 80 = 250 \qquad \text{Mbps}$$

which is larger than the link capacity. The initial fair rate f is given by

$$f = 155/5 = 31 \qquad \text{Mbps}$$

Flow 1 has the minimal rate and is assigned the flow

$$\lambda_1' = \min(10, 31) = 10 \qquad \text{Mbps}$$

We adjust the link capacity shared among the remaining four users as

$$C = 155 - 10 = 145 \qquad \text{Mbps}$$

The fair rate among four remaining users becomes

$$f = 145/4 = 36.25 \qquad \text{Mbps}$$

Flow 2 has minimal rate and is assigned the flow

$$\lambda_2' = \min(20, 36.25) = 20 \qquad \text{Mbps}$$

We adjust the link capacity shared among the remaining two users as

$$C = 145 - 20 = 125\text{Mbps}$$

The fair rate among three remaining users becomes

$$f = 125/3 = 41.7 \qquad \text{Mbps}$$

Flow 3 has minimal rate and is assigned the flow

$$\lambda'_3 = \min(60, 41.7) = 41.7 \qquad \text{Mbps}$$

We adjust the link capacity shared among the remaining two users as

$$C = 125 - 41.7 = 93.3 \qquad \text{Mbps}$$

The fair rate among two remaining users becomes

$$f = 93.3/2 = 46.7 \qquad \text{Mbps}$$

The bandwidths assigned to the flows become

$$\begin{aligned} \lambda'_1 &= 10 && \text{Mbps} \\ \lambda'_2 &= 20 && \text{Mbps} \\ \lambda'_3 &= 41.7 && \text{Mbps} \\ \lambda'_4 &= 46.7 && \text{Mbps} \\ \lambda'_5 &= 46.7 && \text{Mbps} \end{aligned}$$

Note that the aggregate assigned rates equal the outgoing link capacity.

12.14 Processor Sharing (PS)

Processor sharing is an ideal work-conserving scheduler that treats each flow like a fluid model. Data of a given flow is assumed to be infinitely divisible (even at level smaller than a bit), and all flows are served simultaneously. Processor sharing provides max–min fair service. However, PS is an ideal solution that is not implementable in practice. We study it only to provide the background to other practical algorithms.

Assuming we have $m(t)$ active flows at a given time t, the rate assigned to flow i is given by

$$\lambda_i(t) = \frac{C}{m(t)} \tag{12.33}$$

where C is the outgoing link capacity. Figure 12.4 shows a PS scheduler serving four queues.

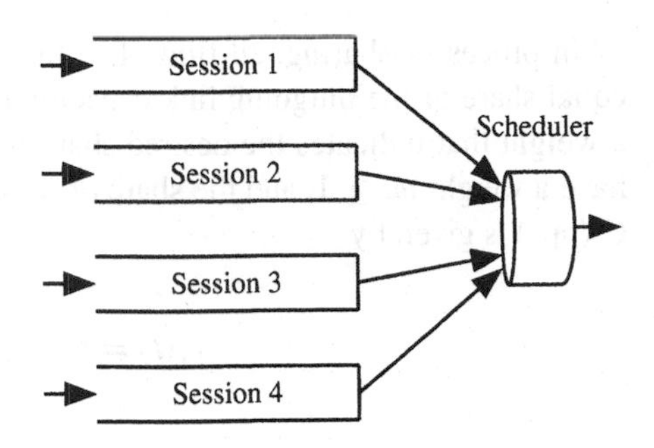

Fig. 12.4 Processor sharing (PS) scheduling in which each queue shares an equal portion of server bandwidth in a fluid flow fashion

Example 3 Assume a PS scheduler that serves three flows. Packets arrive at the scheduler according to the following table:

Time	0	1	2	3	4
Flow 1	0	1	0	0	1
Flow 2	1	1	1	0	1
Flow 3	0	0	1	1	1

A "0" indicates that the flow is inactive and does not require any output bandwidth. A "1" entry indicates an active flow that requires a fair portion of the outgoing link capacity. For simplicity, we assume discrete time intervals T_0, T_1, etc. Calculate the percentage of the outgoing link capacity allocated to each active flow at different times.

According to PS scheduling, the percentage of the bandwidth dedicated to each active flow is shown in the following table:

Time	0	1	2	3	4
Flow 1	0	1	0	0	1
Flow 2	1	1	1	0	1
Flow 3	0	0	1	1	1
% rate	100	50	50	100	33.3

12.15 Generalized Processor Sharing (GPS)

Generalized processor sharing (GPS) is an ideal work-conserving scheduling scheme that is not implemented in practice. It only helps as a reference to compare the performance of other more practical scheduling algorithm. GPS assumes a fluid flow traffic model where the scheduler is able to serve an infinitely small amount of data from each queue at the same time and at different rates.

The bit-by-bit GPS server works on incoming flows in a round robin fashion transmitting one bit from each flow before it moves on to the next flow. When a session is idle, the server skips over to the queue of the next session. Therefore, a single packet requires several rounds of service before it is able to move out of the scheduler but each packet is guaranteed a fair share of the outgoing link capacity.

In processor sharing, all flows had the same weight and all active flows had an equal share of the outgoing link capacity. In GPS, each session or flow is assigned a weight that indicates the desired share of the outgoing link capacity. Flow i will have a weight $w_i \geq 1$, and the share of session i out of the outgoing link bandwidth C (bps) is given by

$$c_i(t) = C \times \frac{w_i}{\sum_{j \in B(t)} w_j} \tag{12.34}$$

where $B(t)$ is the set of backlogged sessions at time t.

The number of bits transmitted from flow i in a time period $t_2 - t_1$ is given by

$$s_i = c_i \, (t_2 - t_1) \quad bits \tag{12.35}$$

where $t_2 \geq t_1$.

The time T required to completely transmit a packet of length A_i in flow i is determined from the expression

$$A_i = \int_{t=0}^{T} c_i(t) \, dt \tag{12.36}$$

This time depends on the number of backlogged sessions which could vary.

Example 4 Assume a GPS scheduler serving four flows with associated weights $w_1 = 1$, $w_2 = 2$, $w_3 = 3$, and $w_4 = 4$. The outgoing link capacity is 1 Mbps and the packet arrival pattern in the different flows is as shown below:

t (ms)	0	10
Session 1	3	
Session 2		1
Session 3	2	
Session 4		7

where the numbers in the rows of each session indicate the length of the packet that arrived at that time in units of kb. Calculate the assigned bandwidth and completion times for the arriving packets.

At $t = 0$ sessions 1 and 3 are backlogged and their combined weights are $1+3 = 4$. The bandwidth assigned to sessions 1 and 3 is

$$c_1(0) = 10^6 \times 1/4 = 0.25 \qquad \text{Mbps}$$
$$c_3(0) = 10^6 \times 3/4 = 0.75 \qquad \text{Mbps}$$

Assuming the system is not changed, thc completion times for the backlogged packets are

$$t_1 = 3 \times 10^3/0.25 = 12 \quad \text{ms}$$

$$t_3 = 2 \times 10^3/0.75 = 2.67 \quad \text{ms}$$

At $t = 10$, packet 3 is gone but a small portion of packet 1 is still left.

The number of bits transmitted from packet 1 of session 1 is given from (12.35) by

$$s_1 = 0.25 \times 10 = 2.5 \quad \text{kb}$$

Thus at $t = 10$ ms, there are 0.5 kb still left to be transmitted for packet $p_1(1)$. The bandwidth assigned to sessions 1, 3, and 4 is

$$c_1(10) = 10^6 \times 1/8 = 0.125 \quad \text{Mbps}$$

$$c_3(10) = 10^6 \times 3/8 = 0.375 \quad \text{Mbps}$$

$$c_4(10) = 10^6 \times 4/8 = 0.5 \quad \text{Mbps}$$

Assuming no more sessions become backlogged, the completion times for the backlogged packets are

$$t_1 = 0.5 \times 10^3/0.125 = 4 \quad \text{ms}$$

$$t_2 = 10^3/0.375 = 2.67 \quad \text{ms}$$

$$t_4 = 7 \times 10^3/0.5 = 14 \quad \text{ms}$$

12.16 Fair Queuing (FQ)

The fair queuing (FQ) algorithm proposed independently in [9, 15] is completely equivalent to the virtual clock (VC) algorithm proposed by Zhang [10] in which the individual sessions are assigned separate queues. The queues are serviced in a round robin manner which prevents a source from arbitrarily increasing its share of the bandwidth. It is called "fair" because it allocates an equal share of the output bandwidth to each traffic flow or queue.

Figure 12.5 schematically shows queue serving sequence in FQ. In the figure, it was assumed that the incoming flows are divided among m queues. Fair queuing is used on a per-flow basis. Note, however, that the algorithm works on a packet-by-packet basis with no consideration to the separate end-to-end connections carried in each flow.

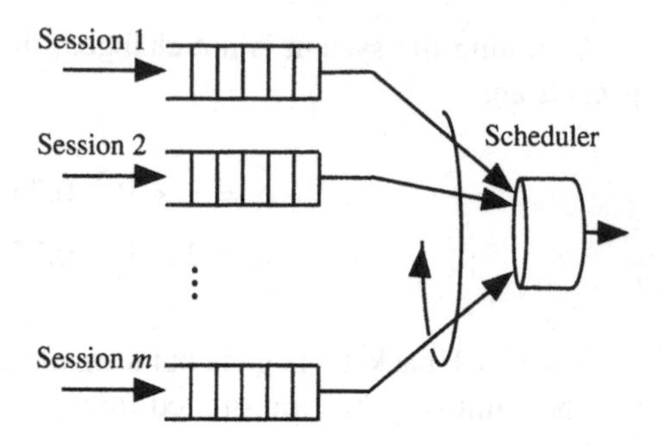

Fig. 12.5 Fair queuing scheduling in which each queue shares an equal portion of server bandwidth in a round robin fashion

Assume flow i has an arrival rate λ_i. The bandwidth allocated to flow i is determined according to max–min scheduling strategy discussed in Section 12.13:

$$\lambda_i' = \min(\lambda_i, f) \tag{12.37}$$

where f is the *fair rate* assigned by the algorithm to each flow as follows:

$$f = \frac{C}{K} \tag{12.38}$$

where C is the outgoing link rate and K is the number of backlogged sessions. f is calculated such that when the switch is congested, the aggregate flow rate equals the switch capacity C.

Equation (12.37) indicates that f is calculated recursively by removing the user with the minimal λ_i and reducing the link capacity accordingly

$$C \leftarrow C - \lambda_{\text{min}} \tag{12.39}$$

Example 5 Assume four sessions are being served by a fair queuing scheduler. The system parameters (in units of Mbps) are as follows:

$$C = 20$$
$$\lambda_1 = 1$$
$$\lambda_2 = 3$$
$$\lambda_3 = 8$$
$$\lambda_4 = 10$$

Find the rates assigned to each flow.

The sum of the flows due to all backlogged sessions is

$$1 + 3 + 8 + 10 = 22$$

which is larger than the link capacity. The initial fair share f is give by

$$f = 20/4 = 5$$

Flow 1 has the minimal rate and is assigned the flow:

$$\lambda_1' = \min(1, 5) = 1$$

We adjust the link capacity shared among the remaining three users as

$$C = 20 - 1 = 19$$

The fair share becomes

$$f = 19/3 = 6.33$$

Flow 2 has minimal rate and is assigned the flow:

$$\lambda_2' = \min(3, 56.33) = 3$$

We adjust the link capacity shared among the remaining two users as

$$C = 19 - 3 = 16$$

The fair share becomes

$$f = 16/2 = 8$$

The bandwidths assigned to the flows become

$$\lambda_1' = 1$$
$$\lambda_2' = 3$$
$$\lambda_3' = 8$$
$$\lambda_4' = 8$$

The sum of all the assigned rates equals the outgoing link capacity. ■

Greedy flows that exceed the fair rate will have similar flow rate at the output equal to the fair rate assigned by the scheduler and will succeed only in filling their buffer which increases their cell loss probability.

Fair queuing is not completely satisfactory because it does not distinguish among long versus short queues or high-priority versus low-priority queues. Thus when bursty traffic is encountered, some queues will become full and their packets will be lost even if some of the other queues are far from full.

Switches or routers employing FQ have to classify each incoming packet to assign it to the proper queue. Furthermore, the switch or router has to perform some operations on each queue to determine its instantaneous flow λ_i.

12.17 Packet-by-Packet GPS (PGPS)

Packet-by-packet GPS (PGPS) is a packetized approximation of the GPS algorithms for fluid flow [34, 35]. This algorithm is also known as weighted fair queuing (WFQ). Thus in PGPS, data are served in terms of complete packets and only one packet can be served at a time.

A PGPS/WFQ server works on incoming flows in a static priority fashion based on a *timestamp* calculation. The flows or sessions are assigned separate queues based on packet header information. The scheduler scans the backlogged queues or sessions to select a packet for service. The packet with the highest priority (least timestamp) is selected and the output link capacity is dedicated to sending that packet without sharing the resource with any other queued packets. After a packet has moved out of a FIFO queue, all packet in that queue move ahead by one position and the head of the line (HOL) packet enters the pool of selection from among all other HOL packets belonging to other sessions.

PGPS requires the computation of three quantities:

1. *Virtual time*, $V(t)$: Indicates the share of the outgoing link capacity for each backlogged session.
2. *Finish number*, F_i: Determines the priority of serving the packet in flow i. The packet with the least finish number is the one that will be served by the scheduler.
3. *Completion time*, T_i: Determines the service time required by the packet based on the packet length and outgoing link bandwidth.

12.17.1 Virtual Time Calculation for PGPS/WFQ

Assuming $B(t)$ to the set of backlogged sessions at time t, the *virtual time* $V(t)$ is defined using the differential equation

$$V(0) = 0 \tag{12.40}$$

$$\frac{\mathrm{d}V(t)}{\mathrm{d}t} = \frac{1}{\sum_{i \in B} w_i} \tag{12.41}$$

where $w_i \geq 1$ is the weight assigned to session i. Figure 12.6 shows the time development of $V(t)$ as packets arrive and sessions become backlogged.

When all sessions are idle, the virtual time is reset to zero

$$V(t) = 0 \qquad \text{when all sessions are idle} \tag{12.42}$$

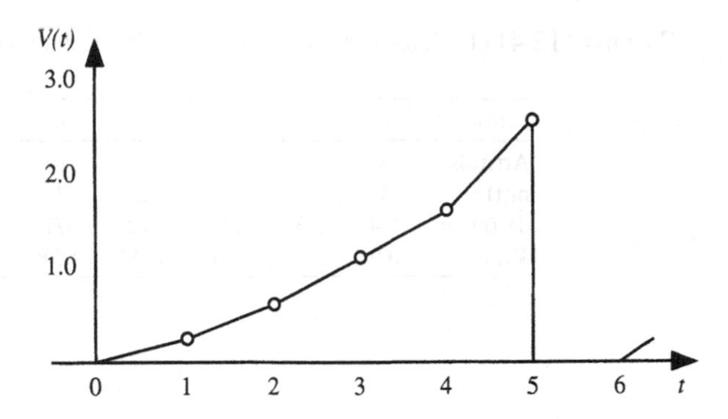

Fig. 12.6 Development of the virtual time $V(t)$ in PGPS/WFQ scheduling

Example 6 Assume a PGPS/WFQ scheduler serving flows of equal weights. The outgoing link capacity is 1 Mbps and all arriving packets equal lengths of 1 kb. The packet arrival pattern is shown below where it is assumed that packets arrived at the idle sessions:

t (ms)	0	1	2	3	4	5	6
Arrivals	4			1			2

Determine the number of backlogged sessions versus t and obtain the values for $V(t)$.

A packet will require 1 ms to be transmitted.

At $t = 0$, four packets arrived and the number of backlogged sessions is $m(0) = 4$.

At $t = 1$, no packets arrive and number of active sessions is reduced by one. Therefore, $m(1) = 3$ since one packet is guaranteed to be serviced.

At $t = 2$, no packets arrive and number of active sessions is decremented by one $m(2) = 2$.

At $t = 3$, one packet arrives and one packet leaves which leaves $m(3) = m(2) = 2$.

At $t = 4$, no packets arrive and number of active sessions is decremented by one.

At $t = 5$, no packets arrive and number of active sessions is decremented by one and we get $m(5) = 0$.

At $t = 6$, two packets arrive and $m(6) = 2$.

The following table shows the development of $m(t)$:

t (ms)	0	1	2	3	4	5	6
Arrivals	4	0	0	1	0	0	2
$m(t)$	4	3	2	2	1	0	2

We use (12.41) to determine the value of $V(t)$ at each time:

t (ms)	0	1	2	3	4	5	6
Arrivals	4	0	0	1	0	0	2
m(t)	4	3	2	2	1	0	2
$dV(t)/dt$	1/4	1/3	1/2	1/2	1/1	0	1/2
$V(t)$	0.0	0.25	0.58	1.08	1.58	2.58	0

12.17.2 Finish Number Calculation for PGPS/WFQ

Assuming packet k has arrived at the queue for session i, then the finish number for that packet is calculated as

$$F_i(k) = \max\left[F_i(k-1),\, V(t)\right] + \frac{A_i}{w_i} \tag{12.43}$$

where $F_i(k-1)$ is the finish number for preceding packet in queue i and A_i is the length of the arriving packet. An empty queue will have a zero finish number associated with it.

The first term on the RHS ensures that for a backlogged queue, an arriving packet will have bigger finish number compared to packets already in queue i so that each queue functions as first-in/first-out (FIFO).

The second term on the RHS ensures that the finish number for packet k of queue i takes into account the size of that packet and the weight associated with the session. This ensures that sessions with lots of bits to send are slightly penalized to ensure fairness to other users that might have smaller packets to send.

Notice that a long packet or a greedy session will be characterized by large finish numbers and will receive lower service priority. On the other hand, short packets belonging to a conforming session with short queue will by characterized by small finish numbers and will be served more frequently. A greedy or bursty user traffic will only succeed in filling its buffer and losing packets while other users will still access their fair share of the link capacity.

The scheduler orders all the finish numbers for the packets at the head of all the queues. The packet with the least finish number is served first.

One last remark is worth making here. The finish number equation (12.43) guarantees that packets already in the system at time t will be served *before* any packets that arrive after t.

When session i is idle, its finish number is reset to zero

$$F_i = 0 \qquad \text{whensessionisidle} \tag{12.44}$$

12.17.3 Completion Time Calculation for PGPS/WFQ

The completion time for a packet is simple to calculate in PGPS/WFQ since the outgoing resources are completely dedicated to the selected packet.

Assuming packet k in session i has a length A_k, then the time T required to transmit it is given by

$$T = \frac{A_k}{C} \quad \text{s} \tag{12.45}$$

where C is the outgoing link capacity in bps.

Example 7 Assume a PGPS/WFQ scheduler serving four flows of equal weights. The outgoing link capacity is $C = 1$ Mbps and the packet arrival pattern is shown below:

t (ms)	0	1	2	3
Session 1	3		2	
Session 2		1		4
Session 3	2		7	
Session 4				5

where the numbers in the rows of each session indicate the length of the packet in units of kb. Calculate the system virtual time and finish numbers for the arriving packets of each session.

A packet of unit length (1 kb) takes 1 ms to transmit. At the start, virtual time and all the finish numbers are reset to 0.

At $t = 0$, our table will be

t (ms)	0	1	2	3
Session 1*	3		2	
Session 2		1		4
Session 3*	(2)		7	
Session 4				5
dV/dt	0.5			
$V(t)$	0.0			

Sessions 1 and 3 are backlogged, as indicated by the asterisk (*), and the finish numbers for the arriving packets are

$$F_1(1) = \max(0, 0) + 3 = 3$$
$$F_3(1) = \max(0, 0) + 2 = 2$$

Packet $p_3(1)$ will be served first and will require 2 ms to transmit. This is indicated by the brackets round the entry for this packet.

At $t = 1$, our table will be

t (ms)	0	1	2	3
Session 1*	3			
Session 2*		1		4
Session 3*	(2)		7	
Session 4				5
dV/dt	0.5	0.3		
$V(t)$	0.0	0.5		

The finish numbers for all congested sessions (i.e., 1, 2, and 3) are given by

$$F_1(1) = 2$$
$$F_2(1) = \max(0, 0.5) + 1 = 1.5$$
$$F_3(1) = 3$$

Since $F_2(1)$ has the least finish number, we could have chosen the packet in session 2 for transmission. However, this would mean that we stop the transmission of packet in session 3 which is not finished yet. This is a form of *preemptive scheduling*. We choose a nonpreemptive scheduling scheme and continue transmitting packet out of session 3 as indicated by the brackets surrounding the packet of session 3.

At $t = 2$, our table will be

t (ms)	0	1	2	3
Session 1*	3		2	
Session 2*		(1)		4
Session 3*	2		7	
Session 4				5
dV/dt	0.5	0.3	0.3	
$V(t)$	0.0	0.5	0.8	

Packet $p_2(1)$ is chosen for transmission since it has the lowest finish number among all the backlogged sessions. The finish numbers for the new packets are

$$F_1(2) = \max(3, 0.8) + 2 = 5$$
$$F_3(2) = \max(2, 0.8) + 7 = 9$$

Since finish number for packet in session 2 is 1.5, we choose this packet for transmission as indicated by the brackets surrounding that packet.

At $t = 3$, our table will be

t (ms)	0	1	2	3
Session 1*	(3)		2	
Session 2*		1		4
Session 3*	2		7	
Session 4*				5
dV/dt	0.5	0.3	0.3	0.25
$V(t)$	0.0	0.5	0.8	1.1

The finish numbers for the new packets are

$$F_2(2) = \max(1.5, 1.1) + 4 = 5.5$$
$$F_4(1) = \max(0, 1.1) + 5 = 6.1$$

Since the finish number for the HOL packet in session 1 is 3, we pick this packet for transmission. ■

12.18 Frame-Based Fair Queuing (FFQ)

Frame-based fair queuing (WFQ) belongs to the general class of rate-proportional servers (RPS) proposed in [36]. This type of schedulers are claimed to offer similar delay and fairness bounds as PGPS/WFQ but with much simpler computations of the packet priorities.

FFQ server works on incoming flows in a static priority fashion based on a *timestamp* calculation. The flows or sessions are assigned separate queues based on packet header information. The scheduler scans the backlogged queues or sessions to select a packet for service. The packet with the highest priority (least timestamp) is selected and the output link capacity is dedicated to sending that packet without sharing the resource with any other queued packets. After a packet has moved out of a FIFO queue, all packets in that queue move ahead by one position and the head of the line (HOL) packet enters the pool of selection from among all other HOL packets belonging to other sessions.

FFQ requires the computation of three quantities:

1. *System potential*, $P(t)$: Indicates the amount of data transferred through the outgoing link up to time t.
2. *Timestamp*, S_i: Determines the priority of serving the packet in flow i. The packet with the least timestamp is the one that will be served by the scheduler.
3. *Completion time*, T_i: Time required by the packet to be fully transmitted. It depends on the packet length and outgoing link bandwidth.

12.18.1 System Potential Calculation for FFQ

Assume that the server started serving a packet at time t_s. At a later time $t \geq t_s$, the *system potential* $P(t)$ is defined as

$$P(t) \leftarrow P + \frac{C(t - t_s)}{F} \tag{12.46}$$

where C (bps) is the outgoing link capacity and F is the *frame size* in bits. The system potential $P(t)$ measures the amount of data transferred up to time t relative to the frame size F. The system potential is updated each time a packet starts service and when all sessions are idle, the system potential is reset to zero.

$$P(t) = 0 \qquad \text{when all sessions are idle} \tag{12.47}$$

The frame size F is chosen so that at least the maximum length packet from any session can be sent during one frame period, i.e.,

$$F > A_{\text{max}} \qquad \text{bits} \tag{12.48}$$

The *frame period* T corresponding to the chosen frame size is given by

$$T = \frac{F}{C} \qquad \text{s} \tag{12.49}$$

12.18.2 Timestamp Calculation for FFQ

When a k packet arrives at session i, a timestamp is associated with it according to the following formula

$$S_i(k) = \max\left[S_i(k-1), P(t)\right] + \frac{A_i}{\lambda_i} \tag{12.50}$$

where $S_i(k-1)$ is the timestamp of the previous packet in the queue, A_i is the packet length in bits, and λ_i is the reserved rate for session i. An empty queue will have a zero timestamp associated with it.

A long packet will be penalized by having a large timestamp value and users with higher reserved bandwidth will consistently receive lower timestamp values so as to obtain their reserved rate in a fair manner.

One issue remains to be resolved which is determining when the current frame is to be completed and a new frame is to be started. We mentioned above that a new frame starts when all sessions are idle. When one or more sessions are backlogged, a new frame is started when the accumulated bits transferred approaches the frame size F. To keep track of the number of bits transferred, a bit counter could be used. When a packet is selected for transmission, the counter contents are updated

$$B(k) = B(k-1) + A_k \tag{12.51}$$

If $B(k) \leq F$, the current frame is continued and the packet is sent. If $B(k) > F$, a new frame is started and the packet is sent during the new frame.

12.18.3 Completion Times Calculation for FFQ

The completion time for a packet is simple to calculate in FFQ since the outgoing resources are completely dedicated to the selected packet.

Assuming packet k in session i has a length A_k, then the time T required to transmit it is

$$T = \frac{A_k}{C} \quad \text{s} \tag{12.52}$$

where C is the outgoing link capacity in bps.

Example 8 Assume a FFQ scheduler serving flows of equal weights. The outgoing link capacity is 1 Mbps and the frame size is chosen as $F = 10^4$ bits. The packet arrival pattern is shown below:

t (ms)	0	1	2	3
Session 1	3		2	
Session 2		1		4
Session 3	2		7	
Session 4				5

where the numbers in the rows of each session indicate the length of the packet in units of kb. Calculate the system potential and timestamps for the arriving packets of each session.

A packet of unit length (1 kb) takes 1 ms to transmit. At the start, system potential and all the timestamps are reset to 0.

At $t = 0$, our table will be

t (ms)	0	1	2	3
Session 1*	3		2	
Session 2	0	1		4
Session 3*	(2)		7	
Session 4	0			5
$B(t)$	0			
$P(t)$	0.0			

where the session entries at $t = 0$ are *timestamp values*. Sessions 1 and 3 are backlogged, as indicated by the asterisk ($*$), and the timestamp values are

$$S_1(1) = \max(0, 0) + 3 = 3$$
$$S_3(1) = \max(0, 0) + 2 = 2$$

Packet $p_3(1)$ will be served first and will require 2 ms to transmit. This is indicated by the brackets a round the entry for this packet.

At $t = 1$, our table will be

t (ms)	0	1	2	3
Session 1*	3			
Session 2*		1		4
Session 3*	(2)		7	
Session 4				5
$B(t)$	0	1		
$P(t)$	0.0	0.1		

where the entry for $B(t)$ is in kbits. The timestamp for $p_2(1)$ is

$$S_2(1) = \max(0, 0.1) + 1 = 1.1$$

At $t = 2$, our table will be

t (ms)	0	1	2	3
Session 1*	3		2	
Session 2*		(1)		4
Session 3*	2		7	
Session 4				5
$B(t)$	0	1	2	
$P(t)$	0.0	0.1	0.2	

Packet $p_2(1)$ is chosen for transmission since it has the lowest timestamp among all the backlogged sessions. The timestamps for the new packets are

$$S_1(2) = \max(3, 0.2) + 2 = 5$$
$$S_3(2) = \max(2, 0.2) + 7 = 9$$

At $t = 3$, our table will be

t (ms)	0	1	2	3
Session 1*	(3)		2	
Session 2*		1		4
Session 3*	2		7	
Session 4*				5
$B(t)$	0	1	2	3
$P(t)$	0.0	0.1	0.2	0.3

The timestamps for the new packets are

$$S_2(2) = \max(1.1, 0.3) + 4 = 5.1$$
$$S_4(1) = \max(0, 0.3) + 5 = 5.3$$

12.19 Core-Stateless Fair Queuing (CSFQ)

The problem with the schedulers discussed so far is the need to maintain a separate state for each flow at all routers in the path of the packets. Such schedulers allow the system to provide firm QoS guarantees for each flow. However, they are complex to implement and their performance is limited by the number of flows that can be supported [37].

Core-stateless fair queuing (CSFQ) attempts to simplify matters by dividing the routers in the network into two categories: edge routers and core routers as shown in Fig. 12.7 [37].

Edge routers maintain a per-flow state and label incoming packets accordingly. They also regulate the incoming flows such that flow i receives a fair service rate λ_i determined by

$$\lambda_i = \min[\lambda_i, f] \tag{12.53}$$

where λ_i is the arrival rate for flow i and f is the fair share rate determined in the following section.

Figure 12.8 shows the functions performed by each edge router. The rate of each incoming flow is estimated based on the timing of arriving packets. The estimated arrival rates for all incoming traffic are used to obtain an estimated value for the fair share f. The edge router also decides whether to accept or drop the arriving packet based on the arrival rate and the fair share estimate as discussed below. The figure

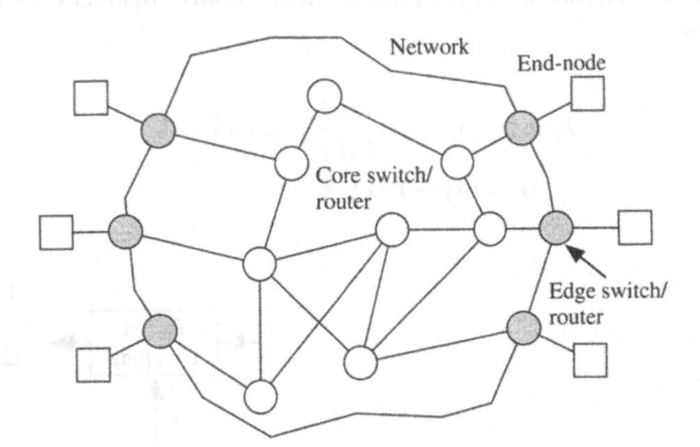

Fig. 12.7 Routers in core-stateless fair queuing (CSFQ) are divided into edge routers (*grey circles*) and core routers (*empty circles*). End-nodes (*squares*) are connected to edge routers

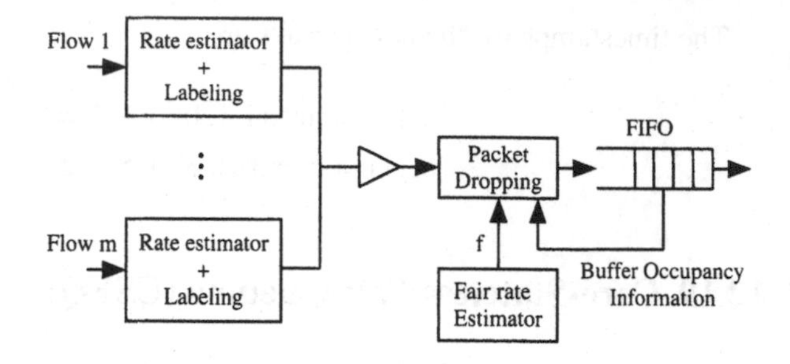

Fig. 12.8 The architecture of an edge router implementing CSFQ scheduling

shows that the packet drop probability depends on the arrival rate, estimated fair share, and the state of the FIFO buffer occupancy.

Core routers do not maintain a per-flow state but use the per-flow label in each packet to decide whether to drop an incoming packet or to accept it. The probability that a packet is dropped is calculated by each core router based on the packet label and on the estimated fair rate at the core router. Accepted packets are placed in a simple FIFO buffer for transmission.

Figure 12.9 shows the functions performed by each core router. The rate of each incoming flow is extracted from the header of arriving packets. The arrival rates for all incoming traffic are used to obtain an estimated value for the fair share f. The core router also decides whether to accept or drop the arriving packet based on the arrival rate and the fair share estimate as discussed below. The figure shows that the packet drop probability depends on the arrival rate, estimated fair share, and the state of the FIFO buffer occupancy.

12.19.1 Determination of Packet Arrival Rate λ_i

Central to CSFQ is obtaining an accurate estimate of the average packet arrival rate for each flow. The arrival rate λ_i is estimated in an iterative manner each time a new packet arrives:

$$\lambda_i(k) = (1 - \alpha)\,\frac{A_i(k)}{T_i(k)} + \alpha\,\lambda_i(k-1) \qquad (12.54)$$

$$\alpha = \exp -T_i(k)/K \qquad (12.55)$$

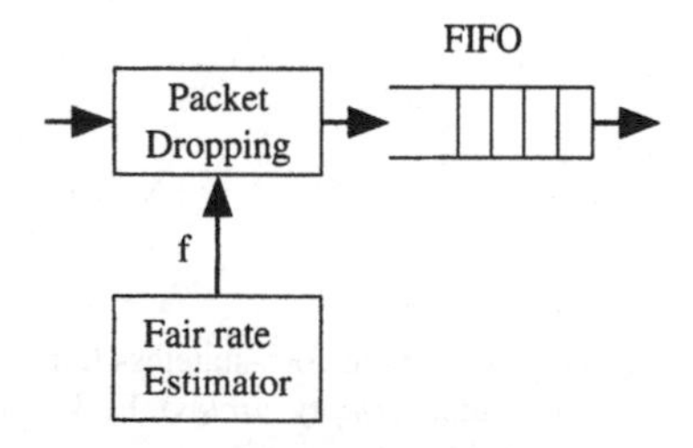

Fig. 12.9 The architecture of a core router implementing CSFQ scheduling

where $T_i(k)$ is the packet interarrival time for flow i at time k, K is some constant, and $A_i(k)$ is the length of the arriving packet.

A conforming user will have a high value for the interarrival time $T_i(k)$. This will result in an exponentially low value for α which increases the assigned rate $\lambda_i(k)$.

12.19.2 Determination of Fair Rate f

The fair share f is determined based on whether the link is congested or uncongested. The update operation is performed at regular intervals of time determined by the system administrator.

The link is congested when the total arrival rate exceeds the outgoing link capacity C:

$$C \leq \sum_{i=1}^{m} \min(\lambda_i, f) \qquad \text{linkcongested} \tag{12.56}$$

and in that case, the fair share is set equal to its old value

$$f(k) = f(k-1) \tag{12.57}$$

The link is uncongested when the total arrival rate is smaller than the outgoing link capacity C:

$$C > \sum_{i=1}^{m} \min(\lambda_i, f) \qquad \text{linkuncongested} \tag{12.58}$$

and in that case, the fair share is determined as follows

$$f(k) = f(k-1) \frac{C}{\max(\lambda_i)} \tag{12.59}$$

The estimated fair share also is slightly modified based on the level of occupancy of the FIFO buffer but this is a minor detail that the reader could check in reference [37].

12.19.3 Bit Dropping Probability p_b

When the traffic rate for flow i is such that $\lambda_i < f$, then that session is conforming and no bit dropping is necessary. The packets of that flow can pass through the network.

A misbehaving session i is characterized by $\lambda_i > f$ and the packet must be dropped with a bit drop probability given by

$$p_b = \max\left(0, 1 - \frac{f}{\lambda_i}\right) \tag{12.60}$$

12.20 Random Early Detection (RED)

The schedulers we discussed in the previous sections emphasized techniques for selecting the next packet for service. Selecting which packet to drop when the buffer overflows was simple. The last packet to arrive is dropped when the buffer is full. This type of packet drop is called *tail drop* strategy. To achieve max–min buffer sharing, the scheduler must assign a buffer to each service class or each user.

Random early drop detection (RED) belongs to the class of *early drop schedulers* where a packet is dropped even when the buffer is not full. There are several plausible reasons why early dropping of packets is beneficial:

1. In tail drop schedulers, misbehaving users occupy precious buffer space and packets belonging to conforming users would be dropped if they arrive when the buffer is full.
2. Dropping packets from some sessions or classes sends a message to end points to reduce their rate before the network becomes congested and more packets would then be lost.

In random early detection, the switch calculates the average queue size at each time step. Based on this estimate, the switch decides whether to drop the packet or label it with a probability that is a function of the queue size [38]. The switch calculates the average queue size Q_a using a low-pass filter. Calculating an average queue length based on filtering is better than using the instantaneous queue length since this allows small temporary traffic bursts to go through unharmed.

The average queue size is compared to two thresholds Q_{min} and Q_{max}:

1. When $Q_a < Q_{min}$, packets are not marked.
2. When $Q_a > Q_{max}$, all packets are marked.
3. When $Q_{min} \leq Q_a \leq Q_{max}$, packets are marked with a probability p_a given by

$$p_a = \frac{p_b}{1 - Q\, p_b} \tag{12.61}$$

where Q is the number of packets received since the last marked packet and p_b is given by

$$p_b = \frac{Q_a - Q_{min}}{Q_{max} - Q_{min}} \tag{12.62}$$

One of the problems of RED is the amount of operations that have to be done at each time step on each packet. This is one reason why FIFO with tail drop method is still in use.

Another related algorithm that is similar to RED is flow random early drop (FRED) where the switch drops each arriving packet with a fixed drop probability p_b when the queue size exceeds a certain drop threshold. FRED is classified as a *early random drop* algorithm where arriving packets are randomly dropped. The reasoning being that misbehaving users send more packets and will have a higher probability that their packets are dropped. However, it has been shown that this drop policy is not very successful [10].

12.21 Packet Drop Options

We discussed above two advantages for dropping packets in a router. Packets are dropped to reduce network congestion and to improve its efficiency [39]. There are several options for *selecting* the next packet to drop and for determining the *times* when the drop policies are implemented. Below, we discuss some of the packet drop policies that could be implemented alone or in combinations, depending on the transmission and scheduler protocols being used.

The simplest packet drop policy is to drop incoming packets when the shared buffer or queue is full. This drop policy does not offer protection against misbehaving users. A full buffer most probably has been caused by a misbehaving user and the packets dropped might belong to conforming users. Of course, such simple policy does not require maintaining any state information. There is only one state to maintain here which is the level of occupancy of the buffer. Attempting to protect conforming users requires defining state variables for each user indicating the amount of packets present in the system and belonging to each session. During periods of congestion, users that have high buffer occupancy states are eligible to have their packets dropped. While this offers protection against misbehaving users, the system must maintain many state variables.

An intermediate solution is to group or aggregate the users into classes of service and maintain state information for these classes only. This solution offers some protection and isolation and also reduces the amount of state variables required.

If the scheduler implements PGPS/WFQ or FFQ algorithms, then a simple packet drop strategy is to drop the packet with the highest finish number or timestamp value. The reasoning for this is that large values for these parameters indicate either long packets or many packets belonging to this session.

Sometimes the packet header contains information that can help with the packet drop policy. For example, in ATM, the AAL layer contains information about missing cells. When a switch or router detects that a connection has a missing cell, it drops all subsequent cells until the last cell of the frame [39]. This frees the network from supporting a cell stream that will be discarded by the receiver since the frame will be retransmitted anyway.

Problems

Scheduler Functions

12.1 Explain the main functions performed by a scheduler.

12.2 What are the main switch resources shared among the different users?

12.3 What are the scheduler performance measures most important to the following applications: electronic mail, file transfer, web browsing, voice communications, and one-way video.

12.4 Which is the important QoS parameter for best-effort applications and CBR traffic.

12.5 Explain the functions performed by the scheduler at the different contention points for input, output, and shared buffer switches.

12.6 One solution to solve the HOL problem in input queued switches is to use virtual output queuing (VOQ). Explain how VOQ works then explain how the scheduler will work in such a scheme.

12.7 Explain what is meant by fairness and what is meant by protection from a scheduler perspective.

12.8 It was explained that to provide deterministic QoS guarantees, the scheduler must maintain separate queues for the separate sessions. Explain how this can be done in input, output, and shared buffer switches. Discuss the pros and cons of each scheme from the point of view of the implementation of the scheduler in each case.

Scheduler Performance Measures

12.9 Explain what is meant by protection in terms of outgoing link bandwidth utilization.

12.10 Explain what is meant by protection in terms of switch buffer utilization.

12.11 Investigate how a scheduler might be able to reduce delay jitter for the different sessions.

Scheduler Classifications

12.12 Explain the difference between work-conserving and nonwork-conserving schedulers.

12.13 Explain what is meant by degree of aggregation and the advantages and disadvantages of this strategy.

12.14 Explain the different packet drop policies that could be used by schedulers.

Max–Min Fairness

12.15 Explain max–min fairness as it applies to outgoing link bandwidth.

12.16 Explain max–min fairness as it applies to outgoing shared buffer space in a switch.

12.17 Assume an outgoing link is being shared among six channels. The system parameters (in units of Mbps) are as follows:

$$C = 622$$
$$\lambda_1 = 200$$
$$\lambda_2 = 30$$
$$\lambda_3 = 100$$
$$\lambda_4 = 50$$
$$\lambda_5 = 180$$
$$\lambda_6 = 180$$

Find the rates assigned to each flow according to the max–min algorithm.

12.18 Assume a 1000 byte buffer is being shared among five sessions. The buffer requirements (in units of bytes) for each session are as follows:

$$B_1 = 250$$
$$B_2 = 250$$
$$B_3 = 300$$
$$B_4 = 400$$
$$B_5 = 150$$

Find the buffer space assigned to each flow according to the max–min algorithm.

FIFO (or FCFS) Scheduling

12.19 Assume a FIFO scheduler where the output link rate is C and the arrival rate for each session is λ, the number of arriving sessions is m and all flows have equal packet lengths L. Find the performance of the system assuming a fluid flow model.

12.20 Assume a FIFO scheduler where there are m users accessing the buffer but one of the users has an arrival probability that is different from that of the other users. Derive the transition matrix for such a system.

12.21 Assume a FIFO scheduler where there are m users accessing the buffer but one of the users produces packets with different lengths compared to those of the other users. Derive an expression for the average length of the queue and the average queuing delay of such a system.

Static Priority Scheduling

12.22 Consider queue i in the static priority scheduler. Assume the arrival probability for this queue a_i and its priority is i, where lower values of i indicate higher priority. Write down the transition matrix for this queue and comment on methods to find its steady-state distribution vector.

12.23 Repeat Problem 12.22 for the case when all queues have the same size B and have the same arrival probability a.

12.24 Consider a static priority protocol serving four users where the packet arrival probabilities for all users are equal (i.e., $a_i = 0.3$ for all $1 \leq i \leq 4$) and all users have the same buffer size (i.e., $B_i = 8$ for all $1 \leq i \leq 4$). Estimate the performance of each user.

12.25 Consider a static priority protocol serving four users where the packet arrival probabilities for all users are equal (i.e., $a_i = 0.6$ for all $1 \leq i \leq 4$) and all users have the same buffer size (i.e., $B_i = 4$ for all $1 \leq i \leq 4$). Estimate the performance of each user.

12.26 Repeat Problem 12.25 when the probability of the queue being empty becomes high $e = 0.9$.

Round Robin Scheduler

12.27 Assume a round robin scheduler in which all packets have equal lengths. Obtain expressions for the maximum scheduler delay and the fraction of the bandwidth assigned to any session.

12.28 Assume a round robin scheduler in which all queues have identical traffic characteristics (arrival probability, packet length, etc.) Obtain the transition matrix for one queue and obtain expressions for its performance parameters: queue length, throughput, and loss probability.

12.29 Assume a round robin scheduler in which all sessions are identical with the following parameters:

$$\begin{aligned} m &= 8 \text{ sessions} & C &= 10 \text{ Mbps} \\ B &= 8 \text{ packets} & L &= 512 \text{ bits} \\ \lambda &= 100 \text{ kbps} & \sigma &= 500 \text{ kbps} \end{aligned}$$

Determine the transition matrix and determine the system performance parameters.

Generalized Processor Sharing

12.30 Assume a GPS scheduler serving four flows with associated weights $w_1 = 4$, $w_2 = 2$, $w_3 = 3$, and $w_4 = 1$. The outgoing link capacity is 10 Mbps and the packet arrival pattern in the different flows is as shown below:

t (ms)	0	1	2
Session 1	3		2
Session 2		1	3
Session 3	2		
Session 4		7	

where the numbers in the rows of each session indicate the length of the packet that arrived at that time in units of kb. Calculate the assigned bandwidth and completion times for the arriving packets.

12.31 What is the longest delay bound experienced by the packet in session i in GPS?

Fair Queuing (FQ)

12.32 Assume four sessions are being served by a fair queuing scheduler. The system parameters (in units of Mbps) are as follows:

$$C = 40$$
$$\lambda_1 = 6$$
$$\lambda_2 = 2$$
$$\lambda_3 = 20$$
$$\lambda_4 = 16$$

Find the rates assigned to each flow.

Packet-by-Packet GPS (PGPS)/Weighted Fair Queuing (WFQ)

12.33 Assume a PGPS/WFQ scheduler serving flows of equal weights. The outgoing link capacity is 1 Mbps and all arriving packets have equal lengths of 1 kb. The packet arrival pattern is shown below:

t (ms)	0	1	2	3	4	5	6
Arrivals	2			2		4	

Determine the number of backlogged sessions versus t and obtain the values for $V(t)$.

12.34 Assume a PGPS/WFQ scheduler serving flows of equal weights. The outgoing link capacity is 1 Mbps and the packet arrival pattern is shown below:

t (ms)	0	1	2	3
Session 1	4		2	
Session 2		2		3
Session 3	1		5	
Session 4				4

where the numbers in the rows of each session indicate the length of the packet in units of kb. Calculate the system virtual time and finish numbers for the arriving packets of each session.

12.35 Assume a PGPS/WFQ scheduler serving flows of equal weights. The outgoing link capacity is 1 Mbps and the packet arrival pattern is shown below:

t (ms)	0	1	2	3
Session 1			2	
Session 2	3	2	1	
Session 3		1		2
Session 4	1		5	

where the numbers in the rows of each session indicate the length of the packet in units of kb. Calculate the system virtual time and finish numbers for the arriving packets of each session.

12.36 Assume a PGPS/WFQ scheduler serving flows of equal weights. The outgoing link capacity is 2 Mbps and the packet arrival pattern is shown below:

t (ms)	0	1	2	3
Session 1	7		2	
Session 2		3		2
Session 3	3		7	
Session 4				8

where the numbers in the rows of each session indicate the length of the packet in units of kb. Calculate the system virtual time and finish numbers for the arriving packets of each session.

Frame-Based Fair Queuing (FFQ)

12.37 Assume a FFQ scheduler serving flows of equal weights. The outgoing link capacity is 1 Mbps and the frame size is chosen as $F = 10^4$ bits. The packet arrival pattern is shown below:

t (ms)	0	1	2	3
Session 1	1		2	
Session 2		1		4
Session 3	2		7	
Session 4				5

where the numbers in the rows of each session indicate the length of the packet in units of kb. Calculate the system potential and timestamps for the arriving packets of each session.

12.38 Assume a FFQ scheduler serving flows of equal weights. The outgoing link capacity is 1 Mbps and the frame size is chosen as $F = 10^4$ bits. The packet arrival pattern is shown below.

t (ms)	0	1	2	3
Session 1	2		2	
Session 2		1		4
Session 3	1		7	
Session 4				5

where the numbers in the rows of each session indicate the length of the packet in units of kb. Calculate the system potential and timestamps for the arriving packets of each session.

References

1. N. McKeown and T.E. Anderson, "A quantitative comparison of iterative scheduling algorithms for input-queued switches", *Computer Networks and ISDN Systems*, vol. 30, pp. 2309–2326, 1998.
2. T. Nasser and F. Gebali, "A new scheduling technique for packet switching networks using the transportation problem optimization model", 1st International Symposium on Signal Processing and Information Technology (ISSPIT 2001), Cairo, Egypt, December 2001, pp. 360–363.
3. G.B. Dantzig, *Linear Programming*, Springer-Verlag, New York, 1985.
4. S.S. Rao, *Engineering Optimization*, Princeton University Press, Princeton, New Jersey, USA, 1996.
5. J.A. cobb, M.G. Gouda, and A. El-Nahas, Time-shift scheduling—fair scheduling of flows in high-speed networks. *IEEE/ACM Transactions on Networking*, vol. 6, no. 3, pp. 274–285, 1998.
6. K. Nichols, S. Blake, F. Baker, and D.L. Blake, "Definition of the differentiated services field (DS Field) in the IPv4 and IPv6 headers", December 1999, IETF RFC 2474.
7. D. Stiliadis and A. Varma, "Latency-rate servers: A general model for analysis of traffic scheduling algorithms", *IEEE/ACM Transactions on Networking*, vol. 6, no. 5, pp. 611–624, 41 pp, 1995.
8. A.K. Parekh and R.G. Gallager, "A generalized processor sharing approach to flow control - the single node case," *Proceedings of IEEE INFOCOM*, vol. 2, pp. 915–924, May 1992.
9. A. Demers, S. Keshav, and S. Shenker, "A generalized processor sharing approach to flow control in integrated services networks: the multiple node case", A.K. Parekh, and R.G. Gallager, Eds., *IEEE/ACM Transactions on Networking*, Vol. 2, no. 2, pp. 137–150, April 1994.
10. L. Zhang, "Virtual Clock: A new traffic control algorithm for packet switching networks," *ACM Transactions on Computer Systems*, vol. 9, pp. 101–124, May 1991.

11. M. Datevenis, S. Sidiropoulos, and C. Courcoubetis, "Weighted round-robin cell multiplexing in a general-purpose ATM switch chip", *IEEE Journal on Selected Areas in Communications*, vol. 9, pp. 1265–1279, 1991.
12. D. Ferrari and D. Verma, "A scheme for real-time channel establishment in wide-area networks", *IEEE Journal on Selected Areas in Communications*, vol. 8, pp. 368–379, 1990.
13. M. Shreedhar and G. Varghese, "Efficient fair queuing using deficit round robin", *SIGCOMM'95*, September 1995.
14. D.C. Verma, H. Zhang, and D. Ferrari, "Delay jitter control for real-time communication in a packet switching network", *Proceedings of Tricomm'91*, pp. 35–43, Chapel Hill, North Carolina, April 1991.
15. J. Nagle, "On packet switches with infinite storage", *IEEE Transactions on Communications*, vol. 35, pp. 435–438, 1987.
16. C.R. Kalmanek, H. Kanakia, and S. Keshav, "Rate Controlled Servers for Very High-Speed Networks", *Conference on Global Communications (GLOBECOM)*, San Diego, California, USA, pp. 12–20, December 2–5, 1990.
17. S.J. Golestani, "A Framing Strategy for Congestion Management", *IEEE Journal of Selected Areas in Communications*, vol. 9, no. 7, pp. 1064–1077, September 1991.
18. S.J. Golestani, "Congestion-Free Transmission of Real-Time Traffic in Packet Networks", *Proceedings of IEEE INFOCOM'90*, San Francisco, California, USA, pp. 527–536, June 1990.
19. S.J. Golestani, "A Stop-and-Go Queuing Framework for Congestion Management", *Proceedings of the ACM Symposium on Communications Architectures and Protocols (SIGCOMM'90)*, ACM Press, Philadelpia, Pennsylvania, USA, vol. 20, pp. 8–18, September 24–27, 1990.
20. S.J. Golestani, "Duration-Limited Statistical Multiplexing of Delay-Sensitive Traffic in Packet Networks", *Proceedings of the ACM Symposium on Communications Architectures and Protocols (SIGCOMM'91)*, ACM Press, Zurich, Switzerland, pp. 323–332, September 3–6, 1991.
21. H. Zhang and D. Ferrari, "Rate-Controlled Static-Priority Queueing", *Conference on Computer Communications (IEEE INFOCOM)*, San Francisco, California, USA, pp. 227–236, March 28–April 1, 1993.
22. M. Aras, J.F. Kurose, D.S. Reeves, and H. Schulzrinne, "Real-time Communication in Packet-Switched Networks", *Proceedings of the IEEE*, vol. 82, no. 1, pp. 122–139, January 1994.
23. D. Ferrari and D.C. Verma, "Scheme for Real-Time Channel Establishment in Wide-Area Networks", *IEEE Journal on Selected Areas in Communications*, vol. 8, no. 3, pp. 368–379, April 1990.
24. D.C. Verma, H. Zhang, and D. Ferrari, "Delay Jitter Control for Real-Time Communications in a Packet Switching Network", *Proceedings of TRICOMM'91*, pp. 35–46, Chapel Hill, North Carolina, April 1991.
25. D.D. Kandlur and K.G. Shin and D. Ferrari, "Real-Time Communication in Multi-Hop Networks", *11th International Conference on Distributed Computing Systems (ICDCS)*, Arlington, TX, pp. 300–307, May 1991.
26. H. Fattah, "Frame-based fair queuing: Analysis and design", M.A.Sc. Thesis, Department of Electrical and Computer Engineering, University of Victoria, Victoria, British Columbia, 111 pp., 2000.
27. D. Ferrari, "Client requirements for real-time communication services", *IEEE Communications Magazine*, vol. 28, no. 11, November 1990. also RFC 1193.
28. S.J. Glestani, "Congestion-free transmission of real-time traffic in packet networks", *ACM SIGCOMM'90*, pp. 527–542, June 1990.
29. R. Guerin and V. Peris, "Quality of service in packet networks: Basic mechanisms and directions", *Proceedings of INFOCOM'97*, vol. 31, pp. 169–189, 1999.
30. L. Bic and A.C. Shaw, *The Logical Design of Operating Systems*, pp. 275–280, 1995.
31. F. Elguibaly, "Analysis and design of arbitration protocols", *IEEE Transactions on Computers*, vol. 38, no. 2, pp. 1168–1175, 1989.

32. J. Crowcroft and J. Oechslin, "Differentiated end-to-end Internet services using a weighted proportional fair sharing TCP", *ACM SIGCOMM*, vol. 28, no. 3, pp. 53–67, 1998.
33. D. Ferrari and D. Verma, "A scheme for real-time channel establishment in wide-area networks", *IEEE Journal on Selected Areas in Communications*, vol. 8, pp. 368–379, 1990.
34. S. Floyd and V. Jacobson, "Link-sharing and resource management models for packet networks", *IEEE/ACM Transactions on Networking*, vol. 3, 1995.
35. M. Grossglauser, S. Keshav, and D. Tse, "RCBR: A simple and efficient service for multiple time-scale traffic", *Proceedings of ACM SIGCOMM'95*, 1995.
36. D. Stiliadis and A. Varma, "Efficient fair queuing algorithms for packet-switched networks", *IEEE/ACM Transactions on Networking*, vol. 6, no. 2, pp. 175–185, 1998.
37. I. Stoica, S. Shenkar, and H. Zhang, "Core-stateless fair queuing: A scalable architecture to approximate fair bandwidth allocation in high speed networks", *SIGCOMM'98*, pp. 118–130.
38. S. Floyd and V. Jackobson, "Random early detection for congestion avoidance", *IEEE/ACM Transactions on Networking*, vol. 1, pp. 397–413, 1993.
39. S. Keshav, *An Engineering Approach to Computer Networking*, Addison Wesley, Reading, Massachusetts, 1997.

Chapter 13
Switches and Routers

13.1 Introduction

The three main building blocks of high-performance networks are the links, the switching equipment connecting the links together, and the software layers (comprising protocols and applications) implemented at the switching equipments and end-nodes. These components enable transferring information between the users of the network [1]. Figure 13.1 shows the principal network hardware elements which are end-nodes, links, and switches.

The *end-nodes*, shown as squares in the figure, are the sources and sinks of information and users access the network through these end-nodes. Examples of end-nodes could be a personal computer, a file server, a printer, or a scanner.

The *links* provide the paths to carry the information streams from one end-node to another. Examples of links could be the telephone copper wires, TV cable, wireless transmitters/receivers, optical fiber cables, and satellite links through dish antennas. As can be expected, different channels offer different service characteristics such as the available data rate, the bit error rate to be expected, and the propagation delay encountered in each hop. It is often mentioned that the optical fiber channel is the most ideal one since it has the highest data rate and least amount of bit error rate. However, the increasing data rates on these channels lead to increased bit error rate due to channel imperfection (dispersion and attenuation) and detection problems. If I am allowed to digress a bit here, I note that the current prevalent way to move data on a fiber channel is to send optical pulses. This is the same technique that was used to send data on the telegraph channel (dots and dashes) about 200 years ago at the time when Custer hastily sheared his hair during the heat of battle, so he would not be recognized during the battle of Little-Big-Horn!

The links or channels comprising the network need not be identical. For example, some end-nodes could be connected to the network through TV cables while other end-nodes could be connected through telephone wires.

Not only need the links of the network in Fig. 13.1 not be identical, but also the data rates supported by the different links are different. For example, an end-node might access the network using an old-style modem operating at 56.6 kbps while another user could be accessing the network using a TV cable connection offering

F. Gebali, *Analysis of Computer and Communication Networks*,
DOI: 10.1007/978-0-387-74437-7_13,

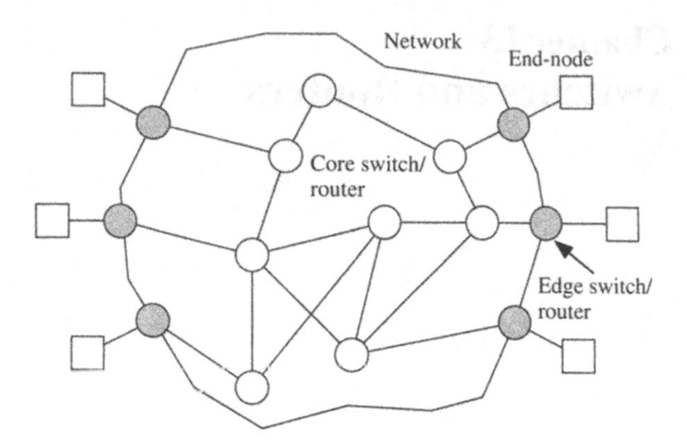

Fig. 13.1 The main network elements are end-nodes (*squares*), links, and switches. The links connect the end-nodes and switches together

a maximum (advertised) data rate of 1 Mbps. Some of the links of that network might also be optical fiber lines operating at OC-192 rate, which corresponds to 9.9533 Gbps.

The *switches* connect several incoming and outgoing links together. The function of the switch is to route the data flows or streams from one incoming link to another outgoing link. The switch determines the proper outgoing link for each incoming packet based on the information in its header. We see that such decisions require quick *classification* of data which is done electronically at electronic data rates even when incoming data are optical pulses. Figure 13.1 identifies two types of switches: edge switches (grey circles) and core switches (empty circles). An edge switch connects one or more end-nodes to the network. On the other hand, core switches connect to other core switches.

The *software* enables error-free transfer of data between the network users. That software spans many layers of the network. Table 13.1 shows some examples of networking software components and their location in the TCP/IP and ISO models of the network.

Table 13.1 Examples of networking software components and their location in the TCP/IP and ISO models of the network

Software	TCP/IP layer	OSI layer
E-mail	Application layer	Layer 7
FTP	Application layer	Layer 7
SNMP	Application layer	Layer 7
DNS	Application layer	Layer 7
Voice over IP	Application layer	Layer 7
World-Wide Web	Application layer	Layer 7
TCP	Transport layer	Layer 4
UDP	Transport layer	Layer 4
ATM	Transport and Internet layer	Layers 4 and 3
Frame relay	Internet layer	Layer 3
IP	Internet	Layer 3
HDLC	Internet layer	Layer 3
Ethernet	Host-to-network	Layer 2
SONET/SDH	Physical layer	Layer 1

The *performance* of a communication network is determined by the links, the routers, and the switches that tie the network together. We will find that message switching through the network is the fundamental activity done by the different components of the network. The network performance parameters such as capacity, bandwidth, and delay, are basically determined by the switches and routers. We start this chapter by reviewing the concepts of switching networks and identifying their main components. Different switch design strategies are introduced which include input queuing, output queuing, and shared buffering. The impact of each design option on the switch performance is discussed such as packet loss probability, speed, and ability to support broadcast and hot-spot traffic.

13.2 Networking

Communication networks are used to achieve connectivity between users through shared communication links. Users could be physically close or they could be at different continents. For any reasonable number of users, it is not possible to establish one-to-one connection since this has many disadvantages such as cost and the need to reconfigure the network every time someone joins or leaves the group. Communication networks aim to achieve the desired connectivity without the disadvantages of one-to-one connections by using shared communication links. The cost to each user of the network drops as the network size grows. This is known as the *economies of scale* which we are all familiar with from the classic telephone network and in the dropping prices of computers and related hardware and software products. The natural downside to sharing is a reduced *quality of service* (QoS) such as increased delay to reach the intended user or occasional loss of data. The dropping of price per user in networks allows for establishing global high-performance networks whose price would otherwise be beyond the reach of small groups of users. Local-area networks (LANs) are examples of communication networks where a group of users is connected at a reasonable cost.

Wide and metropolitan area networks are constructed in a hierarchical fashion from groups of LANs through the use of switches[1] that connect the different parts together. The result is called a switched network since switches are used as the glue that connects the different parts of the network together [2]. In effect, switches enable users with few links to access a large high-capacity network and be able to connect to any location in the globe at a very reasonable cost.

13.3 Media Access Techniques

We mentioned earlier that the links in a network could be wire based, point-to-point wireless, broadcast wireless, or optical. Irrespective of the nature of the medium, the techniques used for sharing it among several users are similar and we discuss them

[1] Most network components such as bridges and routers contain switches in their hardware.

briefly here. Access to the medium is done through a sharing scheme commonly referred to as media access control (MAC) .

13.3.1 Time Division Multiple Access (TDMA)

In time division multiple access, a single link is shared among many users but each user has sole access to the link at any given time. This is similar to staying in a hotel room. The room is shared among many guests but each guest has sole access to the room at any given time. Figure 13.2 shows two different TDMA schemes where time is divided into frames and each frame is in turn divided into time slots. Figure 13.2(a) shows TDMA using *fixed assignment* where each user has a reserved time slot in each frame. Figure 13.2(b) shows TDMA using *random assignment* where each user has to compete to use any time slot in each frame.

An example of TDMA is the use of a wireless broadcast channel within one geographical cell of a cellular phone system. In that case, several users in the geographical cell are allocated a portion of the time to use the channel. Another example of TDMA is of course the backplane bus in computers where the peripherals access the shared channel (bus) one at a time using an interrupt mechanism.

13.3.2 Space Division Multiple Access (SDMA)

In space division multiple access (SDMA), several physical links are provided and the users get access to them through some arbitration protocol, which could be considered as media access control (MAC) protocol. An example of SDMA is the telephone switching network where separate lines are used to connect the users to the central office. Another example is in cellular telephones where a certain region or city is divided into several cells. The users in the different cells can use the network using the same frequency band since there is no chance of interference. Thus SDMA refers to the simultaneous use of several communication media in parallel.

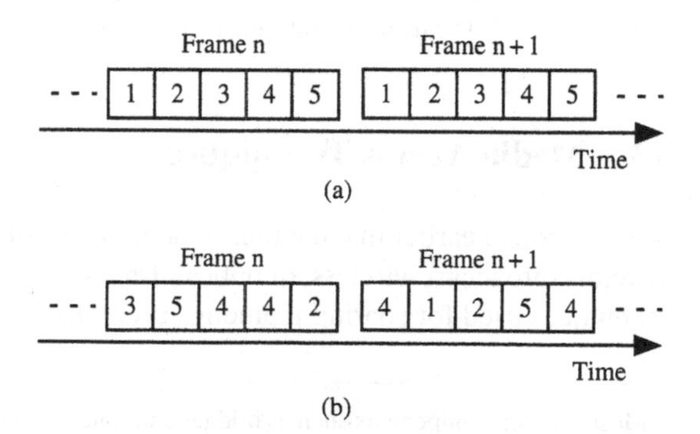

Fig. 13.2 Time division switching: (**a**) fixed assignment, (**b**) random assignment

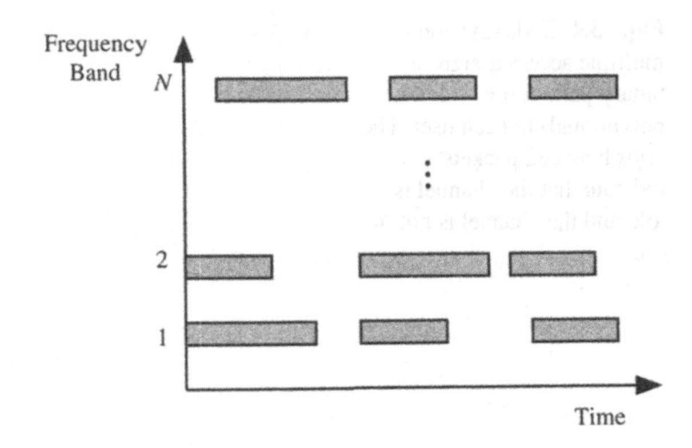

Fig. 13.3 Frequency division multiple access assigns a frequency band to each user. The *gaps* between packets indicate that the channel is idle and the frequency band is not in use

Similar to TDMA, there are two ways that SDMA could be used. Fixed assignment is called *circuit switching* and is discussed in Section 13.4.1. Random assignment is called *packet switching* which is discussed in Section 13.4.2.

13.3.3 Frequency Division Multiple Access (FDMA)

In this technique, each user is assigned a unique frequency band in the frequency spectrum. Radio and analog TV broadcasting use this technique where each station is allocated a frequency band. Figure 13.3 shows the assignment of the frequency bands among users versus time.

This technique is also used in optical fibers by assigning a particular optical wavelength (light color) to each channel. This is known as wavelength division multiple access (WDMA).

13.3.4 Code Division Multiple Access (CDMA)

In code division multiple access technique, data from each user are used to modulate a unique binary pattern that belongs to the user. The binary patterns (called polynomials) are designed such that no interference takes place when the signals are sent. At the receiver end, information carried on each pattern can be extracted with no interference from the other signals. Figure 13.4 shows an example of CDMA where N codes are used to transmit data from N users.

13.4 Circuit and Packet Switching

In digital communications, there are two main techniques for establishing a path through the network: circuit switching and packet switching.

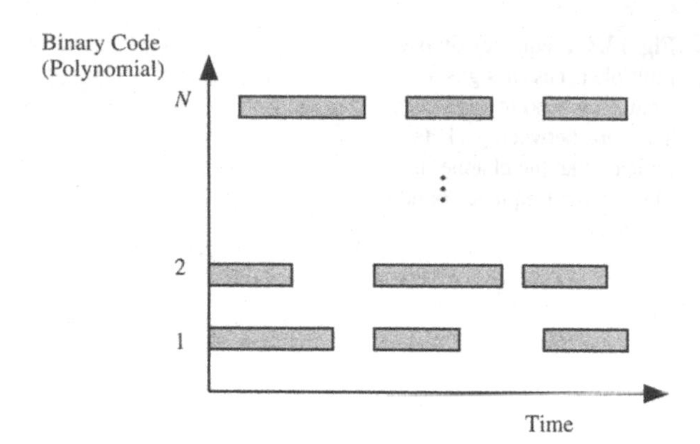

Fig. 13.4 Code division multiple access assigns a binary pattern (or polynomial) to each user. The *gaps* between packets indicate that the channel is idle and the channel is not in use

13.4.1 Circuit Switching

In circuit switching, a unique path is used to move data between the two end users. Thus, the path or the resource is reserved for the duration of a particular session. This is a waste of resources when there is no activity on the channel while other users are waiting to access the channel. Constant bit rate traffic does well with circuit switching since the path is dedicated and there are always data sent on the channel. Computer communication, however, is bursty and does not do well under circuit-switched techniques. The telephone network is circuit switched and the path is held up by the user until it is released.

The important characteristics of circuit switching are as follows:

1. Overhead of establishing the path is only spent at the start of the session.
2. Data format is simple since the path is already established beforehand.
3. In-sequence delivery of data is possible since all packets will arrive using the same route.
4. The channel bandwidth is guaranteed and this suites constant bit rate traffic.
5. The channel is better utilized carrying constant bit rate traffic and not bursty traffic.
6. Traffic information is easily observable since switching nodes maintain information about each call.

To establish a path in circuit switching, three consecutive phases are required:

1. *Connection establishment*: For data to be transferred between two end users, a connection (path) must be established. The intermediate switches exchange information related to the availability of a path and inform the end users when the path is available. This is similar to using the phone by waiting for the dial tone, then dialing the desired number.
2. *Data transfer*: Data can now be moved between the users over the connection that was established. The type of connection will determine the data rate, average delay, and data loss rate. This is similar to talking on the phone once the other party picks up the phone.

3. *Connection tear down*: After data transfer is complete, the end users inform the intermediate switches to free the connection. This is similar to placing the phone on the hook to inform the phone company that the circuit can be released.

13.4.2 Packet Switching

In packet switching, data are broken into packets of fixed or variable size, depending on the protocol used. There are two approaches for packet switching: datagram and virtual circuit.

In *datagram switching*, each packet carries with it all the routing information it needs to reach its destination and each packet is treated independently of the ones before it or after it. In fact, it is possible that different packets might travel down different paths. This is actually an advantage since the resulting network is immune to faults, and packets can be dynamically routed on any available links. This was the main reason for using packet communication in the first place in the 1970s to build a highly fault-tolerant communication network that is immune to enemy attack. Since each packet carries all the information it needs to determine its destination, there is no need to establish a connection before transmitting the data. The connection establishment phase of circuit switching is removed. However, packet switching carries the following penalties:

1. In-sequence delivery of data is not possible since different packets might travel down different paths.
2. Packet routing will be complex since the router has to choose the optimum route to send each packet.
3. Packet delay will vary depending on the route chosen for each packet.
4. Packet loss cannot be detected by the network. The end-nodes are able to detect lost packets.
5. Traffic information is not easily observable since switching nodes do not have information about the state of each stream.

In *virtual circuit switching*, a unique path is used to move data between the two end users. This is similar to the circuit switching approach. However, the channel or path could be used by other data when there are no packets to send. Virtual circuit switching has several advantages over circuit switching and datagram switching:

1. Since packets belonging to a certain session are distinct, it is easy to associate quality of service guarantees to the session and deliver those guarantees.
2. The ability to identify individual sessions enables the service provider to guarantee the service, monitor the user traffics, and charge for the services provided.
3. Different classes of service are easily established for the different sessions. The cost of these services is definitely cheaper compared to leasing a private telephone line.
4. The packets have a simple format since the routing information has been determined during path, or connection, establishment phase.

5. Packet routing is simple since there is only one route associated with all the packets belonging to a call.
6. In-sequence delivery of data is possible since the packets from a certain call traverse the same route.
7. The channel is well utilized since different calls share the same channel. This results in an overall gain in the network capacity, which is usually referred to as *statistical gain.*
8. Packet loss can be detected by the network since the packets follow each other in sequence and each carries a unique sequence number.
9. Virtual circuit switching is suited to both bursty and constant bit rate traffic.

To establish a path in virtual circuit switching, the same steps are followed as in circuit switching: connection establishment, data transfer, then connection tear down.

13.5 Packet Switching Hardware

We review some of the hardware components that are commonly used in communication networks. We should point out that the following definitions are not very strict since the capabilities of the components become more sophisticated with advances in technology. The component capabilities are implemented in software or hardware and there are grey areas where a device bridges the gap between two classifications or partially implements the functionality of a given class of devices. We discuss below several important networking components in the order of their level of complexity. Figure 13.5 shows the ISO and TCP/IP reference model layers implemented by the different components: hubs, bridges, switches, routers, and gateways.

13.5.1 End-Node

An end-node is a device that is attached to the network. The user accesses the network through the end-node, which could be a workstation, a PC, a printer, a file server, etc.

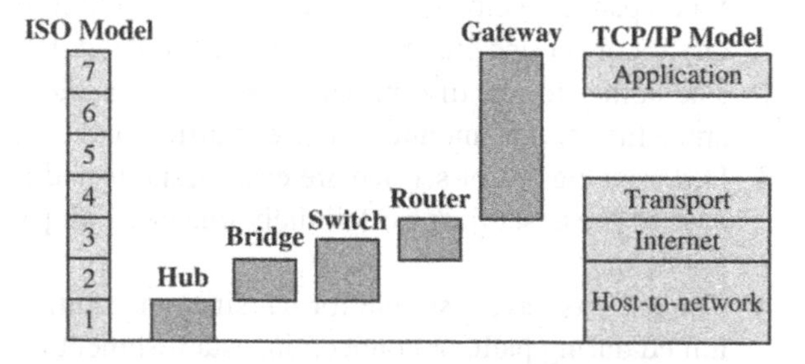

Fig. 13.5 Hubs, bridges, switches, routers, and gateways and the layers implemented by each component

13.5.2 Hub

A hub connects several segments of a LAN. A hub has several ports such that when a packet arrives at a port, it is copied to *all* the other ports so that all segments of the LAN see the packet. In that sense, a hub acts as a repeater to repeat one message on one LAN to all the other LANs [3].

A *passive hub* serves simply as a conduit for the data, enabling it to go from one device (or network segment) to another. An *intelligent hub* includes additional features that enable an administrator to monitor the traffic passing through the hub and to configure each port in the hub. A *switching hub* actually reads the destination address of each packet and then forwards the packet to the correct port based on the header information. A hub does not improve network capacity or performance. It only acts as the "wiring" between the network segments. Thus we can think of the hub as operating mainly in Layer 1 of the ISO reference model since it only enables the operation of the physical layer of the ISO reference model. Figure 13.5 shows the location of the hub and the other main network connectivity components in relation to the ISO reference model layers.

13.5.3 Bridge

A bridge is a device that connects two local-area networks (LANs) or two segments of the same LAN. The two LANs being connected can be alike or dissimilar. For example, a bridge can connect an Ethernet with a token-ring network. Unlike routers, bridges are protocol independent. They simply forward packets without analyzing and re-routing messages. Consequently, they are faster than routers, but less versatile. A bridge uses the packet header to determine whether to pass the packet to the other LAN or not. A bridge operates at Layer 2 of the ISO reference model since it supports the operation of the data link (network access) layer . A bridge is simpler than a router but still requires a switch for delivering the packets to the correct output port.

13.5.4 Switch

A switch connects several LANs. A switch has several ports such that when a packet arrives at a port, it is forwarded *only* to the appropriate output port based on the header information. In that sense, a switch provides a temporary dedicated connection between an input port and an output port. A switch improves network performance by dividing the network into several independent segments, thereby increasing the overall capacity. In that sense, the switch is smarter than a hub. Typically, switches work in Layer 2 of the ISO reference model, which is equivalent to a bridge. The new trend is to move to Layer 3 switching, which is capable of switching millions of packets per second. It should be mentioned that a switch is protocol specific. However, newer switches are able to handle several protocols such as multi-protocol label switching (MPLS).

13.5.5 Router

A router connects networks that may or may not be similar. A router uses the packet header and a forwarding table to determine the best way a packet should go between the networks. Routers use ICMP[2] to communicate with each other and determine the best route between any two hosts. Very little filtering of data is done through routers. Routers know how the whole network is connected and how to move information from one part of the network to another. They free the end-nodes from having to do these tasks. Routers enhance the network by connecting networks that use different protocols. Note that a router requires the use of a switch to perform its packet routing (switching) functions. Routers are smarter than hubs and switches. The router operates at Layer 3 of the ISO reference model since its supports the operation of the network layer (such as data transmission and switching) [2] .

13.5.6 Gateway

A gateway is a computer that uses a combination of hardware and software to link two different types of networks using different protocols. Gateways between e-mail systems, for example, allow users on different e-mail systems to exchange messages. Thus a gateway translates between two different protocols and sometimes topologies. For example, a gateway is needed to translate between TCP/IP over Ethernet and ATM over SONET. We can think of the gateway as operating mainly *above* Layer 3 of the ISO model since its enables the operation of the application layer and above.

From the above definitions, we see that a switch is the basic component of networking, and the design of an efficient high-speed switch is absolutely necessary to obtain networks capable of meeting customer demands for ever-increasing capacity and reduced delay.

13.6 Basic Switch Components

As was explained above, network routers rely on switches to perform their functions. Thus it is worthwhile to study the construction of switches in more detail. A switch is a hardware device that accepts packets at its inputs and routes them to its outputs according to the routing information provided in the packet header and the switch

[2] Internet Control Message Protocol. ICMP supports packets containing error, control, and informational messages.

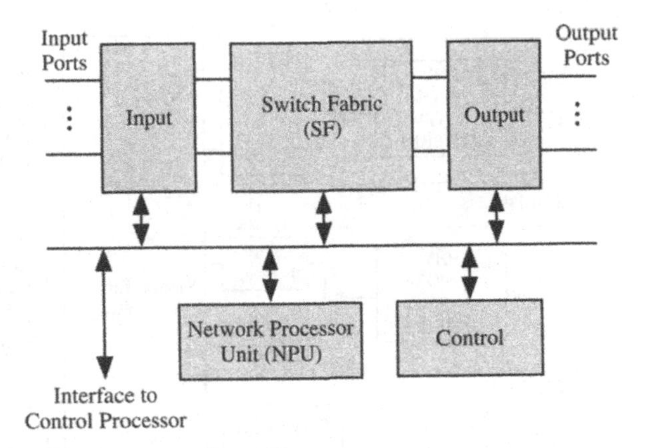

Fig. 13.6 Basic components of a switch

routing table. Figure 13.6 is a block diagram showing the main components of a switch.

The switch has three main architectural components: a network processor unit (NPU), a controller, a datapath comprising input/output ports, and switch fabric. The functionalities of each are explained in the following sections. In general, the functions of intensive data processing and control should be distributed, whenever possible, among the different switch components to reduce the delay and be able to handle many inputs/outputs simultaneously. The components of the switch that have the most impact on its performance are the storage buffers/queues and the switch fabric. For example, packet loss in the switch results due to buffer overflow or inability to establish a path from an input to an output port through the switching fabric. Also, the maximum line rate depends on the memory access speed.

13.6.1 Network Processing Unit (NPU)

The network processing unit is required to do compute-intensive tasks to enable the switch to process data at high speeds. An NPU is a programmable processor with special instructions or hardware components that are specifically designed to efficiently perform networking tasks. A general-purpose processor proves too slow to do the required networking tasks. However, specialized hardware meets the demands but has the disadvantages that once the tasks or protocols are upgraded, a major redesign is required and this takes time. By the time a specialized hardware is available, the protocols might have changed already. Thus the major motives for an NPU are flexibility/programmability and efficiency in executing networking tasks.

Figure 13.7 shows a switch with NPU-specialized hardware components that do specific tasks. Examples of the tasks that the NPU might be required to perform include the following:

1. Implementation of security protocols for firewalls, network security, data encryption/decryption, etc

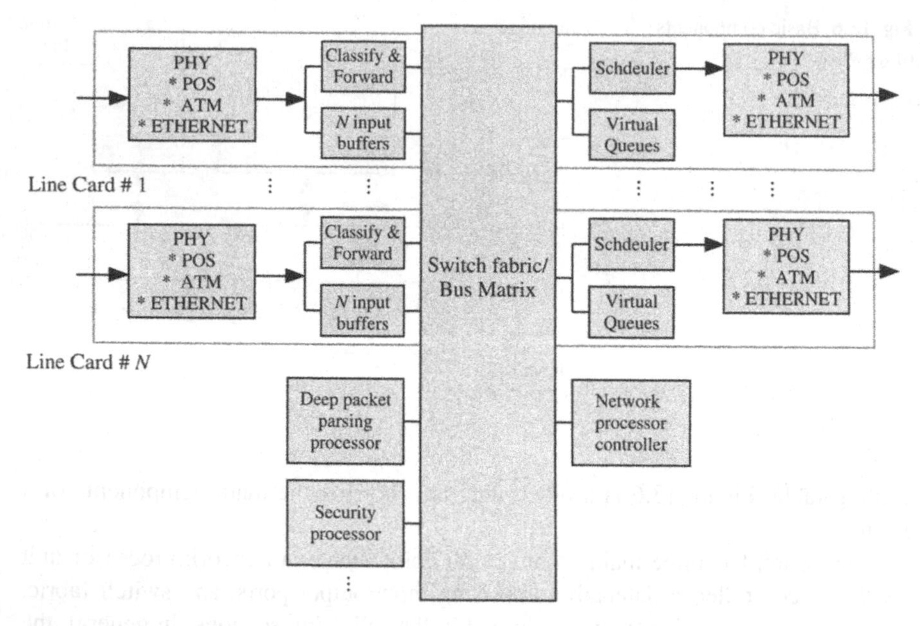

Fig. 13.7 A switch incorporating NPU components

2. Deep packet search which is essentially string search for pattern matching using the packet payload as input
3. Implementation of quality of service protocols
4. Traffic shaping
5. Packet routing using routing tables
6. Interface

13.6.2 Control Section

The control section deals with packet streams and establishing and tearing down circuits and is composed of the bottom two components of the switch block diagram in Fig. 13.6. The control section performs the following functions:

1. Maintains the contents of the routing table, which determines the proper destination output port of an incoming packet
2. For the case of circuit switching, decides whether to accept new connections or not based on current utilization of switch resources
3. Provides congestion control for the switch by continually monitoring the switch resources (e.g., buffer occupancy) and issuing proper actions
4. Assigns switch resources to the established connections based on the class of service

It should be mentioned that these functionalities could be centralized or they could be distributed between the input and output ports. The latter option is desirable to ensure high-speed operation.

13.6.3 Datapath Section

The datapath deals with individual packets or cells and is composed of the top three components of the switch block diagram in Fig. 13.6. The datapath performs the following functions:

1. Accepts incoming packets on any of its N input ports and stores them in temporary buffers for processing their header information
2. Establishes a path to the desired output port through the switch fabric (SF)
3. Stores the routed packets in the output queues and schedules the stored packets for transmission based on some scheduling protocol

13.6.4 Switch Fabric

The switch fabric (SF) establishes the required paths between pairs of input and output ports. The switch fabric must also support any type of connection such as single cast (one input to one output), multicast (one input to many outputs), broadcast (one input to all outputs), or hot point (many inputs to one output). A detailed discussion of switch fabrics and their performance is found in Chapter 14.

13.6.5 Lookup Table Design

The *routing* or *lookup table* stores information about which output ports should receive a packet that arrives at an input port. This information is maintained and updated by the control processor. The data in the lookup table could be organized using any of the following approaches:

1. *Heap storage*: Store the data in a heap with no particular order using a random-access memory (DRAM or SRAM). This approach is simple but searching for a particular data item requires searching the entire memory. RAMs suffer from I/O port limitations (typically one!) and large cycle time. Thus accessing the database is done one item at a time using a slow search strategy which is not practical in high-speed switches.
2. *Content-addressable memory* (*CAM*): This is sometimes called associative memory or associative storage. CAM could be a small memory but complex in design and slow in performance. Searching for a data item is done in a distributed fashion *within* the memory which speeds up the search operation. The design of an

efficient and fast CAM is certainly a challenging and exciting hardware design problem.

3. *Hash tables*: Here the input header information is compressed to a smaller number of bits (e.g., from 48 bits to 16 bits) using a hashing function to reduce the RAM size [4–6]. The hashing function should be simple to implement without requiring much processing time. Typically, hashing functions are implemented using a linear feedback shift register (LFSR) that performs polynomial division. The input to the LFSR is treated as a polynomial and the LFSR contents are the "signature" that corresponds to the input. See Appendix F for a discussion of hashing.
4. *Balanced tree* (*B-tree*): The B-tree is a data structure that is used to efficiently build large dictionaries and implement the algorithms used to search, insert, and delete keys from it. The advantage of using B-trees for constructing lookup tables is the small time required to search for the routing information. However, complex hardware design is required to implement the other lookup table functions such as data insertion and deletion. Again, the design of B-trees is a very challenging and exciting hardware design project. See Appendix F for a discussion of B-trees.

In an ATM switch, the routing table uses direct lookup to determine the destination output port and to update the VPI/VCI bytes. For an Ethernet switch, the lookup table uses associative lookup (content-addressable memory or CAM). For an IP router, the routing table could be based on the new CIDR or cache memory.

13.7 Switch Functions

A switch or a router has to perform several functions beside simply routing packets from its inputs to its outputs.

13.7.1 Routing

The switch or router must be able to read the header of each incoming packet to determine which output link must be used to move the packet to its destination. It is obvious that an arriving packet cannot be routed until packet *classification* based on its header information has been performed. These routing decisions are done using *routing tables*. Techniques for building routing tables are discussed in more detail in Section 13.6.5 and in Appendix F.

13.7.2 Traffic Management

When too many packets are present in the network, *congestion* is said to have taken place. When congestion occurs in a router, the buffers become filled and packets start getting lost. This situation typically gets worse since the receiver will start sending negative acknowledgments and the sender will start retransmitting the lost frames. This leads to more packets in the network and the overall performance starts to deteriorate [7]. *Traffic management* protocols must be implemented at the routers to ensure that users do not tax the resources of the network. Examples of traffic management protocols are leaky bucket, token bucket, and admission control techniques. Chapter 8 discusses the traffic management protocols in more detail.

13.7.3 Scheduling

Routers must employ scheduling algorithms for two reasons: to provide different quality of service (QoS) to the different types of users and to protect well-behaved users from misbehaving users that might hog the system resources (bandwidth and buffer space). A scheduling algorithm might operate on a per-flow basis or it could aggregate several users into broad service classes to reduce the workload. Chapter 12 discusses scheduling techniques in more detail.

13.7.4 Congestion Control

The switch or router drops packets to reduce network congestion and to improve its efficiency. The switch must select which packet to drop when the system resources become overloaded. There are several options such as dropping packets that arrive after the buffer reaches a certain level of occupancy (this is known as tail dropping). Another option is to drop packets from any place in the buffer depending on their tag or priority. Chapter 12 discusses packet dropping techniques in more detail.

13.8 Switch Performance Measures

Many switch designs have been proposed and our intention here is not to review them but to discuss the implications of the designer's decisions on the overall performance of the switch. This, we believe, is crucial if we are to produce novel switches capable of supporting terabit communications.

The basic performance measures of a switch are as follows:

1. Maximum data rate at the inputs.
2. The number of input and output ports.
3. The number of independent logical channels or calls it can support.
4. The average delay a packet encounters while going through the switch.
5. The average packet loss rate within the switch.
6. Support of different quality of service (QoS) classes, where QoS includes data rate (bandwidth), packet delay, delay variation (jitter), and packet loss.
7. Support of multicast and broadcast services.

8. Scalability of the switch, which refers to the ability of the switch to work satisfactorily if the line data rates are scaled up or if the number of inputs and outputs is scaled up.
9. The capacity (packets/s or bits/s) of the switch is defined as

$$\text{Switch capacity} = \text{input line rate} \times \text{number of input ports} \tag{13.1}$$

10. Flexibility of the switch architecture to be able to upgrade its components. For example, we might be interested in upgrading the input/output line cards or we might want to replace only the switch fabric.

13.9 Switch Classifications

The two most important components of a switch are its buffers and its switching fabric. Based on that we can describe the architecture of a switch based on the following two criteria:

- The type of switch fabric (SF) used in the switch to route packets from the switch inputs (ingress points) to the switch outputs (egress points). Detailed qualitative and quantitative discussion of switch fabrics, also known as interconnection networks, is found in Chapter 14.
- Location of the buffers and queues within the switch. An important characteristic of a switch is the number and location of the buffers used to store incoming traffic for processing. The placement of the buffers and queues in a switch is of utmost importance since it will impact the switch performance measures such as packet loss, speed, ability to support differentiated services, etc. The following sections discuss the different buffering strategies employed and the advantages and disadvantages of each option. Chapter 15 provides qualitative discussion of the performance parameters of the different switch types.

The remainder of this chapter and Chapter 16 discuss the different design options for switch buffering. Chapter 14 discusses the different possible design options for interconnection networks.

13.10 Input Queuing Switch

Figure 13.8 shows an input queuing switch. Each input port has a dedicated FIFO buffer to store incoming packets. The arriving packets are stored at the tail of the queue and only move up when the packet at the head of the queue is routed through the switching fabric to the correct output port. A controller at each input port classifies each packet by examining its header to determine the appropriate path through the switch fabric. The controller must also perform traffic management functions. In one time step, an input queue must be able to support one write and one read

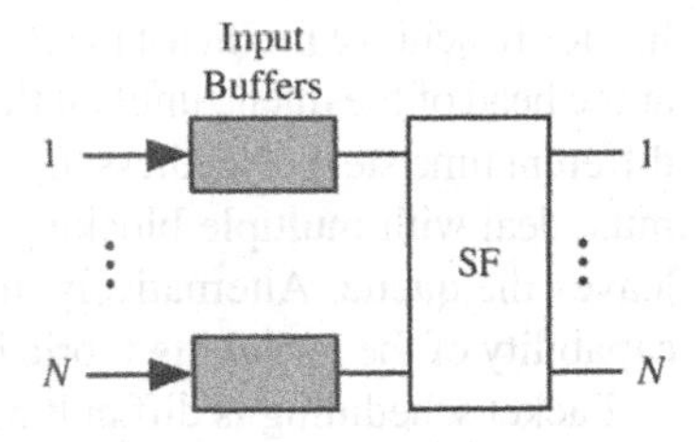

Fig. 13.8 Input queuing switch. Each input has a queue for storing incoming packets

operations which is a nice feature since the memory access time is not likely to impose any speed bottlenecks.

Assuming an $N \times N$ switch, the switch fabric (SF) must connect N input ports to N output ports. Only a space division $N \times N$ switch can provide simultaneous connectivity.

The main advantages of input queuing are

1. Low memory speed requirement.
2. Distributed traffic management at each input port.
3. Distributed table lookup at each input port.
4. Support of broadcast and multicast does not require duplicating the data.

Distributed control increases the time available for the controller to implement its functions since the number of sessions is limited at each input. Thus input queuing is attractive for very high bandwidth switches because all components of the switch can run at the line rate [8].

The main disadvantages of input queuing are

1. Head of line (HOL) problem, as discussed below
2. Difficulty in implementing data broadcast or multicast since this will further slow down the switch due to the multiplication of HOL problem
3. Difficulty in implementing QoS or differentiated services support, as discussed below
4. Difficulty in implementing scheduling strategies since this involves extensive communications between the input ports

HOL problem arises when the packet at the head of the queue is blocked from accessing the desired output port [9]. This blockage could arise because the switch fabric cannot provide a path (*internal blocking*) or if another packet is accessing the output port (*output blocking*). When HOL occurs, other packets that may be queued behind the blocked packet are consequently blocked from reaching possibly idle output ports. Thus HOL limits the maximum throughput of the switch [10]. A detailed discrete-time queuing analysis is provided in Section 15.2 and the maximum throughput is not necessarily limited to this figure.

The switch throughput can be increased if the queue service discipline examines a window of w packets at the head of the queue, instead of only the HOL packet. The first packet out of the top w packets that can be routed is selected and the queue size decreases by one such that each queue sends only one packet to the switching fabric. To achieve multicast in an input queuing switch, the HOL packet must remain at the head of the queue until all the multicast ports have received their own copies at different time steps. Needless to say, this aggravates the HOL problem since now we must deal with multiple blocking possibilities for the HOL packet before it finally leaves the queue. Alternatively, the HOL packet might make use of the multicast capability of the switching fabric if one exists.

Packet scheduling is difficult because the scheduler has to scan all the packets in all the input ports. This requires communication between all the inputs which limits the speed of the switch. The scheduler will find it difficult to maintain bandwidth and buffer space fairness when all the packets from different classes are stored at different buffers at the inputs. For example, packets belonging to a certain class of service could be found in different input buffers. We have to keep a tally of the buffer space used up by this service class.

In input queuing, there are three *potential* causes for packet loss:

1. Input queue is full. An arriving packet has no place in the queue and is discarded.
2. Internal blocking. A packet being routed within the switch fabric is blocked inside the SF and is discarded. Of course, this type of loss occurs only if the input queue sends the packet to the SF without waiting to verify that a path can be provided.
3. Output blocking. A packet that made it through the SF reaches the desired output port but the port ignores it since it is busy serving another packet. Again, this type of loss occurs only if the input queue sends the packet to the output without waiting to verify that the output link is available.

13.11 Output Queuing Switch

To overcome the HOL limitations of input queuing, the standard approach is to abandon input queuing and place the buffers at the output ports as shown in Fig. 13.9. Notice, however, that an output queuing switch must have small buffers at its inputs to be able to temporarily hold the arriving packets while they are being classified and processed for routing.

An incoming packet is stored at the input buffer and the input controller must read the header information to determine which output queue is to be updated. The packet must be routed through the switch fabric to the correct output port. The controller

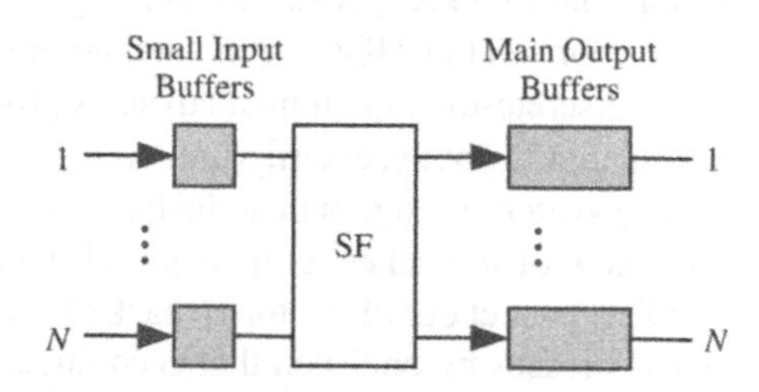

Fig. 13.9 Output queuing switch. Each output has a queue for storing the packets destined to that output. Each input must also have a small FIFO buffer for storing incoming packets for classification

must also handle any contention issues that might arise if the packet is blocked from leaving the buffer for any reason.

A controller at each input port classifies each packet by examining the header to determine the appropriate path through the switch fabric. The controller must also perform traffic management functions.

In one time step, the small input queue must be able to support one write and one read operations which is a nice feature since the memory access time is not likely to impose any speed bottlenecks. However, in one time step, the main buffer at each output port must support N write and one read operations.

Assuming an $N \times N$ switch, the switch fabric (SF) must connect N input ports to N output ports. Only a space division $N \times N$ switch can provide simultaneous connectivity.

The main advantages of output queuing are

1. Distributed traffic management
2. Distributed table lookup at each input port
3. Ease of implementing QoS or differentiated services support
4. Ease of implementing distributed packet scheduling at each output port

The main disadvantages of output queuing are

1. High memory speed requirements for the output queues.
2. Difficulty of implementing data broadcast or multicast since this will further slow down the switch due to the multiplication of HOL problem.
3. Support of broadcast and multicast requires duplicating the same data at different buffers associated with each output port.
4. HOL problem is still present since the switch has input queues.

The switch throughput can be increased if the switching fabric can deliver more than one packet to any output queue instead of only one. This can be done by increasing the operating speed of the switch fabric which is known as *speedup*. Alternatively, the switch fabric could be augmented using duplicate paths or by choosing a switch fabric that inherently has more than one link to any output port. When this happens, the output queue has to be able to handle the extra traffic by increasing its operating speed or by providing separate queues for each incoming link.

As we mentioned before, output queuing requires that each output queue must be able to support one read and N write operations in one time step. This, of course, could become a speed bottleneck due to cycle time limitations of current memory technologies.

To achieve multicast in an output queuing switch, the packet at an input buffer must remain in the buffer until all the multicast ports have received their own copies at different time steps. Needless to say, this leads to increased buffer occupancy since now we must deal with multiple blocking possibilities for the packet before it finally leaves the buffer. Alternatively, the packet might make use of the multicast capability of the switching fabric if one exists.

In output queuing, there are four *potential* causes for packet loss:

1. Input buffer is full. An arriving packet has no place in the buffer and is discarded.
2. Internal blocking. A packet being routed within the switch fabric is blocked inside the SF and is discarded.
3. Output blocking. A packet that made it through the SF reaches the desired output port but the port ignores it since it is busy serving another packet.
4. Output queue is full. An arriving packet has no place in the queue and is discarded.

13.12 Shared Buffer Switch

Figure 13.10 shows a shared buffer switch design that employs a single common buffer in which all arriving packets are stored. This buffer *queues* the data in separate queues that are located within one common memory. Each queue is associated with an output port. Similar to input and output queuing, each input port needs a local buffer of its own in which to store incoming packets until the controller is able to classify them.

A flexible mechanism employed to construct queues using a regular random-access memory is to use the *linked list* data structure. Each linked list is dedicated to an output port. In a linked list, each storage location stores a packet and a pointer to the next packet in the queue as shown. Successive packets need not be stored in successive memory locations. All that is required is to be able to know the address of the next packet though the pointer associated with the packet. This pointer is indicated by the solid circles in the figure. The lengths of the linked lists need not be equal and depend only on how many packets are stored in each linked list. The memory controller keeps track of the location of the last packet in each queue, as shown by the empty circles. There is no need for a switch fabric since the packets are effectively "routed" by being stored in the proper linked list.

When a new packet arrives at an input port, the buffer controller decides which queue it should go to and stores the packet at any available location in the memory,

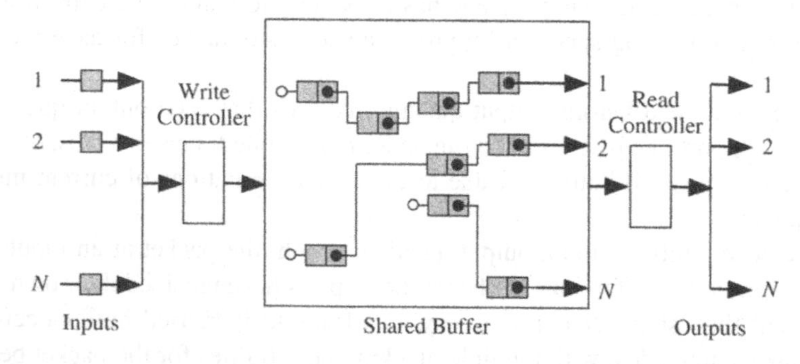

Fig. 13.10 Shared buffer switch. *Solid circles* indicate next packet pointers. *Empty circles* indicate pointers to the tail end of each linked list

then appends that packet to the linked list by updating the necessary pointers. When a packet leaves a queue, the pointer of the next packet now points to the output port and the length of the linked list is reduced by one.

The main advantages of shared buffering are

1. Ability to assign different buffer space for each output port since the linked list size is flexible and limited only by the amount of free space in the shared buffer.
2. A switching fabric is not required.
3. Distributed table lookup at each input port.
4. There is no HOL problem in shared buffer switch since each linked list is dedicated to one output port.
5. Ease of implementing data broadcast or multicast.
6. Ease of implementing QoS and differentiated services support.
7. Ease of implementing scheduling algorithms at each linked list.

The main disadvantages of shared buffering are

1. High memory speed requirements for the shared buffer.
2. Centralized scheduler function implementation which might slow down the switch.
3. Support of broadcast and multicast requires duplicating the same data at different linked lists associated with each output port.
4. The use of a single shared buffer makes the task of accessing the memory very difficult for implementing scheduling algorithms, traffic management algorithms, and QoS support.

The shared buffer must operate at a speed of at least $2N$ since it must perform a maximum of N write and N read operations at each time step.

To achieve multicast in a shared buffer switch, the packet must be duplicated in all the linked lists on the multicast list. This needlessly consumes storage area that could otherwise be used. To support differentiated services, the switch must maintain several queues at each input port for each service class being supported.

In shared buffering, there are two *potential* causes for packet loss:

1. Input buffer is full. An arriving packet has no place in the buffer and is discarded.
2. Shared buffer is full. An arriving packet has no place in the buffer and is discarded.

13.13 Multiple Input Queuing Switch

To overcome the HOL problem in input queuing switch and still retain the advantages of that switch, m input queues are assigned to each input port as shown in Fig. 13.11. If each input port has a queue that is dedicated to an output port (i.e., $m = N$), the switch is called *virtual output queuing* (VOQ) switch. In that case, the input controller at each input port will classify an arriving packet and place it in

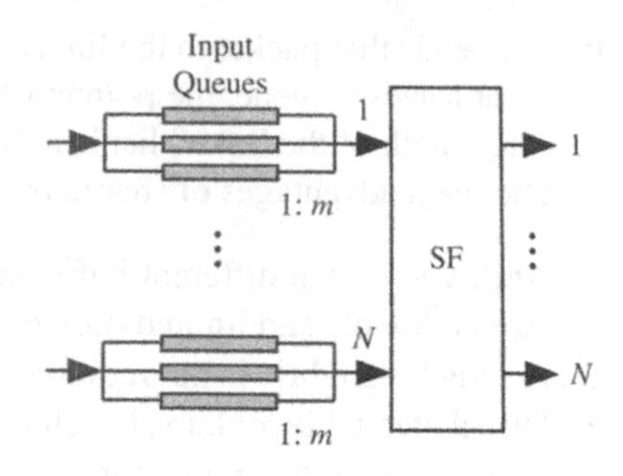

Fig. 13.11 Multiple input queue switch. Each input port has a bank of FIFO buffers. The number of queues per input port could represent the number of service classes supported or it could represent the number of output ports

the FIFO buffer belonging to the destination output port. In effect, we are creating output queues at each input and hence the name "virtual output queuing".

This approach removes the HOL problem and the switch efficiency starts to approach 100% depending only on the efficiency of the switch fabric and the scheduling algorithm at each output port. Multicast is also very easily supported since copies of an arriving packet could be placed at the respective output queues. Distributed packet classification and traffic management are easily implemented in that switch also.

There are, however, several residual problems with this architecture. Scheduling packets for a certain output port becomes a major problem. Each output port must choose a packet from N virtual queues located at N input ports. This problem is solved in the VRQ switch that is discussed in Section 13.16 and Chapter 16. Another disadvantage associated with multiple input queues is the contention between all the queues to access the switching fabric. Dedicating a direct connection between each queue and the switch fabric results in a huge SF that is of dimension $N^2 \times N$ which is definitely not practical.

In multiple input queuing, there are three *potential* causes for packet loss:

1. Input buffer is full. An arriving packet has no place in the buffer and is discarded.
2. Internal blocking. A packet being routed within the switch fabric is blocked inside the SF and is discarded.
3. Output blocking. A packet that made it through the SF reaches the desired output port but the port ignores it since it is busy serving another packet.

13.14 Multiple Output Queuing Switch

To support sophisticated scheduling algorithms, n output queues are assigned to each output port as shown in Fig. 13.12. If each output port has a queue that is dedicated to an input port (i.e., $n = N$), the switch is called *virtual input queuing* (VIQ) switch. In that case, the output controller at each output port will classify an arriving packet and place it in the FIFO buffer belonging to the input port it came on. In effect, we are creating input queues at each output and hence the name "virtual input queuing". Another advantage of using several output queues is that the FIFO

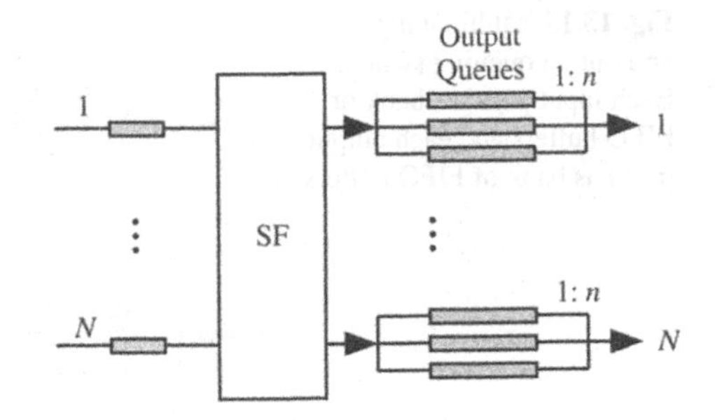

Fig. 13.12 Multiple output queuing switch. Each output port has bank of FIFO buffers. The number of queues per output port could represent the number of service classes supported or it could represent the number of connections supported

speed need not be N times the line rate as was the case in output queuing switch with a single buffer per port.

Several disadvantages are not removed from output queue switch using this approach. The HOL problem is still present and packet broadcast still aggravates the HOL problem. Another disadvantage associated with multiple output queues is the contention between all the queues to access the switching fabric. Dedicating a direct connection between each queue and the switch fabric results in a huge SF that is of dimension $N \times N^2$, which is definitely not practical. This problem is solved in the VRQ switch that is discussed in Section 13.16 and Chapter 16.

In multiple output queuing, there are four *potential* causes for packet loss:

1. Input buffer is full. An arriving packet has no place in the buffer and is discarded.
2. Internal blocking. A packet being routed within the switch fabric is blocked inside the SF and is discarded.
3. Output blocking. A packet that made it through the SF reaches the desired output port but the port ignores it since it is busy serving another packet.
4. Output queue is full. An arriving packet has no place in the queue and is discarded.

13.15 Multiple Input/Output Queuing Switch

To retain the advantages of multiple input and multiple output queuing and avoid their limitations, multiple queues could be placed at each input and output port as shown in Fig. 13.13. An arriving packet must be classified by the input controller at each input port to be placed in its proper input queue. Packets destined to a certain output port travel through the switch fabric (SF), and the controller at each output port classifies them, according to their class of service, and places them in their proper output queue.

The advantages of multiple queues at the input and the output are removal of HOL problem, distributed table lookup, distributed traffic management, and ease of implementation of differentiated services. Furthermore, the memory speed of each queue could match the line rate.

The disadvantage of the multiple input and output queue switch is the need to design a switch fabric that is able to support a maximum of $N^2 \times N^2$ connections

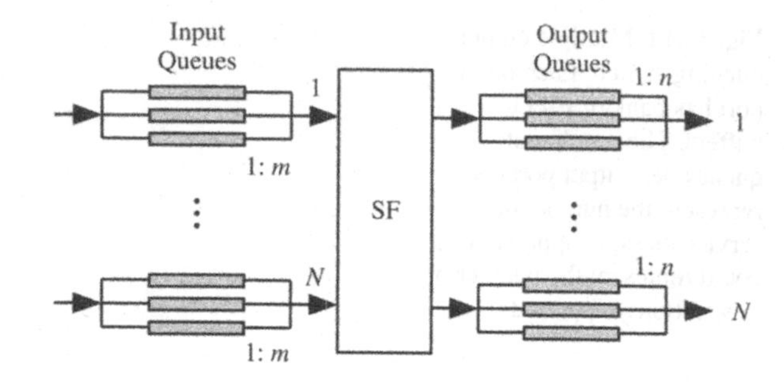

Fig. 13.13 Multiple input and output queuing switch. Each input port has bank of FIFO buffers and each output port has bank of FIFO buffers

simultaneously. This problem is solved in the VRQ switch that is discussed in Section 13.16 and Chapter 16.

In multiple input and output queuing, there are four *potential* causes for packet loss:

1. Input buffer is full. An arriving packet has no place in the buffer and is discarded.
2. Internal blocking. A packet being routed within the switch fabric is blocked inside the SF and is discarded.
3. Output blocking. A packet that made it through the SF reaches the desired output port but the port ignores it since it is busy serving another packet.
4. Output queue is full. An arriving packet has no place in the queue and is discarded.

13.16 Virtual Routing/Virtual Queuing (VRQ) Switch

We saw in the previous sections the many alternatives for locating and segmenting the buffers. Each design had its advantages and disadvantages. The virtual routing/virtual queuing (VRQ) switch has been proposed by the author such that it has all the advantages of earlier switches but none of their disadvantages. In addition, the design has extra features such as low power, scalability, etc. [11–13].

A more detailed discussion of the design and operation of the virtual routing/virtual queuing (VRQ) switch is found in Chapter 16. Figure 13.14 shows the main components of that switch. Each input port has N buffers (not queues) where incoming packets are stored after being classified. Similarly, each output port has K FIFO queues, where K is determined by the number of service classes or sessions that must be supported. The switch fabric (SF) is an array of backplane buses. This gives the best throughput compared to any other previously proposed SF architecture including crossbar switches.

The input buffers store incoming packets which could be variable in size. The input controller determines which output port is desired by the packet and sends a *pointer* to the destination output port. The pointer indicates to the output port the location of the packet in the input buffer, which input port it came from, and any other QoS requirements. The output controller queues that pointer—The packet

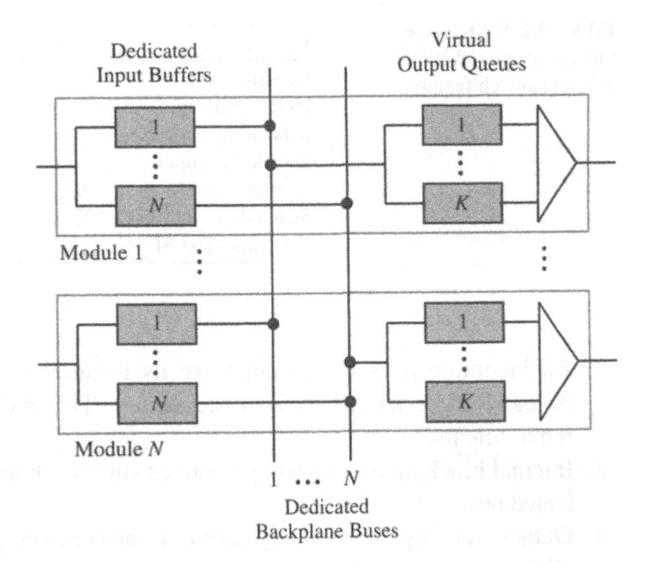

Fig. 13.14 The virtual routing/virtual queuing (VRQ) high-performance switch

itself remains in the input buffer. The buffer storage requirements for the output queues are modest since they store pointer information, which is small in size compared to the size of the packets stored in the input buffers.

When a pointer is selected from an output queue, the location of the corresponding packet is determined and the packet is selected to access the SF. We call this mode of operation *output-driven routing*, which never leads to SF contention. The classic or usual way of accessing the SF is called *input-driven routing*, which is guaranteed to lead to contention as we have seen in each switch design we have studied so far.

Let us see how the VRQ switch is able to overcome all the limitations of earlier designs:

1. Traffic management, scheduling, and congestion control are all distributed among the input and output ports. This allows more time for the algorithms to complete their operations and for the designer to implement more sophisticated algorithms.
2. The HOL problem is completely eliminated because the VRQ switch is output driven and not input driven.
3. The input buffers operate at the line rate and each output queue needs to process at most N pointers which is much more simpler than processing N packets.
4. Packets are stored at the inputs in regular memory, not FIFO memory, which is more simpler to implement.
5. There is great freedom in configuring the output queues. The queues could be constructed based on a per-connection basis, per-input basis, or per-service class basis.
6. Data broadcast is very simple to implement and no extra copies of a packet need to be stored.

Table 13.2 Switch types capable of supporting the different switch features

Feature	Input	Output	Shared	VRQ
QoS support		X	X	X
HOL elimination			X	X
Scheduling support		X	X	X
Broadcast support			X	X
Memory speed	X			X
Scalability	X			X
Contentionless SF			X	X

7. An incoming packet does not leave its location in the input buffer until it is ready to be moved through the switch. This reduces power and storage requirements.
8. Internal blocking is completely removed since each input port has its own dedicated bus.
9. Output blocking is completely removed since each output port is able to process all the pointers that arrive to it.
10. The backplane buses operate at the line rate in a bit-serial fashion with no need whatsoever for internal speedup or use of parallel data lines.
11. The switch fabric is contentionless since it is based on a matrix of *dedicated buses* that are *output driven*.

Table 13.2 summarizes the desirable features to be supported by a switch and switch type that can support these features. From the table, we see that both the shared buffer switch and the VRQ switch can easily implement most of the functionalities of a high-speed switch. The performance results in Chapter 15 show that indeed both the shared buffer switch and the VRQ switch perform the best compared to all the other switch types.

Problems

Networking

13.1 What are the main components of a computer network?
13.2 Give examples of communication links used in computer networks and in other types of networks (e.g., cable TV, radio, telephone, etc.)
13.3 Give examples of network software used at the different layers of the network.
13.4 What is a core switch and an edge switch?
13.5 What are the main performance parameters of a link in a computer network?
13.6 Why should users in a network share the links even though this reduces the quality of service offered?

Media Access Techniques

13.7 What are the main multiplexing techniques that allow users to share links?
13.8 Explain the two main types of TDMA and state the advantages and disadvantages of each.
13.9 What types of random assignment MAC protocols could be used to guarantee fair access to the medium in a reasonable time delay?
13.10 Develop a model for the access probability of a user in a TDMA system supporting N users where all users have equal priority and each time frame has m available slots. What is the throughput of each user and the average time delay?
13.11 Assume a TDMA system with a fixed priority protocol in which each user has a priority equal to its index. Derive an expression for the probability of accessing the medium for any user. Based on that, find the average delay that each user experiences and the average throughput for each user.
13.12 Explain the two main types of SDMA and state the advantages and disadvantages of each.
13.13 Explain how cellular phone technology uses both SDMA and FDMA or SDMA and CDMA.
13.14 Explain how fixed assignment and random assignment could be used in FDMA.
13.15 Explain CDMA and explain how fixed assignment and random assignment could be used.

Circuit and Packet Switching

13.16 Explain the two main types of packet switching.
13.17 What are the advantages and disadvantages of circuit switching?
13.18 What are the important characteristics of circuit switching?
13.19 Explain the steps necessary to establish a path in circuit switching.
13.20 What are the advantages and disadvantages of packet switching?
13.21 Derive expressions for channel usage efficiency for the case of packet switching using datagram versus virtual circuit switching. Assume each packet for datagram is composed of n_1 header/trailer bytes and n_2 payload bytes. For virtual circuit switching, assume n_1 header/trailer bytes, n_2 payload bytes, and n_3 bytes required for call establishment and tear down.
13.22 Explain what is meant by virtual circuit switching.
13.23 What are the advantages and disadvantages of virtual circuit switching?

Packet Switching Hardware

13.24 Explain the functions performed by a hub.
13.25 What is a passive hub?
13.26 What is an intelligent hub?

13.27 What is a switching hub?
13.28 Explain the functions performed by a switch.
13.29 Explain the functions performed by a bridge.
13.30 Explain the functions performed by a router.
13.31 Explain the functions performed by a gateway.

Switch Components

13.32 What are the two main architectural components in a switch?
13.33 Explain the functions performed by the datapath section of a switch.
13.34 Explain the functions performed by the control section of a switch.
13.35 Explain the main components of a switch and the functions performed by each component.
13.36 One of the necessary components of a switch is the FIFO. How can this be implemented using an ordinary one-ported random-access memory?
13.37 One of the necessary components of a switch is a CAM (content-addressable memory). Discuss how a CAM could be implemented using specialized hardware or using a random-access memory or groups of random-access memories.

Switch Functions

13.38 What are the main tasks performed by a switch?

Input Queuing Switch

13.39 Explain the basic operation of input queuing switch.
13.40 What is the head of line (HOL) problem in input queuing switches?
13.41 In input queuing switch, it was mentioned that HOL problem leads to reduced throughput and increased packet loss probability. Suppose a windowing scheme is used to examine w packets at the head of the queue instead of only one. How can this scheme preserve the packet sequence?
13.42 What are the advantages of input queuing switches?
13.43 What are the disadvantages of input queuing switches?
13.44 What are the potential sources of packet loss in input queuing switches?
13.45 Explain how a per-connection scheduling scheme can be implemented in an input queuing switch.

Output Queuing Switch

13.46 Explain the basic operation of output queuing switch.
13.47 What are the advantages of output queuing switches?
13.48 What are the disadvantages of output queuing switches?
13.49 What are the potential sources of packet loss in output queuing switches?

13.50 Explain how a per-connection scheduling scheme can be implemented in an output queuing switch.

Shared Buffer Switch

13.51 Explain the basic operation of shared buffer switch.
13.52 What are the advantages of shared buffer switches?
13.53 What are the disadvantages of shared buffer switches?
13.54 What are the potential sources of packet loss in shared buffer switches?
13.55 In shared buffer switches, it was mentioned that the shared buffer must perform many read and write operations per time step. Discuss how this can be implemented using a bank of one-ported memories that is accessible by all inputs and outputs of the switch.
13.56 Explain how a per-connection scheduling scheme can be implemented in a shared buffer switch.
13.57 Assume a protocol is devised to evenly distribute packets between two buffers (A and B) as follows:

1. Start with n packets (assume n even).
2. Pick a packet at random and move it to the other buffer.
3. The state of the system is the number of packets in buffer A.

Study the long-term performance of this protocol.

Multiple Input Queuing Switches

13.58 Explain how a per-connection scheduling scheme can be implemented in a multiple input queuing switch.
13.59 Explain the basic operation of a virtual output queuing (VOQ) switch.
13.60 What are the advantages of virtual output queuing switch?
13.61 What are the disadvantages of virtual output queuing switch?

Multiple Output Queuing Switches

13.62 Explain the basic operation of a multiple output queuing switch.
13.63 What are the advantages of multiple output queuing switch?
13.64 What are the disadvantages of multiple output queuing switch?
13.65 Explain how a per-connection scheduling scheme can be implemented in an multiple output queuing switch.

Virtual Routing/Virtual Queuing (VRQ) Switch

13.66 Explain the basic operation of virtual routing/virtual queuing (VRQ) switch.
13.67 Discuss the advantages and disadvantages of the VRQ switch.
13.68 What is meant by virtual routing in the VRQ switch?

13.69 How is broadcast operation implemented in the VRQ switch?
13.70 What is meant by "input-driven" and "output-driven" operation in a switch?
13.71 Why does input-driven switching suffer from contention?
13.72 Why is output-driven switching contentionless?

References

1. J. Warland and R. Varaiya, *High-Performance Communication Networks*, Moragan Kaufmann, San Francisco, 2000.
2. W. Stallings, *Data and Computer Communications*, Fourth edition, Prentice Hall, New Jersey, 1994.
3. W. Stallings, *Data and Computer Communications*, Fifth edition, Prentice Hall, New Jersey, 1998.
4. A.V. Aho, J.E. Hopcroft, and J.D. Ullman, *Data Structures and Algorithms*, Addison-Wesley, Reading, Massachusetts, 1983.
5. R. Sedgewick, *Algorithms*, Addison-Wesley, Reading, Massachusetts, 1984.
6. S. Baase, *Computer Algorithms*, Addison-Wesley, Reading, Massachusetts, 1983.
7. A.S. Tanenbaum, *Computer Networks*, Prentice Hall PTR, Upper Saddle River, New Jersey, 1996.
8. N. McKeown and T.E. Anderson, "A quantitative comparison of iterative scheduling algorithms for input-queued switches", *Computer Networks and ISDN Systems*, vol. 30, pp. 2309–2326, 1998.
9. M. Karol, M. Hluchyj, and S. Morgan, "Input versus output queuing on a space division switch", *IEEE Transactions on Communications*, vol. 35, pp. 1347–1356, 1987.
10. M.G. Hluchyj and M.J. Karol, "Queuing in high-performance packet switching", *IEEE J. Selected Areas in Communications*, vol. 6, pp. 1587–1597, 1988.
11. F. Elguibaly, A. Sabaa, and D. Shpak, "A new shift-register based ATM switch", *The First Annual Conference on Emerging Technologies and Applications in Communications (ETACOM)*, Portland, Oregon, pp. 24–27, May 7–10, 1996.
12. F. Elguibaly and S. Agarwal, "Design and performance analysis of shift register-based ATM switch", *IEEE Pacific Rim Conference on Communications, Computers and Signal Processing*, Victoria, British Columbia, pp. 70–73, August 20–22, 1997.
13. A. Sabaa, F. Elguibaly, and D. Shpak, "Design and modeling of a nonblocking input-buffer ATM switch", *Canadian Journal of Electrical & Computer Engineering*, vol. 22, no. 3, pp. 87–93, 1997.

Chapter 14
Interconnection Networks

14.1 Introduction

Interconnection networks form the switching fabric part of any switch. The capabilities and characteristics of the interconnection network have a direct influence on the resulting performance of the switch. We review in this chapter the different types of switching networks and derive analytic expression for their performance.

14.2 Network Design Parameters

An interconnection network (IN) is also known as the switching fabric (SF). We shall use the two terms interchangeably. The main purpose of the SF is to provide full connectivity between the inputs and the outputs of the switch with reasonable hardware and delay. In that sense, the input and output ports share the SF as a common communication resource and, as such, conflicts may arise that have to be resolved using some contention-resolving technique or arbitration protocol. There are several arbitration protocols and each impacts the performance of the switch. Reference [1] reviews and models five protocols and explains how they can be implemented in hardware.

Because of logistics and office building code limitations, the maximum practical number of ports in a central office switch is 16–128. Backbone or core switches and routers deal with even smaller number of very high-speed links. Thus the switch size in core routers could be between 8 and 32, according to references [2, 3].

14.2.1 Network Performance

Different interconnection networks have varying performance features due to the design decisions employed. *Throughput* depends on the type of network used and the number of input and output ports it can support. Packet *loss probability* also depends on the type of network since this determines the number of paths that can be simultaneously established to reduce the amount of buffer occupancy.

F. Gebali, *Analysis of Computer and Communication Networks*,
DOI: 10.1007/978-0-387-74437-7_14, © Springer Science+Business Media, LLC 2008

Fault tolerance is the ability of the network to operate even in the presence of faulty links or internal components. Gradual degradation of performance is desirable so that the operators have a chance of isolating and replacing the faulty component. *Hot insertion* is the ability to replace faulty components without having to halt, or rest, the entire system.

14.2.2 Network Hardware

Synchronization and clock speed determine whether data will move within the SF in a synchronous or an asynchronous manner. As the switching fabric gets bigger and the operating speed increases, it becomes very difficult to synchronize the whole system to a single clock with little skew. Further, the clock lines supporting the fast clock transitions will consume power and will result in system noise. On the other hand, asynchronous data transmission will require more control lines and more sophisticated data transfer protocols.

Complexity includes several aspects such as: (a) how many bits of a packet will move from one location to another within the SF in once clock cycle; (b) the number and type of packages or boards required to implement the SF; (c) the number of I/O pins per chip which impact the overall price and reliability of the system; (d) the routing algorithm required to establish a path between an input and an output. This routing algorithm might be simple and distributed among all the components of the interconnection network or it could require complex operations and need to be centralized thereby slowing down the whole system.

Power consumption is intimately dependent on the level of chip integration used, amount of inter-chip communication, and the operating speed. Total power consumed will determine the size, price, cooling, and weight of the required power supply.

Modularity implies the SF is built from a small number of modules.

Hierarchy and *partitionability* are the ability to divide the SF into subsystems that might simplify the task of building and operating the system.

Communication between the modules and the chips will have an impact on the speed and power consumption of the SF and also on the size and reliability of the switch.

14.2.3 Scalability Issues

Number of input and output ports might have an upper limit due to the way the SF was built. Increasing that number might require complete rewiring and use of other basic modules. Alternatively, increasing the number of I/O ports in some SFs might degrade the performance or might not be feasible from a hardware point of view due, for example, to pin limitations of the chips.

Operating speed of the SF might have an upper limit dictated by the way the SF operates. For example, if communication is required between all the input ports to decide which one gains access to the shared memory, say, then the larger the network, the slower this operation will become which will in turn reduce the overall throughput.

14.3 Classification of Networks

A detailed classifications for interconnection networks can be found in [4]. A broad classification of networks could be based on the protocol used to provide access to the users to the shared medium—this is also called media access control (MAC). Section 13.3 on page 479 provided a detailed discussion of the possible media access control schemes. However, as in [5] and [6], we classify a switch fabric based on time division multiplexing or space-division multiplexing which we discuss in the following sections.

14.4 Time Division Multiple Access (TDMA)

In time division multiple access (TDMA), the communication medium gives sole access to the communication medium for a short amount of time. Time division requires a communication medium be shared among all the users (packets in our case). The shared medium could be a bus that joins all the inputs and outputs or it could be a wireless channel, as in the case of satellite communications and cellular telephones. Simultaneous requests lead to contention or corruption of data and a contention resolution protocol is used to resolve these conflicts. A shared medium has a maximum bandwidth that is determined by the channel used. For the case of a shared bus, the maximum bandwidth is set by the electrical drive capability of the bus drivers and the physical length of the bus. For the case of a wireless channel, the channel bandwidth is determined by the frequency and power restrictions of wireless and any channel impairment such as noise and fading. Many routers and multiport bridges are shared media switches with bandwidth limitations of 800 Mbits/s to 3.2 Gbits/s [2].

We classify TDMA into two types: static assignment and random assignment, which we discuss next.

14.4.1 Static-Assignment TDMA

Constant bit rate traffic is efficiently served by the static-assignment protocol. In that scheme, time is broken into frames and each frame is divided into slots. Each user is assigned a certain slot in each frame. Figure 14.1(a) shows such a scheme where five users are assumed to share the medium. Each user is guaranteed access to the medium even when there is no data to send. Static-assignment techniques do not work well for bursty traffic or mixed traffic with long idle times since either the

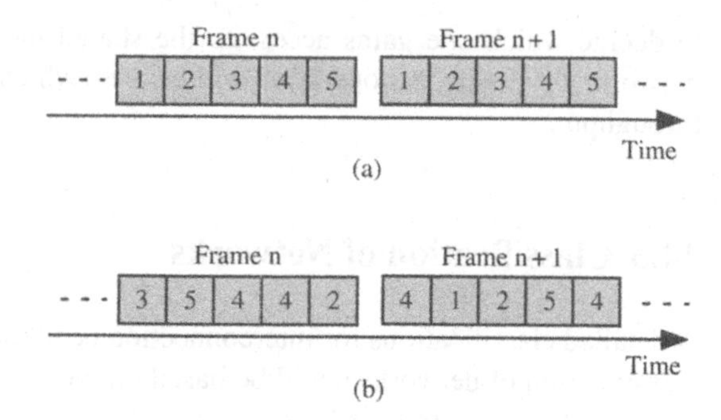

Fig. 14.1 Time division switching: (**a**) Static assignment. (**b**) Random assignment

medium will be grossly underutilized or the access time will be large. Examples of static-assignment networks include the token ring and the telephone network trunk lines.

14.4.2 Random Assignment TDMA

This is also known as *statistical multiplexing* in packet switching. Bursty traffic is efficiently served by the random-assignment protocol. Figure 14.1(b) shows such a scheme where again five users are assumed to share the medium. At a given time step, all users with data to send compete to access that slot and only one winner gains access and the others wait for the next slot. As a result, idle sources do not interfere with active users but access time now is not guaranteed and could be long at times of heavy traffic. Random assignment techniques do not work well for constant bit rate traffic since the access time will be needlessly large. An important aspect of random assignment schemes is the arbitration protocol used to resolve contention among the competing users and assign the channel to one user. Reference [1] dealt with five types of common arbitration protocols and their distributed hardware implementation.

From the above, we see that time division multiplexing is characterized by low bandwidth assigned for each user. On the other hand, it requires little hardware resources since the interconnection network comprises one shared medium which requires a small wiring area, or a narrow frequency band. As such, it might not be well suited to high-speed communications, but can still be used in the switch for communicating the control information between the different modules. Examples of random assignment networks include Ethernet LANs and common computer buses.

14.5 Space Division Switching

Space division switching is the method of choice to implement high-performance switches since they provide high capacity and low access time delay. Space division switching can be broadly classified as crossbar networks and multistage interconnection networks (MIN). We discuss each in the following sections.

14.6 Crossbar Network

Crossbar switches have not been well represented in the literature, with the exception of [7], perhaps due to the original article by Clos [8], in which he claimed that a crossbar network is very expensive to implement. With the current state of VLSI technology, it is possible to place several switching elements and their state registers on a single chip with the only limitation being the number of I/O pins and pad size [2]. Happily, several companies already produce high-speed network switches built around a crossbar network [9–11].

An $N \times N$ crossbar network consists of N inputs and N outputs. It can connect any input to any free output without blocking. Figure 14.2 shows a 6×6 crossbar network. The network consists of an array of crosspoints (CP) connected in a grid fashion. CP(i, j) lies at the intersection of row i with column j. Each CP operates in one of two configurations as shown in Fig. 14.3. The X-configuration is the default configuration where the SE allows simultaneous data flow in the vertical and horizontal directions without interference. If CP(3,5) was in the X-configuration, then data flowing horizontally originates at input 3 and is sent to all the intersection points at this row. Data flowing vertically in column 5 could have originated from any input above or below row 3.

In the T-configuration, the CP allows data flow in the horizontal direction and interrupts data flow in the vertical direction. Data flowing vertically at its output is a copy of the horizontal data. For example, if CP(3,5) was in the T-configuration, then data flowing horizontally originates at input 3. Data flowing downward at the output

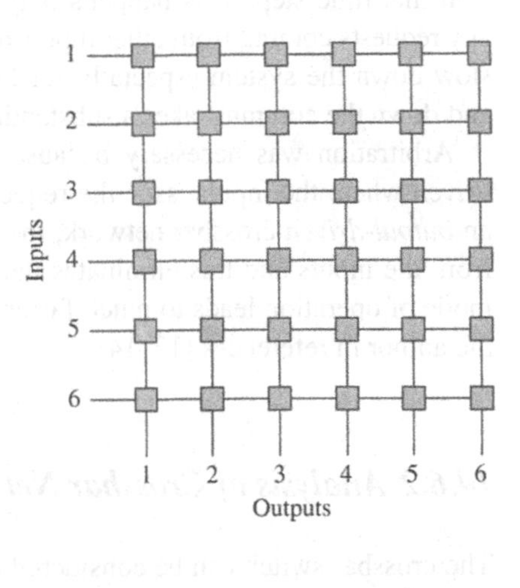

Fig. 14.2 A 6×6 crossbar interconnection network

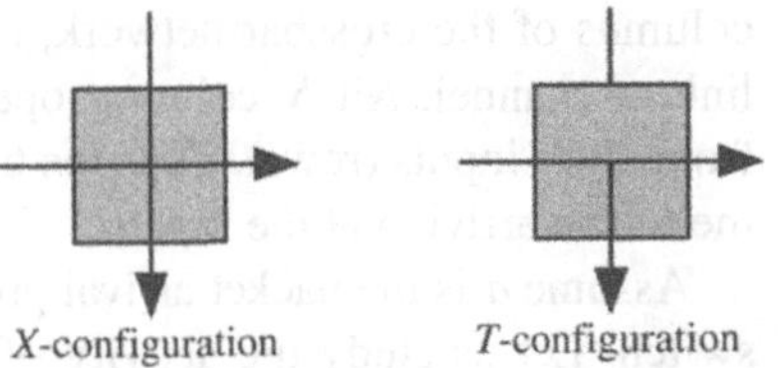

Fig. 14.3 States of the crosspoint (CP) in a crossbar network

is a copy of the horizontal data coming from row 3. This way, output at column 5 sees a copy of the data that was moving in row 3.

A crossbar network supports high capacity due to the N simultaneous connections it can provide. This comes at the expense of the number of CPs that grows as N^2. This is one reason why a crossbar network is used mainly for demanding applications that requires a relatively small values of N (about 10). However, advances in VLSI technology and electro-optics make crossbar switches a viable switching alternative.

Data multicast in a crossbar network can be easily accomplished. Suppose that input 3 requests to multicast its data to outputs 1, 3, and 5. Input 3 would then request to configure CP(3,1), (3,3), and (3, 5) into the T-configuration and all other CPs in row 3 would remain in the default X-configuration.

14.6.1 Crossbar Network Contention and Arbitration

Suppose that two or more inputs request access to the same output. In that case, contention arises and some arbitration mechanism has to be provided to settle this dispute. In fact, we have to provide N arbiters such that each one is associated with a column in the crossbar network. For example, when input 1 requests to communicate with output 3, it requests to configure CP(1,3) into the T-configuration and must wait until the arbiter in column 3 issues a grant to that input. At the same time, the arbiter in column 3 must inform all other inputs that they cannot access column 3 in that time step. This happens only after the arbiter checks to see if there are any requests coming from other inputs demanding access to output 3. These arbiters slow down the system especially for large networks where signal propagation up and down the columns takes a substantial amount of time.

Arbitration was necessary because access to the crossbar network was *input driven* where the inputs issue the requests to configure the crosspoint switches. In an *output-driven* crossbar network, the outputs initiate issuing of requests for data from the inputs and this eliminates the need for the arbiters. Needless to say, this mode of operation leads to much faster switch operation and was first proposed by the author in references [12–14].

14.6.2 Analysis of Crossbar Network

The crossbar switch can be considered as a collection of N shared media, viz., the columns of the crossbar network, since each column is associated with an output link or channel. All N columns operate in parallel and each output accepts traffic from all N inputs (rows). Thus the traffic arriving at each output port is a fraction of the traffic arriving at the inputs.

Assume a is the packet arrival probability at an input port of an $N \times N$ crossbar switch. Let us study the activity of a certain output port. We call this the *tagged output port*. The probability that a packet arrives at any input port such that it is destined to the tagged output port is given by

$$a' = \frac{a}{N} \tag{14.1}$$

This is because an arriving packet has an equal probability of requesting one of N output ports. Essentially, each output port deals with N users but the probability of packet arrival is reduced to a/N. The probability that i requests arrive at a time slot addressed to the tagged output port is given by

$$p(i) = \binom{N}{i} \left(\frac{a}{N}\right)^i \left(1 - \frac{a}{N}\right)^{N-i} \tag{14.2}$$

The input traffic $N_a(\text{in})$ that is destined for the tagged output port per time slot is given by

$$N_a(\text{in}) = E\left[i\ p\left(i\right)\right] = \sum_{i=0}^{N} ip\left(i\right) = a \tag{14.3}$$

The throughput of the tagged output port is equal to the output traffic of the tagged output port and is given by the expression

$$\text{Th} = \sum_{i=1}^{N} p(i) = 1 - p(0) \tag{14.4}$$

After substituting the value for a' we could write

$$\begin{aligned} \text{Th} &= N_a(\text{out}) \\ &= 1 - \left(1 - \frac{a}{N}\right)^N \qquad \text{packets/time step} \end{aligned} \tag{14.5}$$

The second term in the RHS is simply the probability that no packets are destined to the tagged output port. For light loading, we get Th $\approx a$ which indicates that most of the arriving packets are transmitted.

In the limit for large networks ($N \to \infty$), we get

$$\begin{aligned} \text{Th} &= \lim_{N \to \infty} \left[1 - \left(1 - \frac{a}{N}\right)^N\right] \\ &= 1 - e^{-a} \qquad \text{packets/time step} \end{aligned} \tag{14.6}$$

For very large crossbar network at very light loads $a \ll 1$, we get Th $\to a$ which indicates that almost all the arriving packets make it through the switch due to the light loading.

What is really exciting is the performance of very large crossbar networks under extremely heavy loading. The maximum throughput Th(max) occurs at very heavy traffic ($a = 1$):

$$\text{Th(max)} = 1 - e^{-1} = 63.21\% \qquad \text{packets/time step} \tag{14.7}$$

Thus the crossbar network is characterized by very high throughput even at the highest load and for large networks.

From (14.3) and (14.5), the packet acceptance probability is given by

$$p_{\mathrm{a}} = \frac{N_{\mathrm{a(out)}}}{N_{\mathrm{a(in)}}} = \frac{1}{a}\left[1 - \left(1 - \frac{a}{N}\right)^N\right] \tag{14.8}$$

For light traffic ($a \ll 1$), we get $p_{\mathrm{a}} \approx 1$ which indicates very high efficiency. In the limit for large networks, we get

$$\begin{aligned} p_{\mathrm{a}} &= \lim_{N\to\infty} \frac{1}{a}\left[1 - \left(1 - \frac{a}{N}\right)^N\right] \\ &= \frac{1 - e^{-a}}{a} \end{aligned} \tag{14.9}$$

The minimum acceptance probability p_{a}(min) occurs at very heavy traffic ($a = 1$):

$$p_{\mathrm{a}}(\text{min}) = 1 - e^{-1} = 63.21\% \tag{14.10}$$

Thus the crossbar network is characterized by very high acceptance probability even at the highest load and for large networks.

The delay of the network is defined as the average number of attempts to access the desired output. The probability that the packet succeeds after k tries is given by the geometric distribution

$$p(k) = p_{\mathrm{a}}\,(1 - p_{\mathrm{a}})^k \tag{14.11}$$

The average delay time is

$$\begin{aligned} n_{\mathrm{a}} &= \sum_{k=0}^{\infty} k\, p(k) \\ &= \sum_{k=0}^{\infty} k\, p_{\mathrm{a}}\,(1 - p_{\mathrm{a}})^k \\ &= \frac{1 - p_{\mathrm{a}}}{p_{\mathrm{a}}} \qquad \text{time steps} \end{aligned} \tag{14.12}$$

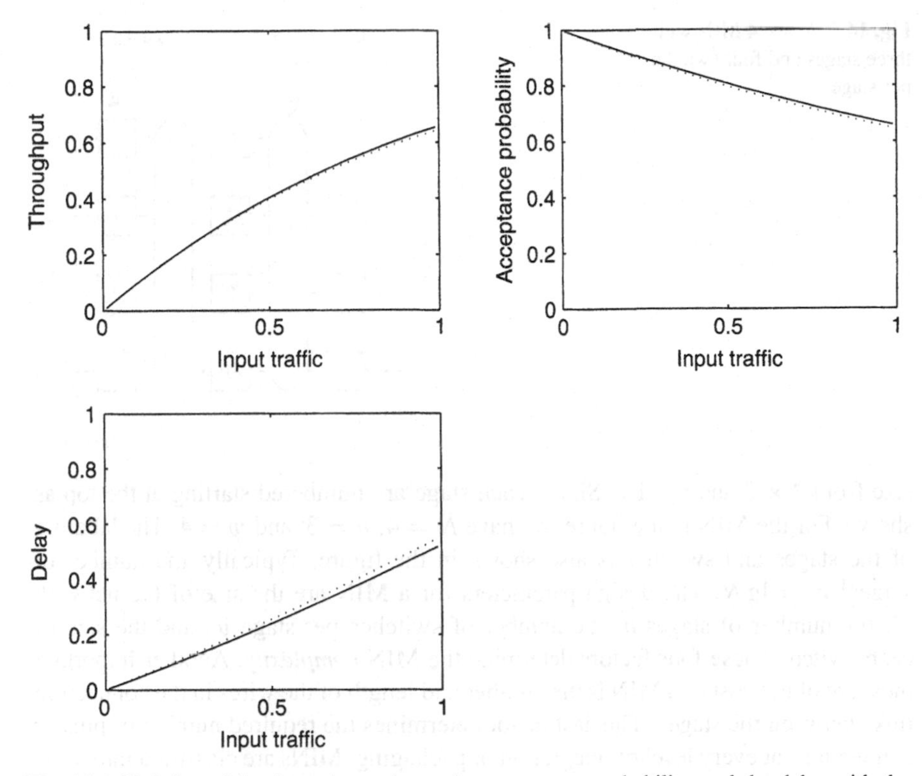

Fig. 14.4 Variation of the throughput, the packet acceptance probability, and the delay with the input traffic for the crossbar network when $N = 8$ (*solid line*) and $N = 16$ (*dotted line*)

For light traffic ($a \approx 0$), we get $n_a \approx 0$ which indicates that arriving packets get through on their first attempt. For heavy traffic, $a = 1$ and the maximum number of attempts becomes 0.582 on the average.

Example 1 Plot the throughput (Th), the access probability (p_a), and the average delay (n_a) versus the input traffic for the crossbar network when the size of the network is $N = 8$ and $N = 16$.

We evaluate the expressions for throughput, packet acceptance probability, and delay when N has the values 8 and 16 and a is varied. Figure 14.4 shows the variation of throughput, the packet acceptance probability, and the delay with the input traffic when $N = 8$ (solid line) and $N = 16$ (dotted line). We see that the minimum throughput is around 63.66% and occurs at $a = 1$. What is striking is the good overall performance of the crossbar network and the little dependence on N. ■

14.7 Multistage Interconnection Networks

As Fig. 14.5 shows, an $N \times N$ multistage interconnection network consists of n stages with stage i connected to stages $i - 1$ and $i + 1$ through some pattern of connection lines. Each stage has w crossbar switching elements (SE) that vary in

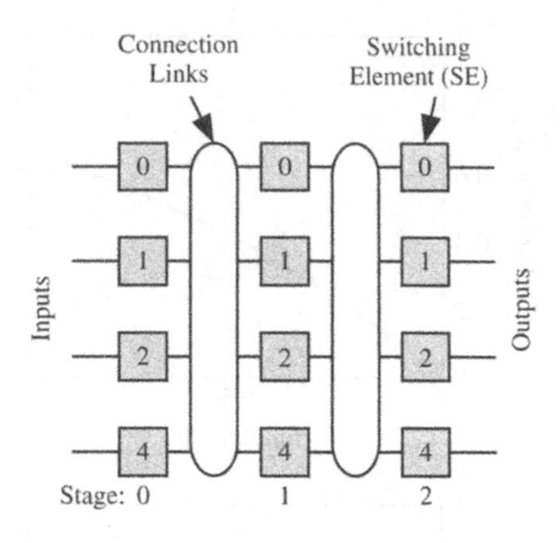

Fig. 14.5 A 4 × 4 MIN with three stages and four switches per stage

size from 2 × 2 and up. The SEs in each stage are numbered starting at the top as shown. For the MIN in the figure, we have $N = 4$, $n = 3$, and $w = 4$. The labeling of the stages and switches is also shown in the figure. Typically, the number of stages[1] $n = \lg N$. The design parameters for a MIN are the size of the network N, the number of stages n, the number of switches per stage w, and the size of each switch. These four factors determine the MIN *complexity*. Another important measure of the cost of a MIN is the number and length of the wires in the connection links between the stages. This last factor determines the required number of pins or connections at every level of integration or packaging. MINs are classified into three classes according to the ability of the network to establish a path between an input and an output [4, 15]:

Blocking MIN: A connection between a free input/output pair cannot be established due to conflicts with already-existing connections. This blocking is called internal blocking since the internal wiring of the network prevented path establishment. This type of MIN usually has a single path connecting any input/output pair which leads to hardware economy at the cost of reducing the fault tolerance and the throughput due to internal blocking.

Nonblocking MIN: A connection between a free input/output pair can be established independent of already-existing connections. This type of MIN has many alternative routes connecting any input/output pair which makes them expensive to implement although they are not as expensive as a crossbar network.

Rearrangeable MIN: Any input can be connected to any free output port by rearranging existing connections. Rearrangeable networks should not be confused with nonblocking networks that do not rearrange existing paths to establish new connections. Rearrangeable networks are less expensive than nonblocking or crossbar networks.

---

[1] We use Knuth's [11] notation "lg N" to denote the base-2 logarithm $\log_2 N$.

14.7.1 Definitions

We start our study of MINs by defining some functions that are commonly used in the field of multistage interconnection networks:

Shuffle function: Assume the label for a row in the MIN is represented as binary number A. The perfect shuffle function S performs a circular left shift on the bits of A:

$$\begin{aligned} A &= a_{n-1} \quad a_{n-2} \quad \cdots \quad a_1 \quad a_0 \\ S(A) &= a_{n-2} \quad a_{n-3} \quad \cdots \quad a_0 \quad a_{n-1} \end{aligned}$$

Inverse shuffle function: The inverse perfect shuffle function S^{-1} performs a circular right shift on the bits of A:

$$\begin{aligned} A &= a_{n-1} \quad a_{n-2} \quad \cdots \quad a_1 \quad a_0 \\ S^{-1}(A) &= a_0 \quad a_{n-1} \quad \cdots \quad a_2 \quad a_1 \end{aligned}$$

Exchange function: The exchange function $E^{i,j}$ exchanges the bits at positions i and j leaving all other bits intact:

$$\begin{aligned} A &= a_{n-1} \quad \cdots \quad a_i \quad \cdots \quad a_j \quad \cdots \quad a_0 \\ E^{i,j}(A) &= a_{n-1} \quad \cdots \quad a_j \quad \cdots \quad a_i \quad \cdots \quad a_0 \end{aligned}$$

Butterfly function: The butterfly function B^i exchanges the least significant bit (a_0) and the ith bit (a_i) of the binary number leaving all other bits intact:

$$\begin{aligned} A &= a_{n-1} \quad \cdots \quad a_{i+1} \quad a_i \quad a_{i-1} \quad \cdots \quad a_0 \\ B^i(A) &= a_{n-1} \quad \cdots \quad a_{i+1} \quad a_0 \quad a_{i-1} \quad \cdots \quad a_i \end{aligned}$$

Cube function: The cube function C^i complements the ith bit (a_i) of the binary number leaving all other bits intact:

$$\begin{aligned} A &= a_{n-1} \quad \cdots \quad a_{i+1} \quad a_i \quad a_{i-1} \quad \cdots \quad a_0 \\ C^i(A) &= a_{n-1} \quad \cdots \quad a_{i+1} \quad \overline{a_i} \quad a_{i-1} \quad \cdots \quad a_0 \end{aligned}$$

Plus–Minus 2^i (PM2I) function: The plus–minus 2^i (PM2I) function adds $\pm 2^i$ to the row address of a packet. The result is reduced using the *modulo* function as follows:

$$\begin{aligned} \text{PM2}_{+i}(A) &= \left(A + 2^i\right) \bmod N \\ \text{PM2}_{-i}(A) &= \left(A - 2^i\right) \bmod N \end{aligned}$$

where i varies between 0 and $n = \lg N$. Notice that the C^i function could be implemented using the PM2I function:

$$C^i(A) = A + \overline{a_i}\,\mathrm{PM2}_{+i}(A) - a_i\,\mathrm{PM2}_{-i}(A)$$

Below we show examples of several MINs that were proposed for applications in communications.

14.8 Generalized-Cube Network (GCN)

Figure 14.6 shows an 8×8 generalized-cube network [16, 17]. The interconnection pattern of the network is based on the cube and shuffle functions. For an $N \times N$ network, the number of stages is n where $n = \lg N$ and the number of SEs in each stage is $N/2$. Each SE is a 2×2 crossbar switch and the number of links between stages is N. This network is equivalent to many other networks that were proposed such as the omega [18], banyan [19], delta, and baseline. The generalized-cube network is a blocking network and provides only one path from any input to any output. As such, it possesses no tolerance for faults.

Switching element SE(i, j) at stage i and position j is connected to SE($i + 1$, k) such that k is given by

$$k = \begin{cases} j & \text{straight connection} \\ C^i(j) & \text{cube connection} \end{cases} \tag{14.13}$$

where $0 \le i < n$ and $0 \le j < N/2$. Note that the SE row label j requires only $n-1$ bits for its representation. Thus when $N = 8$, the switch row label is composed only of two bits.

As an example, SE(1,2) at stage 1 is connected to switches SE(2,2), the straight connection, and switch SE(2,0), the $C^1(2)$ connection.

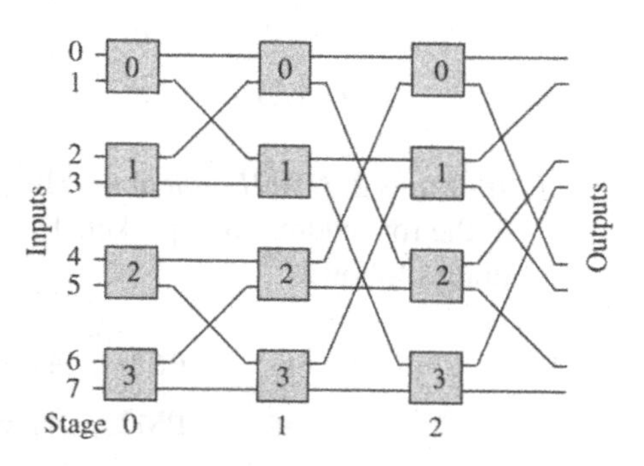

Fig. 14.6 Generalized-cube network for $N = 8$

At the last stage, output stage, switch SE(2,j) is connected to row k such that k is given by

$$k = \begin{cases} j & \text{straight connection} \\ C^2(j) & \text{cube connection} \end{cases} \tag{14.14}$$

where $0 \leq j < N$. Note that the output row label j requires only $n - 1$ for its representation. Thus when $N = 8$, the switch row label is composed of three bits.

As an example, SE(2,3) is connected to output rows 3 and 7 since we have $j = 3$ which has the binary equivalent 011 and applying the straight and cube connection $C^2(3)$ gives output rows 3 and 7, respectively.

The GCN provides one unique path from any input to any output based on the input row address and the destination address. The packet path is established by properly configuring the connections within each SE that the packet goes through. Figure 14.7 shows the two types of connections that could be *simultaneously* established for the two inputs of an SE at stage i:

Straight connection: The packet enters and exits at the same row location.
Cube connection: The packet enters at row location R and exits at row location $C^i(R)$.

Depending on the SE design, one SE input could simultaneously establish the straight and cube connections for itself, while the other input will not be able to access the two outputs. This situation is useful to broadcast packets to two or more outputs.

14.8.1 Routing Algorithm for GCN Network

The routing algorithm in GCN is distributed among the SEs and is based on performing a bitwise XOR operation on the source row address (S) and destination row address D. S indicates the row location of the input port and D indicates the row

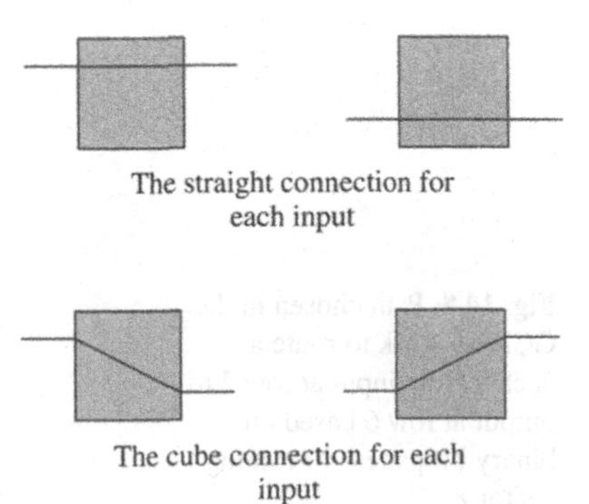

The straight connection for each input
The cube connection for each input

Fig. 14.7 The straight and cube connections for each input of an SE in a GCN network

location of the desired output port. The routing vector r is used to determine the path of the packet through the network. r is obtained as follows:

$$r = \text{XOR}(S, D) \tag{14.15}$$

where the n-bit *routing vector* r carries the information about the desired path. Bit i of that vector specifies the type of connection of the SE at stage i:

$$r_i = \begin{cases} 0 & \text{straight connection} \\ 1 & \text{cube connection} \end{cases} \tag{14.16}$$

For an 8×8 GCN network $n = \lg 8 = 3$ and the routing vector r will have three bits. If an incoming packet arrives at row 2 (binary 010) and is destined to row 6 (binary 110), then we have

$$S = \begin{bmatrix} 0\,1\,0 \end{bmatrix}^t \tag{14.17}$$

$$D = \begin{bmatrix} 1\,1\,0 \end{bmatrix}^t \tag{14.18}$$

$$r = \begin{bmatrix} 1\,0\,0 \end{bmatrix}^t \tag{14.19}$$

The path selected is explained in Table 14.1.

Figure 14.8 shows the path chosen to route a packet from input at row 2 to output at row 6 based on binary bit pattern of routing vector r.

Table 14.1 SE settings for path in GCN network from input 2 to output 6

Stage	$i = 0$	$i = 1$	$i = 2$
Bit scanned	r_0	r_1	r_2
Bit value	0	0	1
Connection type	Straight	Straight	Cube

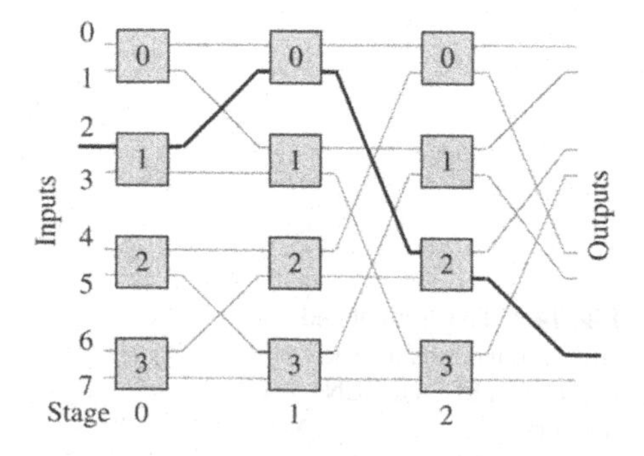

Fig. 14.8 Path chosen in the GCN network to route a packet from input at row 2 to output at row 6 based on binary bit pattern of routing vector r

14.8.2 Analysis of GCN Network

An $N \times N$ GCN network is built using 2×2 crossbar switching elements. We use the analysis for the $N \times N$ crossbar network to study each 2×2 SE, then extend the analysis to the whole network [20, 21].

Consider a 2×2 SE at stage i and define $a_i(\text{in})$ as the probability that a packet appears at one of its two inputs and $a_i(\text{out})$ as the probability that a packet appears at one of the two outputs. Both these probabilities are equal to the input and output traffic, respectively.

For any SE at the input stage ($i = 0$), we can adapt the crossbar throughput given by (14.5) when $N = 2$:

$$a_0(\text{in}) = a \tag{14.20}$$

$$a_0(\text{out}) = 1 - \left(1 - \frac{a}{2}\right)^2 \tag{14.21}$$

Now we know the throughput at the output of all SEs at stage 0.

For the SEs at the other stages, we have the recursive expression

$$a_i(\text{in}) = a_{i-1}(\text{out}) \tag{14.22}$$

$$a_i(\text{out}) = 1 - \left[1 - \frac{a_i(\text{in})}{2}\right]^2 \tag{14.23}$$

for $0 < i < n$.

According to (14.3), we can write the traffic at the input of the SE at the first stage $i = 0$ as

$$N_{\text{a}}(\text{in}) = a \tag{14.24}$$

and the output traffic at the tagged output of the network is

$$N_{\text{a}}(\text{out}) = a_{n-1}(\text{out}) \tag{14.25}$$

The throughput of the GCN network is given by

$$\text{Th} = N_{\text{a}}(\text{out}) = a_{n-1}(\text{out}) \tag{14.26}$$

A simple expression for the throughput is not possible because of the nonlinear nature of (14.23).

The packet acceptance probability of the GCN network is given by

$$p_{\text{a}} = \frac{\text{Th}}{N_{\text{a}}(\text{in})} \tag{14.27}$$

The delay of the network is defined as the average number of attempts to access the desired output. The probability that the packet succeeds after k tries is given by the geometric distribution

$$p(k) = p_{\mathrm{a}}\,(1 - p_{\mathrm{a}})^k \tag{14.28}$$

The average delay time is

$$\begin{aligned} n_{\mathrm{a}} &= \sum_{k=0}^{\infty} k\, p(k) \\ &= \sum_{k=0}^{\infty} k\, p_{\mathrm{a}}\,(1 - p_{\mathrm{a}})^k \\ &= \frac{1 - p_{\mathrm{a}}}{p_{\mathrm{a}}} \end{aligned} \tag{14.29}$$

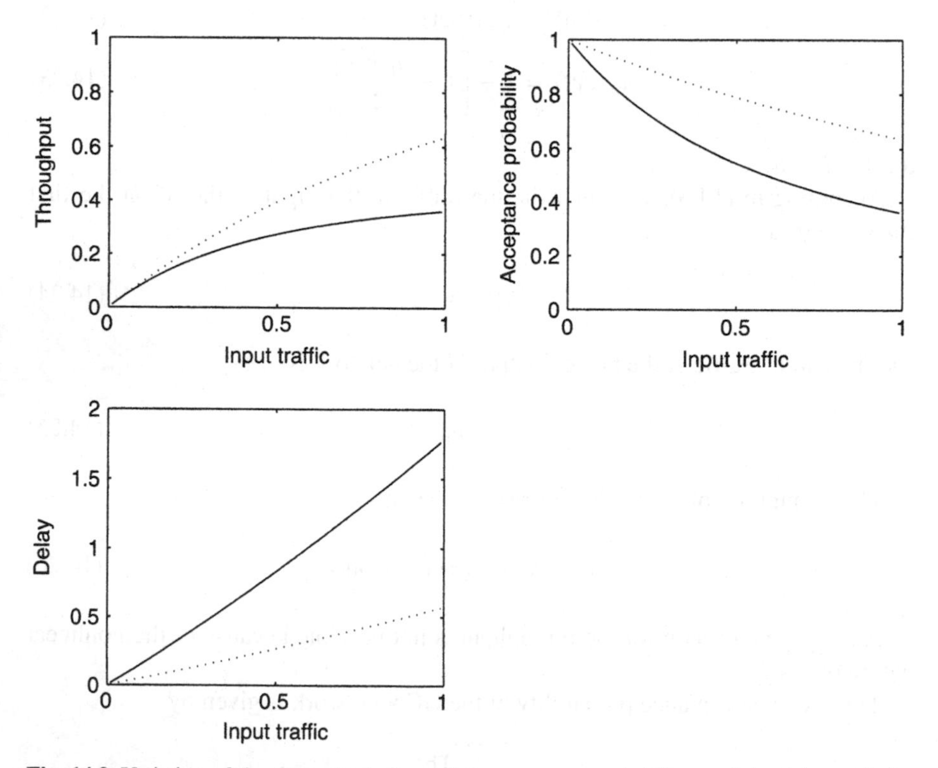

Fig. 14.9 Variation of the throughput, the packet acceptance probability, and the delay with the input traffic when $N = 64$ for the GCN network (*solid line*) and the similar-sized crossbar network (*dotted line*).

Example 2 Plot the throughput (Th), the access probability (p_a), and the average delay (n_a) for the GCN network versus the input traffic when the size of the network is $N = 64$, and compare this with the similar-sized crossbar network.

We evaluate the expressions for throughput, packet acceptance probability, and delay when $N = 64$, and a is varied for both the GCN network and the crossbar network. Figure 14.9 shows the variation of throughput, the packet acceptance probability, and the delay with the input traffic when $N = 64$ for the GCN network (solid line) and the crossbar network (dotted line). We see that the crossbar network shows superior performance for the same number of input and output ports as expected. ■

14.9 The Banyan Network

Figure 14.10 shows an 8×8 banyan network. For an $N \times N$ network, the number of stages is $n + 1$, where $n = \lg N$, and the number of SEs in each stage is N. Each SE is a 2×2 crossbar switch and the number of links between the stages is $2N$.

An $N \times N$ banyan network is built using 1-to-2 selectors in the input stage ($i = 0$), 2×2 crossbar SEs in the $n - 1$ internal stages ($0 < i < n$), and 2-to-1 concentrators in the output stage ($i = n$). However, the banyan network is a blocking network and provides only one path from any input to any output. As such, it possesses no tolerance for faults.

Switching element SE(i, j) at stage i and row position j is connected to SE($i + 1, k$) such that k is given by

$$k = \begin{cases} j & \text{straight connection} \\ C^i(j) & \text{cube connection} \end{cases} \tag{14.30}$$

where $0 \le i < n$. Thus, at stage 1, we see that SE(1, 2) is connected to switches SE(2, 2), the straight connection, and switch SE(2, 0), the C^1 (2) connection.

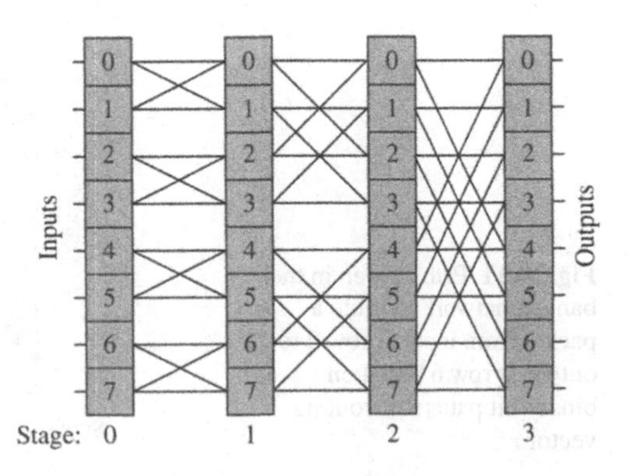

Fig. 14.10 An 8×8 banyan network

The banyan network provides one unique path from any input to any output based on the input row address and the destination address. Figure 14.7 shows the two types of connections that could be established for the two inputs of an SE at stage i:

Straight connection: The packet enters and exits at the same row location.
Cube connection: The packet enters at row location R and exits at row location $C^i(R)$.

14.9.1 Routing Algorithm for Banyan Network

The routing algorithm for the banyan network is identical to that of the GCN network.

Figure 14.11 shows the path chosen to route a packet from input at row 2 to output at row 6 based on binary bit pattern of routing vector r.

14.9.2 Analysis of Banyan Network

An $N \times N$ banyan network is built using 1-to-2 selectors in the input stage ($i = 0$), 2×2 crossbar SEs in the $n - 1$ internal stages ($0 < i < n$), and 2-to-1 concentrators in the output stage ($i = n$). Therefore, we can use the results we obtained for the crossbar network after some modifications.

Because we have 1-to-2 selectors at stage 0, the packet arrival probabilities at the input and the outputs of each selector are given by

$$a_0(\text{in}) = a \tag{14.31}$$

$$a_0(\text{out}) = \frac{a}{2} \tag{14.32}$$

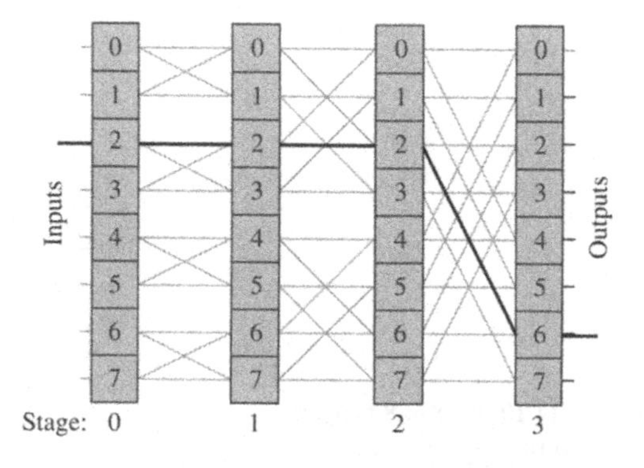

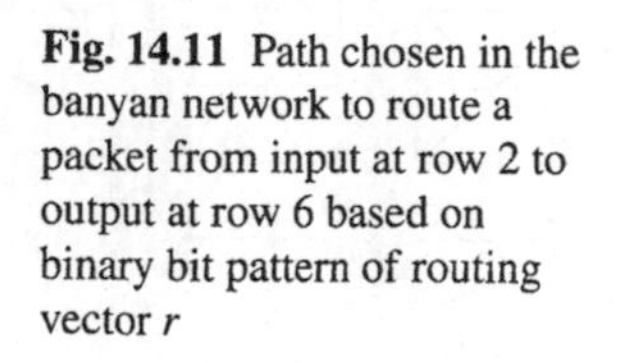
Fig. 14.11 Path chosen in the banyan network to route a packet from input at row 2 to output at row 6 based on binary bit pattern of routing vector r

For the SEs at the internal stages ($0 < i < n$), we use the expression for the throughput in (14.5) with $N = 2$

$$a_i(\text{in}) = a_{i-1}(\text{out}) \tag{14.33}$$

$$a_i(\text{out}) = 1 - \left[1 - \frac{a_i(\text{in})}{2}\right]^2 \tag{14.34}$$

where $0 < i < n$.

At the output stage ($i = n$), we have 2-to-1 concentrators. Depending on the design, an output concentrator could accept one packet only per time step. In other designs, the output concentrator could accept two packets coming on the two input links at the same time step.

For the output concentrator that accepts only one packet per time step, the packet arrival and departure probabilities at the output are given by

$$a_n(\text{in}) = a_{n-1}(\text{out}) \tag{14.35}$$

$$a_n(\text{out}) = x_1 + x_2 \tag{14.36}$$

where x_j is the probability that j packets arrived

$$x_j = \binom{2}{j} a_n^j(\text{in}) \; [1 - a_n(\text{in})]^{2-j} \tag{14.37}$$

From the definition of x_j, we can write $a_n(\text{out})$ as

$$a_n(\text{out}) = 2\, a_n(\text{in}) - a_n^2(\text{in}) \tag{14.38}$$

For the output concentrator that accepts all packets arriving at its two inputs, the packet arrival and departure probabilities are given by

$$a_n(\text{in}) = a_{n-1}(\text{out}) \tag{14.39}$$

$$a_n(\text{out}) = x_1 + 2x_2 \tag{14.40}$$

From the definition of x_j, we can write $a_n(\text{out})$ as

$$a_n(\text{out}) = 2\, a_n(\text{in}) \tag{14.41}$$

According to (14.3), we can write the traffic at the input of the banyan network as

$$N_a(\text{in}) = a \tag{14.42}$$

and the output traffic at the tagged output of the network is

$$N_a(\mathrm{out}) = a_n(\mathrm{out}) \tag{14.43}$$

where $a_n(\mathrm{out})$ is given by (14.38) or (14.41) depending on the details of the design.
The throughput of the banyan network is given by

$$\mathrm{Th} = N_a(\mathrm{out}) = a_n(\mathrm{out}) \tag{14.44}$$

The packet acceptance probability for the banyan network is given by

$$p_a = \frac{\mathrm{Th}}{N_a(\mathrm{in})} \tag{14.45}$$

The delay of the banyan network is defined as the average number of attempts to access the desired output. The probability that the packet succeeds after k tries is given by the geometric distribution

$$p(k) = p_a \, (1 - p_a)^k \tag{14.46}$$

The average delay time is

$$\begin{aligned} n_a &= \sum_{k=0}^{\infty} k \; p(k) \\ &= \sum_{k=0}^{\infty} k \; p_a \, (1 - p_a)^k \\ &= \frac{1 - p_a}{p_a} \end{aligned} \tag{14.47}$$

Example 3 Plot the throughput (Th), the access probability (p_a), and the average delay (n_a) for the banyan network versus the input traffic when the size of the network is $N = 64$ and compare this with the similar-sized crossbar network. Assume an output concentrator that accepts only one packet.

We evaluate the expressions for throughput, packet acceptance probability, and delay when $N = 64$ and a is varied for both the banyan network and the crossbar network. Figure 14.12 shows the variation of throughput, the packet acceptance probability, and the delay with the input traffic when $N = 64$ for the banyan network (solid line) and the crossbar network (dotted line). We see that the crossbar network shows superior performance for the same number of input and output ports as expected. ■

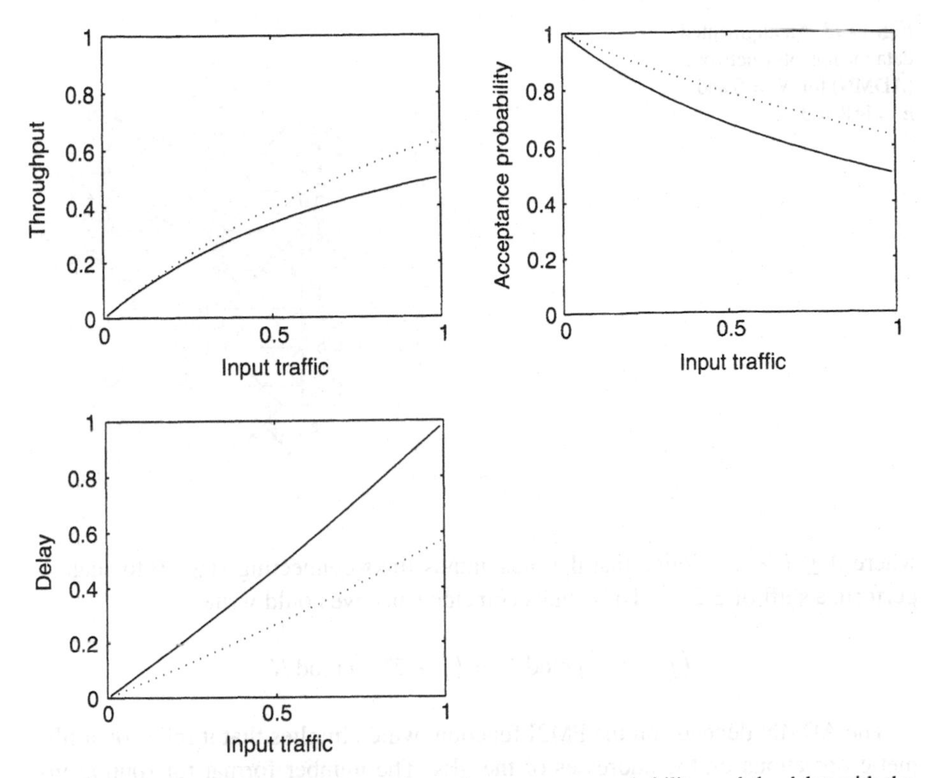

Fig. 14.12 Variation of the throughput, the packet acceptance probability, and the delay with the input traffic when $N = 64$ for the banyan network (*solid line*) and the similar-sized crossbar network (*dotted line*). The output concentrator accepts only one packet

14.10 Augmented Data Manipulator Network (ADMN)

Figure 14.13 shows an 8×8 ADMN network. For an $N \times N$ network, the number of stages is $n + 1$ where $n = \lg N$, and the number of SEs in each stage is N. Each SE is a 3×3 crossbar switch and the number of links between the stages is $3N$.

An $N \times N$ ADMN network is built using 1-to-3 selectors in the input stage ($i = 0$), 3×3 crossbar SEs in the $n - 1$ internal stages ($0 < i < n$), and 3-to-1 concentrators in the output stage ($i = n$) [17]. The ADMN network is a nonblocking network and provides two paths from any input to any output. As such, it is 1-fault tolerant.

As the packet travels through the stages of the network, each SE is capable of shifting the path of the packet among the rows. The shifting distance decreases as packets traverse the network.

Switching element SE(i, j) is connected to SE($i + 1$, k) such that k is given by

$$k = \begin{cases} \left(j - 2^{n-i-1}\right) \bmod N & -2I : \text{up connection} \\ j & s : \text{straight connection} \\ \left(j + 2^{n-i-1}\right) \bmod N & +2I : \text{down connection} \end{cases} \tag{14.48}$$

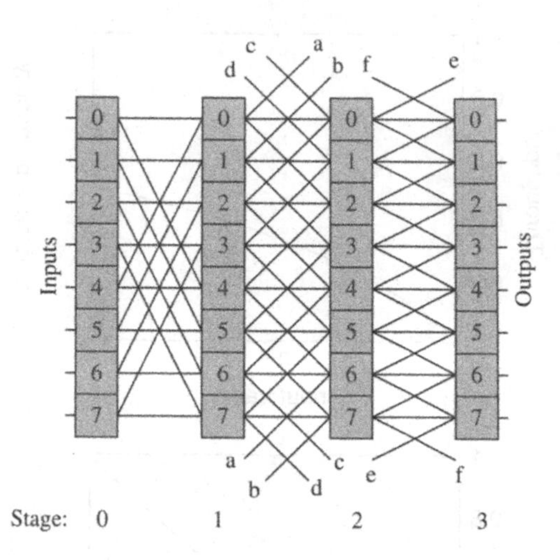

Fig. 14.13 An augmented data manipulator network (ADMN) for $N = 8$ and $n = \lg 8 = 3$

where $0 \leq i < n$. Notice that the plus–minus links connecting stage 0 to stage 1 perform a shift of $\pm 2^{n-1}$. These links coincide since we could write

$$\left(j - 2^{n-1}\right) \bmod N = \left(j + 2^{n-1}\right) \bmod N$$

The ADMN depends on the PM2I function, which implies that it relies on arithmetic operations on the addresses of the SEs. The number format for routing information could be decimal notation, $(n+1)$-bit 2's complement notation, or the n-bit radix-2 redundant signed-digit (RSD) number notation. In the RSD system, any n-bit number x can be represented as

$$x = \sum_{i=0}^{n-1} x_i \times 2^i \tag{14.49}$$

where $x_i \in \{-1, 0, +1\}$. The range of an n-bit RSD number is

$$-2^n + 1, -2^n + 2, \cdots, -1,\ 0,\ 1, \cdots, 2^n - 2, 2^n - 1 \tag{14.50}$$

or

$$-N + 1, -N + 2, \cdots,\ -1,\ 0,\ 1,\ \cdots,\ N - 2,\ N - 1 \tag{14.51}$$

where $n = \lg N$.

One bit location in that number system is represented not in binary format (0, 1) but in ternary format $(\overline{1},\ 0,\ 1)$, where $\overline{1}$ denotes the value -1. For example, a number $r = 3$ could be represented as

$$r = \left[0\,\overline{1}\,\overline{1}\right] \tag{14.52}$$

14.10.1 Routing Algorithms for ADMN Network

We discuss here three routing algorithms for the ADMN network. The algorithms are distributed among the SEs. The first routing algorithm uses the decimal value of the routing vector, which is calculated once at the input stage but has to be updated at all the internal stages. The second routing algorithm uses the $(n + 1)$-bit 2's complement of the routing vector, which has to be calculated once at the input stage. The other stages merely scan the bits of the routing vector to make configure their SEs. The third routing algorithm uses the (n)-bit RSD bit pattern of the routing vector which is calculated once at the input stage. The other stages merely scan the bits of the routing vector to make configure their SEs.

14.10.2 First ADMN Routing Algorithm

This algorithm updates the routing vector value at each stage and uses current value to make its routing decisions. To establish a path from a source address at an input port location S to a destination address at an output port location D, we initialize the routing vector r as

$$r_0 = D - S \tag{14.53}$$

where r_0 is the routing vector at the input of the SE at stage 0. A packet that has been routed as far as stage i will have a routing vector r_i. Upon exiting stage i, the routing vector r_i is updated according to the equation

$$r_{i+1} = r_i + \mu_i\, \delta_i \qquad 0 \leq i < n \tag{14.54}$$

where

$$\delta_i = 2^{n-i-1} \tag{14.55}$$

$$\mu_i = \begin{cases} -1 & r_i \geq \delta_i \\ 0 & -\delta_i < r_i < \delta_i \\ +1 & r_i \leq -\delta_i \end{cases} \tag{14.56}$$

Our objective in Equation (14.54) is to reduce r_i to zero. When $r_i = 0$, we know that the packet has reached the desired destination row D.

With the values of δ_i and μ_i determined, the packet will move to row location R_{i+1} at stage $i + 1$ where R_{i+1} is given by

$$R_{i+1} = (R_i - \mu_i\, \delta_i) \bmod N, \qquad 0 \leq i < n \tag{14.57}$$

Table 14.2 SE settings for path in ADMN network from input 6 to output 1 based on the decimal value of the routing vector r

Stage	$i=0$	$i=1$	$i=2$	$i=3$
$\delta_i = 2^{n-i-1}$	4	2	1	NA
μ_i	+1	0	+1	NA
Input routing vector (r_i)	−5	−1	−1	0
Output routing vector (r_{i+1})	−1	−1	0	0
Entry row location of packet (R_i)	6	2	2	1
Exit row location of packet (R_{i+1})	2	2	1	1

As an example, assume a packet arrives at input port 6 and is destined to output port 1. In that case, we have

$$S = 6 \tag{14.58}$$

$$D = 1 \tag{14.59}$$

$$r_0 = 1 - 6 = -5 \tag{14.60}$$

The routing decisions at each stage are explained in Table 14.2

Figure 14.14 shows the path chosen to route a packet from input at row 6 to output at row 1 based on decimal value of routing vector r.

Finding the Alternative Route

To find an alternative route for the incoming packet, we calculate the routing vector, as in the previous section, then we find the new routing vector obtained from the operation

$$\overline{r}_0 = (r_0 \pm N) \bmod N \tag{14.61}$$

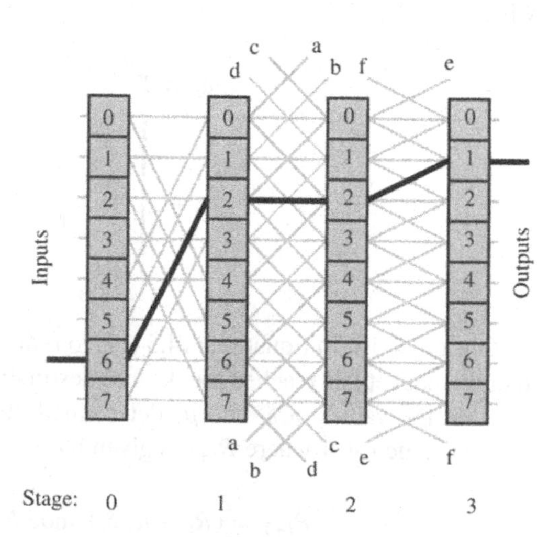

Fig. 14.14 Path chosen in the ADMN network to route a packet from input at row 6 to output at row 1 based on decimal value of routing vector r

Table 14.3 SE settings for path in ADMN network from input 6 to output 1 based on decimal value of routing vector $\overline{r}_0 = (r_0 \pm N) \bmod N$

Stage	$i = 0$	$i = 1$	$i = 2$	$i = 3$
$\delta_i = 2^{n-i-1}$	4	2	1	NA
μ_i	0	-1	-1	NA
Input routing vector ($\overline{r}_i$)	3	3	1	0
Output routing vector ($\overline{r}_{i+1}$)	3	1	0	0
Entry row location of packet (R_i)	6	6	0	1
Exit row location of packet (R_{i+1})	6	$8 \equiv 0$	1	1

For our case, the new routing vector will be

$$\overline{r}_0 = (-5 + 8) \bmod 8 = 3 \tag{14.62}$$

Let us illustrate this routing algorithm by finding how a path is established between input at row 6 and output at row 1 when $\overline{r}_0 = 3$. The routing decisions at each stage are explained in Table 14.3.

Figure 14.15 shows the path chosen to route a packet from input at row 6 to output at row 1 based on modulo operation on decimal value of routing vector $\overline{r}_0$.

The maximum shift that a packet experiences using this network is $\pm(N - 1)$. This implies that there is only one possible route when $R = D$. This problem could be solved by introducing double links for the straight connection and changing our SEs to 4×4 switches instead of 3×3 ones.

14.10.3 Second ADMN Routing Algorithm

The first routing algorithms discussed in the previous section required addition and subtraction operations—and these operations were repeated at each stage. Needless

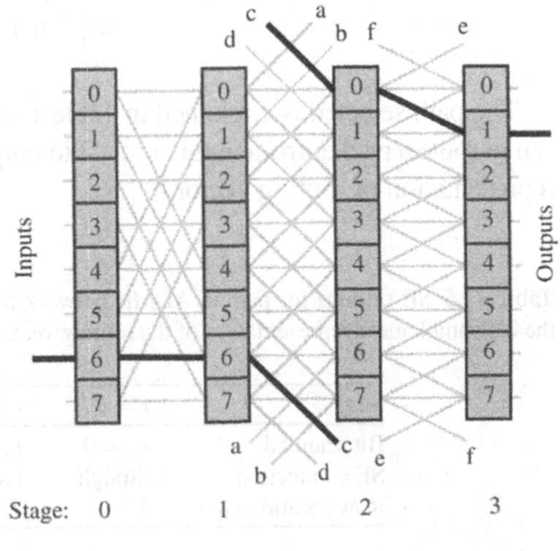

Fig. 14.15 Path chosen in the ADMN network to route a packet from input at row 6 to output at row 1 based on modulo operation on routing vector $\overline{r}_0 = 3$

Table 14.4 SE settings for path establishment in ADMN network based on 2's complement number format

Sign bit value	Bit value	Connection type
0	0	Straight
0	1	Up
1	0	Straight
1	1	Down

to say, these operations are complex and slow compared to the operations required of the second routing algorithm discussed here. This routing algorithm is also distributed among the SEs and relies on 2's complement representation of the routing vector. Calculating the routing vector is done only once at the input stage. The rest of the stages merely scan the bits of the routing vector to make up their routing decisions.

To establish a path from a source address at an input port location S to a destination address at an output port location D, we calculate r as

$$r = D - S \tag{14.63}$$

where r is represented in $(n + 1)$-bit 2's complement notation.

A switching element at stage i will have to scan two bits of the routing vector: the sign bit (which is bit n) and bit $n - i - 1$. Therefore, an SE at stage 0 scans the sign bit and bit $n - 1$. An SE at stage 1 scans the sign bit and bit $n - 2$, and so on. Finally, an SE at stage $n - 1$ scans the sign bit and bit 0. The SE settings for path establishment are based on the rules explained in Table 14.4.

As an example, assume $S = 7$ and $D = 2$. In that case, r is given as

$$r = D - S = -5 \tag{14.64}$$

The routing vector in $n + 1$-bit 2's complement notation will be

$$r = \begin{bmatrix} 1 & 0 & 1 & 1 \end{bmatrix} \tag{14.65}$$

The path selected is explained in Table 14.5, and Fig. 14.16 shows the path chosen to route a packet from input at row 7 to output at row 2 based on 2's complement representation of routing vector r.

Table 14.5 SE settings for path in ADMN network from input $S = 7$ to output $D = 2$ based on the 2's complement representation of the routing vector r. The sign bit for routing vector is $r_n = 1$

Stage	$i = 0$	$i = 1$	$i = 2$	$i = 3$
Bit scanned	$r_2 = 0$	$r_1 = 1$	$r_0 = 1$	NA
SE connection	Straight	Down	Down	NA
Row location (R_i)	7	7	1	2

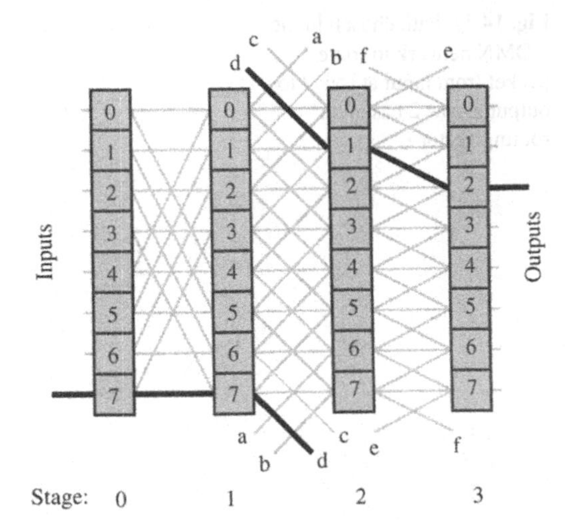

Fig. 14.16 Path chosen in the ADMN network to route a packet from input at row 7 to output at row 2 based on 2's complement representation of routing vector

Finding the Alternative Route

To find an alternative route for the incoming packet, we calculate the routing vector, as in the previous section, then we find the 2's complement of the routing vector. For our case, the original routing vector is

$$r = \begin{bmatrix} 1\,0\,1\,1 \end{bmatrix} \tag{14.66}$$

The 2's complement of this vector is

$$\overline{r} = \begin{bmatrix} 0\,1\,0\,1 \end{bmatrix} \tag{14.67}$$

The path selected is explained in Table 14.6, and Fig. 14.17 shows the path chosen to route a packet from input at row 7 to output at row 2 based on routing vector $\overline{r}$.

14.10.4 Third ADMN Routing Algorithm

The third ADMN routing algorithm is also distributed among the SEs and relies on combining RSD with modulo operation. Calculating the routing vector is done only

Table 14.6 SE settings for path in ADMN network from input $S = 7$ to output $D = 2$ based on the 2's complement representation of the routing vector $\overline{r}$. The sign bit for routing vector is $\overline{r}_n = 0$

Stage	$i = 0$	$i = 1$	$i = 2$	$i = 3$
Bit scanned	$\overline{r}_2 = 1$	$\overline{r}_1 = 0$	$\overline{r}_0 = 1$	NA
SE connection	Up	Straight	Up	NA
Row location (R_i)	7	3	3	2

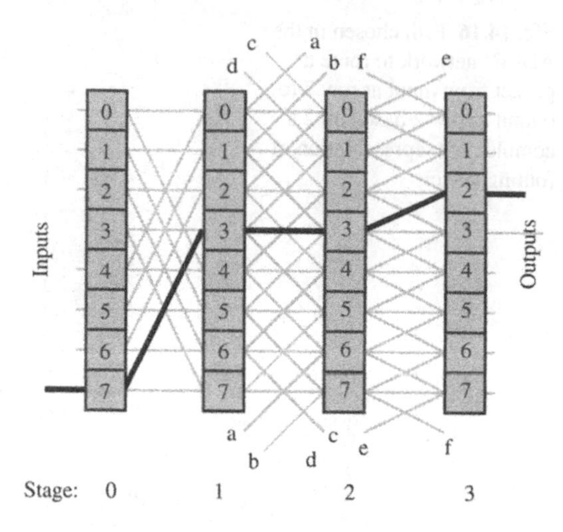

Fig. 14.17 Path chosen in the ADMN network to route a packet from input at row 7 to output at row 2 based on routing vector $\overline{r}$

once at the input stage. The rest of the stages merely scan the bits of the routing vector to make up their routing decisions.

To establish a path from a source address at an input port location S to a destination address at an output port location D, we need to calculate the routing vector r as given by the formula

$$r = D - S \tag{14.68}$$

where r is now represented in an n-bit RSD notation. The bits of the resulting RSD number are examined starting with the MSB first. So bit $n-1$ is examined by SEs at stage 0. Bit $n-2$ is examined by SEs at stage 1, and so on. Finally, bit 0 is examined by SEs at stage $n-1$. The SE settings for path establishment are based on the rules explained in Table 14.7.

As an example, assume $S = 5$ and $D = 2$. In that case r is given as

$$r = D - S = -3 \tag{14.69}$$

The routing vector in n-bit RSD notation will be

$$r = \left[0\,\overline{1}\,\overline{1}\right] \tag{14.70}$$

Table 14.7 SE settings for path establishment in ADMN network based on RSD number format

Bit value	Connection type
$\overline{1}$	Up
0	Straight
1	Down

Table 14.8 SE alternative settings for path in ADMN network from input 5 to output 2 based on RSD number format

Stage	0	1	2
Bit scanned	r_2	r_1	r_0
Bit value	0	$\overline{1}$	$\overline{1}$
Connection type	Straight	Up	Up

The path selected is explained in Table 14.8.

Figure 14.18 shows the path chosen to route a packet from input at row 5 to output at row 2 based on RSD number format of the routing vector r.

Finding the Alternative Route

The alternative route using the RSD number format is obtained by finding the value of $\overline{r}$ according to the formula

$$\overline{r} = (r \pm N) \bmod N \tag{14.71}$$

The alternate path $\overline{r}$ is

$$\overline{r} = (-3 + 8) \bmod 8 = 5 \tag{14.72}$$

which corresponds to the routing vector

$$\overline{r} = \begin{bmatrix} 1 & 0 & 1 \end{bmatrix} \tag{14.73}$$

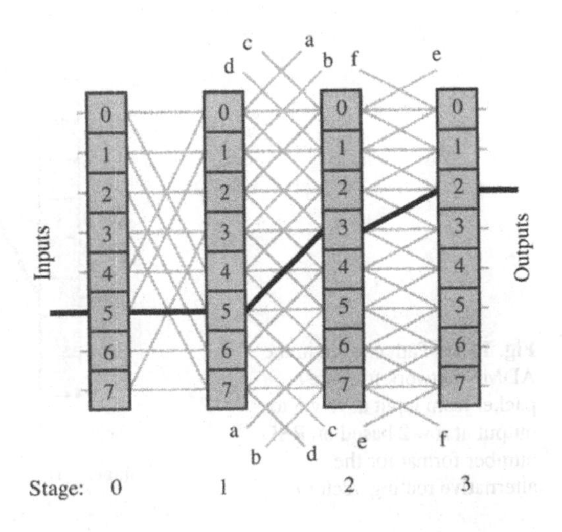

Fig. 14.18 Path chosen in the ADMN network to route a packet from input at row 5 to output at row 2 based on RSD number format of the routing vector r

Table 14.9 SE alternative settings for path in ADMN network from input 5 to output 2 based on RSD number format

Stage	$i = 0$	$i = 1$	$i = 2$
Bit scanned	r_2	r_1	r_0
Bit value	1	0	1
Connection type	Down	Straight	Down

The path selected is explained in Table 14.9. Note that each stage scans one bit in descending order.

Figure 14.19 shows the path chosen to route a packet from input at row 5 to output at row 2 based on RSD value of routing vector $\bar{r}$.

The maximum shift that a packet experiences using this network is $\pm(N - 1)$. This implies that there is only one possible route when $R = D$. This problem could be solved by introducing double links for the straight connection and changing our SEs to 4×4 switches instead of 3×3 ones.

14.10.5 Analysis of ADMN Network

An $N \times N$ ADMN network is built using 1-to-3 selectors in the input stage ($i = 0$), 3×3 crossbar SEs in $n - 1$ internal stages ($0 < i < n$), and 3-to-1 concentrators in the output stage ($i = n$). Therefore, we can use the results we obtained for the banyan network after some modifications.

Because we have 1-to-3 selectors at stage 0, the packet arrival and departure probabilities at each selector are given by

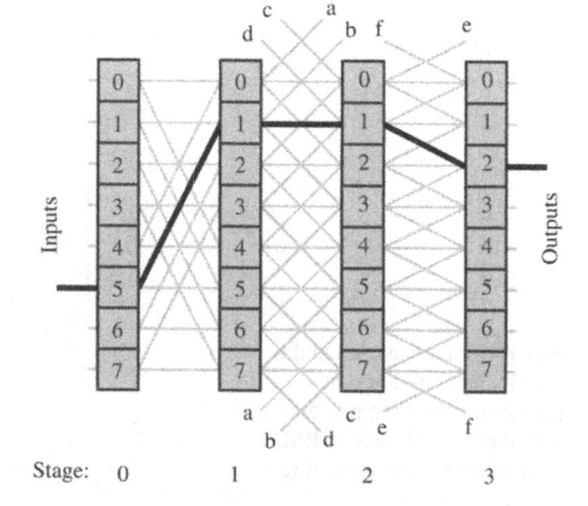

Fig. 14.19 Path chosen in the ADMN network to route a packet from input at row 5 to output at row 2 based on RSD number format for the alternative routing vector $\bar{r}$

$$a_0(\text{in}) = a \tag{14.74}$$

$$a_0(\text{out}) = \frac{a}{3} \tag{14.75}$$

For the SEs at the internal stages ($0 < i < n$), we use the expression for the throughput in (14.5) with $N = 3$

$$a_i(\text{in}) = a_{i-1}(\text{out}) \tag{14.76}$$

$$a_i(\text{out}) = 1 - \left[1 - \frac{a_i(\text{in})}{3}\right]^3 \tag{14.77}$$

At the output stage ($i = n$), we have 3-to-1 concentrators. For an ADMN network whose output concentrator accepts m packets ($1 \le m \le 3$), we can write the arrival and departure probabilities at the output as

$$a_n(\text{in}) = a_{n-1}(\text{out}) \tag{14.78}$$

$$a_n(\text{out}) = \sum_{j=1}^{m-1} j\, x_j + m \sum_{j=m}^{3} x_j \tag{14.79}$$

where $1 \le m \le 3$ is the maximum number of packets that could be accepted at the output in one time step and x_j is the probability that j packets arrived

$$x_j = \binom{3}{j} a_n^j(\text{in})\ [1 - a_n(\text{in})]^{3-j} \tag{14.80}$$

According to (14.3), we can write the traffic at the input of the ADMN network as

$$N_a(\text{in}) = a \tag{14.81}$$

and the output traffic at the tagged output of the network is

$$N_a(\text{out}) = a_n(\text{out}) \tag{14.82}$$

where $a_n(\text{out})$ is given by (14.79).

The throughput of the ADMN network is given by

$$\text{Th} = N_a(\text{out}) = a_n(\text{out}) \tag{14.83}$$

The packet acceptance probability for the ADMN network is given by

$$p_a = \frac{\text{Th}}{N_a(\text{in})} \tag{14.84}$$

The delay of the ADMN network is defined as the average number of attempts to access the desired output. The probability that the packet succeeds after k tries is given by the geometric distribution

$$p(k) = p_a (1 - p_a)^k \tag{14.85}$$

The average delay time is

$$\begin{aligned} n_a &= \sum_{k=0}^{\infty} k\, p(k) \\ &= \sum_{k=0}^{\infty} k\, p_a (1 - p_a)^k \\ &= \frac{1 - p_a}{p_a} \end{aligned} \tag{14.86}$$

Example 4 Plot the throughput (Th), the access probability (p_a), and the average delay (n_a) for the ADMN network versus the input traffic when the size of the network

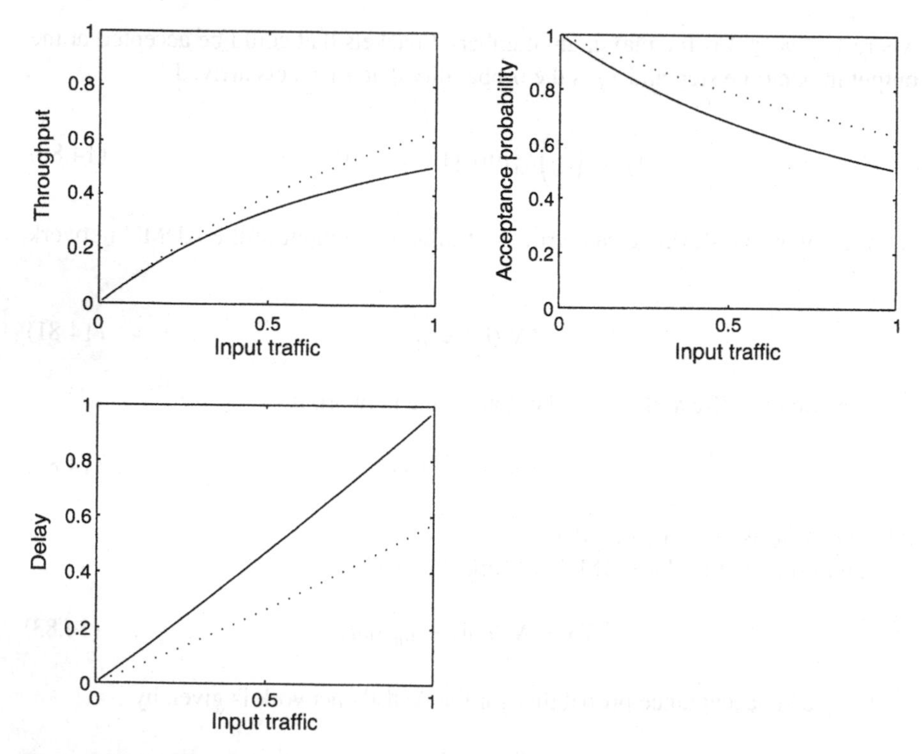

Fig. 14.20 Variation of the throughput, the packet acceptance probability, and the delay with the input traffic when $N = 64$ for the ADMN network (*solid line*) and the similar-sized crossbar network (*dotted line*). The output concentrator accepts only one packet

is $N = 64$ and compare this with the similar-sized crossbar network. Assume an output concentrator that accepts only one packet.

We evaluate the expressions for throughput, packet acceptance probability, and delay when $N = 64$ and a is varied for both the ADMN network and the crossbar network. Figure 14.20 shows the variation of throughput, the packet acceptance probability, and the delay with the input traffic when $N = 64$ for the ADMN network (solid line) and the crossbar network (dotted line). We see that the crossbar network shows superior performance for the same number of input and output ports as expected. ■

14.11 Improved Logical Neighborhood (ILN)

This family of networks is based on the data manipulator network but has much better fault tolerance capability [22]. Figure 14.21 shows an 8×8 ILN network. For an $N \times N$ network, the number of stages is $n + 1$ where $n = \lg N$, and the number of SEs in each stage is N. Each SE is an $(n + 1) \times (n + 1)$ crossbar switch and the number of links between the stages is $(n + 1)N$.

An $N \times N$ ILN network is built using 1-to-$(n + 1)$ selectors in the input stage ($i = 0$) and all internal stages ($0 < i < n$) have identical SEs that are $(n + 1) \times (n + 1)$ crossbar switches. The output stage ($i = n$) has $(n + 1)$-to-1 concentrators.

The ILN network provides $n!$ alternate paths between any input and any output. For the 8×8 case, we have six alternate paths. As such, the network is very fault tolerant and the blocking probability is much smaller than other MIN networks.

SE(i, j) is connected to SE$(i + 1, k)$ such that k is given by

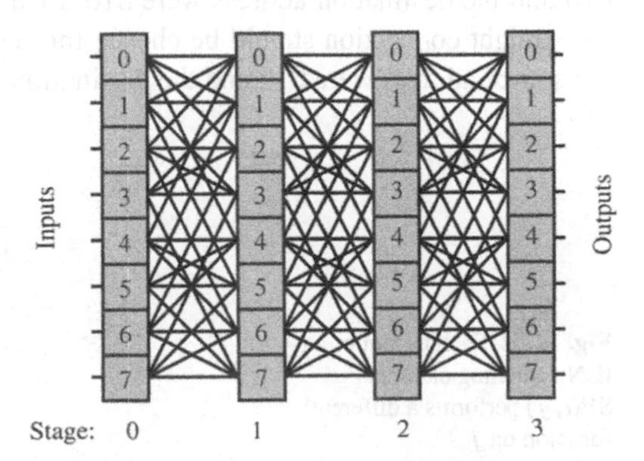

Fig. 14.21 An improved logical neighborhood network (ILN) for $N = 8$ and $n = \lg 8 = 3$

$$k = \begin{cases} j & \text{straight connection} \\ C^0(j) & \text{0-cube connection} \\ C^1(j) & \text{1-cube connection} \\ C^2(j) & \text{2-cube connection} \end{cases} \tag{14.87}$$

where $0 \leq i < n$. Figure 14.22 shows the four outputs of the switch and the operations they do on incoming packets.

14.11.1 Routing Algorithm for ILN Network

We discuss here one distributed routing algorithm that does not involve arithmetic operations and relies only on bitwise XOR operations involving the source address S and the destination address D. This leads to very fast routing decisions at each SE and is very compatible with high-performance switches.

The routing vector r is given by

$$r = \text{XOR}(S, D) \tag{14.88}$$

where the routing vector r carries the information about the desired path. The path selected corrects the routing vector r such that all its bits are zeros to indicate that the packet has reached the destination row address.

Each SE is able to correct a "1" bit located at any position in r. In addition, any SE is also able to *destroy* an already-existing "0" in r by changing it to a "1". This action might be necessary in order to select an alternative route in case of contention. The reason for this strategy will be explained in the next section.

Table 14.10 shows the outputs that should be used to establish the path based on the numerical value of r. The number of bits that need correction by proper selection of the output ports varies between 0 and 3. For example, if the input address were 010 and the destination address were 010, then the routing vector is $r = 000$ and the straight connection should be chosen through all stages. On the other hand, if the source address were 010 and the destination address were 101, then the routing

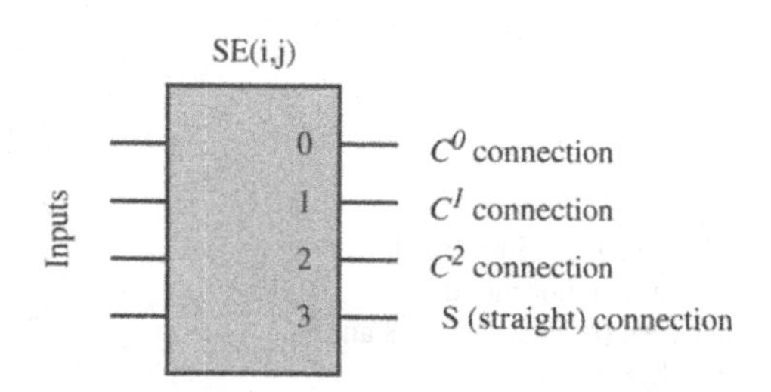

Fig. 14.22 Each output of an ILN switching element SE(i, j) performs a different function on j

Table 14.10 Possible SE outputs based on the routing vector r

Binary value of r	C^2	C^1	C^0	S
000				x
001			x	
010		x		
100	x			
011		x	x	
101	x		x	
110	x	x		
111	x	x	x	

vector is $r = 111$. In that case, any of the cube connections could be used in any order at each stage to eliminate any of the "1"s in any bit location.

The vector r carries information about the mismatch between the incoming address and the destination address. This mismatch should decrease by one bit each time the packet goes through each stage of the network.

As an example, let us attempt to move a packet from input 6 (110) to output 0 (000). The resulting routing vector is $r = [1\ 1\ 0]$. Figure 14.23 shows two possible paths for routing a packet from input at row 6 to output at row 1 based on routing vector r.

Stage 0: We have $r = [1\ 1\ 0]$. Thus we need to access ports C^1 or C^2. Picking the C^1 connection produces the updated routing vector $r' = [1\ 0\ 0]$. This choice makes the packet travel from row 6 to row 4 at stage 1. On the other hand, picking the C^2 connection produces the updated routing vector $r' = [0\ 1\ 0]$. This choice makes the packet travel from row 6 to row 2 at stage 1.
Stage 1: If routing vector at this stage is $r = [1\ 0\ 0]$, we choose port C^2. The updated routing vector becomes $r' = [0\ 0\ 0]$. This choice makes the packet travel from row 4 to row 0 at stage 2. If routing vector at this stage

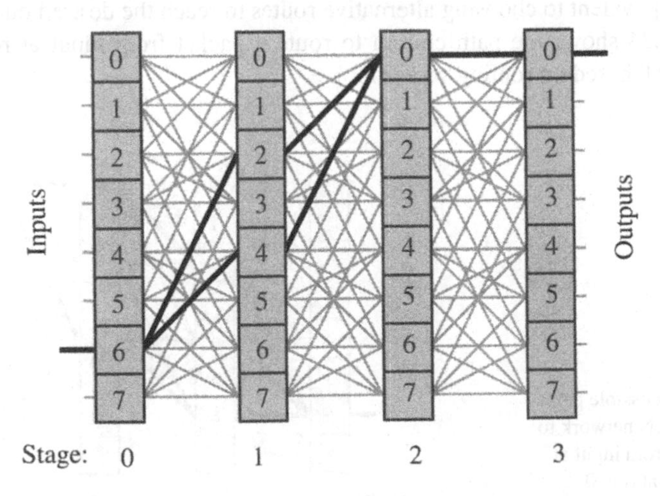

Fig. 14.23 Paths chosen in the ILN network to route a packet from input at row 6 to output at row 0

is $r = [0\ 1\ 0]$, we choose port C^1. The updated routing vector becomes $r' = [0\ 0\ 0]$. This choice makes the packet travel from row 2 to row 0 at stage 2.

Stage 2: Since our routing vector is $r = [0\ 0\ 0]$, we have to choose the straight connection to route the packet to the desired output at row 0 at stage 3.

As an illustration of the richness of paths from any input to any output, Fig. 14.23 shows the paths chosen to route a packet from input at row 6 to output at row 1.

14.11.2 Path Selection Issues

The routing algorithm chosen by each SE in the ILN network is based on choosing a connection that will change an existing "1" in r to a "0". It was also mentioned that contention might force the routing algorithm to change a "0" bit to "1" to choose an alternative path. These newly created "1"s will be corrected at later stages in the network. Let us illustrate that using a simple example. Assume only two packets, a and b, arrive at the inputs 0 and 4, respectively, at a given slot time. Both packets are destined to output 0. In that case, packet a will choose the straight connection throughout switching elements SE(0,0), SE(1,0), and SE(2,0). Packet b will choose the C^2 connection for SE(0,4) and it will require the straight connection for switching elements SE(1,0) and SE(2,0). Thus, only one packet will be routed and the others will be blocked even though there are many alternative routes to output 0.

The routing algorithm at each SE must be able to identify the alternative routes at any stage in order to prevent this unnecessary contention and fully utilize the link redundancy.

The intelligent routing algorithm, in effect, destroys some of the zero bits in r, trusting that they will be returned back to zero value at later stages. The destruction of 0-bits is equivalent to choosing alternative routes to reach the desired output.

Figure 14.24 shows the path chosen to route a packet from input at row 6 to output at row 0 based on routing vector r.

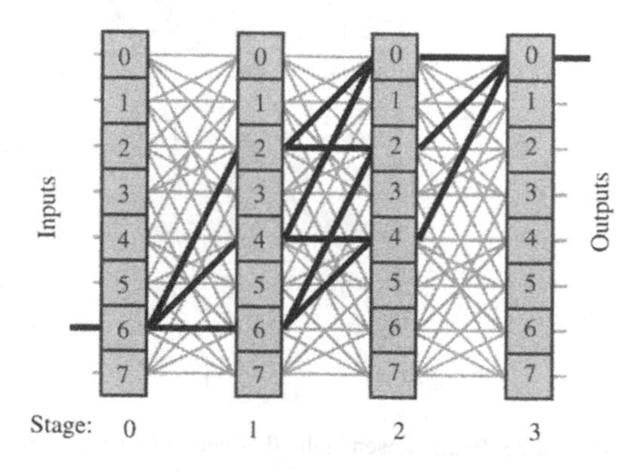

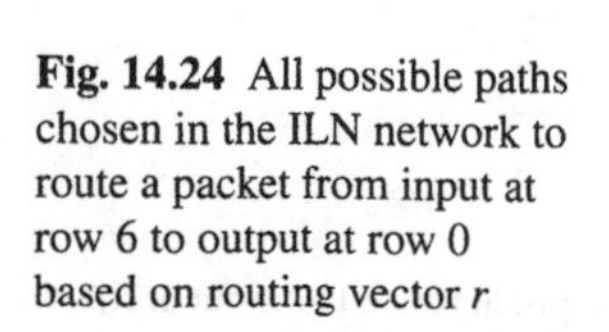

Fig. 14.24 All possible paths chosen in the ILN network to route a packet from input at row 6 to output at row 0 based on routing vector r

14.11.3 Analysis of ILN Network

An $N \times N$ ILN network is built using 1-to-$(n+1)$ selectors in the input stage $(i = 0)$, $(n+1) \times (n+1)$ crossbar SEs in the $n-1$ internal stages $(0 < i < n)$, and $(n+1)$-to-1 concentrators in the output stage $(i = n)$. Therefore, we can use the results we obtained for the banyan and ADMN networks after some modifications.

Because we have 1-to-$(n+1)$ selectors at stage 0, the packet arrival and departure probabilities at the input and the outputs of each selector are given by

$$a_0(\text{in}) = a \tag{14.89}$$

$$a_0(\text{out}) = \frac{a}{n+1} \tag{14.90}$$

For the SEs at the internal stages $(0 < i < n)$, we use the expression for the throughput in (14.5) with $N = n+1$:

$$a_i(\text{in}) = a_{i-1}(\text{out}) \tag{14.91}$$

$$a_i(\text{out}) = 1 - \left[1 - \frac{a_i(\text{in})}{n+1}\right]^{(n+1)} \tag{14.92}$$

At the output stage $(i = n)$, we have $(n+1)$-to-1 concentrators. For an ILN network whose output concentrator accepts m packets $(1 \le m \le n+1)$, we can write the arrival and departure probabilities at the output as

$$a_n(\text{in}) = a_{n-1}(\text{out}) \tag{14.93}$$

$$a_n(\text{out}) = \sum_{j=1}^{m-1} j\, x_j + m \sum_{j=m}^{n+1} x_j \tag{14.94}$$

where $1 \le m \le n+1$ is the maximum number of packets that could be accepted at the output in one time step and x_j is the probability that j packets arrived

$$x_j = \binom{n+1}{j} a_n^j(\text{in})\ [1 - a_n(\text{in})]^{n+1-j} \tag{14.95}$$

According to (14.3), we can write the traffic at the input of the banyan network as

$$N_a(\text{in}) = a \tag{14.96}$$

and the output traffic at the tagged output of the network is

$$N_a(\text{out}) = a_n(\text{out}) \tag{14.97}$$

where $a_n(\text{out})$ is given by (14.94).

The throughput of the ILN network is given by

$$\mathrm{Th} = N_a(\mathrm{out}) = a_n(\mathrm{out}) \tag{14.98}$$

The packet acceptance probability for the ADMN network is given by

$$p_a = \frac{\mathrm{Th}}{N_a(\mathrm{in})} \tag{14.99}$$

The delay of the ILN network is defined as the average number of attempts to access the desired output. The probability that the packet succeeds after k tries is given by the geometric distribution

$$p(k) = p_a\,(1 - p_a)^k \tag{14.100}$$

The average delay time is

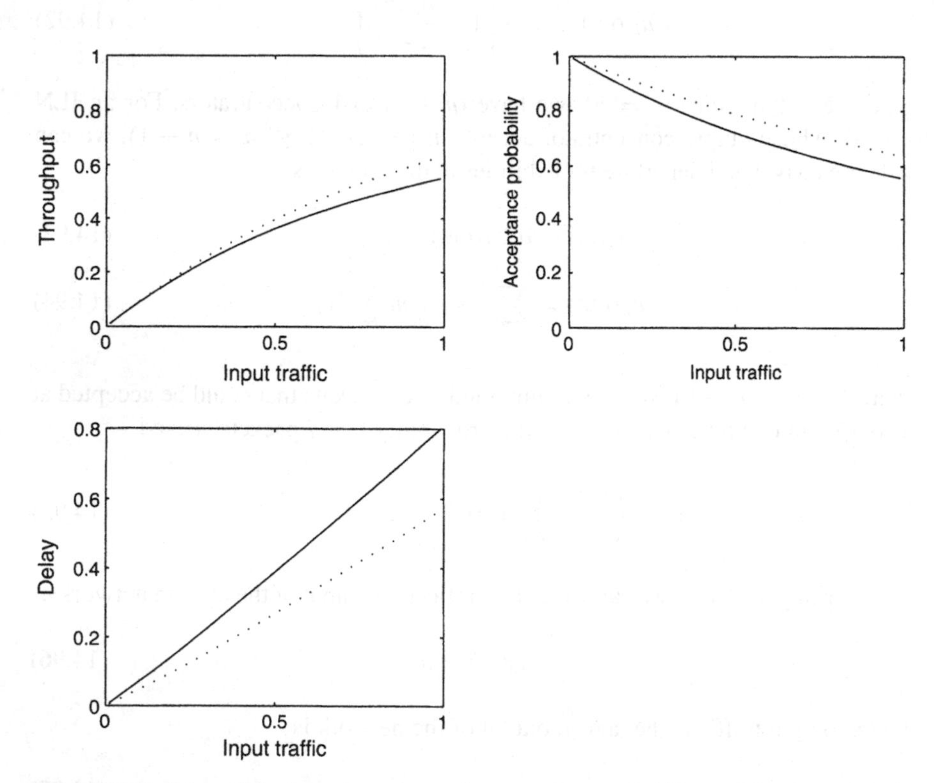

Fig. 14.25 Variation of the throughput, the packet acceptance probability, and the delay with the input traffic when $N = 64$ for the ILN network (*solid line*) and the similar-sized crossbar network (*dotted line*)

$$n_a = \sum_{k=0}^{\infty} k\ p(k)$$
$$= \sum_{k=0}^{\infty} k\ p_a (1 - p_a)^k$$
$$= \frac{1 - p_a}{p_a} \tag{14.101}$$

Example 5 Plot the throughput (Th), the access probability (p_a), and the average delay (n_a) for the ILN network versus the input traffic when the size of the network is $N = 64$ and compare this with the similar-sized crossbar network. Assume an output concentrator that accepts all the packets that arrive at its two inputs.

We evaluate the expressions for throughput, packet acceptance probability, and delay when $N = 64$ and a is varied for both the ILN network and the crossbar network. Figure 14.25 shows the variation of throughput, the packet acceptance probability, and the delay with the input traffic when $N = 64$ for the ILN network (solid line) and the crossbar network (dotted line). We see that the crossbar network shows superior performance for the same number of input and output ports as expected. ■

Problems

Interconnection Networks

14.1 What is the main function of an interconnection network and what are the main performance requirements of the network?

14.2 Discuss the main issues related to synchronous data transfer in the switching fabric and between the main components of a switch.

14.3 Assume two communicating modules in the switch using synchronous clocking scheme. Construct a table explaining (a) the main signal lines required between the two modules; (b) the direction of signal on each line; and (c) the function of each signal.
Using a timing diagram show how a packet could move between the two modules.

14.4 To maintain data integrity using a system-wide clock, clock skew must be controlled and clock signal integrity must be maintained. How can clock skew be controlled?

14.5 Assume two modules exchange data using a common clock. One option is to use a two-phase clock and the other is to use single-phase clock. Explain the two options using timing diagrams and explain the problem of clock skew in both of them.

14.6 Discuss the main issues related to asynchronous data transfer in the switching fabric and between the main components of a switch.

14.7 Assume two communicating modules in the switch using asynchronous clocking scheme. Construct a table explaining (a) the main signal lines required between the two modules; (b) the direction of signal on each line; and (c) the function of each signal.
Using a timing diagram show how a packet could move between the two modules.

14.8 Assume you need two transfer a packet between two blocks in the switch in the least amount of time. What are your options and what are the advantages and disadvantages of each option?

14.9 What are the main sources of power loss in a chip operating at very high data rates? Is there much difference between using GaAs or CMOS logic at high speeds?

Time Division Multiple Access (TDMA)

14.10 Explain the two main types of TDMA.

14.11 Discuss the advantages and disadvantages of static-assignment TDMA.

14.12 Discuss the advantages and disadvantages of random assignment TDMA.

Multiplexing Techniques (TDMA)

14.13 In time division switching, one media access protocol was mentioned, the random assignment. What other types of MAC protocols could be used to guarantee fair access to the medium in a reasonable time delay.

14.14 Perform numerical simulations of the random assignment TDS with $N = 10$ for variable arrival probability a between 0.1 and 1. Plot the average delay for an input and plot the average throughput of any input.

14.15 Assume a TDS with a static priority protocol in which each user has a priority equal to its index. Derive an expression for the probability of accessing the medium for any user. Based on that, find the average delay that each user experiences and the average throughput for each user.

Crossbar Interconnection Network

14.16 Explain how a path is established in the crossbar switch and explain why the switch is nonblocking.

14.17 Discuss the need for arbitration in a crossbar network and propose some techniques for resolving output contention. Discuss the advantages and disadvantages of the arbitration techniques you propose from the point of view of hardware complexity and speed.

14.18 Assume we have N devices attached to a bus and each device has a priority of accessing the bus equal to its index number. Thus device 0 has the least probability, device 1 has higher probability than device 0, and so on. Derive an expression for the probability that device i access the bus. Having done that, derive an expression for the delay.

14.19 Derive expressions for channel usage efficiency of packet switching using datagram versus virtual circuit switching. Assume each packet for datagram is composed of n_1 header/trailer bytes and n_2 payload bytes. For virtual circuit switching, assume n_1 header/trailer bytes, n_2 payload bytes, and n_3 bytes required for call establishment and tear down.

14.20 Prove equation (14.23) by estimating the probability of a packet transmitted to the output of an SE as the sum of two probabilities: (*a*) only one packet arrives on one of the input links; (*b*) packets arrive at both input links.

14.21 One of the necessary components of a switch is the FIFO. How can this be implemented using an ordinary one-ported random-access memory?

14.22 One of the necessary components of a switch is a CAM (content addressable memory). Discuss how a CAM could be implemented using specialized hardware or using a random-access memory or groups of random-access memories.

14.23 In input queuing switch, it was mentioned that HOL problem leads to reduced throughput and increased packet loss probability. Suppose a windowing scheme is used to examine w packets at the head of the queue instead of only one. How can this scheme preserve the packet sequence?

14.24 In shared buffer switches, it was mentioned that the shared buffer must perform many read and write operations per time step. Discuss how this can be implemented using a bank of one-ported memories that is accessible by all inputs and outputs of the switch.

14.25 Perform numerical simulations for the crossbar network of size 8×8 for variable arrival probability a between 0.1 and 1. Plot the average delay for an input and plot the average throughput of any input.

Generalized-Cube Network (GCN)

14.26 Sketch a 4×4 GCN network.

14.27 Sketch a 16×16 GCN network.

14.28 Find the path taken by a packet in an 8×8 GCN network if the source row address is $S = 2$ and the destination row address is $D = 1$.

14.29 Find the path taken by a packet in an 8×8 GCN network if the source row address is $S = 7$ and the destination row address is $D = 0$.

Banyan Interconnection Network

14.30 Prove (14.32) on page 524.

14.31 Prove (14.38) on page 525.

14.32 Prove (14.41) on page 525.

14.33 Sketch a 4×4 banyan network.

14.34 Sketch a 16×16 banyan network.

14.35 Find the path taken by a packet in an 8×8 banyan network if the source row address is $S = 2$ and the destination row address is $D = 1$.

14.36 Find the path taken by a packet in an 8×8 banyan network if the source row address is $S = 7$ and the destination row address is $D = 0$.

14.37 Assume a banyan network is built using 3×3 switching elements instead of 2×2. Plot the resulting network and derive an expression for the recurrence formula.

14.38 Perform numerical simulation for an 8×8 banyan network for variable arrival probability a between 0.1 and 1. Plot the average delay for an input and plot the average throughput of any input.

ADMN Interconnection Network

14.39 Perform numerical simulation for an 8×8 ADMN network for variable arrival probability a between 0.1 and 1. Plot the average delay for an input and plot the average throughput of any input.

14.40 Perform numerical simulation for an 8×8 ILN network for variable arrival probability a between 0.1 and 1. Plot the average delay for an input and plot the average throughput of any input.

14.41 Sketch a 4×4 ADMN network.

14.42 Sketch a 16×16 ADMN network.

14.43 Find the paths taken by a packet in an 8×8 ADMN network if the source row address is $S = 2$ and the destination row address is $D = 1$.

14.44 Find the paths taken by a packet in an 8×8 ADMN network if the source row address is $S = 7$ and the destination row address is $D = 0$.

ILN Interconnection Network

14.45 Explore other alternative routes for the routing example given for the ILN network.

14.46 Sketch a 4×4 ILN network.

14.47 Sketch a 16×16 ILN network.

14.48 Find the paths taken by a packet in an 8×8 ILN network if the source row address is $S = 2$ and the destination row address is $D = 1$.

14.49 Find the paths taken by a packet in an 8×8 ILN network if the source row address is $S = 7$ and the destination row address is $D = 0$.

References

1. F. Elguibaly, "Analysis and design of arbitration protocols", *IEEE Transactions on Computers*, vol. 38, no. 2, pp. 1168–1175, 1989.
2. R.J. Simcoe and T.-B. Pei, "Perspectives on ATM switch architecture and the influence of traffic pattern assumptions on switch design", *Computer Communication Review*, vol. 25, pp. 93–105, 1995.

3. N. McKeown, "The iSLIP scheduling algorithm for input-queued switches", *IEEE/ACM Transactions on Networking*, vol. 7, pp. 188–201, 1999.
4. J. Duato, S. Yalamanchili, and L. Ni, *Interconnection Networks: An Engineering Approach*, IEEE Computer Society Press, Los Alamito, California, 1997.
5. J. Walrand, *Communication Networks: A First Course*, McGraw-Hill, New York, 1998.
6. M. Swartz, *Telecommunication Networks: Protocols, Modeling and Analysis*, Addison Wesley, Reading, Massachusetts, 1987.
7. S.W. Furhmann, "Performance of a packet switch with crossbar architecture", *IEEE Transactions on Communications*, vol. 41, pp. 486–491, 1993.
8. C. Clos, "A study of non-blocking switching networks", *Bell System Technology Journal*, vol. 32, pp. 406–424, 1953.
9. "Performing Internet routing and switching at gigabit speeds", Cisco 12000 Series GSR Technical Product Description, Cisco Systems, San Jose, California [On line]. Available HTTP: http://www.ieng.com/warp/public/cc/pd/rt/12000/ index.shtml
10. N. McKeown, M. Izzard, A. Mekkittikul, B. Ellersick, and M. Horowitch, "The tiny tera: A small high-bandwidth packet switch core", *IEEE Micro*, vol. 17, pp. 26–33, 1997.
11. C. Partridge, P.P. Carvey, E. Burgess, I. Castineyra, T. Clarke, L. Graham, M. Hathaway, P. Herman, A. King, S. Kohalmi, T. Ma, J. Mcallen, T. Mendez, W.C. Milliken, R. Pettyjohn, J. Rokosz, J. Seeger, M. Sollins, S. Storch, B. Tober, G.D. Troxel, D. Waitzman, and S. Winterble, "A 50 Gb/s IP Router", *IEEE/ACM Transactions on Networking*, vol. 6, no. 3, pp. 237–248, 1998.
12. F. Elguibaly, A. Sabaa and D. Shpak, "A new shift-register based ATM switch", *The First Annual Conference on Emerging Technologies and Applications in Communications (ETACOM)*, Portland, Oregon, pp. 24–27, May 7–10, 1996.
13. F. Elguibaly, and S. Agarwal, "Design and performance analysis of shift register-based ATM switch", *IEEE Pacific Rim Conference on Communications, Computers and Signal Processing*, Victoria, B.C., pp. 70–73, August 20–22, 1997.
14. A. Sabaa, F. Elguibaly and D. Shpak, "Design and modeling of a nonblocking input-buffer ATM switch", *Canadian Journal of Electrical and Computer Engineering*, vol. 22, no. 3, pp. 87–93, 1997.
15. V.E. Benes, *Mathematical Theory of Connecting Networks and Telephone Traffic*, Academic Press, New York, 1965.
16. K. Hwang and F.A. Briggs, *Computer Architecture and Parallel Processing*, McGraw-Hill, New York, 1984.
17. H.J. Siegel, *Interconnection Networks for Large-Scale Parallel Processing: Theory and Case Studies*, Lexington Book, Lexington, Massachusetts, 1990.
18. D. Lawrie, "Access and alignment of data in an array processor", *IEEE Transactions on Computers*, vol. C-24, pp. 1145–1155, 1975.
19. L.R. Goke and G.J. Lipvski, "Banyan networks for partitioning multiprocessor systems", *First Annual International Symposium on Computer Architecture*, pp. 21–28, December 1973.
20. K. Pibulyarojana, S. Kimura, and Y. Ebihara, "A study on a hybrid dilated banyan network", *IEICE Transaction on Communications*, vol. E80 B, pp. 116–126, 1997.
21. J.H. Patel, "Performance of processor-memory interconnections for multiprocessors", *IEEE Transaction on Computers*, vol. C-30, pp. 771–780, 1981.
22. M. Abd-El-Barr, K. Al-Tawil, and O. Abed, "Fault-tolerant and reliability analysis of multistage data manipulator networks", *International Conference on Distributed Computing*, pp. 275–280, 1995.

Chapter 15
Switch Modeling

15.1 Introduction

As we saw in Chapter 13, a packet switch must contain input (ingress) ports, output (egress) ports, switch fabric, and control section. When a packet arrives at a switch, it must be processed at the input before it can be routed to the proper switch output port. Processing involves matching the packet header to information in a lookup table to determine the destination and the priority and type of packet. These operations take time and the packets must be temporarily stored in input buffers.

Chapter 13 discusses several types of switches based on the placement of buffers within each switch which includes.

1. Input queuing switch
2. Output queuing switch
3. Shared buffer switch
4. Multiple input queuing switch
5. Multiple output queuing switch
6. Multiple input and output queuing switch
7. Virtual routing/virtual queuing (VRQ) switch

The placement of the buffers and the priority queues in the switch has direct impact on the overall switch performance.

The literature abounds with sophisticated models for all types of switches [1–14]. Most of these models assumed an deal switch fabric in order to simplify the analysis. This assumption is not true in reality. We saw in Chapter 14 that all the interconnection networks in use today have limitations and the ability of a packet to go through the SF decreases with increasing input traffic or increasing network size.

In this chapter, we illustrate how to model the performance of the three main switch types taking into consideration the limitations of the switching fabric (SF). We choose here to present simple and accurate analyses that take into consideration the effect of the interconnection network. Having developed the relevant equations, we derive expressions for the switch performance figures such as throughput, average buffer size, packet loss probability, and packet delay.

F. Gebali, *Analysis of Computer and Communication Networks*,
DOI: 10.1007/978-0-387-74437-7_15, © Springer Science+Business Media, LLC 2008

15.2 Input Queuing Switch

Figure 15.1 shows an input queuing switch where each input port has a queue that accepts all incoming packets for the input line.

At the beginning of each time step, packets arrive at the inputs and are stored in the input FIFO queues. The input controller services the packets at the head of each queue – These packets are usually called head of line (HOL) packets. The controller sets up a path to the desired output port for each HOL packet. Sometimes the path cannot be established due to internal or output blocking as discussed in Chapter 14 and the blocked packets remain in their respective queues.

We make the following assumptions to simplify our analysis:

1. Each input queue has one input and one output.
2. The size of the input queue is B.
3. a is the packet arrival probability for any input of the switch.
4. p_a is the packet departure probability from any input queue where p_a is the packet acceptance probability of the switching fabric.
5. Packets could be served in the same time step at which they arrive.
6. Each packet has equal probability $1/N$ of requesting an output port.
7. Data broadcast or multicast are not implemented.
8. Packets will be lost when a packet arrives at a full input queue whose HOL packet cannot be routed in that time step.

Under these assumptions, we can model each input queue as an $M/M/1/B$ queue. The arrival probability for this queue is a and the departure probability c is given by

$$c = p_a \tag{15.1}$$

where p_a is the packet acceptance probability of the interconnection network used. Notice that if a path is established for the packet, we are assured the output will accept it. This was part of our simplifying assumptions.

Figure 15.2 shows one input queue associated with an input port. The arrival probability a and departure probability p_a are shown at the queue input and output, respectively.

Equation (15.1) poses a problem for us. From the results of Chapter 14, we know that the switching fabric (SF) packet acceptance probability p_a depends on the traffic arriving at the inputs. In other words, we can determine p_a only if we knew

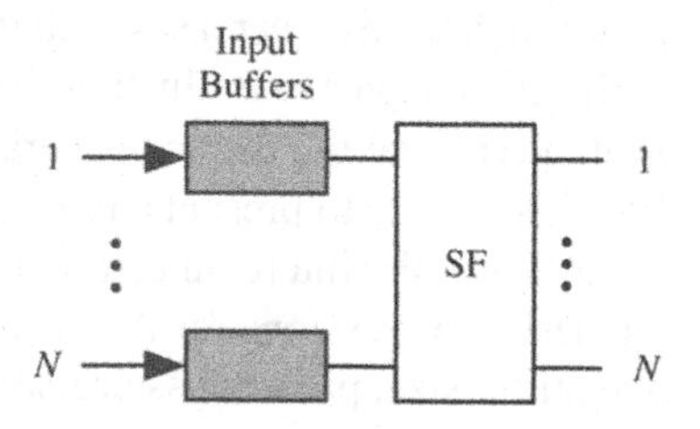

Fig. 15.1 The input queuing switch

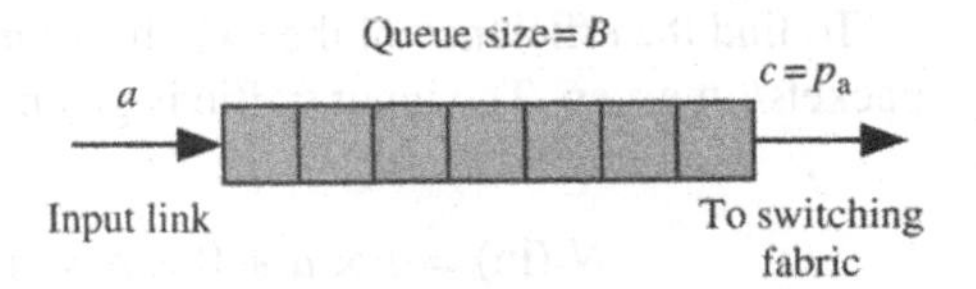

Fig. 15.2 The arrival and departure probabilities for one queue of an input queuing switch

which input queues are occupied and demand to access the network. The state of occupancy of the queues depends on the p_a, and this is the circular argument that is posing the problem.

Researchers and development engineers chose the easy way out by assuming the switch fabric to have $p_a = 1$ and dealt with each input queue in isolation. However, this is a very optimistic assumption. We get around this problem by assuming a fixed value for p_a that we set depending on the type of SF we plan to use.

According to Equation (7.36) on page 235, we have the following expression for the probability that the $M/M/1/B$ queue is empty

$$s_0 = \frac{1-\rho}{1-\rho^{B+1}} \tag{15.2}$$

where B is the size of the input queue and ρ was identified as the distribution index which is given by

$$\rho = \frac{a\,d}{b\,c} = \frac{a\,d}{b\,p_a} \tag{15.3}$$

with $b = 1 - a$ and $d = 1 - p_a$.

The throughput or output traffic for the input queuing switch is given by

$$\begin{aligned} \text{Th} &= N_a(\text{out}) \\ &= a\ p_a\ s_0 + \sum_{i=1}^{B} p_a\ s_i \\ &= a\ p_a\ s_0 + p_a\,(1 - s_0) \qquad \text{packets/time step} \end{aligned} \tag{15.4}$$

where s_i is the probability that the queue has i packets. The first term on the RHS is the probability that a packet arrives while the queue is empty and the second term on the RHS is the probability that the queue has packets to transmit. Simplifying, we get

$$\text{Th} = p_a\,(1 - b s_0) \qquad \text{packets/time step} \tag{15.5}$$

The throughput in units of packets/s is

$$\text{Th}' = \frac{\text{Th}}{T} \qquad \text{packets/s} \tag{15.6}$$

where T is the time step value.

To find the efficiency of the switch, we must estimate the input traffic in units of packets/time step. The input traffic is given by

$$N_a(\text{in}) = 1 \times a + 0 \times b = a \qquad \text{packets/time step} \tag{15.7}$$

The efficiency of the input queuing switch is defined as the ratio between output traffic and input traffic:

$$\begin{aligned}\eta &= \frac{N_a(\text{out})}{N_a(\text{in})} \\ &= \frac{p_a\,(1 - b s_0)}{a}\end{aligned} \tag{15.8}$$

According to our assumptions, packets are lost in an input queuing switch when a packet arrives on a full queue and no packet can leave the queue. To find the lost traffic in the switch, we must first find the probability that the queue becomes full. Using the traffic conservation principle, the average lost traffic $N_a(\text{lost})$ is given by

$$\begin{aligned}N_a(\text{lost}) &= N_a(\text{in}) - N_a(\text{out}) \\ &= a - p_a\,(1 - b s_0) \qquad \text{packets/time step}\end{aligned} \tag{15.9}$$

The average lost traffic per second is given by

$$N_a'(\text{lost}) = \frac{N_a(\text{lost})}{T} \tag{15.10}$$

The packet loss probability L at the input queue is given by

$$\begin{aligned}L &= \frac{N_a(\text{lost})}{N_a(\text{in})} \\ &= 1 - \frac{p_a\,(1 - b s_0)}{a}\end{aligned} \tag{15.11}$$

The average queue size is given by the equation

$$Q_a = \sum_{i=0}^{B} i s_i \qquad \text{packets} \tag{15.12}$$

where s_i is the probability that there are i packets in the input queue. s_i is given by the equation

$$s_i = \frac{(1-\rho)\rho^i}{1-\rho^{B+1}}; \qquad 0 \le i \le B \tag{15.13}$$

The wait time, or delay, of the input queuing switch is mainly due to the input queue. Using Little's result, the switch delay is given by

$$Q_a = W \, \mathrm{Th} \tag{15.14}$$

or

$$W = \frac{Q_a}{\mathrm{Th}} \qquad \text{time steps} \tag{15.15}$$

The delay in seconds is given by

$$W' = W \, T \qquad \text{seconds} \tag{15.16}$$

15.2.1 Congestion in Input Queuing Switch

When the arrival probability starts to increase, congestion takes place due to the unstable situation described below:

1. The increasing input traffic results in $s_0 \rightarrow 0$.
2. The queues start to demand access to the switching fabric.
3. The increased traffic sensed by the SF leads to decreased acceptance probability p_a.
4. Decreased p_a will lead to decreased throughput at the switch outputs.
5. The decreased p_a will lead to more filling of the queues.
6. Ultimately cell loss starts to occur.

15.2.2 Performance Bounds on Input Queuing Switch

Under full load conditions, the buffers of the input queuing switch become full and we can assume

$$a \rightarrow 1 \tag{15.17}$$

$$s_0 \rightarrow 0 \tag{15.18}$$

$$Q_a \rightarrow B \tag{15.19}$$

The maximum throughput is given, from (15.5), by

$$\text{Th}(\text{max}) = p_a \qquad \text{packets/time step} \tag{15.20}$$

The minimum efficiency of the switch is given, from (15.8), by

$$\eta(\text{min}) = p_a \tag{15.21}$$

The maximum lost traffic is given, from (15.9), by

$$N_a(\text{lost})_{\text{max}} = 1 - p_a \qquad \text{packets/time step} \tag{15.22}$$

Packet loss probability is given, from (15.11), by

$$L(\text{max}) = 1 - p_a \tag{15.23}$$

The maximum wait time is given, from (15.15), by

$$W(\text{max}) = \frac{B}{\text{Th}(\text{max})} \tag{15.24}$$

$$= \frac{B}{p_a} \tag{15.25}$$

We see the importance of designing a good switching fabric (i.e., high packet acceptance probability $p_a \to 1$) as a prerequisite for designing a high-performance switch.

Example 1 Plot the input queuing switch performance for a 10×10 switch that uses two different input buffer sizes $B_1 = 16$ and $B_2 = 64$ and assuming the SF packet acceptance probability is $p_a = 0.7$.

Figure 15.3 shows the switch throughput, efficiency, and loss probability for the switch with two input buffer sizes. The dotted line is for $B = 16$ and the solid line is for $B = 64$.

We note that the switch throughput saturates and becomes equal to the SF acceptance probability p_a as soon as the input traffic a approaches $p_a = 0.7$. This trend is the same irrespective of the buffer size.

The switch efficiency is 100% as long as $a < p_a$. However, when $a \approx p_a$, the efficiency of both switches decreases at approximately the same rate.

The delay of both switches increases with increased input traffic since the input buffers start to fill up. As soon as the arrival probability approaches the SF acceptance probability ($a \approx p_a$), we see that the delay of both switches saturates very rapidly toward its maximum value. Of course, the switch with the smaller buffers has lower delay as shown by the dotted line. Furthermore, the maximum delay of each switch is governed by the expression

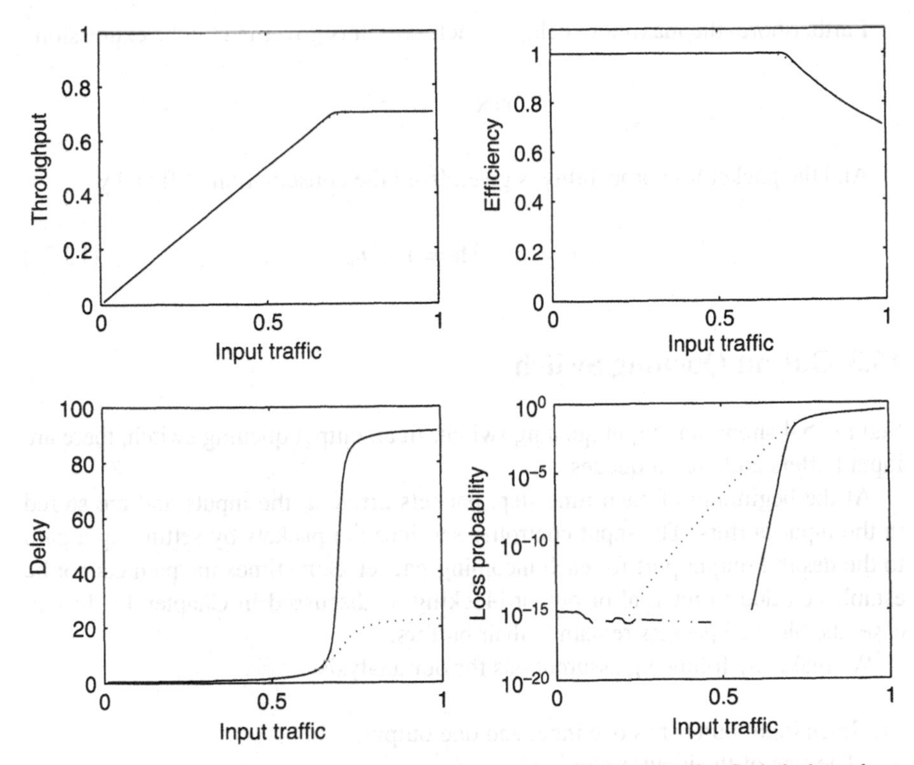

Fig. 15.3 Performance of an input queuing switch versus the input traffic. Switch throughput, efficiency, delay, and loss probability are plotted against input traffic. The *solid line* is for a switch with input queue size $B = 64$ and the *dotted line* is for a switch with input queue size $B = 16$. Switching fabric packet acceptance probability $p_a = 0.7$

$$W(\max) = B/p_a$$

The packet loss probability is very small for both switches as long as the input traffic is small. However, packet loss probability increases early on with increasing traffic for the switch with smaller buffer size. As soon as $a \approx p_a$, both switches will have large packet loss probability equal to its maximum value given by

$$L(\max) = 1 - p_a$$

■

Although the previous example was crude and employed several simplifications, we can make one very important conclusion. The throughput of the input queuing switch is limited by the performance of the switch fabric it is using. The best throughput we can hope to achieve is given by

$$\text{Th}(\max) \leq p_a \tag{15.26}$$

Furthermore, the maximum delay of each switch is governed by the expression

$$W(\max) = B/p_a$$

And the packet loss probability is given from the conservation of flow by

$$L = 1 - \text{Th} = 1 - p_a \tag{15.27}$$

15.3 Output Queuing Switch

Figure 15.4 shows an output queuing switch. In an output queuing switch, there are input buffers and output queues.

At the beginning of each time step, packets arrive at the inputs and are stored in the input buffers. The input controller services the packets by setting up a path to the desired output port for each incoming packet. Sometimes the path cannot be established due to internal or output blocking as discussed in Chapter 13. In that case, the blocked packets remain in their buffers.

We make the following assumptions for our analysis.

1. Each input buffer has one input and one output.
2. The size of the input buffer is B_1.
3. a is the packet arrival probability for any input of the switch.
4. p_a is the packet departure probability from any input buffer, where p_a is the packet acceptance probability of the switching fabric.
5. Packets could be served in the same time step at which they arrive.
6. Each arriving packet has equal probability $1/N$ of requesting an output port.
7. Each output queue has m inputs and one output. m depends on the specific design of the switch fabric.
8. The size of the output queue is B_2.
9. The packet departure probability from the output queue is 1 assuming the queue is not empty.
10. Data broadcast or multicast are not implemented.
11. Packets will be lost when a packet arrives at full input buffer *or* a full output queue.

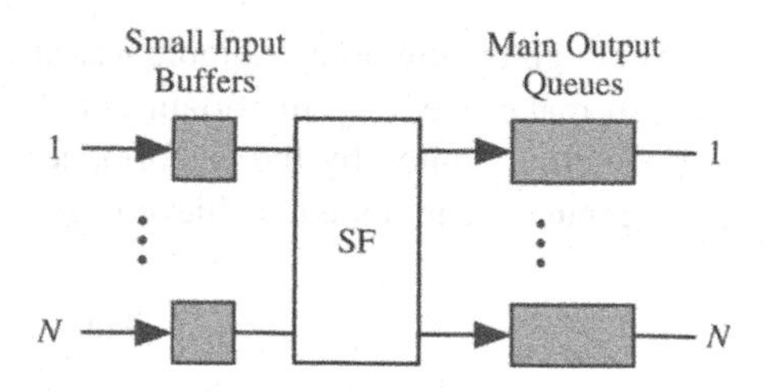

Fig. 15.4 The output queuing switch

The last assumption implies that HOL packets travel through the interconnection network to the output queue even if it is full. This assumption effectively decouples the analysis of input buffers and output queue models.

Figure 15.5 shows one output queue associated with an input port and one output queue associated with an output port. The arrival and departure probabilities for each queue are shown.

15.3.1 Modeling the Input Buffer

Queuing analysis of the buffers at each input port is very similar to the analysis used for the input queue switch previously discussed. The assumptions imply the input buffer can be treated as $M/M/1/B$ queue since only one packet can arrive per time step and only one packet can depart. The arrival probability at an input buffer is given by

$$a_1 = a \tag{15.28}$$

where a is the packet arrival probability at one of the input ports. The departure probability from an input buffer c_1 is given by

$$c_1 = p_a \tag{15.29}$$

where p_a is the packet acceptance probability of the interconnection network used.

The probability that the input buffer is empty is given by

$$e_1 = s_0 = \frac{1 - \rho_1}{1 - \rho_1^{B_1+1}} \tag{15.30}$$

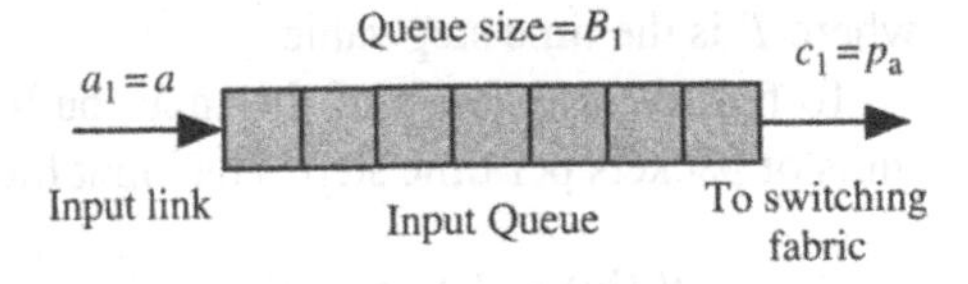

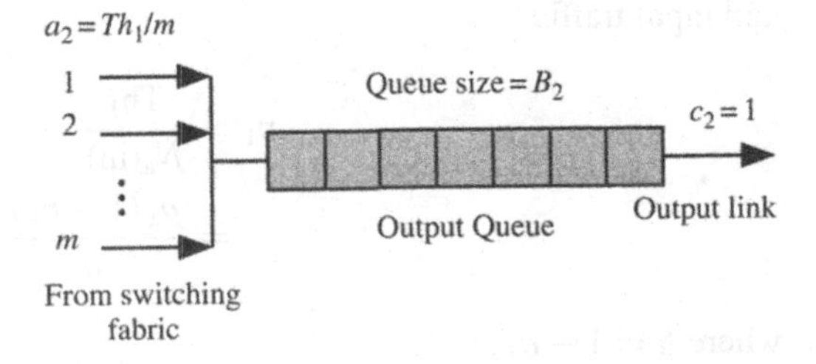

Fig. 15.5 The arrival and departure probabilities for one input buffer and one output queue of an output queuing switch

where e_l is the probability that the input buffer is empty, B_l is the size of the input buffer, and ρ_l was identified as the distribution index which is given by

$$\rho_l = \frac{a_l\, d_l}{b_l\, c_l} \tag{15.31}$$

with $b_l = 1 - a_l$ and $d_l = 1 - c_l$. Substituting the values of a_l and c_l, we get

$$\rho_l = \frac{a\, d}{b\, p_a} \tag{15.32}$$

where $b = 1 - a$ and $d = 1 - p_a$.

The throughput or output traffic for the input buffer is given by

$$\begin{aligned} \text{Th}_l &= N_{a,l}(\text{out}) \\ &= a_l\, c_l\, e_l + \sum_{i=1}^{B_l} c\, s_i \\ &= a\, p_a\, e_l + p_a\,(1 - e_l) \qquad \text{packets/time step} \end{aligned} \tag{15.33}$$

where s_i is the probability that the buffer has i packets. The first term on the RHS is the probability that a packet arrives while the buffer is empty and the second term on the RHS is the probability that the buffer has packets to transmit. Simplifying, we get

$$\text{Th}_l = p_a\,(1 - b\, e_l) \qquad \text{packets/time step} \tag{15.34}$$

The throughput in units of packets/s is

$$\text{Th}'_l = \frac{\text{Th}_l}{T} \qquad \text{packets/s} \tag{15.35}$$

where T is the time step value.

To find the efficiency of the input buffer, we must estimate the input traffic in units of packets per time step. The input traffic is given by

$$N_a(\text{in}) = 1 \times a_l + 0 \times b_l = a \qquad \text{packets/time step} \tag{15.36}$$

The efficiency of the input buffer is defined as the ratio between output traffic and input traffic:

$$\begin{aligned} \eta_l &= \frac{\text{Th}_l}{N_a(\text{in})} \\ &= \frac{p_a\,(1 - b e_l)}{a} \end{aligned} \tag{15.37}$$

where $b = 1 - a$.

The average lost traffic in the input buffer is given by

$$\begin{aligned} N_{a,1}(\text{lost}) &= N_a(\text{in}) - \text{Th}_1 \\ &= a - p_a(1 - be_1) \qquad \text{packets/time step} \end{aligned} \tag{15.38}$$

The packet loss probability at the input buffer is given by

$$\begin{aligned} L_1 &= 1 - \eta_1 \\ &= 1 - \frac{p_a(1 - be_1)}{a} \end{aligned} \tag{15.39}$$

The average buffer size is given by the equation

$$Q_1 = \sum_{i=0}^{B_1} i\ s_i \qquad \text{packets} \tag{15.40}$$

where s_i is the probability that there are i packets in the input buffer. s_i is given by the equation

$$s_i = \frac{(1 - \rho_1)\rho_1^i}{1 - \rho_1^{B_1+1}}; \qquad 0 \le i \le B_1 \tag{15.41}$$

The distribution index ρ_1 is obtained as usual for a $M/M/1/B$ queue.
Using Little's result, the input buffer delay is given by

$$Q_1 = W_1\ \text{Th}_1 \qquad \text{packets} \tag{15.42}$$

or

$$W_1 = \frac{Q_1}{\text{Th}_1} \qquad \text{time steps} \tag{15.43}$$

The delay in seconds is given by

$$W_1' = W_1\ T \qquad \text{seconds} \tag{15.44}$$

15.3.2 Modeling the Output Queue

The assumptions we used for the switch imply that the output queue can be treated as $M^m/M/1/B$ queue since a maximum of m packets can arrive per time step and only one packet can depart. The value of m depends on the switching fabric (SF) used and the maximum number of packets that could be accepted in one time step by the output queue. For example, if we had a GCN network, then $m = 1$. If we had a banyan network, then $m = 2$. If we had an ADMN network, then $m = 3$. If we had an ILN network, then $m = \lg N + 1$, where N is the number of input ports of the switch.

The packet arrival probability at each of the m links for an output queue is denoted by a_2. We obtain a value for a_2 using the flow conservation principle which states that

$$\text{Total traffic into switching fabric} \approx \text{Total traffic out of switching fabric}$$

The above equation is true only if we ignore packet loss flow in the switch, which is reasonable for most high-performance switches. Mathematically, we could write this as

$$N \times \text{Th}_1 \approx N m a_2 \tag{15.45}$$

Therefore, a_2 is given by

$$\begin{aligned} a_2 &= \frac{\text{Th}_1}{m} \\ &= \frac{p_a (1 - b\, e_1)}{m} \end{aligned} \tag{15.46}$$

Thus, if we are able to determine the throughput of the input buffers, we will be able to proceed with the study of the output queues.

The probability of i packets arriving at an output queue is given by

$$r_i = \binom{m}{i} a_2^i \,(1 - a_2)^{m-i} \tag{15.47}$$

where m is the maximum number of packets that could arrive at an output queue in one time step. The departure probability of the output queue is really simple

$$c_2 = 1 \tag{15.48}$$

Output Queue State Transition Matrix

Having found the probability r_i that i packets arrive at each output port, we are now able to write down the state transition matrix for the tagged output queue. This is a lower Hessenberg matrix of dimension $(B_2 + 1) \times (B_2 + 1)$ with m subdiagonals.

For the special case when $m = 3$ and $B_2 = 6$, $\mathbf{P}_2$ is 7×7 and is given by

$$\mathbf{P}_2 = \begin{bmatrix} r_0 + r_1 & r_0 & 0 & 0 & 0 & 0 & 0 \\ r_2 & r_1 & r_0 & 0 & 0 & 0 & 0 \\ r_3 & r_2 & r_1 & r_0 & 0 & 0 & 0 \\ 0 & r_3 & r_2 & r_1 & r_0 & 0 & 0 \\ 0 & 0 & r_3 & r_2 & r_1 & r_0 & 0 \\ 0 & 0 & 0 & r_3 & r_2 & r_1 & r_0 \\ 0 & 0 & 0 & 0 & r_3 & p_2 & p_1 \end{bmatrix} \quad (15.49)$$

where the terms p_i at the bottom row are such that the sum of each column must add to unity. It was assumed that packets could be served in the same time step at which they arrive. This explains the double term for element p_{11} of the matrix.

We are now able to derive the output queue performance once we solve for the equilibrium distribution vector. The equation we have to solve is

$$\mathbf{P}_2 \, \mathbf{s} = \mathbf{s} \quad (15.50)$$

This is an eigenvalue problem or we can convert it to a system of homogeneous linear equations:

$$(\mathbf{P}_2 - \mathbf{I}) \, \mathbf{s} = 0 \quad (15.51)$$

Since we have a lower Hessenberg matrix, we use forward substitution by assuming a value for $s_0 = 1$, say. The above matrix gives us all the other components of $\mathbf{s}$. To get the true value for the state vector $\mathbf{s}$, we use the normalizing equation

$$\sum_{i=0}^{5} s_i = 1 \quad (15.52)$$

Let us assume that the sum of the components that we obtained for the vector s gives

$$\sum_{i=0}^{5} s_i = x \quad (15.53)$$

Then we must divide each value of $\mathbf{s}$ by x to get the true normalized vector that we desire.

Output Queue Performance

The throughput of the output queue Th_2 (packets/time step) is obtained from the equation

$$\text{Th}_2 = (1 - e_2) + e_2 \, (1 - r_0) \quad (15.54)$$

where e_2 is the probability that the output queue is empty. The first term on the RHS is the probability that a packet leaves the queue, given that the output queue is not empty. In that case, the statistics of packet arrival do not matter. The second term on the right-hand side is the probability that a packet leaves the queue, given that it was empty and one or more packets arrived at the output queue. Simplifying, we get

$$\mathrm{Th}_2 = 1 - r_0\, e_2 \qquad \text{packets/time step} \tag{15.55}$$

To find the efficiency of the output queue, we must estimate the input traffic at that queue in units of packets/time step. The average number of packets arriving at the tagged output is given by

$$N_{\mathrm{a},2}(\mathrm{in}) = \sum_{i=0}^{m} i\, r_i \tag{15.56}$$

$$= a_2\, m \tag{15.57}$$

$$= \mathrm{Th}_1 \qquad \text{packets/time step} \tag{15.58}$$

The efficiency of the output queue is given by

$$\eta_2 = \frac{\mathrm{Th}_2}{N_{\mathrm{a},2}(\mathrm{in})} = \frac{\mathrm{Th}_2}{\mathrm{Th}_1} = \frac{1 - r_0\, e_2}{p_{\mathrm{a}}\,(1 - b\, e_1)} \tag{15.59}$$

According to our assumptions, packets are lost at an output queue if more than one packet arrive when the queue is starting to fill. The maximum number of arriving packets depends on the details of the interconnection network. To find the lost traffic at the tagged output queue, we must first find the traffic at its input. It is a lot more simpler to estimate the lost traffic using the traffic conservation principle:

$$N_{\mathrm{a},2}(\mathrm{lost}) = N_{\mathrm{a},2}(\mathrm{in}) - Th_2 = \mathrm{Th}_1 - \mathrm{Th}_2 \qquad \text{packets/time step} \tag{15.60}$$

The packet loss probability is

$$\begin{aligned} L_2 &= \frac{N_{a,2}(\text{lost})}{N_{a,2}(\text{in})} \\ &= 1 - \frac{\text{Th}_2}{\text{Th}_1} \\ &= 1 - \frac{1 - r_0 e_2}{p_a (1 - b\, e_1)} \end{aligned} \quad (15.61)$$

The average queue size is given by the equation

$$Q_2 = \sum_{i=0}^{B_2} i\, s_i \qquad \text{packets} \quad (15.62)$$

where s_i is the probability that there are i packets in the output queue.

We can invoke Little's result to estimate the average wait time (number of time steps) a packet spends in the input queue before it is routed as

$$Q_2 = W_2 \times \text{Th}_2 \qquad \text{packets} \quad (15.63)$$

where W_2 is the average number of time steps that a packet spends in the queue.

$$W_2 = \frac{Q_2}{\text{Th}_2} \qquad \text{time steps} \quad (15.64)$$

15.3.3 Putting It All Together

The throughput of the switch per output port equals the throughput of the output queue

$$\begin{aligned} \text{Th} &= \text{Th}_2 \\ &= 1 - r_0\, e_2 \qquad \text{packets/time step} \end{aligned} \quad (15.65)$$

The efficiency of the switch is given by

$$\begin{aligned} \eta &= \frac{N_{a,2}(\text{out})}{N_{a,1}(\text{in})} \\ &= \frac{\text{Th}_2}{a} \\ &= \frac{1 - r_0 e_2}{a} \end{aligned} \quad (15.66)$$

The lost traffic for the switch is given by

$$\begin{aligned} N_{\rm a}(\text{lost}) &= N_{\rm a,1}(\text{in}) - N_{\rm a,2}(\text{out}) \\ &= a - \text{Th}_2 \\ &= a - 1 + r_0 e_2 \qquad \text{packets/time step} \end{aligned} \tag{15.67}$$

Because the events of packet loss in the input buffer and output queue are not mutually exclusive, the total packet loss probability is given by

$$\begin{aligned} L &= L_1 + L_2 - L_1\, L_2 \\ &= \frac{N_{\rm a}(\text{lost})}{N_{\rm a}(in)} \\ &= 1 - \frac{1 - r_0 e_2}{a} \end{aligned} \tag{15.68}$$

The average queue length is given by

$$Q_{\rm a} = Q_1 + Q_2 \qquad \text{packets} \tag{15.69}$$

The total delay of packets within the switch is given by

$$W = W_1 + W_2 \qquad \text{time steps} \tag{15.70}$$

15.3.4 Performance Bounds on Output Queuing Switch

Under full load conditions, the input queue switch becomes full and we can assume

$$a \rightarrow 1 \tag{15.71}$$

$$e_1 \rightarrow 0 \tag{15.72}$$

$$a_2 \rightarrow p_{\rm a}/m \tag{15.73}$$

$$e_2 \rightarrow \epsilon \tag{15.74}$$

$$Q_1 \rightarrow B_1 \tag{15.75}$$

$$Q_2 \rightarrow B_2 \tag{15.76}$$

where $0 < \epsilon \ll 1$.

The maximum throughput is found from (15.65) when $e_2 = \epsilon$ and $a = 1$:

$$\begin{aligned} \text{Th(max)} &= 1 - \epsilon \left(1 - \frac{p_{\rm a}}{N}\right)^N \\ &\approx 1 - \epsilon\, e^{-p_{\rm a}} \qquad \text{packets/time step} \end{aligned} \tag{15.77}$$

The minimum efficiency of the switch is given from (15.66) by

$$\begin{aligned}\eta(\text{min}) &= \frac{\text{Th(max)}}{N_a(\text{in})_{\text{max}}} \\ &= 1 - \epsilon\left(1 - \frac{p_a}{m}\right)^m \\ &\approx 1 - \epsilon\, e^{-p_a}\end{aligned} \tag{15.78}$$

The maximum lost traffic is given from (15.67) by

$$\begin{aligned}N_a(\text{lost})_{\text{max}} &= N_a(\text{in})_{\text{max}} - \text{Th(max)} \\ &= \epsilon\left(1 - \frac{p_a}{m}\right)^m \\ &\approx \epsilon\, e^{-p_a} \qquad \text{packets/time step}\end{aligned} \tag{15.79}$$

where we assumed at maximum load $N_a(\text{in})_{\text{max}} = 1$.

Maximum packet loss probability in (15.68) is be given by

$$\begin{aligned}L(\text{max}) &= \frac{N_a(\text{lost})_{\text{max}}}{N_a(\text{in})_{\text{max}}} \\ &= \epsilon\, e^{-p_a}\end{aligned} \tag{15.80}$$

The maximum queue size is given by

$$Q(\text{max}) = B_1 + B_2 \tag{15.81}$$

The maximum wait time is given from (15.70) by

$$W(\text{max}) = \frac{B_1}{p_a} + \frac{B_2}{1 - \epsilon\, e^{-p_a}} \tag{15.82}$$

We see the importance of designing a good switching fabric (i.e., high packet acceptance probability $p_a \rightarrow 1$) as a prerequisite for designing a high-performance switch.

Example 2 Plot the output queuing switch performance for switches; one has $B_1 = 16$ and $B_2 = 64$ and the other has $B_1 = 4$ and $B_2 = 16$. Assume the SF is an ADMN network with packet acceptance probability $p_a = 0.7$.

For an ADMN network, $m = 3$. Figure 15.6 shows the switch throughput, efficiency, and loss probability for the switch with two input queue sizes. The solid line is for a switch with $B_1 = 16$ and $B_2 = 64$. The dotted line is for a switch with $B_1 = 4$ and $B_2 = 16$.

We note that the switch throughput saturates at a maximum value Th $= p_a$ as soon as a approaches p_a. This trend is the same irrespective of the buffer size.

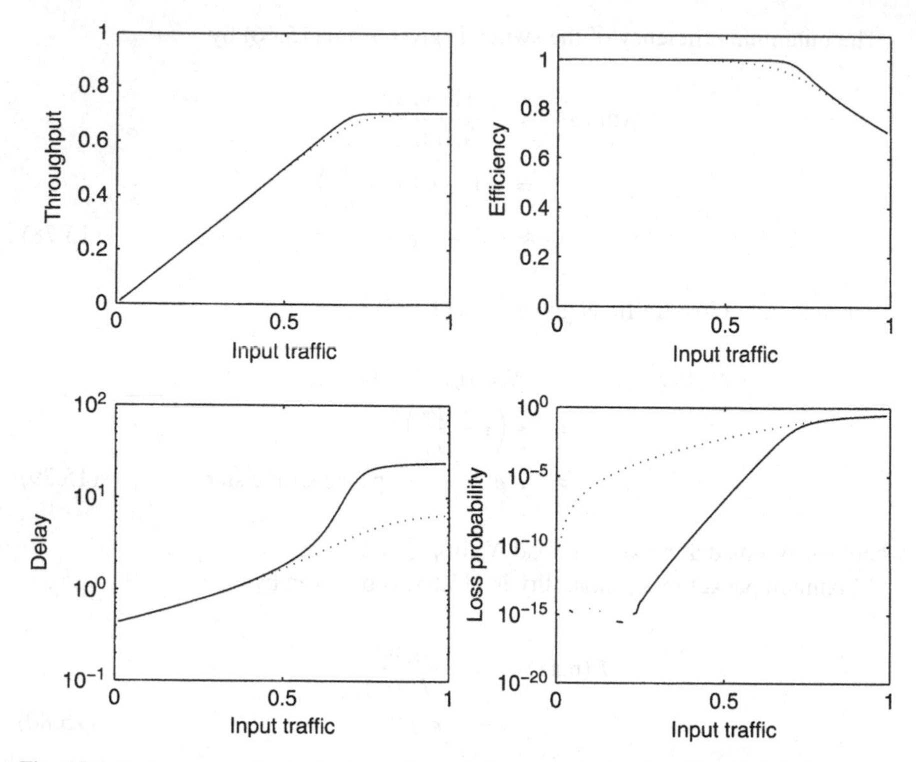

Fig. 15.6 Performance of an output queuing switch versus the input traffic. Switch throughput, efficiency, delay, and loss probability are plotted against input traffic. The *solid line* is for a switch with $B_1 = 16$ and $B_2 = 64$. The *dotted line* is for a switch with $B_1 = 4$ and $B_2 = 16$. An ADMN network is assumed with $p_a = 0.7$

The switch efficiency is 100% for most of the input traffic as long as $a \leq p_a$. When the input traffic is $a \approx p_a$, the efficiency of both switches decreases at approximately the same rate. However, the efficiency of the switch with smaller input and output queues starts its decrease slightly before a approaches p_a.

The delay of both switches increases with increased input traffic since the input queues start to fill up. As soon as the arrival probability approaches the SF acceptance probability ($a \approx p_a$), we see that the delay of both switches saturates at its maximum value. The switch with larger buffers shows higher delay as shown.

The switch with smaller input and output queues shows higher packet loss probability even for very small values of a. As soon as $a \approx p_a$, both switches give very high and equal cell loss probabilities. ■

Although the previous example was crude and employed several simplifications, we can make one very important conclusion. The throughput of the input queuing switch is limited by the performance of the switch fabric it is using. The best throughput we can hope to achieve is given by

$$\text{Th(max)} \leq p_a \tag{15.83}$$

And the packet loss probability is given from the conservation of flow by

$$L = 1 - \text{Th} = 1 - p_a \tag{15.84}$$

15.4 Shared Buffer Switch

Figure 15.7 shows a shared buffer switch. In a shared buffer switch, there is one common memory that is accessed by all input and output ports. The memory is organized, using linked lists, into several FIFO queues such that each output port has associated with it at least one queue. At the beginning of each time step, packets arrive at the inputs and are temporarily stored in a small buffer at each input port until the shared memory controller services them. The shared memory controller scans each input port in turn and appends incoming packets to the correct FIFO queue associated with each output port.

We make the following assumptions to simplify our analysis:

1. The shared buffer is divided into N linked lists (or queues) such that each linked list is associated with an output port.
2. The maximum size of each linked list is B and the total size of the shared memory is NB.
3. Each queue has N inputs and one output.
4. a is the packet arrival probability at any input of the switch.
5. Packet departure probability from any queue is 1.
6. Packets could be served in the same time step at which they arrive.
7. Each arriving packet has equal probability $1/N$ of being appended at the end of any linked list associated with an output port.
8. Data broadcast or multicast are not implemented.
9. Packets will be lost when more than one packet are destined to an output port whose linked list is full.

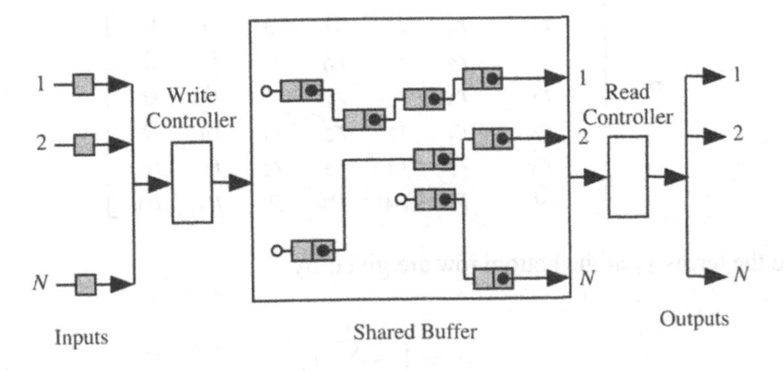

Fig. 15.7 The shared buffer switch

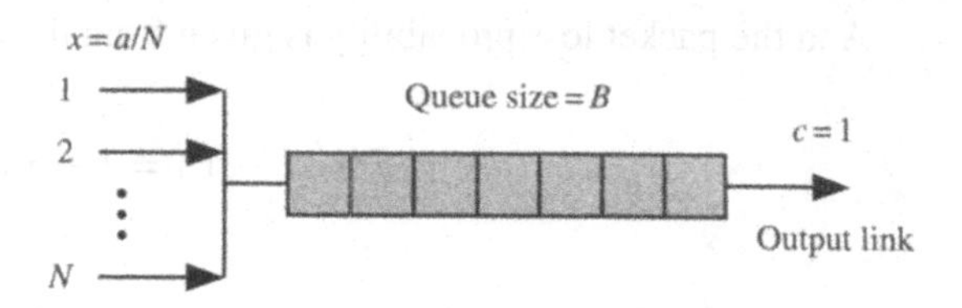

Fig. 15.8 The arrival and departure probabilities for one linked list associated with an output port. The arrival and departure probabilities are shown

Figure 15.8 shows one linked list associated with an output port. The arrival and departure probabilities are shown.

Under these assumptions, we can model the shared buffer as a *collection of independent* $M^N/M/1/B$ queues. We study one linked list belonging to an output port. We call this the tagged output port. The probability that an input port has a packet destined to the tagged output is

$$x = \frac{a}{N} \tag{15.85}$$

The probability that the tagged output port receives i packets in one time slot is given by

$$r_i = \binom{N}{i} x^i (1-x)^{N-i} \tag{15.86}$$

where $0 \le i \le N$.

According to our assumptions, the departure probability from the linked list is

$$c = 1 \tag{15.87}$$

Having found the arrival and departure probabilities, we are now able to write down the state transition matrix for the tagged linked list. This is a lower Hessenberg matrix of dimension $(B+1) \times (B+1)$ with B subdiagonals.

For the special case when $B = 6$, $\mathbf{P}$ is given by

$$\mathbf{P} = \begin{bmatrix} r_0 + r_1 & r_0 & 0 & 0 & 0 & 0 & 0 \\ r_2 & r_1 & r_0 & 0 & 0 & 0 & 0 \\ r_3 & r_2 & r_1 & r_0 & 0 & 0 & 0 \\ r_4 & r_3 & r_2 & r_1 & r_0 & 0 & 0 \\ r_5 & r_4 & r_3 & r_2 & r_1 & r_0 & 0 \\ r_6 & r_5 & r_4 & r_3 & r_2 & r_1 & r_0 \\ 0 & p_5 & p_4 & p_3 & p_2 & p_1 & p_0 \end{bmatrix} \tag{15.88}$$

where the terms p_i at the bottom row are given by

$$p_i = 1 - \sum_{j=0}^{i} r_j \tag{15.89}$$

This ensures that that the sum of each column of the transition matrix is unity. For example, p_2 is given by

$$p_2 = 1 - (r_0 + r_1 + r_2) \tag{15.90}$$

The solution to the state vector is simply found by assuming a value for s_0 and then using forward substitution to find all the other components. The normalizing condition is invoked to obtain the true value for the state vector. Once the state vector is determined using numerical techniques, the switch performance can be calculated.

Shared buffer performance
The throughput of the tagged output Th is given by

$$\text{Th} = (1 - s_0) + s_0\,(1 - r_0) \tag{15.91}$$

where s_0 is the probability that the linked list is empty. The first term on the RHS is the probability that a packet leaves the queue given that the queue is not empty. In that case, the statistics of packet arrival do not matter. The second term on the RHS is the probability that a packet leaves the queue, given that it was empty and one or more packets arrived at the queue input. Simplifying, we get

$$\text{Th} = 1 - s_0 r_0 \qquad \text{packets/time step} \tag{15.92}$$

To find the efficiency of the shared buffer switch, we must first find the traffic arriving at the tagged linked list. The average number of packets arriving at the tagged output is given by

$$\begin{aligned} N_{\mathrm{a}}(\mathrm{in}) &= \sum_{i=0}^{N} i\, r_i \\ &= x\,N \\ &= a \qquad \text{packets/time step} \end{aligned} \tag{15.93}$$

The efficiency of the input queuing switch is defined as the ratio between output traffic and input traffic:

$$\begin{aligned} \eta &= \frac{N_a(out)}{N_a(in)} \\ &= \frac{1 - s_0\, r_0}{a} \end{aligned} \tag{15.94}$$

According to our assumptions, packets are lost in a shared buffer switch if more than one packet arrive when the tagged linked list is starting to fill. To find the lost traffic at the tagged output linked list, we must first find the traffic at its input. It is a lot more simpler to estimate the lost traffic using the traffic conservation principle

$$\begin{aligned} N_a(lost) &= N_a(in) - Th \\ &= a - (1 - s_0\, r_0) \qquad \text{packets/time step} \end{aligned} \tag{15.95}$$

The packet loss probability is given by

$$L = 1 - \eta = 1 - \frac{1 - s_0\, r_0}{a} \tag{15.96}$$

The average queue size is given by the equation

$$Q_a = \sum_{i=0}^{B} i\ s_i \qquad \text{packets} \tag{15.97}$$

where s_i is the probability that there are i packets in the queue.

We can invoke Little's result to estimate the average wait time (number of time steps) a packet spends in the input buffer before it is routed as

$$Q_a = W \times Th \tag{15.98}$$

where W is the average number of time steps that a packet spends in the queue.

$$W = \frac{Q_a}{Th} \qquad \text{time steps} \tag{15.99}$$

15.4.1 Performance Bounds on Shared Buffer Switch

Under full load conditions, the buffer of the shared buffer switch becomes full and we can assume

$$a \rightarrow 1 \tag{15.100}$$

$$s_0 \rightarrow \epsilon \tag{15.101}$$

$$Q_a \rightarrow B \tag{15.102}$$

where $\epsilon \ll 1$. The maximum throughput is found from (15.92) when $s_0 = \epsilon$ and $a = 1$:

$$\begin{aligned} \text{Th(max)} &= 1 - \epsilon \left(1 - \frac{1}{N}\right)^N \\ &\approx 1 - \epsilon\, e^{-1} \qquad \text{packets/time step} \end{aligned} \tag{15.103}$$

The minimum efficiency of the switch is given from (15.94) by

$$\begin{aligned}\eta(\min) &= 1-\epsilon\left(1-\frac{1}{N}\right)^N \\ &\approx 1-\epsilon\, e^{-1}\end{aligned} \tag{15.104}$$

The switch efficiency is very close to 100% even at full load. Maximum packet loss probability in (15.96) will be given by

$$\begin{aligned}L(\max) &= \epsilon\left(1-\frac{1}{N}\right)^N \\ &= \epsilon\, e^{-1}\end{aligned} \tag{15.105}$$

The maximum lost traffic is

$$\begin{aligned}N_a(\text{lost})_{\max} &= L \times N_a(\text{in})_{\max} \\ &= \epsilon\left(1-\frac{1}{N}\right)^N \\ &= \epsilon\, e^{-1} \qquad \text{packets/time step}\end{aligned} \tag{15.106}$$

where we assumed at maximum load $N_a(\text{in})_{\max} = 1$. The maximum wait time is given by the approximate formula

$$\begin{aligned}W(\max) &= \frac{B}{\text{Th}(\max)} \\ &= \frac{B}{1-\epsilon\left(1-\frac{1}{N}\right)^N} \\ &= \frac{B}{1-\epsilon\, e^{-1}}\end{aligned} \tag{15.107}$$

Example 3 Find the performance of a 10×10 shared buffer switch for the two cases when the size of the linked list per output port is limited to $Q = 16$ and $B = 4$.

Figure 15.9 shows the switch throughput, efficiency, and loss probability for the switch with two buffer sizes. The solid line is for a switch with $B = 16$ and the dotted line is for a switch with $B = 4$.

We note that the switch throughput shows little signs of saturation. However, the smaller buffer size results in slightly reduced throughput at high input traffic.

The switch efficiency is 100% for most of the operating range. However, when $a \approx 1$, the efficiency of the switch with the smaller buffer decreases slightly.

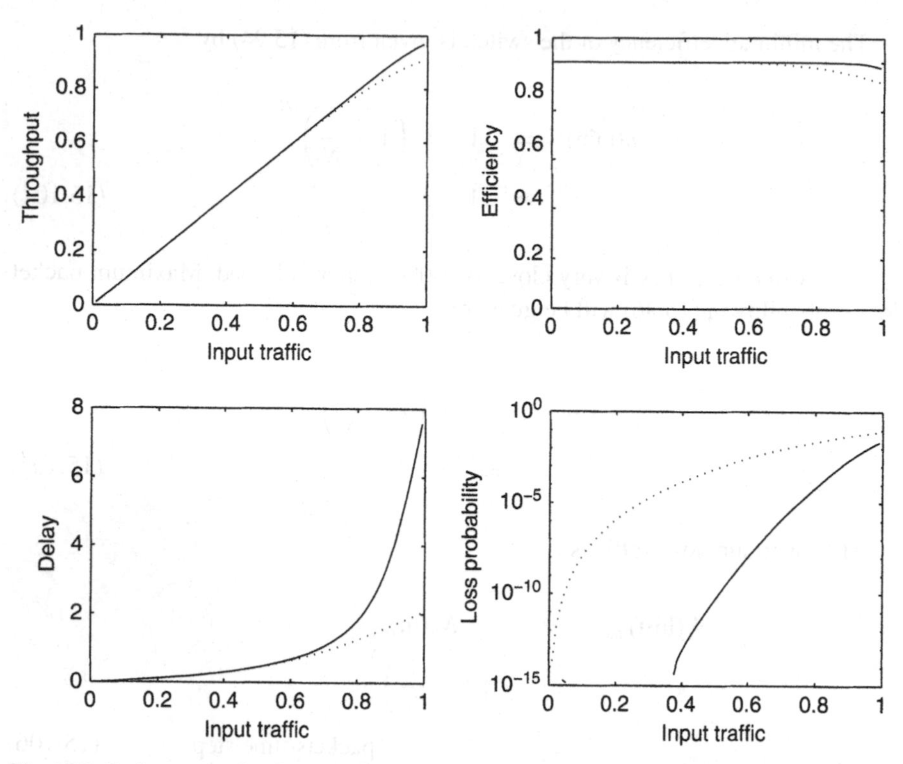

Fig. 15.9 Performance of a shared buffer switch versus the input traffic. Switch throughput, efficiency, delay, and loss probability are plotted against input traffic. The *solid line* is for a switch with $B = 16$ and the *dotted line* is for a switch with $B = 4$

The delay of both switches increases with increased input traffic since the buffers start to fill up. The switch with the larger buffer shows increased delay compared to the switch with the smaller buffer.

The packet loss probability increases with input traffic. However, the switch with the smaller buffer shows higher packet loss probability even for very small input traffic. ■

15.5 Comparing the Three Switches

In this section, we attempt to compare the performance of the three types of switches discussed earlier. For a full comparison, the switch parameters should be changed over their practical ranges. However, our aim here is simplicity.

We assume that the input queuing switch has an input buffer of size $B = 16$ and the packet acceptance probability of the switching fabric is $p_a = 0.7$.

We assume that the output queuing switch has an input buffer of size $B_1 = 4$ and an output queue of size $B_2 = 16$. The packet acceptance probability of the switching fabric is $p_a = 0.7$. The switch fabric is such that only three packets can be accepted by the output buffer at any time step (i.e., $m = 3$).

We assume that the shared buffer switch such that each output port has a queue of size $B = 16$ and the switch is of size $N = 10$.

Under these conditions, the simplified models for the three switches produce the performance curves shown in Fig. 15.10.

The throughput of all switch types are identical until the input traffic approaches $a \approx p_a$. After that, the shared buffer switch shows increasing throughput but the input and output queuing switches show signs of throughput saturation.

The efficiency of all switch types are identical and equal to unity until the input traffic approaches $a \approx p_a$. After that, the shared buffer switch still has unity efficiency but the input and output queuing switches show signs of decreasing efficiency.

The shared buffer switch shows the least delay among the three switch types for most of the input traffic range. On the other hand, the input and output queuing switches show almost identical delays until $a \approx p_a$. After that, the input buffer switch shows larger delay.

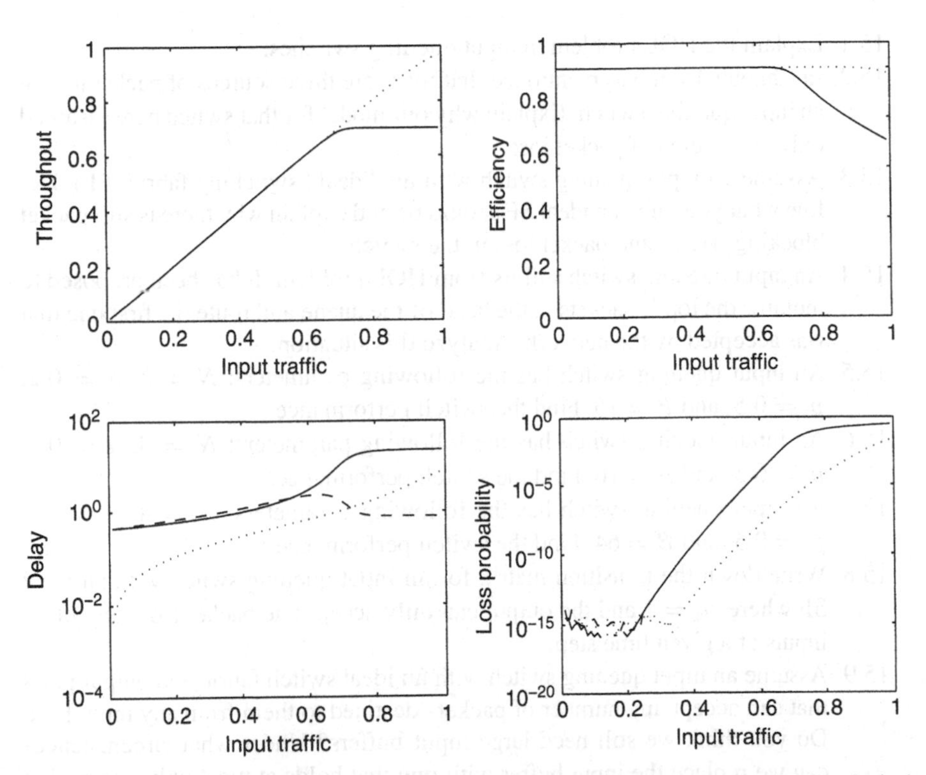

Fig. 15.10 Performance of an input buffered switch (*solid lines*), output buffer switch (*dashed lines*), and shared buffer switch (*dotted lines*) versus the input traffic. Switch throughput, efficiency, delay, and loss probability are plotted against input traffic

The packet loss probability is lowest for the shared buffer switch under the assumed parameters. The output and input queuing switches showed the highest packet loss probability.

15.6 Modeling Other Switch Types

It is possible to use the techniques developed in the previous sections to model other switch types such as multiple input queuing, multiple output queuing, and multiple input and output queuing. We leave this to the problems at the end of this chapter.

Problems

Input Queuing Switch

15.1 Explain the HOL problem in input queuing switches.

15.2 In Chapter 13, it was mentioned that there are three sources of packet loss in an input queuing switch. Explain why our model for that switch here assumed only one source of packet loss.

15.3 Assume an input queuing switch with an "ideal" switching fabric (SF). Define what you think an ideal SF should be and explain why there is still packet blocking, HOL, and packet loss in the switch.

15.4 An input queuing switch suffers from HOL problem. It has been proposed to monitor the top L packets at the head of the queue and route the first one that was accepted by the network. Analyze this situation.

15.5 An input queuing switch has the following parameters: $N = 8$, $a = 0.2$, $p_a = 0.5$, and $B = 16$. Find the switch performance.

15.6 An input queuing switch has the following parameters: $N = 4$, $a = 0.5$, $p_a = 0.5$, and $B = 16$. Find the switch performance.

15.7 An input queuing switch has the following parameters: $N = 4$, $a = 0.8$, $p_a = 0.5$, and $B = 64$. Find the switch performance.

15.8 Write down the transition matrix for an input queuing switch with an ideal SF where $p_a = 1$ and the output can only accept one packet from any of its inputs at a given time step.

15.9 Assume an input queuing switch with an ideal switch fabric and output ports that can accept any number of packets destined to them from any input port. Do you think we still need large input buffers? Under what circumstances can we replace the input buffer with one that holds at most only one packet so that the input port can process it.

Output Queuing Switch

15.10 In the output queuing switch we assumed that the output queue can accept a maximum of L packets per time step. Based on that, do you think output queuing has HOL problems just like input queuing switches? Explain your answer.

15.11 In Chapter 13, it was mentioned that there are four sources of packet loss in an output queuing switch. Explain why our model for that switch here assumed only two sources of packet loss.

15.12 Equation (15.61) on page 565 for packet loss probability in the output queue assumed that the maximum number of arriving packets is smaller than the size of the buffer ($m < B_2$). Write down the equation when $m > B_2$.

15.13 Write down the transition matrix for an output queuing switch with an ideal SF where $p_a = 1$ and the output can only accept one packet from any of its inputs at a given time step.

15.14 An output queuing switch has the following parameters: $N = 8$, $a = 0.2$, $p_a = 0.5$, $B_1 = 4$, and $B_2 = 16$. Find the switch performance.

15.15 An output queuing switch has the following parameters: $N = 4$, $a = 0.5$, $p_a = 0.5$, $B_1 = 4$, and $B_2 = 16$. Find the switch performance.

15.16 An output queuing switch has the following parameters: $N = 4$, $a = 0.8$, $p_a = 0.5$, $B_1 = 16$, and $B_2 = 64$. Find the switch performance.

15.17 In the output queue switch analysis, we effectively decoupled the input buffer from the output queue by insisting that packets at the output buffer can move to an output queue even if it were full. Can you remove this assumption and analyze the switch?

Shared Buffer Switch

15.18 In Chapter 13, it was mentioned that there are two sources of packet loss in a shared buffer switch. Explain why our model for that switch here assumed only one source of packet loss.

15.19 Equation (15.96) on page 572 for packet loss probability in the linked list assumed that the maximum number of arriving packets is smaller than the size of the buffer ($N < B$). Write down the equation when $N > B$.

15.20 In the shared buffer switch, we assumed that the linked list for each output can accept a maximum of N packets per time step. Based on that, do you think shared buffering has HOL problems just like input queuing switches? Explain your answer.

15.21 In a shared buffer switch, the linked list for each output can accept a maximum of one packet only per time step. Based on that, do you think shared buffering has HOL problems just like input queuing switches? Explain your answer.

15.22 Assume that the linked list for each output of a shared buffer switch can accept a maximum of m packets only per time step ($1 \leq m \leq N$). Write down an expression for the transition matrix and derive the performance equations.

Other Switch Structures

15.23 Develop a Markov chain analysis of the multiple input queuing switch.
15.24 Develop a Markov chain analysis of the multiple output queuing switch.
15.25 Develop a Markov chain analysis of the multiple input and output queuing switch.

References

1. H.J. Siegel, R.J. McMillen and P.T. Mueller, "A survey of Interconnection methods for reconfigurable parallel processing systems", *Proc. AFIPS 1979*, vol. 48, pp. 529–542, 1979.
2. F.A. Tobagi, "Fast packet switch architectures for broadband integrated services digital networks", *Proc. IEEE*, vol. 78, pp. 133–166, 1990.
3. S.E. Butner and R. Chivukula, "On the limits of electronic ATM switching", *IEEE Networks*, vol. 10, no. 6, pp. 26–31, Nov./Dec. 1996.
4. A. Sabaa, F. Elguibaly and D. Shpak, "Design and modeling of a nonblocking input-buffer ATM switch", *Canadian Journal of Electrical and Computer Engineering*, vol. 22, pp. 87–93, 1997.
5. D. Present, C. Fayet, and G. Pujolle, "An optimal solution for ATM switches", *Computer Networks and ISDN Systems*, vol. 29, pp. 2039–2052, 1998.
6. R.Y. Awdeh and H.T. Muftah, "Survey of ATM switch architectures", *Computer Networks and ISDN Systems*, vol. 27, pp. 1567–1613, 1995.
7. J. Garcia-Haro and A. Jajszczyk, "ATM shared-memory switching architectures", *IEEE Networks*, vol. 8, no. 4, pp. 18–26, Jul./Aug. 1994.
8. M. Murata, "Requirements on ATM switch architectures for quality-of-service guarantees", *IEICE Transactions on Communications.*, vol. E81-B, pp. 138–151, 1998.
9. A. Patavinal, "Nonblocking architecture for ATM switching", *IEEE Communications Magazine*, pp. 38–48, Feb. 1993.
10. B. Patel, F. Schaffa, and M. Willebeek-LeMair, "The helix switch: A single chip cell switch design", *Computer Networks and ISDN Systems*, vol. 28, pp. 1791–1807, 1996.
11. "Design and evaluation of scalable shared-memory ATM switches", *IEICE Transactions on Communications.*, vol. E81-B, pp. 224–236, 1998.
12. N. Endo, T. Kozaki, T. Ohuchi, H. Kuwahara, and S. Gohara, "Shared buffer memory switch for an ATM exchange", *IEEE Trans. Commun.*, vol. 41, pp. 237–245, 1993.
13. J.-F. Lin and S.-D. Wang, "A high performance fault-tolerant switching network for ATM", *IEICE Transactions on Communications.*, vol. E781-B, pp. 1518–1528, 1995.
14. S.-C. Yang and J.A. Silvester, "A reconfigurable ATM switch fabric for fault tolerance and traffic balancing", *IEEE Journal on Selected Areas in Communications*, vol. 9, pp. 1205–1217, 1991.

Chapter 16
Examples of Switches

16.1 Introduction

The different switch architectures discussed in Chapter 13 attempts to provide simultaneous paths between any input to any output. The switch performance varied depending on the location of the buffers/queues and depended on the type of switching fabric employed. As a result, different switches produced different throughput, delay, and packet loss probability. In addition to high performance, a switch must be able to provide extra features such as multiple service classes (differentiated service) and multicast and broadcast capabilities.

Traffic through the switch encounters multiple contention points before a path to the desired output could be found. The main contention points in a switch occur at these locations:

1. At the input ports
2. Within the switching fabric
3. At the output port

Contentions lead to delay, packet loss, congestion, and reduced throughput. Like a chain, the performance of the switch is as good as the performance of the worst-managed contention point [1]. Multiple contention points may also introduce head of line (HOL) blocking, a state in which a packet at the head of a queue loses contention and so blocks packets behind it from traversing the switch fabric. In an attempt to avoid HOL blocking, some architectures use feedback flow control. This strategy references the available capacity of output buffers to control the flow rate from input buffers. Multiple contention points may add complex overhead and queue-management tasks, introducing the potential for performance degradation.

To counteract the effect of performance loss due to multiple contention points, the switch designer must consider these approaches:

1. Increase buffer and queue sizes and place the queues at the output only.
2. Create multiple queues for each connection (per VC in ATM jargon) or service class.
3. Employ selective packet discard based on the connection or service class.

F. Gebali, *Analysis of Computer and Communication Networks*,
DOI: 10.1007/978-0-387-74437-7_16,

4. Use a switching fabric that does not offer as much internal contention. Typically, this is achieved through switching fabrics with multiple paths, dilating each link, and speedup of the switch fabric operation.
5. Employ sophisticated routing algorithms through the switch fabric if possible.
6. Employ large buffers or queues at each contention point. This might not be practical if buffers are used within the switch fabric.
7. Use traffic shaping or flow control to reduce congestion by limiting the number of packets coming from a certain input.

In an effort to support multiple service classes and certain quality of service guarantees, a design might provide several queues at each input port. Each queue is dedicated to a virtual circuit (VC) or flow. The total number of queues Q required in the switch is given by

$$Q = NV \tag{16.1}$$

where N is the number of ports and V is the number of virtual circuits, or flows, supported per port. It is possible to have $Q = 64 \times 4000$ or more.

Good switch architectures eliminate unnecessary points of contention by combining a contentionless switch fabric with buffering on output which confines contention to the output ports. And good switch performance is ensured only if the output queues are managed with sophisticated traffic scheduling algorithms.

Next-generation switches are characterized by their lack of traditional switch fabric and use of output queuing. These two features lead to single contention point during accessing output ports. This chapter presents two next-generation designs: the VRQ switch and N.E.T.'s Promina 4000® ATM switch.

16.2 Promina 4000 Switch

A block diagram of the Promina 4000 switch is shown in Fig. 16.1 [1, 2]. The switch is an output queue switch since the queues are located at the output ports. This is one of the main requirements of a high-performance switch. The switch fabric is contention-free (contentionless) and is based on a broadcast matrix of dedicated backplane buses.

The switch is composed of N input/output port cards and N backplane buses. The input block in each module is a four-to-one multiplexer such that the output of the multiplexer is connected to a dedicated backplane bus. The output block in each module has N inputs and four outputs, as indicated. Each backplane bus is dedicated to carry the traffic of one input port and is connected to one of the N inputs of each output port.

Fig. 16.1 Block diagram for N.E.T.'s Promina 4000 ATM switch

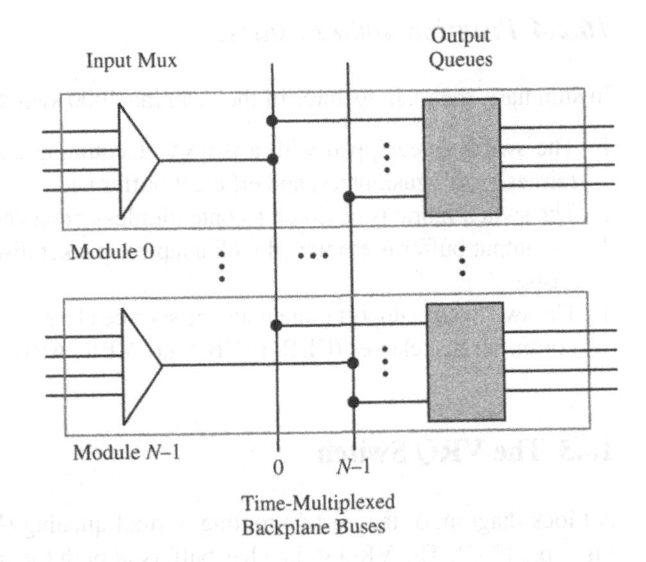

16.2.1 Input Port Operation

The input port accepts data from four 155 Mbps lines. A TDMA multiplexer places those packets on the bus dedicated to the input port. The input data are broadcast to all the outputs and each output selects the data destined to it. Note that the switch does not require any input buffers, thereby circumventing the HOL problem. Input buffers are not needed since each incoming packet is broadcast to all the output ports and each output port decides whether to accept the packet or not.

16.2.2 Backplane Bus Operation

Each bus is dedicated to an input port. The bus operates at 622 Mbps and sends its data to all the output ports. Thus, each bus can carry one 622 Mbps channel or four 155 Mbps TDMA channels.

16.2.3 Output Port Operation

Each input port accepts data from N buses at a maximum rate of $622 \times N$ Mbps. The output port has a common buffer that is used to queue the packets. However, the switch maintains information about buffer occupancy for each connection (VC). Thus, packet discard is performed on a per-VC basis which is fair to all other connections and all the service classes supported by the switch.

We see that the switch is inherently modular in architecture since the inputs do not communicate among themselves to do their operation. Similarly, we see that the outputs do not communicate among themselves to do their operation.

16.2.4 Promina 4000 Features

In summary, the main features of the Promina 4000 switch are the following:

1. The switch is equipped with a per-VC accounting capability to meet goals of fairness, QoS guarantees, and efficient buffer use.
2. The switch fabric is based on a contentionless broadcast matrix.
3. An output buffer is employed with adaptive packet discard scheme on a per-VC basis.
4. The switch can support more than one service class for each of the standard ATM Forum service classes (CBR, rt-VBR, nrt-VBR, ABR, and UBR).

16.3 The VRQ Switch

A block diagram of the virtual routing, virtual queuing (VRQ) switch is shown in Fig. 16.2 [3–5]. The VRQ switch has buffers at both the input and output modules. However, the switch could be classified as an output queue switch since all decisions about packet transmission and scheduling are carried out at the output ports. This is one of the main requirements of a high-performance switch. The switch fabric is contention-free (contentionless) and is based on a broadcast matrix of dedicated backplane buses.

An $N \times N$ VRQ switch is composed of N stacked modules where each module contains an input port and an output port, as shown. The modules are connected together using N backplane buses. Each bus is dedicated to an output port but can take its data from any input port. The reader can immediately see that contention

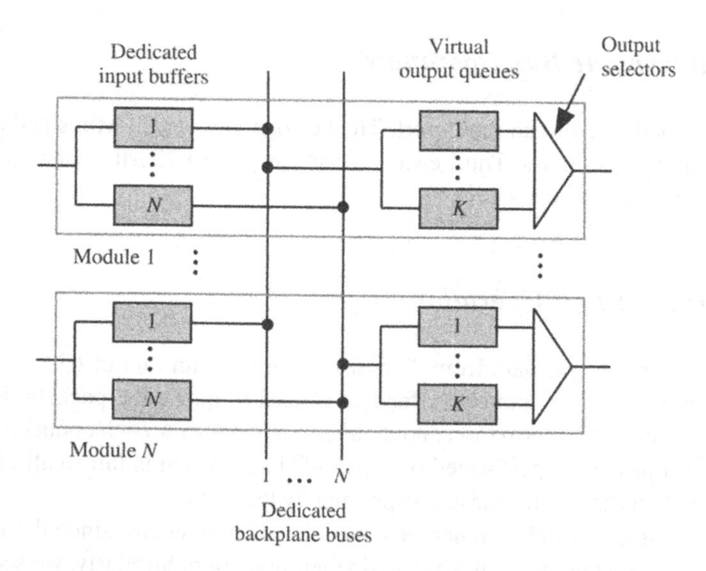

Fig. 16.2 The virtual routing, virtual queuing (VRQ) high-performance switch

could potentially occur when two or more inputs want to access a certain output. This is prevented here since the bus is *output driven*, not *input driven*, in the sense that each output dictates which input is to access the bus. This choice is based on the scheduling algorithm employed at each output port.

16.3.1 Input Port Operation

The input port contains N buffers and each buffer is dedicated to an output port. This in itself presents certain advantages and disadvantages. The main advantage is that a buffer has to perform only one read/write operation per time step, irrespective of the switch size.

Clever memory design ensures that each input buffer has a cycle time that is the least possible through elimination of memory address decoders and bit line drivers. This design ensures that buffer speed will present no bottleneck to switch performance.

When a packet arrives, the input port controller routes it to one of the input buffer in the module, based on the header information. The input controller also informs the destination output port that a packet has arrived and also provides to the output an *address*, or a *pointer*, indicating where this packet is located in the input buffer. Thus packet routing has been virtually accomplished through this pointer exchange, hence the name *virtual routing*. The packet will actually get routed out of the switch when the output port selects it based on its scheduling algorithm.

16.3.2 Backplane Bus Operation

Each bus is dedicated to an output port and could be driven by any input port, as shown in the figure. The output port controls access to the bus, and we say that the bus is output driven. Thus, only one packet moves across the bus per time step and the bus speed matches the line speed without the need to use any speedup factor or multiplexing schemes, such as TDMA. Again, bus speed will present no bottleneck to switch performance. In effect, the switch fabric is contentionless since it is based on a matrix of *dedicated buses* that are *output driven*.

16.3.3 Output Port Operation

The output port contains K queues. The queues store the packet pointers and not the actual packets, hence the name *virtual queuing*. The queues could be constructed based on a per-connection basis, per-input basis, or per-service class basis. In either case, the size of these queues is minimal since they do not store actual packets. They only store pointer to the packets. This gives the designer much freedom in configuring the queues based on actual traffic requirements.

16.3.4 VRQ Features

In summary, the main features of the VRQ switch are as follows:

1. *Fast input buffers*: Each input module of the switch contains input buffers to store incoming packets. Those buffers are shift register based, as opposed to RAM based. The use of shift registers dramatically increases buffer speed due to removal of address decoding circuitry, word and bit line drivers, and sense amplifier delays. A shift register acts in effect as a pipelined memory with operation speed dictated only by the speed of moving data between adjacent flip-flops. Thus, memory delay is the sum of setup and delay times of one-bit flip-flop. This is a much higher memory speed compared to a standard memory where the delay time is determined by the sum of delays of the address decoder, word line, bit line, and sense amplifier.
2. *Distributed routing*: The routing operation is localized in each input port.
3. *Distributed scheduling algorithm:* Packet selection among the different output queues is localized in each output port. The small size of output queues enables supporting several scheduling algorithms.
4. *Contentionless switch fabric*: The switching fabric is output driven which prevents any contention and its operating speed matches the line rate.
5. *Modularity*: The switch is very modular at the input ports, output ports, and at the backplane switch matrix itself.
6. *Local communication*: The input and output modules need not communicate any state information. This removes the need for global communication between the input ports.
7. *Point-to-point bussing*: The backplane buses between each output port and associated buffers at the inputs can be implemented by point-to-point optical fiber lines. This results in fast operation and ability to directly drive output optical fiber links.
8. *Low packet loss probability*: Packets can only be lost due to filled input buffers because internal and output blocking are eliminated.

16.4 Comparing Promina 4000 with VRQ Switch

As a final note, we can compare the two switches since superficially both seem to possess very similar structures. The main differences between the two switches are:

1. In the VRQ switch, each backplane bus is dedicated to an *output* port. In the Promina 4000 switch, each backplane bus is dedicated to an *input* port.
2. The buses in the VRQ switch are output driven to prevent contention and collisions. Thus, contention is removed in the VRQ switch without having to resort to time multiplexing, as in the Promina 4000 switch. The buses in the Promina switch are input driven and contention, or collision, is avoided by using TDMA

multiplexing. This implies high bus speed which could prove to be a bottleneck for higher line rates.

3. The backplane bus speed in the VRQ switch exactly matches the line rate (e.g., 155 Mbps), while the bus speed in the Promina 4000 switch is four times the input line rate. This might prove to be a bottleneck if higher line rates are contemplated unless space division switching (SDM), instead of TDMA, is employed. Of course, this will only increase the area, power, and pin count of the system.
4. The buffer location is different in the two switches. In the VRQ switch, packets are stored in input buffers. In one time step, only one read/write operation is required as a maximum. This relaxes to a great extent the requirements on memory cycle time. In the Promina 4000 switch, the buffer is located at each output port. In one time step, a maximum of N write and one read operations are required. This naturally is a severe restraint on the memory cycle time. The option, of course, is not to share the buffer among all the connections or service classes but to partition the memory so that the number of access requests is reduced.
5. Buffer partitioning is also different in the two switches. In the VRQ switch, a two-level partitioning scheme is employed. First, the buffers are distributed among the input ports. Second, the buffers in each input port are further partitioned such that each local partition stores packets destined to a particular output port. In the 4000 switch, a one-level partitioning scheme is employed. The buffers are distributed among the output ports. In one time step, a maximum of N write and one read operations are required. This naturally is a severe restraint on the memory cycle time. The option of course is not to share the buffer among all the connections or service classes but to partition the memory so that the number of access requests is reduced.
6. Traffic flow in the VRQ switch is output driven which eliminates contention altogether. In the Promina 4000 switch, traffic flow is input driven which normally gives rise to contention. This is avoided by use of time-division multiplexing and dedicated backplane buses dedicated to each input port.

16.5 Modeling the VRQ Switch

We provide in this section a simple queuing model for the VRQ switch to illustrate its performance and to provide more examples of applying queuing theory to another switch architecture. Figure 16.3 is a simplified version of the VRQ switch. The diagram shows a VRQ switch with N input ports and each input port has N buffers for storing the packets destined to the N output ports. Each output port has N virtual queues such that there is one virtual output queue dedicated to each input port. Let us look at a particular tagged output port. That port will have associated with it N input buffers in each of the N input ports as shown in Fig. 16.4. The figure shows the backplane bus connecting all the input buffers in each input to the output virtual queue at the tagged output port.

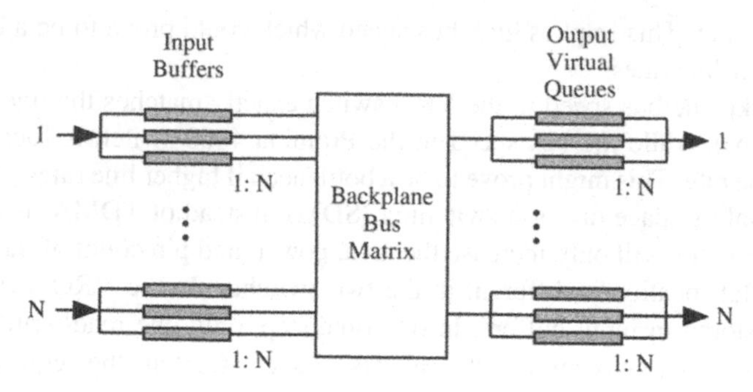

Fig. 16.3 A simple implementation of the virtual routing, virtual queuing (VRQ) high-performance switch

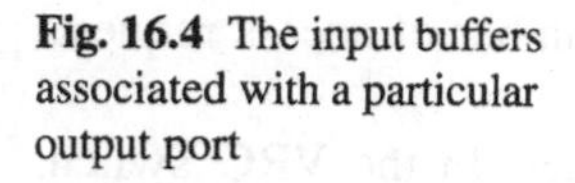
Fig. 16.4 The input buffers associated with a particular output port

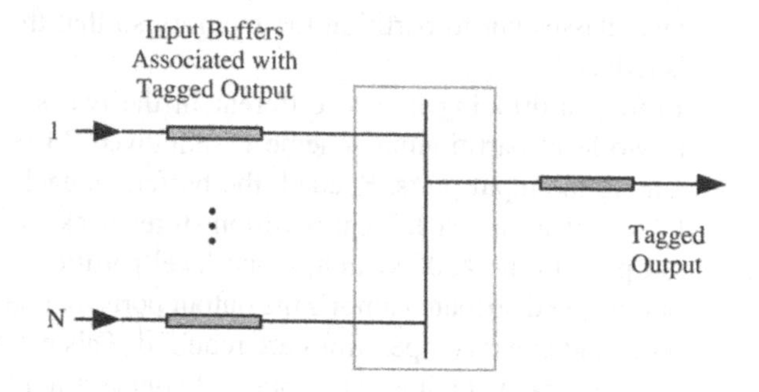

We make the following assumptions for our analysis.

1. There are N buffers at each input port.
2. Each input buffer has one input and one output.
3. The size of each input buffer is B_1.
4. a is the packet arrival probability at any input of the switch.
5. An arriving packet is served at the same time step at which it arrives.
6. Each arriving packet has equal probability $1/N$ of requesting an output port.
7. There is only one service class supported by the switch. Support of different service classes requires more elaborate modeling or can be done accurately only using numerical techniques.
8. There are N queues at each output port such that each output queue is dedicated to a particular input port.
9. Each output queue has N inputs and one output.
10. The size of each output queue is B_2.
11. When a packet arrives at an input port, the corresponding output queue is also updated.
12. When a packet leaves an output port, the corresponding input buffer is also updated.
13. Data broadcast or multicast are not implemented.

14. Packet loss occurs only in the input buffers.
15. The backplane bus is contentionless.

Based on the above assumptions, we model the input buffer in each input port as an $M/M/1/B$ queue and we model each output queue as an $M/M/1/B$ queue. Furthermore, each input buffer has an identical output queue and the buffer size for both queues are equal and is given by

$$B_1 = B_2 = B$$

16.5.1 Analysis of the Input Buffer

Queuing analysis of the buffers at each input port is very similar to the analysis used for the input buffer switch discussed in Chapter 15. The assumptions imply that the input buffer can be treated as $M/M/1/B$ queue since only one packet can arrive per time step and only one packet can depart. The arrival probability at an input buffer is given simply by

$$a_1 = \frac{a}{N} \tag{16.2}$$

and the departure probability from an input buffer c_1 is given by

$$c_1 = c_2 \tag{16.3}$$

where c_2 is the departure probability from an output queue. This equation shows that there is close and unavoidable coupling between the input and output queues in the VRQ switch.

The transition matrix for the input buffer is of dimension $(B+1) \times (B+1)$ and is given by

$$\mathbf{P}_1 = \begin{bmatrix} 1-a_1d_1 & b_1c_1 & 0 & \cdots & 0 & 0 & 0 \\ a_1d_1 & f & b_1c_1 & \cdots & 0 & 0 & 0 \\ 0 & a_1d_1 & f & \cdots & 0 & 0 & 0 \\ 0 & 0 & a_1d_1 & \cdots & 0 & 0 & 0 \\ \vdots & \vdots & \vdots & \ddots & \vdots & \vdots & \vdots \\ 0 & 0 & 0 & \cdots & b_1c_1 & 0 & 0 \\ 0 & 0 & 0 & \cdots & f & b_1c_1 & 0 \\ 0 & 0 & 0 & \cdots & a_1d_1 & f & b_1c_1 \\ 0 & 0 & 0 & \cdots & 0 & a_1d_1 & 1-b_1c_1 \end{bmatrix} \tag{16.4}$$

where

$$
\begin{aligned}
b_1 &= 1 - a_1 \\
d_1 &= 1 - c_1 \\
f &= a_1 c_1 + b_1 d_1
\end{aligned}
$$

The throughput or output traffic for the input buffer is given by

$$
\begin{aligned}
\mathrm{Th}_1 &= N_{\mathrm{a},1}(\mathrm{out}) \\
&= a_1\, c_1\, s_0 + \sum_{i=1}^{B} c_1\, s_i \qquad \text{packets/time step} \qquad (16.5)
\end{aligned}
$$

where s_i is the probability that the queue has i packets. The first term on the RHS is the probability that a packet arrives while the queue is empty and the second term on the RHS is the probability that the queue has packets to transmit. Simplifying, we get

$$
\mathrm{Th}_1 = c_1 \times (1 - b_1\, s_0) \qquad \text{packets/time step} \qquad (16.6)
$$

The throughput in units of packets/s is

$$
\mathrm{Th}'_1 = \frac{\mathrm{Th}_1}{T} \qquad \text{packets/s} \qquad (16.7)
$$

where T is the time step value.

To find the efficiency of the input buffer, we must estimate the input traffic in units of packets per time step. The input traffic is given by

$$
\begin{aligned}
N_{\mathrm{a}}(\mathrm{in}) &= 1 \times a_1 + 0 \times b_1 \\
&= a_1 \\
&= \frac{a}{N} \qquad \text{packets/time step} \qquad (16.8)
\end{aligned}
$$

The efficiency of the input buffer is defined as the ratio between output traffic and input traffic

$$
\begin{aligned}
\eta_1 &= \frac{\mathrm{Th}_1}{N_{\mathrm{a}}(\mathrm{in})} \\
&= \frac{c_1\,(1 - b_1 s_0)}{a_1} \qquad (16.9)
\end{aligned}
$$

The average lost traffic in the input buffer is given by

$$\begin{aligned} N_{a,1}(\text{lost}) &= N_a(\text{in}) - N_{a,1}(\text{out}) \\ &= a_1 - c_1\,(1 - b_1\, s_0) \qquad \text{packets/time step} \end{aligned} \tag{16.10}$$

The packet loss probability at the input buffer is given by

$$\begin{aligned} L_1 &= 1 - \eta_1 \\ &= 1 - \frac{c_1\,(1 - b_1 s_0)}{a_1} \end{aligned} \tag{16.11}$$

The average queue size is given by the equation

$$Q_1 = \sum_{i=0}^{B} i\ s_i \qquad \text{packets} \tag{16.12}$$

where s_i is the probability that there are i packets in the input queue.

Using Little's result, the input buffer delay is given by

$$Q_1 = W_1\ \text{Th}_1 \qquad \text{packets} \tag{16.13}$$

or

$$W_1 = \frac{Q_1}{\text{Th}_1} \qquad \text{time steps} \tag{16.14}$$

The delay in seconds is given by

$$W_1' = W_1\ T \qquad \text{seconds} \tag{16.15}$$

16.5.2 Analysis of the Output Queue

The VRQ dedicates at least one output queue for each output port. The output scheduler selects a packet from one of the queues for service based on any of the scheduling techniques discussed in Chapter 12.

For the case when the VRQ supports differentiated services with K classes, each output port would dedicate K queues for each input port. Thus, each output port would have a total of KN queues. The output scheduler selects a packet from one of the queues for service based on any of the scheduling techniques discussed in Chapter 12. However, we assume that $K = 1$ here for simplicity.

The assumptions we used for the VRQ switch imply that the output queues are $M/M/1/B$ type. The VRQ works on the principle that an arriving packet at an input port updates the contents of the virtual queue of its destination output port.

Therefore, the arrival probability at the input of a tagged output queue is given by

$$a_2 = a_1 = \frac{a}{N} \tag{16.16}$$

The departure probability of the output queue is really simple

$$c_2 = c_1 = \frac{1}{N} \tag{16.17}$$

We conclude, therefore, that the output queue dedicated to an input port behaves exactly like the input buffer we studied in the previous section.

We can write

$$\begin{aligned} \mathrm{Th}_2 &= \mathrm{Th}_1 \\ &= \frac{1}{N} \times (1 - b_1 s_0) \qquad \text{packets/time step} \end{aligned} \tag{16.18}$$

$$\eta_2 = \eta_1 = \frac{1 - b_1 s_0}{a} \tag{16.19}$$

$$\begin{aligned} N_{a,2}(\text{lost}) &= N_{a,1}(\text{lost}) \\ &= \frac{1}{N} \times (a - 1 + b_1 s_0) \qquad \text{packets/time step} \end{aligned} \tag{16.20}$$

$$L_2 = L_1 = 1 - \frac{1 - b_1 s_0}{a} \tag{16.21}$$

$$W_2 = W_1 = \frac{Q_1}{\mathrm{Th}_1} \qquad \text{time steps} \tag{16.22}$$

16.5.3 Putting It All Together

The throughput of the switch per output port equals the throughput of the output queue

$$\begin{aligned} \mathrm{Th} &= \mathrm{Th}_1 \\ &= \frac{1}{N} \times (1 - b_1 s_0) \qquad \text{packets/time step} \end{aligned} \tag{16.23}$$

The lost traffic for the switch is given by

$$\begin{aligned} N_a(\text{lost}) &= N_{a,1}(\text{lost}) \\ &= \frac{1}{N} \times (a - 1 + b_1 s_0) \qquad \text{packets/time step} \end{aligned} \tag{16.24}$$

The total packet loss probability is given by

$$\begin{aligned} L &= L_1 \\ &= 1 - \frac{1 - b_1 s_0}{a} \end{aligned} \tag{16.25}$$

and the total delay of packets within the switch is given by

$$\begin{aligned} W &= W_1 \\ &= \frac{Q_1}{\text{Th}_1} \qquad \text{time steps} \end{aligned} \tag{16.26}$$

16.5.4 Performance Bounds on VRQ Switch

Under full load conditions ($a \to 1$), the packet arrival probability in (16.2) becomes

$$a_1 = \frac{1}{N} \tag{16.27}$$

Thus, the packet arrival and departure probabilities from any input buffer become equal

$$a_1 = c_1 = \frac{1}{N} \tag{16.28}$$

The $M/M/1/B$ queue distribution index ρ becomes unity

$$\rho = \frac{a_1\ d_1}{b_1\ c_1} = 1 \tag{16.29}$$

Thus, the probability of the queue being in state i is given from the $M/M/1/B$ queue results in 7.17 on page 230 by

$$s_i = \frac{1}{B} \qquad 0 \le i \le B \tag{16.30}$$

The throughput in (16.18) under full load conditions will become

$$\text{Th(max)} = \frac{1}{N} \times \left[1 - \frac{1}{B}\left(1 - \frac{1}{N} \right) \right] \qquad \text{packets/time step} \tag{16.31}$$

When $N > 10$ and $B > 10$, we can approximate the above equation as

$$\begin{aligned} \text{Th(max)} &= \frac{1}{N}\left(1 - \frac{1}{B}\right) \\ &\approx \frac{1}{N} \qquad \text{packets/time step} \end{aligned} \tag{16.32}$$

The efficiency in (16.19) of the VRQ switch will be

$$\eta(\text{min}) = 1 - \frac{1}{B} \approx 1 \tag{16.33}$$

We note that the minimum efficiency of the VRQ switch depends on the local buffer size and not on the switch size N. This is evidence that the VRQ switch provides isolation between users.

The lost traffic in (16.20) will be

$$N_a(\text{lost}) = \frac{1}{NB} \qquad \text{packets/time step} \tag{16.34}$$

Maximum packet loss actually decreases with the buffer size *and* the switch size. It is obvious that the VRQ switch actually performs better as its size increases.

Packet loss probability in (16.21) is given by

$$L = \frac{1}{B} \tag{16.35}$$

The maximum packet loss probability depends only on the buffer size. This is, again, further evidence of the user isolation properties of the VRQ switch.

The maximum delay is given from (16.22) by

$$W(\text{max}) = \frac{NB}{2} \qquad \text{time steps} \tag{16.36}$$

Example 1 Plot the VRQ switch performance when the switch size is 10×10. Choose two different buffer sizes. One switch has $B = 64$ and the other has $B = 16$.

Figure 16.5 shows the switch throughput, efficiency, and loss probability for the VRQ switch with two input buffer sizes. The solid line is for a switch with $B = 64$ and the dotted line is for a switch with $B = 16$.

We note that the switch throughput does not saturate even at full load. The two switches show almost identical throughput but the one with smaller input buffers shows slight throughput saturation near full load.

The switch efficiency is flat at its maximum value of 100% for most of the input traffic range. The efficiency of the switch with bigger buffer size hardly shows any

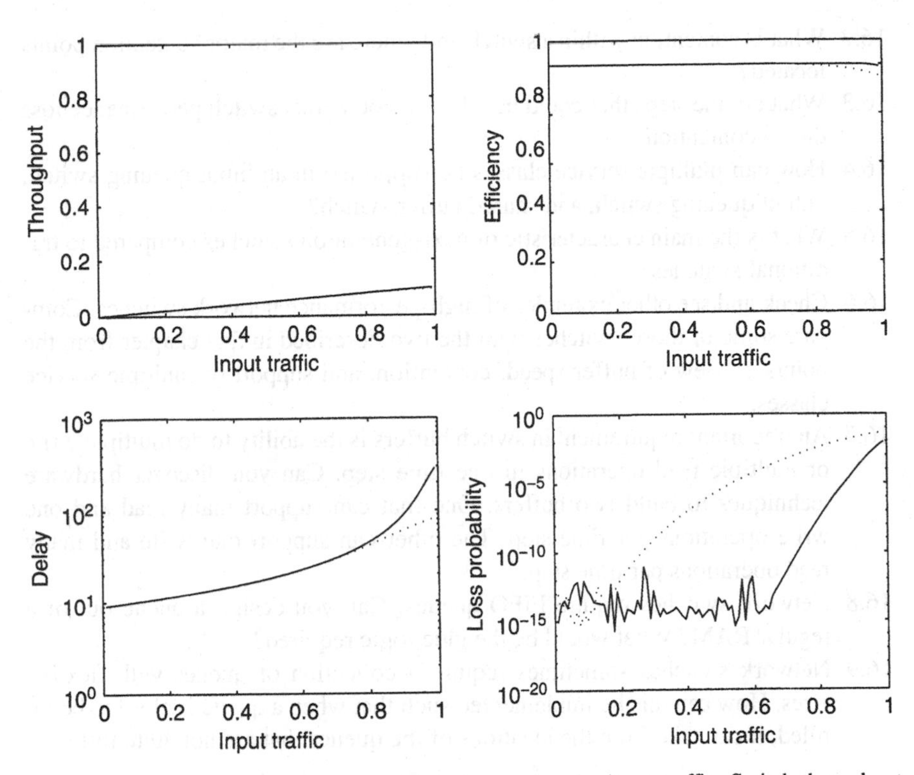

Fig. 16.5 Performance of a 10 × 10 VRQ switch versus the input traffic. Switch throughput, efficiency, delay, and loss probability are plotted against input traffic. The *solid line* is for a switch with $B = 64$ and the *dotted line* is for a switch with $B = 16$

signs of decreasing. The switch with smaller buffer shows slightly decreased efficiency only at full load.

The delay of both switches increases slowly with increased input traffic since the input buffers start to fill up. Near full load conditions, the switch with bigger buffer shows higher delay because of its larger buffer size.

The switch packet loss performance is very impressive even when buffer size is small. The switch with smaller buffer shows higher packet loss probability even for very small values of input traffic. The switch with larger buffer hardly shows any buffer loss for over one half of the range of input traffic. ■

Problems

Switch Design Strategies

16.1 What is congestion and how does it lead to degradation of throughput, packet loss, delay, and efficiency?

16.2 What is contention within a switch and where are the major contention points located?

16.3 What are the steps that could be taken to counteract switch performance loss due to contention?

16.4 How can multiple service classes be supported in an input queuing switch, output queuing switch, and shared buffer switch?

16.5 What is the main characteristic of next-generation switches compared to traditional switches?

16.6 Check and see other examples of high-performance network switches. Compare some of those switches with the two described in this chapter from the points of view of buffer speed, contention, and support of multiple service classes.

16.7 An important requirement in switch buffers is the ability to do multiple write or multiple read operations in one time step. Can you discover hardware techniques to build two buffers. One that can support many read and one write operations per time step. The other can support one write and many read operations per time step.

16.8 Network switches require FIFO queues. Can you design a queue out of a regular RAM? What would be the glue logic required?

16.9 Network switches sometimes require a collection of queues with flexible sizes. How can this be implemented such that when a queue is close to being filled, it can take from the locations of the queues that are not quite full?

Promina 4000 Switch

16.10 Explain how the Promina 4000 switch works.

16.11 Explain why the inputs of the Promina 4000 switch do not require any input buffers, thereby circumventing the HOL problem.

16.12 Is there any limitation on the Promina 4000 scaling (size and speed) based on input port operation?

16.13 How is the switch fabric of the Promina 4000 switch contentionless?

16.14 Is there any limitation on the Promina 4000 scaling (size and speed) based on backplane bus operation?

16.15 Is there any limitation on the Promina 4000 scaling (size and speed) based on output port operation?

16.16 Explain how the scheduling algorithm is implemented in the Promina 4000 switch.

VRQ Switch

16.17 Explain how the VRQ switch works.

16.18 Explain why the input buffers of the VRQ switch do not suffer from the HOL problem.

16.19 Is there any limitation on the VRQ switch scaling (size and speed) based on input port operation?
16.20 How is the switch fabric of the VRQ switch contentionless?
16.21 Is there any limitation on the VRQ switch scaling (size and speed) based on backplane bus operation?
16.22 Is there any limitation on the VRQ switch scaling (size and speed) based on output port operation?
16.23 Explain how the scheduling algorithm is implemented in the VRQ switch.

References

1. http://internet.net.com/techtop/adtmatm_wp/home.html.
2. http://internet.net.com/products/.
3. F. Elguibaly, A. Sabaa, and D. Shpak, "A new shift-register based ATM switch", *The First Annual Conference on Emerging Technologies and Applications in Communications (ETACOM)*, Portland, Oregon, pp. 24–27, May 7–10, 1996.
4. F. Elguibaly and S. Agarwal, "Design and performance analysis of shift register-based ATM switch", *IEEE Pacific Rim Conference on Communications, Computers and Signal Processing*, Victoria, British Columbia, pp. 70–73, August 20–22, 1997.
5. A. Sabaa, F. Elguibaly, and D. Shpak, "Design and modeling of a nonblocking input-buffer ATM switch", *Canadian Journal of Electrical & Computer Engineering*, vol. 22, no. 3, pp. 87–93, 1997.

Appendix A
Summation of Series

A.1 Arithmetic Series

$$\sum_{i=0}^{n-1}(a + id) = \frac{n}{2}(a + l)$$

where $l = a + (n - 1)d$

A.2 Geometric Series

$$\sum_{i=0}^{n-1} ar^i = \frac{a(1 - r^n)}{1 - r}$$

$$\sum_{i=0}^{\infty} ar^i = \frac{a}{1 - r}$$

where $r \neq 1$ in the above two equations.

A.3 Arithmetic–Geometric Series

$$\sum_{i=0}^{n-1}(a + id)\, r^i = \frac{a(1 - r^n)}{1 - r} + \frac{rd\left[1 - nr^{n-1} + (n - 1)\, r^n\right]}{(1 - r)^2}$$

where $r \neq 1$ in the above two equations. If $-1 < r < 1$, the series converge, and we get

$$\sum_{i=0}^{\infty} (a + id) r^i = \frac{a}{1-r} + \frac{rd}{(1-r)^2}$$

$$\sum_{i=0}^{\infty} i^2 r^i = \frac{r + r^2}{(1-r)^3}$$

A.4 Sums of Powers of Positive Integers

$$\sum_{i=1}^{n} i = \frac{n(n+1)}{2} \tag{A.1}$$

$$\sum_{i=1}^{n} i^2 = \frac{n(n+1)(2n+1)}{6} \tag{A.2}$$

$$\sum_{i=1}^{n} i^3 = \frac{n^2(n+1)^2}{4} \tag{A.3}$$

A.5 Binomial Series

$$\sum_{i=0}^{n} \binom{n}{i} a^{n-i} b^i = (a+b)^n$$

special cases

$$\sum_{i=0}^{n} \binom{n}{i} = 2^n$$

$$\sum_{i=0}^{n} (-1)^i \binom{n}{i} = 0$$

$$\sum_{i=0}^{n} \binom{n}{i}^2 = \binom{2n}{n}$$

$$(1+x)^q = 1 + qx + \frac{q(q-1)}{2!} x^2 + \frac{q(q-1)(q-2)}{3!} x^3 \cdots$$

$$(1+x)^{-1} = 1 - x + x^2 - x^3 + x^4 - \cdots$$

where $x < 1$ in the above equations.

A.5.1 Properties of Binomial Coefficients

$$\binom{n}{i} = \binom{n}{n-i}$$
$$\binom{n}{i} = \frac{n}{i}\binom{n-1}{i-1}$$
$$\binom{n}{i} + \binom{n}{i-1} = \binom{n+1}{i}$$
$$\binom{n}{n} = \binom{n}{0} = 1$$
$$n! \sim \sqrt{2\pi}\, n^{n+\frac{1}{2}} \mathrm{e}^{-n} \quad \text{Stirling's formula}$$

A.6 Other Useful Series and Formulas

$$\sum_{i=0}^{n} i \binom{n}{i} a^{i-1} b^{n-i} = na(a+b)^{n-1}$$
$$\sum_{i=0}^{n} i^2 \binom{n}{i} a^{i} b^{n-i} = n(n-1)a^2(a+b)^{n-2} + na(a+b)^{n-1}$$
$$\sum_{i=1}^{n} \binom{n-1}{i-1} \frac{1}{i} a^{i} (1-a)^{n-i} = \frac{1}{n}\left[1-(1-a)^n\right]$$
$$\sum_{i=0}^{\infty} \frac{a^i}{i!} = \mathrm{e}^a$$
$$\sum_{i=0}^{\infty} (i+1)\frac{a^i}{i!} = \mathrm{e}^a\,(a+1)$$
$$\lim_{n\to\infty}\left(1 \pm \frac{x}{n}\right)^n = \mathrm{e}^{\pm x}$$
$$\lim_{n\to\infty} x^n \binom{n}{i} = 0 \quad |x| < 1$$
$$\binom{n}{i} + \binom{n}{i-1} = \binom{n+1}{i}$$
$$\lim_{n\to\infty}\left(1 - \frac{a}{n}\right)^n = \mathrm{e}^{-a}$$
$$\lim_{n\to\infty} \binom{n}{i} a^i\,(1-a)^{n-i} = \frac{a^i \mathrm{e}^{-a}}{i!} \quad a < 1$$

The last equation is used to derive the Poisson distribution from the binomial distribution.

Appendix B
Solving Difference Equations

B.1 Introduction

Difference equations describe discrete-time systems just as differential equations describe continuous-time systems. We encounter difference equations in many fields in telecommunications, digital signal processing, electromagnetics, civil engineering, etc.

In Markov chains and many queuing models, we often get a special structure for the state transition matrix that produces a recurrence relation between the system states. This appendix is based on the results provided in [1], [2], and [3]. We start first by exploring simple approaches for simple situations; then we deal with the more general situation.

B.2 First-Order Form

Assume we have the simple recurrence equation

$$s_i = as_{i-1} + b \tag{B.1}$$

where a and b are given. Our task is to find values for the unknown s_i for all values of $i = 0, 1, \cdots$ that satisfy the above equation. This is a first-order form since each sample depends on the immediate past value only.

Since this is a linear relationship, we assume that the solution for s_i is composed of two components, a constant component c and a variable component v that depend on i. Thus, we write the trial solution for s_i as

$$s_i = v_i + c \tag{B.2}$$

Substitute this into our recursion to get

$$v_i + c = a\,(v_{i-1} + c) + b \tag{B.3}$$

We can group the constant parts and the variable parts together to get

$$c = ac + b \tag{B.4}$$

$$v_i = av_{i-1} \tag{B.5}$$

and the value of the constant component of s is

$$c = \frac{b}{1-a} \tag{B.6}$$

Assume a solution for v_i in (B.5) of the form

$$v_i = \lambda^i \tag{B.7}$$

Substitute this solution in the recursion formula (B.5) for v_i to get

$$\lambda^i = a\,\lambda^{i-1} \tag{B.8}$$

which gives

$$\lambda = a \tag{B.9}$$

and v_i is given by

$$v_i = \lambda^i = a^i \tag{B.10}$$

Thus, s_i is given from (B.2) as

$$s_i = a^i + \frac{b}{1-a} \tag{B.11}$$

This is the desired solution to the difference equation.

B.3 Second-Order Form

Assume that we have the simple recurrence equation

$$s_{i+1} + a\,s_i + b\,s_{i-1} = 0 \tag{B.12}$$

This is a second-order form since each sample depends on the two most recent past samples. Assume a solution for s_i of the form

$$s_i = \lambda^i \tag{B.13}$$

The recursion formula gives

$$\lambda^2 + a\,\lambda + b = 0 \tag{B.14}$$

There are two possible solutions (roots) for λ, which we denote as α and β, and there are three possible situations.

B.3.1 Real and Different Roots $\alpha \neq \beta$

When the two roots are real and different, s_i becomes a linear combination of these two solutions

$$s_i = A\alpha^i + B\beta^i \tag{B.15}$$

where A and B are constants. The values of A and B are determined from any restrictions on the solutions for s_i, such as given initial conditions. For example, if s_i represent the different components of the distribution vector in a Markov chain, then the sum of all the components must equal unity.

B.3.2 Real and Equal Roots $\alpha = \beta$

When the two roots are real and equal, s_i is given by

$$s_i = (A + \mathrm{i}B)\,\alpha^i \tag{B.16}$$

B.3.3 Complex Conjugate Roots

In that case, we have

$$\alpha = \gamma + \mathrm{j}\theta \tag{B.17}$$

$$\beta = \gamma - \mathrm{j}\theta \tag{B.18}$$

s_i is given by

$$s_i = \gamma^i\,[A\cos(\mathrm{i}\,\theta) + B\sin(\mathrm{i}\,\theta)] \tag{B.19}$$

B.4 General Approach

Consider the N-order difference equation given by

$$s_i = \sum_{k=0}^{N} a_k\, s_{i-k}, \qquad i > 0 \tag{B.20}$$

where we assumed $s_i = 0$ when $i < 0$.

We define the one-sided z-transform of s_i as

$$S(z) = \sum_{i=0}^{\infty} s_i \ z^{-i} \tag{B.21}$$

Now take the z-transform of both sides of (B.20) to obtain

$$S(z) - s_0 = \sum_{k=0}^{N} a_k \, z^{-k} \times \left[\sum_{i=0}^{\infty} s_{i-k} \, z^{-(i-k)} \right] \tag{B.22}$$

We assume that $a_k = 0$ when $k > N$ and we also assume that $s_i = 0$ when $i < 0$. Based on these two assumptions, we can change the upper limit for the summation over k and we can change the variable of summation of the term in square brackets as follows.

$$S(z) - s_0 = \sum_{k=0}^{\infty} a_k \ z^{-k} \times \left[\sum_{m=0}^{\infty} s_m \ z^{-m} \right] \tag{B.23}$$

where we introduced the new variable $m = i - k$. Define the z-transform of the coefficients a_k as

$$A(z) = \sum_{k=0}^{\infty} a_k \ z^{-k} \tag{B.24}$$

Thus, we get

$$\begin{aligned} S(z) - s_0 &= A(z) \times \left[\sum_{m=0}^{\infty} s_m \ z^{-m} \right] \\ &= A(z) \times S(z) \end{aligned} \tag{B.25}$$

Notice that the two summations on the right-hand side are now independent. Thus, we finally get

$$S(z) = \frac{s_0}{1 - A(z)} \tag{B.26}$$

We can write the above equation in the form

$$S(z) = \frac{s_0}{\mathcal{D}(z)} \tag{B.27}$$

where the denominator polynomial is

$$\mathcal{D}(z) = 1 - A(z) \tag{B.28}$$

MATLAB allows us to find the inverse z-transform of $S(z)$ using the command

$$\texttt{RESIDUE}(a, b)$$

where a and b are the coefficients of the nominator and denominator polynomials $A(z)$ and $B(z)$, respectively, in descending powers of z^{-1}.

The function RESIDUE returns the column vectors r, p, and c which give the residues, poles, and direct terms, respectively.

The solution for s_i is given by the expression

$$s_i = c_i + \sum_{j=1}^{m} r_j (p_j)^{(i-1)} \quad i > 0 \tag{B.29}$$

where m is the number of elements in r or p vectors.

References

1. J.R. Norris, *Markov Chains*, Cambridge University Press, New York, 1997.
2. V.K. Ingle and J.G. Proakis, *Digital Signal Processing Using MATLAB*, Brooks/Cole, Pacific Grove, 2000.
3. A. Antoniou, *Digital Filters: Analysis, Design, and Applications*, McGraw-Hill, New York, 1993.

Appendix C
Finding s(n) Using the z-Transform

When the transition matrix **P** of a Markov chain is not diagonalizable, we could use the z-transform technique to find the value of the distribution vector at any time instance $\mathbf{s}(n)$. An alternative, and more appealing technique, is to use the Jordan canonic form—however, we will explain the z-transform here. The z-transform of the distribution vector **s** is given by

$$\mathbf{S}(z) = \sum_{n=0}^{\infty} \mathbf{s}(n)\, z^{-n} \tag{C.1}$$

We express $\mathbf{s}(n)$ in the above equation in terms of the transition matrix **P** using the relation

$$\mathbf{s}(n) = \mathbf{P}^n\, \mathbf{s}(0) \tag{C.2}$$

Alternatively, $\mathbf{s}(n)$ can be written as

$$\mathbf{s}(n) = \mathbf{P}\, \mathbf{s}(n-1) \tag{C.3}$$

From (C.1), the z-transform of $\mathbf{s}(n)$ can be written as

$$\mathbf{S}(z) = \mathbf{s}(0) + \sum_{n=1}^{\infty} \mathbf{P}\, \mathbf{s}(n-1) z^{-n} \tag{C.4}$$

Thus we have

$$\mathbf{S}(z) - \mathbf{s}(0) = z^{-1}\mathbf{P} \sum_{n=1}^{\infty} \mathbf{s}(n-1)\, z^{-(n-1)} \tag{C.5}$$

Changing the index of summation, we get

$$\mathbf{S}(z) - \mathbf{s}(0) = z^{-1}\, \mathbf{P}\mathbf{S}(z) \tag{C.6}$$

We can thus obtain an expression for the z-transform of the distribution vector as

$$\mathbf{S}(z) = \left(\mathbf{I} - z^{-1}\mathbf{P}\right)^{-1} \mathbf{s}(0) \tag{C.7}$$

Denoting the transform pair

$$\mathbf{S}(z) \Leftrightarrow \mathbf{s}(n) \tag{C.8}$$

then we can write (C.7), using (C.2), in the form

$$\left(\mathbf{I} - z^{-1}\mathbf{P}\right)^{-1} \mathbf{s}(0) \Leftrightarrow \mathbf{P}^n \mathbf{s}(0) \tag{C.9}$$

Since $\mathbf{s}(0)$ is arbitrary, we have the z-transform pair

$$\mathbf{P}^n \Leftrightarrow \left(\mathbf{I} - z^{-1}\mathbf{P}\right)^{-1} \tag{C.10}$$

WWW: We have defined the MATLAB function `invz(B,a)` which accepts the matrix B whose elements are the nominator polynomials of $\left(\mathbf{I} - z^{-1}\mathbf{P}\right)^{-1}$ and the polynomial a which is the denominator polynomial of $\left(\mathbf{I} - z^{-1}\mathbf{P}\right)^{-1}$. The function returns the residue matrices r that correspond to each pole of the denominator.

The following two examples illustrate the use of this function.

Example 1 Consider the Markov chain matrix

$$\mathbf{P} = \begin{bmatrix} 0.5 & 0.8 & 0.4 \\ 0.5 & 0 & 0.3 \\ 0 & 0.2 & 0.3 \end{bmatrix}$$

Use the z-transform technique to find a general expression for the distribution vector at step n and find its value when $n = 3$ and 100 assuming an initial distribution vector $\mathbf{s}(0) = \begin{bmatrix} 1 & 0 & 0 \end{bmatrix}^t$

First, we must form

$$\mathbf{I} - z^{-1}\mathbf{P} = \begin{bmatrix} 1 - 0.5z^{-1} & -0.8z^{-1} & -0.4z^{-1} \\ -0.5z^{-1} & 1 & -0.3z^{-1} \\ 0 & -0.2z^{-1} & 1 - 0.3z^{-1} \end{bmatrix}$$

We have to determine the inverse of this matrix using determinants or any other technique to obtain

$$\left(\mathbf{I} - z^{-1}\mathbf{P}\right)^{-1} = \frac{1}{D} \times$$

$$\begin{bmatrix} 1 - 0.3z^{-1} - 0.06z^{-2} & 0.8z^{-1} - 0.16z^{-2} & 0.4z^{-1} + 0.24z^{-2} \\ 0.5z^{-1} - 0.15z^{-2} & 1 - 0.8z^{-1} + 0.15z^{-2} & 0.3z^{-1} + 0.05z^{-2} \\ 0.10z^{-2} & 0.2z^{-1} - 0.10z^{-2} & 1 - 0.5z^{-1} - 0.40z^{-2} \end{bmatrix}$$

where

$$D = \left(1 - z^{-1}\right)\left(1 + 0.45z^{-1}\right)\left(1 - 0.25z^{-1}\right)$$

We notice that the poles of this matrix are the eigenvalues of **P**. We now have to find the inverse z-transform of $\left(\mathbf{I} - z^{-1}\mathbf{P}\right)^{-1}$ which can be done on a term-by-term basis [1] or by using the MATLAB function `invz(B,a)` [1]:

$$\mathbf{P}^n = \begin{bmatrix} 0.59 & 0.59 & 0.59 \\ 0.32 & 0.32 & 0.32 \\ 0.09 & 0.09 & 0.09 \end{bmatrix} +$$

$$(-0.45)^n \begin{bmatrix} 0.27 & -0.51 & 0.06 \\ -0.37 & 0.70 & -0.08 \\ 0.10 & -0.19 & 0.02 \end{bmatrix} +$$

$$(0.25)^n \begin{bmatrix} 0.14 & -0.08 & -0.64 \\ 0.05 & -0.02 & -0.24 \\ -0.19 & 0.10 & 0.88 \end{bmatrix} \qquad n = 0, 1, \ldots$$

We notice that the matrix corresponding to the pole $z^{-1} = 1$ is column stochastic and all its columns are equal. For the two matrices corresponding to the poles $z^{-1} = -0.45$ and 0.25, the sum of each column is exactly zero.

s(3) is given from the above equation by substituting $n = 3$ in the expression for $\mathbf{P}^n$:

$$\mathbf{s}(3) = \mathbf{P}^3\mathbf{s}(0) = \mathbf{s}(3) = \begin{bmatrix} 0.57 \\ 0.36 \\ 0.08 \end{bmatrix}^t$$

s(100) is given by

$$\mathbf{s}(100) = \mathbf{P}^{100}\mathbf{s}(0) = \begin{bmatrix} 0.59 \\ 0.32 \\ 0.09 \end{bmatrix}^t$$

■

Example 2 Consider the Markov chain matrix

$$\mathbf{P} = \begin{bmatrix} 0.2 & 0.4 & 0.4 \\ 0.8 & 0.1 & 0.4 \\ 0 & 0.5 & 0.2 \end{bmatrix}$$

Use the z-transform technique to find a general expression for the distribution vector at step n and find its value when $n = 7$ for an initial distribution vector $\mathbf{s}(0) = \begin{bmatrix} 1 & 0 & 0 \end{bmatrix}^t$

First, we must form

$$\mathbf{I} - z^{-1}\mathbf{P} = \begin{bmatrix} 1-0.2z^{-1} & -0.4z^{-1} & -0.4z^{-1} \\ -0.8z^{-1} & 1-0.1z^{-1} & -0.4z^{-1} \\ 0 & -0.5z^{-1} & 1-0.2z^{-1} \end{bmatrix}$$

We have to determine the inverse of this matrix using determinants or any other techniques to obtain

$$(\mathbf{I} - z^{-1}\mathbf{P})^{-1} = \frac{1}{D} \times$$
$$\begin{bmatrix} 1-0.3z^{-1}-0.18z^{-2} & 0.4z^{-1}+0.12z^{-2} & 0.4z^{-1}+0.12z^{-2} \\ 0.8z^{-1}-0.16z^{-2} & 1-0.4z^{-1}+0.40z^{-2} & 0.4z^{-1}+0.24z^{-2} \\ 0.40z^{-2} & 0.5z^{-1}-0.10z^{-2} & 1-0.3z^{-1}-0.30z^{-2} \end{bmatrix}$$

where

$$D = \left(1 - z^{-1}\right)\left(1 + 0.3z^{-1}\right)\left(1 + 0.2z^{-1}\right)$$

We notice that the poles of this matrix are the eigenvalues of $\mathbf{P}$. We now have to find the inverse z-transform of $\left(\mathbf{I} - z^{-1}\mathbf{P}\right)^{-1}$ which can be done on a term-by-term basis [1] or by using the function `invz(B,a)`:

$$\mathbf{P}^n = \begin{bmatrix} 0.33 & 0.33 & 0.33 \\ 0.41 & 0.41 & 0.41 \\ 0.26 & 0.26 & 0.26 \end{bmatrix} +$$
$$(-0.3)^n \begin{bmatrix} 0.0 & 0.0 & 0.0 \\ -3.08 & 1.92 & 0.92 \\ 3.08 & -1.92 & -0.92 \end{bmatrix} +$$
$$(-0.2)^n \begin{bmatrix} 0.67 & 0.67 & 0.67 \\ 2.67 & 2.67 & 2.67 \\ -3.33 & -3.33 & -3.33 \end{bmatrix} \qquad n = 0, 1, \ldots$$

We notice that the matrix corresponding to the pole $z^{-1} = 1$ is column stochastic and all its columns are equal. For the two matrices corresponding to the poles $z^{-1} = -0.3$ and 0.2, the sum of each column is exactly zero.

$\mathbf{s}(7)$ is given from the above equation as

$$\mathbf{s}(7) = \mathbf{P}^7 \mathbf{s}(0)$$

or

$$\mathbf{s}(7) = \begin{bmatrix} 0.33 \\ 0.41 \\ 0.26 \end{bmatrix}$$

■

Problems

C.1 Use the z-transform to find the distribution vector at any time instant n for the transition matrix

$$\begin{bmatrix} 0.2 & 0.4 & 0.4 \\ 0.8 & 0.1 & 0.4 \\ 0 & 0.5 & 0.2 \end{bmatrix}$$

C.2 Use the z-transform to find the distribution vector at any time instant n for the transition matrix

$$\begin{bmatrix} 0.5 & 0.3 \\ 0.5 & 0.7 \end{bmatrix}$$

C.3 Use the z-transform to find the distribution vector at any time instant n for the transition matrix

$$\begin{bmatrix} 0.2 & 0.3 & 0.5 & 0.6 \\ 0.3 & 0.1 & 0.2 & 0.1 \\ 0.5 & 0.1 & 0.1 & 0.2 \\ 0 & 0.5 & 0.2 & 0.1 \end{bmatrix}$$

C.4 Use the z-transform to find the distribution vector at any time instant n for the transition matrix

$$\begin{bmatrix} 0.1 & 0.4 & 0.1 \\ 0.7 & 0 & 0.3 \\ 0.2 & 0.6 & 0.6 \end{bmatrix}$$

Reference

1. V.K. Ingle and J.G. Proakis, *Digital Signal Processing Using MATLAB*, Brooks/Cole, Pacific Grove, CA, 2000.

Appendix D
Vectors and Matrices

D.1 Introduction

The objective of this appendix is to briefly review the main topics related to matrices since we encounter them in most of our work on queuing theory. There are excellent books dealing with this subject and we refer the reader to them for a more comprehensive treatment. Perhaps one of the best books on matrices is [1]. This book is not only easy to read, but the author's writing style and insights make the topic actually enjoyable. The reader that wants to read a comprehensive book, albeit somewhat dry, could consult [2].

D.2 Scalars

A scalar is a real or complex number. We denote scalars in this book by lower-case letters or Greek letters. Sometimes, but not too often, we use upper-case letters to denote scalars.

D.3 Vectors

A vector is an ordered collection of scalars $v_1, v_2, \cdots$, which are its *components*. We use bold lower-case letters to indicate vectors. A subscript on a vector will denote a particular component of the vector. Thus the scalar v_2 denotes the second component of the vector $\mathbf{v}$. Usually, we say $\mathbf{v}$ is a 3-vector to signify that $\mathbf{v}$ has three components.

A vector $\mathbf{v}$ could have its component values change with the time index n. At time n, that vector will be denoted by $\mathbf{v}(n)$.

As an example, a vector that has only three components is written as

$$\mathbf{v} = \begin{bmatrix} v_1 \\ v_2 \\ v_3 \end{bmatrix} \tag{D.1}$$

To conserve space, we usually write a column vector in the following form

$$\mathbf{v} = \begin{bmatrix} v_1 & v_2 & v_3 \end{bmatrix}^t \tag{D.2}$$

where the superscript t indicates that the vector is to be *transposed* by arranging the components horizontally instead of vertically.

A vector is, by default, a *column* vector. A *row* vector $\mathbf{r}$ could be obtained by transposing $\mathbf{v}$

$$\mathbf{r} = \mathbf{v}^t = \begin{bmatrix} v_1 & v_2 & v_3 \end{bmatrix} \tag{D.3}$$

D.4 Arithmetic Operations with Vectors

Two vectors can be added if they have the same number of components

$$\mathbf{v} + \mathbf{w} = \begin{bmatrix} v_1 + w_1 \\ v_2 + w_2 \\ v_3 + w_3 \end{bmatrix} = \begin{bmatrix} (v_1 + w_1) & (v_2 + w_2) & (v_3 + w_3) \end{bmatrix}^t \tag{D.4}$$

A vector can be multiplied by a scalar if all the vector components are multiplied by the same scalar as shown by

$$a\,\mathbf{v} = \begin{bmatrix} av_1 & av_2 & av_3 \end{bmatrix}^t \tag{D.5}$$

A row vector can multiply a column vector as long as the two vectors have the same number of components:

$$\mathbf{r}\,\mathbf{c} = \begin{bmatrix} r_1 & r_2 & r_3 \end{bmatrix} \begin{bmatrix} c_1 \\ c_2 \\ c_3 \end{bmatrix} \tag{D.6}$$

$$= r_1\,c_1 + r_2\,c_2 + r_3\,c_3 \tag{D.7}$$

The *dot product* of a vector is just a scalar that is defined as

$$\mathbf{v} \cdot \mathbf{w} = v_1\,w_1 + v_2\,w_2 + v_3\,w_3 \tag{D.8}$$

where corresponding components are multiplied together.

Two vectors $\mathbf{x}$ and $\mathbf{y}$ are said to be *orthogonal* if their dot product vanishes:

$$\mathbf{x} \cdot \mathbf{y} = 0$$

D.5 Linear Independence of Vectors

The two vectors $\mathbf{x}_1$ and $\mathbf{x}_2$ are said to be *linearly independent* or simply *independent* when

$$a_1\mathbf{x}_1 + a_2\mathbf{x}_2 = \mathbf{0} \tag{D.9}$$

is true if and only if

$$\begin{aligned} a_1 &= 0 \\ a_2 &= 0 \end{aligned}$$

D.6 Matrices

Assume we are given this set of linear equations

$$\begin{aligned} a_{11}\,x_1 + a_{12}\,x_2 + a_{13}\,x_3 &= b_1 \\ a_{21}\,x_1 + a_{22}\,x_2 + a_{23}\,x_3 &= b_2 \\ a_{31}\,x_1 + a_{32}\,x_2 + a_{33}\,x_3 &= b_3 \end{aligned} \tag{D.10}$$

We can write the equations in the form

$$\begin{bmatrix} a_{11} & a_{12} & a_{13} \\ a_{21} & a_{22} & a_{23} \\ a_{31} & a_{32} & a_{33} \end{bmatrix} \begin{bmatrix} x_1 \\ x_2 \\ x_3 \end{bmatrix} = \begin{bmatrix} b_1 \\ b_2 \\ b_3 \end{bmatrix} \tag{D.11}$$

where the first equation is obtained by multiplying the first row and the vector $\mathbf{x}$ and the second equation is obtained by multiplying the second row and the vector $\mathbf{x}$, and so on. A concise form for writing the system of linear equations in (D.10) is to use vectors and matrices:

$$\mathbf{A}\mathbf{x} = \mathbf{b} \tag{D.12}$$

The *coefficient matrix* of this system of equations is given by the array of numbers

$$\mathbf{A} = \begin{bmatrix} a_{11} & a_{12} & a_{13} \\ a_{21} & a_{22} & a_{23} \\ a_{31} & a_{32} & a_{33} \end{bmatrix} \tag{D.13}$$

A matrix with m rows and n columns is called a matrix of order $m \times n$ or simply an $m \times n$ matrix. When $m = n$, we have a square matrix of order n. The elements a_{ii} constitute the *main diagonal* of $\mathbf{A}$.

D.7 Matrix Addition

If **A** is an $m \times n$ matrix and **B** is an $m \times n$ matrix, then we can add them to produce an $m \times n$ matrix **C**

$$\mathbf{C} = \mathbf{A} + \mathbf{B} \tag{D.14}$$

where the elements of **C** are given by

$$c_{ij} = a_{ij} + b_{ij} \tag{D.15}$$

with $1 \leq i \leq m$ and $1 \leq j \leq n$.

D.8 Matrix Multiplication

If **A** is an $l \times m$ matrix and **B** is an $m \times n$ matrix, then we can multiply them to produce an $l \times n$ matrix **C**

$$\mathbf{C} = \mathbf{A}\,\mathbf{B} \tag{D.16}$$

where the elements of **C** are given by

$$c_{ij} = \sum_{k=1}^{m} a_{ik}\, b_{kj} \tag{D.17}$$

with $1 \leq i \leq l$ and $1 \leq j \leq n$.

D.9 Inverse of a Matrix

Given the matrix **A**, its *left inverse* is defined by the equation

$$\mathbf{L}\mathbf{A} = \mathbf{I} \tag{D.18}$$

The *right inverse* is defined by the equation

$$\mathbf{A}\mathbf{R} = \mathbf{I} \tag{D.19}$$

When **A** is square, the left and the right inverses are equal and we denote the *inverse* as $\mathbf{A}^{-1}$, which satisfies the equations

$$\mathbf{A}\mathbf{A}^{-1} = \mathbf{A}^{-1}\mathbf{A} = \mathbf{I} \tag{D.20}$$

The inverse of the matrix is found by treating the above equation as a system of simultaneous equations in the unknown matrix $\mathbf{A}^{-1}$. The matrix $\mathbf{A}^{-1}$ is first written as

$$\mathbf{A}^{-1} = \begin{bmatrix} \mathbf{a}_1 & \mathbf{a}_2 & \cdots & \mathbf{a}_n \end{bmatrix} \quad \text{(D.21)}$$

where n is the dimension of $\mathbf{A}$. The ith column $\mathbf{a}_i$ of $\mathbf{A}^{-1}$ is treated as an unknown vector that is to be found from the equation

$$\mathbf{A}\,\mathbf{a}_i = \begin{bmatrix} 0 & \cdots & 0 & 1 & 0 & \cdots & 0 \end{bmatrix}^t \quad \text{(D.22)}$$

where the vector on the RHS has 1 in the ith location. Gauss elimination is a useful technique for finding the inverse of the matrix.

Not all matrices have inverses. A square matrix has an inverse only if its rank equals the number of rows (rank is explained Section D.11).

D.10 Null Space of a Matrix

Given an $m \times n$ matrix $\mathbf{A}$, a nonzero n-vector $\mathbf{x}$ is said to be a null vector for $\mathbf{A}$ when

$$\mathbf{A}\mathbf{x} = \mathbf{0} \quad \text{(D.23)}$$

All possible solutions of the above equation form the *null space* of $\mathbf{A}$, which we denote by null $(\mathbf{A})$. If n is the number of all possible and independent solutions, then we have

$$n = \text{null}\,(\mathbf{A}) \quad \text{(D.24)}$$

Finding the null vectors $\mathbf{x}$ is done by solving (D.23) as a system of homogeneous linear equations.

MATLAB offers the function `null` to find all the null vectors of a given matrix. The function `null(A)` produces the null space basis vectors. The function `null(A,'r')` produces the null vectors in rational format for presentation purposes. For example

$$A = \begin{bmatrix} 1 & 2 & 3 \\ 1 & 2 & 3 \\ 1 & 2 & 3 \end{bmatrix}$$

$$\text{null}(A,'r') = \begin{bmatrix} -2 & -3 \\ 1 & 0 \\ 0 & 1 \end{bmatrix}$$

D.11 The Rank of a Matrix

The maximum number of linearly independent rows or columns of a matrix **A** is the rank of the matrix. This number r is denoted by

$$r = \text{rank}\,(\mathbf{A}) \tag{D.25}$$

A matrix has only one rank regardless of the number of rows or columns. An $m \times n$ matrix (where $m \leq n$) is said to be *full rank* when $\text{rank}(\mathbf{A}) = m$.

The rank r of the matrix equals the number of pivots of the matrix when it is transformed to its echelon form (see Section D.19.3 for definition of echelon form).

D.12 Eigenvalues and Eigenvectors

Eigenvalues and eigenvectors apply only to *square* matrices. Consider the special situation when we have

$$\mathbf{A}\mathbf{x} = \lambda \mathbf{x} \tag{D.26}$$

The number λ is called the *eigenvalue* and $\mathbf{x}$ is called the *eigenvector*. Math packages help us find all possible eigenvalues and the corresponding eigenvectors of a given matrix.

We can combine all the eigenvectors into the eigenvector matrix **X** whose columns are the eigenvectors and we could write

$$\mathbf{A}\,\mathbf{X} = \mathbf{X}\,\mathbf{D} \tag{D.27}$$

where

$$\mathbf{X} = \begin{bmatrix} \mathbf{x}_1 & \mathbf{x}_2 & \cdots & \mathbf{x}_n \end{bmatrix} \tag{D.28}$$

$$\mathbf{D} = \begin{bmatrix} \lambda_1 & 0 & \cdots & 0 \\ 0 & \lambda_2 & \cdots & 0 \\ \vdots & \vdots & \ddots & \vdots \\ 0 & 0 & \cdots & \lambda_n \end{bmatrix} \tag{D.29}$$

When the inverse $\mathbf{X}^{-1}$ exists, we can diagonalize **A** in the form

$$\mathbf{A} = \mathbf{X}\,\mathbf{D}\,\mathbf{X}^{-1} \tag{D.30}$$

D.13 Diagonalizing a Matrix

We say that the square matrix **A** is diagonalized when it can be written in the form (D.30). A matrix that has no repeated eigenvalues can always be diagonalized. If some eigenvalues are repeated, then the matrix might or might not be diagonalizable.

A general rule for matrix diagonalization is as follows: A matrix is diagonalizable only when its Jordan canonic form (JCF) is diagonal. Section 3.14.1 on page 103 discusses the Jordan canonic form of a matrix.

D.14 Triangularizing a Matrix

Sometimes it is required to change a given matrix **A** to an upper triangular matrix. The resulting matrix is useful in many applications such as

1. Solving a system of linear equations.
2. Finding the eigenvector of a matrix given an eigenvalue.
3. Finding the inverse of a matrix.

Householder or Givens techniques can be used to triangularize the matrix. We illustrate Givens technique here only. The idea is to apply a series of plane rotation, or Givens rotation, matrices on the matrix in question in order to create zeros below the main diagonal. We start with the first column and create zeros below the element a_{11}. Next, we start with the second column and create zeros below the element a_{22} and so on. Each created zero does not disturb the previously created zeros. Furthermore, the plane rotation matrix is very stable and does not require careful choice of the pivot element. Section D.19.7 discusses the Givens plane rotation matrices.

Assume, as an example, a 5×5 matrix **A** and we are interested in eliminating element a_{42}, we choose the Givens rotation matrix $\mathbf{G}_{42}$

$$\mathbf{G}_{42} = \begin{bmatrix} 1 & 0 & 0 & 0 & 0 \\ 0 & c & 0 & s & 0 \\ 0 & 0 & 1 & 0 & 0 \\ 0 & -s & 0 & c & 0 \\ 0 & 0 & 0 & 0 & 1 \end{bmatrix} \tag{D.31}$$

where $c = \cos\theta$ and $s = \sin\theta$. Premultiplying a matrix **A** by $\mathbf{G}_{42}$ modifies only rows 2 and 4. All other rows are left unchanged. The elements in rows 2 and 4 become

$$a_{2j} = c\, a_{2j} + s\, a_{4j} \tag{D.32}$$

$$a_{4j} = -s\, a_{2j} + c\, a_{4j} \tag{D.33}$$

The new element a_{42} is eliminated from the above equation if we have

$$\tan\theta = \frac{a_{42}}{a_{22}} \tag{D.34}$$

The following MATLAB code illustrates how an input matrix is converted to an upper triangular matrix.

```
%File: givens.m
%The program accepts a matrix performs a seriesof
%Givens rotations to transform the matrix into an
%upper-triangular matrix.
%The new matrix is printed after eachzero is created.
%
%Input matrix to be triangularized
A=[-0.6  0.2  0.0
    0.1 -0.5  0.6
    0.5  0.3 -0.6]
n=3;
%iterate for first n-1 columns
for j=1:n-1
  %cancel all subdiagonal elements
  %in column j
  for i=j+1:n
     %calculate theta
     theta=atan2(q(i,j),q(j,j))
     for k=j:n
        temp_x= A(j,k)*cos(theta)+A(i,k)*sin(theta);
        temp_y=-A(j,k)*sin(theta)+A(i,k)*cos(theta);
        %update new elements in rows i and j
        A(j,k) = temp_x;
        A(i,k) = temp_y;
     end
     A %print q after each iteration.
  end
end
```

An example of using the Givens rotation technique is explained in the next section.

D.15 Linear Equations

A system of linear equations has the form

$$\mathbf{Ax} = \mathbf{b} \tag{D.35}$$

where the *coefficient matrix* **A** is a given $n \times n$ nonsingular matrix so that the system possesses a unique solution. The vector **b** is also given. The unknown vector **x** is to be found. The system is said to be *homogeneous* when **b** is zero, otherwise the system is said to be *nonhomogeneous*. Before we discuss methods for obtaining the solution to the above equation, we define first some *elementary row operations* [3], [2]:

1. Interchange of two rows.
2. Multiplication of a row by a nonzero constant.
3. Addition of a constant multiple of one row to another row.

Each of the above elementary row operations is implemented by multiplying the matrix from the left by an appropriate matrix **E** called an *elementary matrix*. The three elementary matrices corresponding to the above three elementary row operations are illustrated below for 4×4 matrices [4]:

$$\begin{bmatrix} 1 & 0 & 0 & 0 \\ 0 & 0 & 1 & 0 \\ 0 & 1 & 0 & 0 \\ 0 & 0 & 0 & 1 \end{bmatrix}, \begin{bmatrix} 1 & 0 & 0 & 0 \\ 0 & a & 0 & 0 \\ 0 & 0 & 1 & 0 \\ 0 & 0 & 0 & 1 \end{bmatrix}, \begin{bmatrix} 1 & 0 & 0 & 0 \\ 0 & 1 & 0 & 0 \\ a & 0 & 1 & 0 \\ 0 & 0 & 0 & 1 \end{bmatrix} \tag{D.36}$$

There are two classes of numerical methods for finding a solution. The *direct* methods guarantee to find a solution in one step. This is the recommended approach for general situations and for small values of n. The *iterative* methods start with an assumed solution then try to refine this assumption until the succeeding estimates of the solution converge to within a certain error limit. This approach is useful for large values of n and for sparse matrices where **A** has a large number of zeros. The advantage of iterative solutions is that they practically eliminate arithmetic roundoff errors and produce results with accuracy close to the machine precision [5]. We discuss the two approaches in the following sections. We refer the reader to Appendix E for a discussion of the techniques used by MATLAB to numerically find the solution.

In the following section, we review the useful techniques for solving systems of linear equations.

D.15.1 Gauss Elimination

Gauss elimination solves a system of linear equations by transforming **A** into upper triangular form using elementary row operations. The solution is then found using back-substitution. To create a zero at position (2,1), we need to multiply row 2 by a_{21}/a_{11}, then subtract this row from row 1. This is equivalent to a row operation matrix of the form

$$\mathbf{E}_{21} = \begin{bmatrix} 1 & 0 & 0 \\ e & 1 & 0 \\ 0 & 0 & 1 \end{bmatrix} \text{ with } e = -\frac{a_{21}}{a_{11}} \tag{D.37}$$

The reader could verify that this matrix performs the desired operation and creates a zero at position (2,1).

We illustrate this using an example of a 3 × 3 system for a matrix with rank 2:

$$\begin{bmatrix} -0.5 & 0.3 & 0.2 \\ 0.3 & -0.4 & 0.5 \\ 0.2 & 0.1 & -0.7 \end{bmatrix} \begin{bmatrix} s_0 \\ s_1 \\ s_2 \end{bmatrix} = \begin{bmatrix} 0 \\ 0 \\ 0 \end{bmatrix} \tag{D.38}$$

Step 1 Create a zero in the (2,1) position using the elementary matrix $\mathbf{E}_{21}$

$$\mathbf{E}_{21} = \begin{bmatrix} 1 & 0 & 0 \\ r & 1 & 0 \\ 0 & 0 & 1 \end{bmatrix} \text{ with } e = -\frac{a_{21}}{a_{11}}$$

which gives

$$\mathbf{T}_1 = \mathbf{E}_{21}\mathbf{A} = \begin{bmatrix} -0.5 & 0.3 & 0.2 \\ 0 & -0.22 & 0.62 \\ 0.2 & 0.1 & -0.7 \end{bmatrix}$$

Step 2 Create a zero in the (31) position using the elementary matrix $\mathbf{E}_{31}$

$$\mathbf{E}_{31} = \begin{bmatrix} 1 & 0 & 0 \\ 0 & 1 & 0 \\ r & 0 & 1 \end{bmatrix} \text{ with } r = -\frac{a_{31}}{a_{11}}$$

which gives

$$\mathbf{T}_2 = \mathbf{E}_{31}\mathbf{T}_1 = \begin{bmatrix} -0.5 & 0.3 & 0.2 \\ 0 & -0.22 & 0.62 \\ 0 & 0.22 & -0.62 \end{bmatrix}$$

Step 3 Create a zero in the (3, 2) position using the elementary matrix $\mathbf{E}_{32}$

$$\mathbf{E}_{32} = \begin{bmatrix} 1 & 0 & 0 \\ 0 & 1 & 0 \\ 0 & e & 1 \end{bmatrix} \text{ with } e = -\frac{a_{32}}{a_{22}}$$

which gives

$$\mathbf{T}_3 = \mathbf{E}_{32}\mathbf{T}_2 = \begin{bmatrix} -0.5 & 0.3 & 0.2 \\ 0 & -0.22 & 0.62 \\ 0 & 0 & 0 \end{bmatrix}$$

Note that the triangularization operation produced a row of zeros, indicating that the rank of this matrix is indeed 2.

Step 4 Solve for the components of **s** assuming $s_3 = 1$

$$\mathbf{s} = \begin{bmatrix} 2.0909 & 2.8182 & 1 \end{bmatrix}$$

After normalization, the true state vector becomes

$$\mathbf{s} = \begin{bmatrix} 0.3538 & 0.4769 & 0.1692 \end{bmatrix}$$

D.15.2 Gauss–Jordan Elimination

Gauss–Jordan elimination is a powerful method for solving a system of linear equations of the form

$$\mathbf{Ax} = \mathbf{b} \tag{D.39}$$

where **A** is a square matrix of dimension $n \times n$ and **x** and **b** are column vectors with n components each. The above equation can be written in the form

$$\mathbf{Ax} = \mathbf{Ib} \tag{D.40}$$

where **I** is the $n \times n$ unit matrix.

We can find the unknown vector **x** by multiplying by the inverse of matrix **A** to get the solution

$$\mathbf{x} = \mathbf{A}^{-1}\mathbf{b} \tag{D.41}$$

It is not recommended to find the inverse as mentioned above. Gauss elimination and Gauss–Jordan elimination techniques are designed to find the solution without the need to find the inverse of the matrix. This solution in the above equation can be written in the form

$$\mathbf{Ix} = \mathbf{A}^{-1}\mathbf{b} \tag{D.42}$$

Comparing (D.40) to (D.42), we conclude that we find our solution if somehow we were able to convert the matrix **A** to **I** using repeated row operation.

Gauss–Jordan elimination does exactly that by constructing the augmented matrix $\begin{bmatrix} \mathbf{A} & \mathbf{I} \end{bmatrix}$ and converting it to the matrix $\begin{bmatrix} \mathbf{I} & \mathbf{A}^{-1} \end{bmatrix}$.

Example 1 Solve the following set of linear equations:

$$\begin{array}{ccccc} 2x_1 & -x_2 & & = & 0 \\ -x_1 & +2x_2 & -x_3 & = & 3 \\ & -x_2 & +2x_3 & = & 2 \end{array}$$

We have

$$\mathbf{A} = \begin{bmatrix} 2 & -1 & 0 \\ -1 & 2 & -1 \\ 0 & -1 & 2 \end{bmatrix}$$

$$\mathbf{b} = \begin{bmatrix} 0 & 3 & 2 \end{bmatrix}^t$$

The augmented matrix is

$$\begin{bmatrix} \mathbf{A} & \mathbf{I} \end{bmatrix} = \begin{bmatrix} 2 & -1 & 0 & 1 & 0 & 0 \\ -1 & 2 & -1 & 0 & 1 & 0 \\ 0 & -1 & 2 & 0 & 0 & 1 \end{bmatrix}$$

$$\rightarrow \begin{bmatrix} 2 & -1 & 0 & 1 & 0 & 0 \\ 0 & 3 & -2 & 1 & 2 & 0 \\ 0 & -1 & 2 & 0 & 0 & 1 \end{bmatrix} \quad \text{row 1} + 2 \text{ row 2}$$

$$\rightarrow \begin{bmatrix} 2 & -1 & 0 & 1 & 0 & 0 \\ 0 & 3 & -2 & 1 & 2 & 0 \\ 0 & 0 & 4 & 1 & 2 & 3 \end{bmatrix} \quad \text{row 2} + 3 \text{ row 3}$$

So far, we have changed our system matrix **A** into an upper triangular matrix. Gauss elimination would be exactly that and our solution could be found by forward substitution.

Gauss–Jordan elimination continues by eliminating all elements not on the main diagonal using row operations:

$$[A\ I] \rightarrow \begin{bmatrix} 6 & 0 & -2 & 4 & 2 & 0 \\ 0 & 3 & -2 & 1 & 2 & 0 \\ 0 & 0 & 4 & 1 & 2 & 3 \end{bmatrix} \quad 3 \text{ row 1} + \text{row 2}$$

$$\rightarrow \begin{bmatrix} 12 & 0 & 0 & 9 & 6 & 3 \\ 0 & 3 & -2 & 1 & 2 & 0 \\ 0 & 0 & 4 & 1 & 2 & 3 \end{bmatrix} \quad 2 \text{ row 1} + \text{row 3}$$

$$\rightarrow \begin{bmatrix} 12 & 0 & 0 & 9 & 6 & 3 \\ 0 & 6 & 0 & 3 & 6 & 3 \\ 0 & 0 & 4 & 1 & 2 & 3 \end{bmatrix} \quad 2 \text{ row 2} + \text{row 3}$$

Now we can simplify to get the unit matrix as follows

$$\begin{bmatrix} \mathbf{I} & \mathbf{A}^{-1} \end{bmatrix} = \begin{bmatrix} 1 & 0 & 0 & \frac{3}{4} & \frac{1}{2} & \frac{1}{4} \\ 0 & 1 & 0 & \frac{1}{2} & 1 & \frac{1}{2} \\ 0 & 0 & 1 & \frac{1}{4} & \frac{1}{2} & \frac{3}{4} \end{bmatrix}$$

As a result, we now know the matrix $\mathbf{A}^{-1}$, and the solution to our system of equations is given by

$$\begin{aligned} \mathbf{x} &= \mathbf{A}^{-1}\mathbf{b} \\ &= \begin{bmatrix} 2 & 4 & 3 \end{bmatrix}^t \end{aligned}$$

D.15.3 Row Echelon Form and Reduced Row Echelon Form

The row echelon forms tells us the important information about a system of linear equations such as whether the system has a unique solution, no solution, or an infinity of solutions.

We start this section with a definition of a *pivot*. A pivot is the first nonzero element in each row of a matrix. A matrix is said to be in *row echelon form* if it has the following properties [7]:

1. Rows of all zeros appear at the bottom of the matrix.
2. Each pivot has the value 1.
3. Each pivot occurs in a column that is strictly to the right of the pivot above it.

A matrix is said to be in *reduced row echelon form* if it satisfies one additional property

4. Each pivot is the only nonzero entry in its column.

MATLAB offers the function `rref` to find the reduced row echelon form of a matrix.

Consider the system of equations

$$\begin{aligned} x_1 \quad &+2x_2 \quad +3x_3 \quad = 1 \\ 2x_1 \quad &+3x_2 \quad +4x_3 \quad = 4 \\ 3x_1 \quad &+3x_2 \quad +5x_3 \quad = 3 \end{aligned}$$

The augmented matrix is given by

$$\begin{bmatrix} \mathbf{A} & \mathbf{b} \end{bmatrix} = \begin{bmatrix} 1 & 2 & 3 & 1 \\ 2 & 3 & 4 & 4 \\ 3 & 3 & 5 & 3 \end{bmatrix}$$

The reduced echelon form is given by

$$\mathbf{R} = \text{rref}(\mathbf{A}, \mathbf{b}) = \begin{bmatrix} 1 & 0 & 0 & 2 \\ 0 & 1 & 0 & 4 \\ 0 & 0 & 1 & -3 \end{bmatrix}$$

Thus the solution to our system of equations is

$$\mathbf{x} = \begin{bmatrix} 1 & 4 & -3 \end{bmatrix}^t$$

Let us now see the reduced echelon form when the system has no solution:

$$\begin{array}{llll} 3x_1 & +2x_2 & +x_3 & = 1 \\ 2x_1 & +x_2 & +x_3 & = 0 \\ 6x_1 & +2x_2 & +4x_3 & = 6 \end{array}$$

The augmented matrix is given by

$$\begin{bmatrix} \mathbf{A} & \mathbf{b} \end{bmatrix} = \begin{bmatrix} 3 & 2 & 1 & 3 \\ 2 & 1 & 1 & 0 \\ 6 & 2 & 4 & 6 \end{bmatrix}$$

The reduced echelon form is given by

$$\mathbf{R} = \text{rref}(\mathbf{A}, \mathbf{b}) = \begin{bmatrix} 1 & 0 & 1 & 0 \\ 0 & 1 & -1 & 0 \\ 0 & 0 & 0 & 1 \end{bmatrix}$$

The last equation implies that

$$0x_1 + 0x_2 + 0x_3 = 1$$

There are no values of x_1, x_2, and x_3 that satisfy this equation and hence the system has no solution.

The following system has an infinite number of solutions.

$$\begin{array}{llll} 3x_1 & +2x_2 & +x_3 & = 3 \\ 2x_1 & +x_2 & +x_3 & = 0 \\ 5x_1 & +3x_2 & +2x_3 & = 3 \end{array}$$

The augmented matrix is given by

$$\begin{bmatrix} \mathbf{A} & \mathbf{b} \end{bmatrix} = \begin{bmatrix} 3 & 2 & 1 & 3 \\ 2 & 1 & 1 & 0 \\ 5 & 3 & 2 & 3 \end{bmatrix}$$

The reduced echelon form is given by

$$\begin{aligned} \mathbf{R} &= \text{rref}(\mathbf{A}, \mathbf{b}) \\ &= \begin{bmatrix} 1 & 0 & 1 & -3 \\ 0 & 1 & -1 & 6 \\ 0 & 0 & 0 & 0 \end{bmatrix} \end{aligned}$$

Thus we have two equations for three unknowns and we could have infinity solutions.

D.16 Direct Techniques for Solving Systems of Linear Equations

Direct techniques for solving systems of linear equations are usually the first one attempted to obtain the solution. These techniques are appropriate for small matrices where computational errors will be small. In the next section, we discuss iterative techniques which are useful for large matrices.

D.16.1 Givens Rotations

Givens rotations operation performs a number of orthogonal similarity transformations on a matrix to make it an upper triangular matrix. Another equivalent technique is the Householder transformation; but we will illustrate the Givens technique here. The technique is very stable and does not suffer from the pivot problems of Gauss elimination. We start with the equilibrium steady-state Markov chain equation

$$\mathbf{P}\,\mathbf{s} = \mathbf{s} \tag{D.43}$$

and write it in the form

$$(\mathbf{P} - \mathbf{I})\;\mathbf{s} = \mathbf{0} \tag{D.44}$$

$$\mathbf{A}\,\mathbf{s} = \mathbf{0} \tag{D.45}$$

where $\mathbf{0}$ is a zero column vector. Thus, finding $\mathbf{s}$ amounts to solving a homogeneous system of n linear equations in n unknowns. The system has a nontrivial solution if the determinant of $\mathbf{A}$ is zero. This is assured here since the determinant of the matrix is equal to the product of its eigenvalues. Here, **A** has a zero eigenvalue, which results in a zero determinant and this guarantees finding a nontrivial solution.

Applying a series of Givens rotations [6] will transform **A** into an upper triangular matrix **T** such that

$$\mathbf{T}\,\mathbf{s} = \mathbf{0} \tag{D.46}$$

We are now able to do back-substitution to find all the elements of **s**. The last row of **T** could be made identically zero since the rank of **T** is $n - 1$. We start our back-substitution by ignoring the last row of **T** and assuming an arbitrary value for $s_n = 1$, then proceed to evaluate s_{n-1}, and so on. There still remains one equation that must be satisfied:

$$\sum_{i=1}^{n} s_i = 1 \tag{D.47}$$

Let us assume that the sum of the components that we obtained for the vector **s** gives

$$\sum_{i=1}^{n} s_i = b \tag{D.48}$$

then we must divide each value of **s** by b to get the true normalized vector that we desire.

Example 2 Use Givens method to find the equilibrium state vector state **s** for the Markov chain with transition matrix given by

$$\mathbf{P} = \begin{bmatrix} 0.4 & 0.2 & 0 \\ 0.1 & 0.5 & 0.6 \\ 0.5 & 0.3 & 0.4 \end{bmatrix}$$

Step 1 Obtain the matrix $\mathbf{A} = \mathbf{P} - \mathbf{I}$

$$\mathbf{A} = \begin{bmatrix} -0.6 & 0.2 & 0.0 \\ 0.1 & -0.5 & 0.6 \\ 0.5 & 0.3 & -0.6 \end{bmatrix}$$

Step 2 Create a zero in the (2, 1) position using the Givens rotation matrix $\mathbf{G}_{21}$

$$\mathbf{G}_{21} = \begin{bmatrix} c & s & 0 \\ -s & c & 0 \\ 0 & 0 & 1 \end{bmatrix} \quad \text{with } \theta = \tan^{-1}\frac{0.1}{-0.6}$$

which gives

$$\mathbf{T}_1 = \mathbf{G}_{21}\mathbf{A} = \begin{bmatrix} 0.6083 & -0.2795 & 0.0986 \\ 0 & 0.4603 & -0.5918 \\ 0.5 & 0.3 & -0.6 \end{bmatrix}$$

Step 3 Create a zero in the (3, 1) position using the Givens rotation matrix $\mathbf{G}_{31}$

$$\mathbf{G}_{31} = \begin{bmatrix} c & 0 & s \\ 0 & 1 & 0 \\ -s & 0 & c \end{bmatrix} \quad \text{with } \theta = \tan^{-1} \frac{0.5}{0.6083}$$

which gives

$$\mathbf{T}_2 = \mathbf{G}_{31}\mathbf{T}_1 = \begin{bmatrix} 0.7874 & -0.0254 & -0.3048 \\ 0 & 0.4603 & -0.5918 \\ 0 & 0.4092 & -0.5261 \end{bmatrix}$$

Step 4 Create a zero in the (3, 2) position using the Givens rotation matrix $\mathbf{G}_{32}$

$$\mathbf{G}_{32} = \begin{bmatrix} 1 & 0 & 0 \\ 0 & c & s \\ 0 & -s & c \end{bmatrix} \quad \text{with } \theta = \tan^{-1} \frac{0.4092}{0.4603}$$

which gives

$$\mathbf{T}_3 = \mathbf{G}_{32}\mathbf{T}_2 = \begin{bmatrix} 0.7874 & -0.0254 & -0.3048 \\ 0 & 0.6159 & -0.7919 \\ 0 & 0 & 0 \end{bmatrix}$$

Step 5 Solve for the components of $\mathbf{s}$ assuming $s_3 = 1$

$$\mathbf{s} = \begin{bmatrix} 0.3871 & 1.2857 & 1 \end{bmatrix}$$

After normalization, the true state vector becomes

$$\mathbf{s} = \begin{bmatrix} 0.1448 & 0.4810 & 0.3741 \end{bmatrix}$$

D.17 Iterative Techniques

Iterative techniques continually refine the estimate of the solution while at the same time suppressing computation noise. Thus, the answer could be accurate within the machine precision. Iterative techniques are used when the system matrix is large and sparse. This is a typical situation in Markov chains and queuing theory.

D.17.1 Jacobi Iterations

We start with the steady-state Markov chain equation

$$\mathbf{P}\,\mathbf{s} = \mathbf{s} \tag{D.49}$$

The matrix **P** is broken down as the sum of three components

$$\mathbf{P} = \mathbf{L} + \mathbf{D} + \mathbf{U} \tag{D.50}$$

where **L** is the lower triangular part of **P**, **D** is the diagonal part of **P**, and **U** is the upper triangular part of **P**. Thus, our steady-state equation can be written as

$$(\mathbf{L} + \mathbf{D} + \mathbf{U})\ \mathbf{s} = \mathbf{s} \tag{D.51}$$

We get Jacobi iterations if we write the above equation as

$$\mathbf{D}\,\mathbf{s} = (\mathbf{I} - \mathbf{L} - \mathbf{U})\ \mathbf{s} \tag{D.52}$$

where **I** is the unit matrix and it is assumed that **P** has nonzero diagonal elements. The technique starts with an assumed solution $\mathbf{s}^0$, then iterates to improve the guess using the iterations

$$\mathbf{s}^{k+1} = \mathbf{D}^{-1}(\mathbf{I} - \mathbf{L} - \mathbf{U})\ \mathbf{s}^k \tag{D.53}$$

Each component of **s** is updated according to the equation

$$s_i^{k+1} = \frac{1}{p_{ii}} \left(s_i^k - \sum_{j=0}^{i-1} p_{ij}\ s_j^k - \sum_{j=i+1}^{n-1} p_{ij}\ s_j^k \right) \tag{D.54}$$

D.17.2 Gauss–Seidel Iterations

We get Gauss–Seidel iterations if we write (D.51) as

$$(\mathbf{D} + \mathbf{L})\ \mathbf{s} = (\mathbf{I} - \mathbf{U})\ \mathbf{s} \tag{D.55}$$

The technique starts with an assumed solution $\mathbf{s}^0$, then iterates to improve the guess using the iterations

$$\mathbf{s}^{k+1} = (\mathbf{D} + \mathbf{L})^{-1}\,(\mathbf{I} - \mathbf{U})\ \mathbf{s}^k \tag{D.56}$$

Each component of **s** is updated according to the equation

$$s_i^{k+1} = \frac{1}{p_{ii}} \left(s_i^k - \sum_{j=0}^{i-1} p_{ij}\, s_j^{k+1} - \sum_{j=i+1}^{n-1} p_{ij}\, s_j^k \right) \tag{D.57}$$

Notice that we make use of previously updated components of **s** to update the next components as is evident by the iteration index associated with the first summation term (compare that with Jacobi iterations).

D.17.3 Successive Overrelaxation Iterations

We explain successive overrelaxation (SOR) technique by rewriting a slightly modified version of (D.57)

$$s_i^{k+1} = s_i^k + \frac{1}{p_{ii}} \left(s_i^k - \sum_{j=0}^{i-1} p_{ij}\, s_j^{k+1} - \sum_{j=i}^{n-1} p_{ij}\, s_j^k \right) \tag{D.58}$$

We can think of the above equation as updating the value of s_i^k by adding the term in brackets. We multiply that term by a *relaxation parameter* ω to get the SOR iterations:

$$s_i^{k+1} = s_i^k + \frac{\omega}{p_{ii}} \left(s_i^k - \sum_{j-0}^{i-1} p_{ij}\, s_j^{k+1} - \sum_{j=i}^{n-1} p_{ij}\, s_j^k \right) \tag{D.59}$$

when $\omega = 1$, we get Gauss–Seidel iterations again of course.

D.18 Similarity Transformation

Assume there exists a square matrix **M** whose inverse is $\mathbf{M}^{-1}$. Then the two square matrices **A** and $\mathbf{B} = \mathbf{M}^{-1}\mathbf{A}\mathbf{M}$ are said to be similar. We say that **B** is obtained from **A** by a *similarity transformation*. Similar matrices have the property that they both have the same eigenvalues. To prove that, assume **x** is the eigenvector of **A**

$$\mathbf{A}\mathbf{x} = \lambda\mathbf{x} \tag{D.60}$$

We can write the above equation in the form

$$\mathbf{A}\,\mathbf{M}\mathbf{M}^{-1}\,\mathbf{x} = \lambda\mathbf{x} \tag{D.61}$$

Premultiplying both sides of this equation by $\mathbf{M}^{-1}$, we obtain

$$\mathbf{M}^{-1}A\,\mathbf{M}\mathbf{M}^{-1}\,\mathbf{x} = \lambda\mathbf{M}^{-1}\mathbf{x} \tag{D.62}$$

But $\mathbf{B} = \mathbf{M}^{-1}\mathbf{A}\mathbf{M}$, by definition, and we have

$$\mathbf{B}\left(\mathbf{M}^{-1}\mathbf{x}\right) = \lambda\left(\mathbf{M}^{-1}\mathbf{x}\right) \tag{D.63}$$

Thus, we proved that the eigenvectors and eigenvalues of $\mathbf{B}$ are $\mathbf{M}^{-1}\mathbf{x}$ and λ, respectively.

Thus, we can say that the two matrices **A** and **B** are similar when their eigenvalues are the same and their eigenvectors are related through the matrix **M**.

D.19 Special Matrices

The following is an alphabetical collection of special matrices that were encountered in this book.

D.19.1 Circulant Matrix

A square circulant matrix has the form

$$\mathbf{P} = \begin{bmatrix} 0 & 0 & 0 & \cdots & 0 & 0 & 1 \\ 1 & 0 & 0 & \cdots & 0 & 0 & 0 \\ 0 & 1 & 0 & \cdots & 0 & 0 & 0 \\ \vdots & \vdots & \vdots & \ddots & \vdots & \vdots & \vdots \\ 0 & 0 & 0 & \cdots & 0 & 0 & 0 \\ 0 & 0 & 0 & \cdots & 1 & 0 & 0 \\ 0 & 0 & 0 & \cdots & 0 & 1 & 0 \end{bmatrix} \tag{D.64}$$

Premultiplying a matrix by a circulant matrix results in circularly shifting all the rows down by one row and the last row will become the first row.

An $m \times m$ circulant matrix **C** has the following two interesting properties:

Repeated multiplication m-times produces the identity matrix

$$\mathbf{C}^k = \mathbf{I} \tag{D.65}$$

where k is an integer multiple of m.

Eigenvalues of the matrix all lie on the unit circle

$$\lambda_i = \exp\left(j2\pi \times \frac{k_i}{m}\right) \qquad k_i = 1,\ 2,\ \cdots,\ m \tag{D.66}$$

where $j = \sqrt{-1}$ and

$$|\lambda_i| = 1 \tag{D.67}$$

D.19.2 Diagonal Matrix

A diagonal matrix has $a_{ij} = 0$ whenever $i \neq j$ and $a_{ii} = d_i$. A 3×3 diagonal matrix $\mathbf{D}$ is given by

$$\mathbf{D} = \begin{bmatrix} d_1 & 0 & 0 \\ 0 & d_2 & 0 \\ 0 & 0 & d_3 \end{bmatrix} \tag{D.68}$$

An alternative way of defining $\mathbf{D}$ is

$$\mathbf{D} = \text{diag}\begin{pmatrix} d_1 & d_2 & d_3 \end{pmatrix} \tag{D.69}$$

or

$$\mathbf{D} = \text{diag}\,(\mathbf{d}) \tag{D.70}$$

where $\mathbf{d}$ is the vector of diagonal entries of $\mathbf{D}$.

D.19.3 Echelon Matrix

An $m \times n$ matrix $\mathbf{A}$ can be transformed using elementary row operations into an upper triangular matrix $\mathbf{U}$, where elementary row operations include

1. Exchanging two rows.
2. Multiplication of a row by a nonzero constant.
3. Addition of a constant multiple of a row to another row.

Such operations are used in Givens rotations and Gauss elimination to solve the system of linear equations

$$\mathbf{Ax} = \mathbf{b} \tag{D.71}$$

by transforming it into

$$\mathbf{Ux} = \mathbf{c} \tag{D.72}$$

then doing back-substitution to find the unknown vector $\mathbf{x}$. The matrix $\mathbf{U}$ that results is in echelon form. An example of a 4×6 echelon matrix is

$$\mathbf{U} = \begin{bmatrix} x & x & x & x & x & x \\ 0 & x & x & x & x & x \\ 0 & 0 & 0 & 0 & x & x \\ 0 & 0 & 0 & 0 & 0 & x \end{bmatrix} \tag{D.73}$$

Notice that the number of leading zeros in each row must increase. The number of pivots[1] equals the rank r of the matrix.

D.19.4 Identity Matrix

An identity matrix has $a_{ij} = 0$ whenever $i \neq j$ and $a_{ii} = 1$. A 3×3 identity matrix $\mathbf{I}$ is given by

$$\mathbf{I} = \begin{bmatrix} 1 & 0 & 0 \\ 0 & 1 & 0 \\ 0 & 0 & 1 \end{bmatrix} \tag{D.74}$$

D.19.5 Nonnegative Matrix

An $m \times n$ matrix $\mathbf{A}$ is nonnegative when

$$a_{ij} \geq 0 \tag{D.75}$$

for all $1 \leq i \leq m$ and $1 \leq j \leq n$

D.19.6 Orthogonal Matrix

An orthogonal matrix has the property

$$\mathbf{A}\mathbf{A}^t = \mathbf{I} \tag{D.76}$$

The inverse of $\mathbf{A}$ is trivially computed as $\mathbf{A}^{-1} = \mathbf{A}^t$. The inverse of an orthogonal matrix equals its transpose. Thus, if $\mathbf{G}$ is an orthogonal matrix, then we have by definition

$$\mathbf{G}^t\mathbf{G} = \mathbf{G}^{-1}\mathbf{G} = \mathbf{I} \tag{D.77}$$

[1] A pivot is the leading nonzero element in each row.

It is easy to prove that the Givens matrix is orthogonal. A matrix **A** is invertible if there is a matrix $\mathbf{A}^{-1}$ such that

$$\mathbf{A}\mathbf{A}^{-1} = \mathbf{A}^{-1}\mathbf{A} = \mathbf{I} \tag{D.78}$$

D.19.7 Plane Rotation (Givens) Matrix

A 5×5 plane rotation (or Givens) matrix is one that looks like the identity matrix except for elements that lie in the locations pp, pq, qp, and qq. Such a matrix is labeled $\mathbf{G}_{pq}$. For example, the matrix $\mathbf{G}_{42}$ takes the form

$$\mathbf{G}_{42} = \begin{bmatrix} 1 & 0 & 0 & 0 & 0 \\ 0 & c & 0 & s & 0 \\ 0 & 0 & 1 & 0 & 0 \\ 0 & -s & 0 & c & 0 \\ 0 & 0 & 0 & 0 & 1 \end{bmatrix} \tag{D.79}$$

where $c = \cos\theta$ and $s = \sin\theta$. This matrix is orthogonal. Premultiplying a matrix **A** by $\mathbf{G}_{pq}$ modifies only rows p and q. All other rows are left unchanged. The elements in rows p and q become

$$a_{pk} = ca_{pk} + sa_{qk} \tag{D.80}$$

$$a_{qk} = -sa_{pk} + ca_{qk} \tag{D.81}$$

D.19.8 Stochastic (Markov) Matrix

A column stochastic matrix **P** has the following properties:

1. $a_{ij} \geq 0$ for all values of i and j.
2. The sum of each column is exactly 1 (i.e., $\sum_{j=1}^{m} p_{ij} = 1$).

Such a matrix is termed *column stochastic* matrix or *Markov* matrix. This matrix has two important properties. First, all eigenvalues are in the range $-1 \leq \lambda \leq 1$. Second, at least one eigenvalue is $\lambda = 1$.

D.19.9 Substochastic Matrix

A column substochastic matrix **V** has the following properties:

1. $a_{ij} \geq 0$ for all values of i and j.
2. The sum of each column is less than 1 (i.e., $\sum_{j=1}^{m} p_{ij} < 1$).

Such a matrix is termed *column substochastic* matrix. This matrix has the important property that all eigenvalues are in the range $-1 < \lambda < 1$.

D.19.10 Tridiagonal Matrix

A tridiagonal matrix is both upper and lower Hessenberg, i.e., nonzero elements exist only on the main diagonal and the adjacent upper and lower subdiagonals. A 5×5 tridiagonal matrix $\mathbf{A}$ is given by

$$\mathbf{A} = \begin{bmatrix} a_{11} & a_{12} & 0 & 0 & 0 \\ a_{21} & a_{22} & a_{23} & 0 & 0 \\ 0 & a_{32} & a_{33} & a_{34} & 0 \\ 0 & 0 & a_{43} & a_{44} & a_{45} \\ 0 & 0 & 0 & a_{54} & a_{55} \end{bmatrix} \tag{D.82}$$

D.19.11 Upper Hessenberg Matrix

An upper Hessenberg matrix has $h_{ij} = 0$ whenever $j < i - 1$. A 5×5 upper Hessenberg matrix $\mathbf{H}$ is given by

$$\mathbf{H} = \begin{bmatrix} h_{11} & h_{12} & h_{13} & h_{14} & h_{15} \\ h_{21} & h_{22} & h_{23} & h_{24} & h_{25} \\ 0 & h_{32} & h_{33} & h_{34} & h_{35} \\ 0 & 0 & h_{43} & h_{44} & h_{45} \\ 0 & 0 & 0 & h_{54} & h_{55} \end{bmatrix} \tag{D.83}$$

A matrix is lower Hessenberg if its transpose is an upper Hessenberg matrix.

Reference

1. G. Strang, *Introduction to Linear Algebra*, Wellesley-Cambridge Press, Wellesley, MA, 1993.
2. F.E. Hohn, *Elementary Matrix Algebra*, MacMillan, New York, 1964.
3. E. Kreyszig, *Advanced Engineering Mathematics*, John Wiley, New York, 1988.
4. R.E. Blahut, *Fast Algorithms for Digital Signal Processing*, Addison-Wesley, Reading, Massachusetts, 1985.
5. W.H. Press, S.T. Teukolsky, W.T. Vetterling, and B.P. Fhannery, *Numerical Recipes in C: The Art of Scientific Computing*, Cambridge University Press, Cambridge, 1992.
6. I. Jacques, and C. Judd, *Numerical Analysis*, Chapman and Hall, New York, 1987.
7. D. Arnold, "MATLAB and RREF", http://online.redwoods.cc.ca.us/instruct/darnold/LinAlg/rref/context-rref-s.pdf

Appendix E
Using MATLAB

E.1 Introduction

MATLAB is the most used software package for many engineering applications [1]. There are other useful packages such as Maple and Mathematica. Another package that is worth mentioning is Scientific Workplace [2] that combines document typesetting with the ability to perform symbolic and numerical computations from within the package using Maple. The results could be plotted with simple menu commands also.

MATLAB is based on vectors and matrices and its usefulness comes from the available commands and functions to manipulate and analyze these quantities. We should mention at the start that MATLAB indexes arrays starting with 1. Throughout this book, we also tried to start our arrays with the index value 1. Examples of functions available are statistical analysis, Fourier transform, matrix operations, and graphics.

E.2 The Help Command

The help command is the most basic way to determine the syntax and behavior of a particular function. Information is displayed directly in the Command Window. For example, to know more about the `EIG` function for finding the eigenvalues and eigenvectors of a matrix, one would type the command `help eig`. The result is shown in Fig. E.1.

E.3 Numbers in MATLAB

MATLAB uses several number types and number formats as shown in Table E.1. For very large or very small numbers, one could use the "e" notation or just multiply the number by 10 raised to the proper value:

$$\begin{aligned} \text{-1.5e-3} &= -1.5 * 10^{(-3)} = \text{-0.0015} \\ \text{3.6e2} &= 3.6 * 10^{2} = 360 \end{aligned}$$

```
help eig

EIG  Eigenvalues and eigenvectors. E = EIG(X) is a
     vector containing the eigenvalues of a square
     matrix X.

     [V,D] = EIG(X) produces a diagonal matrix D of
     eigenvalues and a full matrix V whose columns
     are the corresponding eigenvectors so that X*V =
     V*D.

     [V,D] = EIG(X,'nobalance') performs the
     computation with balancing disabled, which,
     sometimes gives more accurate results for
     certain problems with unusual scaling.

     E = EIG(A,B) is a vector containing the
     generalized eigenvalues of square matrices A
     and B.

     [V,D] = EIG(A,B) produces a diagonal matrix D
     of generalized eigenvalues and a full matrix V
     whose columns are the corresponding eigenvectors
     so that A*V = B*V*D.

     See also CONDEIG, EIGS.
```

Fig. E.1 The result of typing the command `help eig`

Table E.1 Number types in MATLAB

Number type	Example	Comment
Integer	4,-12	Type the number without a decimal point
Real	1.5,-12.0	Type the number with a decimal point
Complex	5.9-4.1*i*	Include *i* or *j* for imaginary part of number

The user can control how the numbers are displayed using the `format` command. Table E.2 shows the common types of formatting instructions.

Table E.2 Formatting numbers in MATLAB

Command	Example of display
format short	31.4162 (4-decimal places)
format short e	3.1416e+01 (4-decimal places)
format short g	Best for displaying fixed or floating point numbers
format bank	3.14 (2-decimal places)

Table E.3 Basic MATLAB operations for scalars

Operation	Symbol	Example
Addition	+	1+2
Subtraction	−	5 − 3
Multiplication	*	2 * 8
Division	/ or \	10/6
Exponentiation	^	2^8

E.4 Basic Arithmetic Operations on Scalars

The basic arithmetic operations on scalars in MATLAB are explained in Table E.3.

E.5 Variables

We can define variables in MATLAB by assigning a value to the name of the variable:

```
x = 3*4
```

In that case, MATLAB will print

```
x =

    12
```

to confirm that x has been assigned the desired value.

The following is a summary of some special features of MATLAB:

- Variable names can contain up to 31 characters and are case-sensitive. Underscore (_) could be used but a variable name must start with a character.
- MATLAB has several special variables such as *pi* (the ratio of circumference of a circle to its diameter) and i or $j = \sqrt{-1}$.
- When a command ends with a semicolon (;), the result of the command is not printed.
- Comments start with the percent sign (%).
- A succession of three periods ($\cdots$) tells MATLAB that the rest of the statement appears on the next line.
- The functions REAL and IMAG are used to extract the real and imaginary components of a complex number. The functions ABS and ANGLE are used to extract its magnitude and angle, respectively.

Table E.4 Special array declarations in MATLAB

ones(3)	Returns a 3 × 3 matrix containing all ones
ones(2,3)	Returns a 2 × 3 matrix containing all ones
zeros(4,5)	Returns a 4 × 5 matrix containing all zeros
zeros(size(A))	Returns a matrix whose dimension matches matrix *A* and containing all zeros

E.6 Arrays

An array is an ordered set of numbers. A row vector **x** can be declared using the command

$$x = [\ 6,\quad 9,\quad 8,\quad 6,\quad 4,\quad 2\ ]$$

The elements of the array can be entered separated by spaces or commas between them. A column vector **y** can be declared by separating the elements using semicolons or entering each element on a separate line:

$$y = [\ 6;\quad 9;\quad 8;\quad 6;\quad 4;\quad 2\ ]$$

To create a matrix **A**, we separate the rows using semicolons or by entering each row on a separate line:

$$A = [\ 1,\quad 2,\quad 3;\quad 4,\quad 5,\quad 6\ ]$$

which will create the 2 × 3 matrix

$$A = \begin{bmatrix} 1 & 2 & 3 \\ 4 & 5 & 6 \end{bmatrix}$$

There are special arrays in MATLAB as explained in Table E.4.

E.7 Neat Tricks for Arrays

Most often, we need to plot (x, y) data where **x** is a vector representing the input data and **y** is the output data. As an example, if we want to obtain a plot of $y = \sqrt{x}$ over the range $0 \leq x \leq 10$, we need a large number of points, say 100, to get a smooth curve. We can do that quickly in MATLAB using the following commands:

```
points = 100; % generate 100 points
lower_limit = 0; % lower limit for data
upper_limit = 10; % upper limit for data
```

```
% Now calculate the step size of input data
step = (upper_limit - lower_limit)/(points-1);

% Initialize a 100-point input data row vector
x = lower_limit:step:upper_limit;

% Initialize a 100-point output data row vector
y = sqrt(x);

% Plot the data
plot(x,y)
```

To find the length of a vector, the command `length` is used. To find the size of an array, the command `size` is used.

E.8 Array–Scalar Arithmetic

Simple arithmetic operations such as multiply, divide, add, or subtract can be done using a scalar and an array as explained below.

Assume we have a matrix **A**:

$$A = \begin{bmatrix} 1 & 2 & 3 \\ 4 & 5 & 6 \end{bmatrix}$$

Multiply **A** by a scalar:

$$2 * A = \begin{bmatrix} 2 & 4 & 6 \\ 8 & 10 & 12 \end{bmatrix}$$

Divide **A** by a scalar:

$$A/2 = \begin{bmatrix} 0.5 & 1 & 1.5 \\ 2 & 2.5 & 3 \end{bmatrix}$$

Add a scalar to **A**:

$$A + 2 = \begin{bmatrix} 3 & 4 & 5 \\ 6 & 7 & 8 \end{bmatrix}$$

Subtract a scalar from **A**:

$$A - 2 = \begin{bmatrix} -1 & 0 & 1 \\ 2 & 3 & 4 \end{bmatrix}$$

E.9 Array–Array Arithmetic

Addition and subtraction operations among two arrays **A** and **B** are defined only when the two arrays have the same dimensions according to matrix addition/subtraction rules. Multiplication of the two arrays is defined only if the two arrays are compatible according to matrix multiplication rules. Matrix division operator (\) will be discussed later in Section E.16.

MATLAB has a special operator called the *dot multiplication* operator (.*). Assume we have two arrays **A** and **B** having the same dimensions and we need to multiply the elements of **A** by the corresponding elements of **B** on an element-by-element basis. We have the operation

$$A \,.\!* B$$

The *dot division operator* is similarly defined:

$$A \,./B$$

Raising the elements of an array to a certain power is possible using the .^ operator

$$A.\hat{}2$$

which will square each element in array **A**. The following operation raises each element in A by the value of the corresponding element in **B**:

$$A\,.\hat{}B$$

E.10 The Colon Notation

A shortcut for generating a row vector is to issue the command

$$m : n$$

which gives the answer $m \; m+1 \; \cdots \; n$.

The command

$$0 : 2 : 10$$

gives the answer 0 2 4 6 8 10.

In general, the command $a : b : c$ specifies a as the start value, b as the step size, and c as the limit (such that the final element will not exceed c).

E.11 Addressing Arrays

The colon notation can be used to pick out selected rows, columns, and elements of vectors, matrices, and arrays.

$A(:)$ arranges all the elements of A as a column vector. This command is really handy to ensure that any vector we generate is a column vector. Thus if v was a vector and we are not sure if it is a column or a row vector, we could simply issue the command

$$v = v(:)$$

Assuming **A** is a 3×3 matrix, we can set its top left element to zero using the command

$$A(1, 1) = 0;$$

The first index indicates the row location (from top to bottom) and the second index indicates the column location (from left to right).

To access several elements at the same time, the colon notation is used. Assuming **A** is a 3×3 matrix, the following command defines a new column vector **c** that contains the first column of **A**:

$$c = A(:, 1)$$

while the command

$$C = A(:, 1 : 2)$$

defines **C** to be a 3×2 matrix equal to the first two columns of **A**.

In general, the statement $A(m : n, k : l)$ generates a matrix that contains rows m to n and columns k to l. Thus the dimension of the new matrix would be (n-m+1) $\times$ $(l - k + 1)$.

To delete the k-column of a matrix, we assign the *null vector* to it using the command

$$A(:, k) = [\,]$$

Similarly, to delete the i-row of a matrix we issue the command

$$A(i, :) = [\,]$$

E.12 Matrix Functions

The *rank* of a matrix can be found using the command

$$\texttt{rank}(A)$$

which returns the number of singular values of **A** that are larger than the default tolerance.

The *eigenvalues* and *eigenvectors* of a matrix can be found using the command

$$d = \texttt{eig}(A)$$

which returns a vector **d** of the eigenvalues of matrix **A**. The command

$$[V, D] = \texttt{eig}(A)$$

produces matrices of eigenvalues (**D**) and eigenvectors (**V**) of matrix **A**, so that $A * V = V * D$. Matrix **D** is the canonical form of **A**—a diagonal matrix with **A**'s eigenvalues on the main diagonal. Matrix **V** is the modal matrix—its columns are the eigenvectors of **A**. The eigenvectors are scaled so that the norm of each is 1.0.

Sometimes it is required to find the *sum* of elements of a matrix:

$$B = \texttt{sum}(A)$$

If **A** is a vector, $sum(A)$ returns the sum of the elements. If **A** is a matrix, $sum(A)$ treats the columns of **A** as vectors, returning a row vector of the sums of each column and

$$B = \texttt{trace}(A)$$

is the trace of **A**.

The *inverse* of a matrix found using the command

$$B = \texttt{inv}(A)$$

returns the inverse of the square matrix **A**. A warning message is printed if **A** is badly scaled or nearly singular. In practice, it is seldom necessary to form the explicit inverse of a matrix. A frequent misuse of `inv` arises when solving the system of linear equations **Ax** = **b**. One way to solve this is with $x = inv(A)*b$. A better way, from both execution time and numerical accuracy standpoints, is to use the matrix division operator $x = A\backslash b$. This produces the solution using Gaussian elimination, without forming the inverse as explained below.

E.13 M-Files (Script Files)

One interacts with MATLAB through the *command window*. When a command or group of commands is to be repeated, the user has to enter the whole sequence again. There is a better way, however, which is to write that sequence of commands and constant definitions in an M-file. An M-file must have a .m extension and commands are written in it using any text editor or the text editor associated with MATLAB.

Before an M-file could be used, the current directory of MATLAB must be where the file is saved. Changing the working directory is done either manually by typing the `cd` command or by typing the path to the file in the *Current Directory* window at the top right corner of the MATLAB command window. To issue the commands contained in the file, it is now sufficient to type the name of the file in the command window.

Here is a list of some useful commands that one could use in the MATLAB command window:

1. To check which is the current working directory of MATLAB, type the print working directory command:

```
pwd
```

2. To check what files are there in the current directory, type the command

```
ls
```

or

```
what
```

3. To open a certain file for editing its contents type,

```
open file_name
```

E.14 Function M-Files

M-files can be used to create user-defined functions. For example, we might want to define a function that finds the radius of a circle if the area is given.

We start by defining the name of the function as `radius` and we create a file named `radius.m`. The contents of the file `radius.m` would be as shown in the listing below. The numbers on the left are not part of the file. They are merely there to help us refer to particular lines in the code.

```
1   function radius r = radius(A)
2   % RADIUS finds the radius of a circle
3   % ifthe area A is given.
```

```
4    if A < 0
5       error('area must be positive');
6    end
7    r = sqrt(A/pi);
```

Line 1 defines the new function radius and indicates that it accepts a value A as input and produces the number r as output. Lines 2 and 3 describe briefly the purpose of this function, and typing the command

```
help radius
```

will print out the commented lines 2 and 3 to remind the user about the function. Line 5 prints out an error message if we attempt to find the radius of a circle with negative area.

We could now use this function in our MATLAB code by typing

$$x = \text{radius}(100)$$

which would return $x = 5.6419$.

Suppose we need to evaluate two properties of the circle, such as the radius r and the associated circumference c when the circle area is given. The above function should be changed to

```
1    function radius [r,c] = radius(A)
2    % RADIUS finds the radius and circumference of
3    % a circle if the area A is given.
4    if A < 0
5       error('area must be positive');
6    end
7    r = sqrt(A/pi);
8    c = 2 * pi * r;
```

We could now use this function in our MATLAB code by typing

$$[\text{r, c}] = \text{radius}(100)$$

which would return $r = 5.6419$ and $c = 35.4491$.

E.15 Statistical Functions

MATLAB has two random number generators RAND and RANDN. The command

$$x = \text{rand}(m, n)$$

Table E.5 The different formats for using the rand function

$Y =$ rand(n)	Returns a square $n \times n$ matrix of random entries
$Y =$ rand(m, n)	Returns an $m \times n$ matrix of random entries
$Y =$ rand($m, n, p, \ldots$)	Generates random arrays
$Y =$ rand($size(A)$)	Returns an array of random entries that is the same size as matrix A

Table E.6 The different formats for using the hist function

HIST(Y)	Draws a histogram of the elements in Y
HIST(Y, x)	Draws a histogram using n bins, where n is length(x). x also specifies the locations on the x-axis where hist places the bins.

returns an $m \times n$ matrix of random numbers having a uniform distribution between 0 and 1. The command

$$x = \texttt{randn}(m, n)$$

returns an $m \times n$ matrix of random numbers having a Gaussian distribution between 0 and 1 with zero mean and unity variance. Table E.5 explains the different ways the rand function could be invoked.

The command

$$x = \texttt{randperm}(n);$$

returns a random permutation of the integers 1 to n.

To plot the pdf of a random number sequence, the hist and stairs functions could be used. Table E.6 explains the different ways the hist function could be invoked. For example, if x is a 5-element vector, HIST distributes the elements of Y into five bins centered on the x-axis at the elements in x.

Example 1 The following MATLAB code generates a Gaussian random variable Y with mean 0 and standard deviation 1. The number of samples will be denoted by n. The resulting data are plotted in one curve and the histogram is plotted in an adjacent window.

```
n=1000; %number of samples
y=randn(1,n);
%plot theresulting dat
subplot(1,2,1)
plot(y);
box off
axis square
xlabel('Sample index')
ylabel('Randomnumber value')
```

```
%generate a histogram of the data to view thepdf
x = -2.9:0.1:2.9;
subplot(1,2,2)
hist(y)
box off
axis square
xlabel('Sample index')
ylabel('Bin count')
```

The output of that code can be found in Fig. 1.9 on page 18.

■

E.16 System of Linear Equations

Solving a system of linear equations is perhaps one of the important problems encountered in linear algebra. We certainly encountered this problem in our queuing analyses. The system of linear equations is expressed as

$$Ax = b \tag{E.1}$$

where **A** is a matrix, **x** is the vector to be found, and **b** is a given vector. The solution to this equation is simply

$$x = A^{-1}b \tag{E.2}$$

where $\mathbf{A}^{-1}$ is the inverse of **A** and in MATLAB this translates to

$$x = \texttt{inv}(A) * b \tag{E.3}$$

This is actually a misuse of the `inv` function. A better way, from both an execution time and numerical accuracy standpoint, is to use the matrix division operator and we should write

$$x = A \backslash b \tag{E.4}$$

This produces the solution using Gaussian elimination without forming the inverse.

E.16.1 A Is Overdetermined

If matrix **A** has more rows than columns, the system of equations is overdetermined since the number of equations is greater than the number of unknowns. In that case the \ and / operators find the solution that minimizes the squared error in $Ax - b$.

E.16.2 A Is Underdetermined

If matrix **A** has more columns than rows, the system of equations is underdetermined since the number of equations is less than the number of unknowns. In that case, the backslash \ and forwardslash / operators find the solution that has the maximum number of zeros in **x**.

E.16.3 **b = 0** *(Homogeneous Equations)*

When **b** = 0, we have the equation

$$Ax = 0 \tag{E.5}$$

and **x** is a null-vector of matrix **A**. A solution exists only if $\det(A) = 0$, which indicates that **A** must be rank deficient. Assuming $\text{rank}(A) = n - 1$, where n is the size of **A**, we can find a solution by assuming a value for one of the components of **x** and solving for the other components since we would have a system of $n - 1$ linear equations that can be solved using the backslash \ and / operators.

E.17 Solution of Nonlinear Equations

MATLAB has functions to obtain a solution for nonlinear equations or systems of nonlinear equations.

FMINBND finds the local minimum of a nonlinear function of one variable within an interval specified by the user

$$x = \texttt{fminbnd(fun}, x1, x2)$$

where FUN is a function that is described in an M-file and $x1$ and $x2$ are the lower and upper values of the interval such that $x1 < x < x2$.

FZERO tries to find the zero of the function near a value specified by the user:

$$x = \texttt{fzero(fun}, x0)$$

where FUN is a function that is described in an M-file and $x0$ is the value near which the zero is expected.

FMINSEARCH is a multidimensional unconstrained nonlinear minimization based on Nedler–Mead technique [3], [4]:

$$x = \texttt{fminsearch}(fun, x0)$$

returns a vector **x** that is a local minimizer of the function that is described in FUN (usually an M-file: fun.m) near the starting vector **x**0.

E.18 Formatting Output

The command `format compact` allows closer presentation of MATLAB results.

Reference

1. The Math Works, Inc. info@mathworks.com.
2. Scientific Work Place. http://www.mackichan.com.
3. J.C. Lagarias, J.A. Reed, M.H. Wright, and P.E. Wright, "Convergence properties of the Nedler-Mead simplex algorithm in low dimensions", *SIAM Journal on Optimization*, vol. 9, no. 1, pp. 112–147, 1998.
4. Nedler, J.A. and Mead, R., "A Simplex Method for Function Minimization", *Computer Journal*, vol. 7, pp. 308–313, 1965.

Appendix F
Database Design

F.1 Introduction

Each switch or router has a routing table or a lookup table, which is basically a database associating the packet header information with routing and control information such as priority, service class, updated VPI/VCI values, and destination output port. Because it is a database, standard database issues are encountered such as

1. The *sorting* scheme used
2. How a *search* is performed
3. *Inserting* data
4. *Deleting* data
5. Average access *time* for insert, search, and delete operations
6. Percentage of *memory* utilized

We shall explain two schemes for implementing the database using hashing functions and trees.

F.2 Hashing

Hashing is a technique used to implement databases where insert, search, and delete operations are performed. Hashing is the transformation of a *search key* into a number by means of mathematical calculations. The search time using hashing is small, which is an advantage over other search algorithms [1–3]. The mathematical operation is usually called *hash function* and the resulting number is called the *hash value*. The number of bits required to represent the hash value is substantially smaller than the number of bits required to represent the key itself.

If x is the key and $h(x)$ is the resulting hash value, then we can divide our memory into B *buckets*. The entries corresponding to key z are stored in the bucket whose label is $h(x)$. Each bucket stores the data in the form of a *heap*, which is nothing but a linked list that links all the data in the bucket with no particular order (i.e., like a heap).

A good hashing function must be able to equally distribute all possible entries of x among the buckets. In other words, we are hoping that the lengths of the linked lists in all buckets be equal so that the search time becomes small. Figure F.1 shows the open hashing data organization where each bucket has associated with it a linked list. Assume N is the number of entries in the set, which could be the number of connections established in an ATM switch or the number of available routes in an IP router.

Some hash functions are better than others at *randomizing* the hash values. The use of linear feedback shift registers (LFSR) for polynomial division is one such function, which is sometimes known as cyclic redundancy check (CRC). If the number of bits in the LFSR is b, then the number of buckets used is $B = 2^b$. Another hashing function that was based on the linear congruential method calculates h using modulo arithmetic:

$$h = (x \times y + 1) \bmod B \tag{F.1}$$

where y is a number with no particular pattern in its digits except that it should end with $\cdots x21$, with x even [2].

Assume N to be the number of data items stored in the database and B to be the number of buckets. The probability that a certain bucket contains i data items is given by

$$p_i = \binom{N}{i} \left(\frac{1}{B}\right)^i \left(1 - \frac{1}{B}\right)^{N-i} \tag{F.2}$$

The average length of the linked list in a bucket is given by

$$L = \sum_{i=0}^{N} i \; p_i \tag{F.3}$$

$$= \frac{N}{B} \tag{F.4}$$

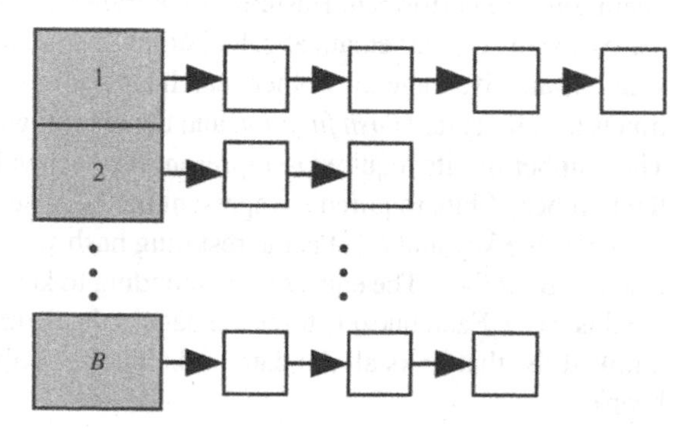

Fig. F.1 Storage of data using hashing

When $N \sim B$, then each linked list will have only one entry and the search time will be small independent of N. The time required to implement insert, search, and delete operations will be on the average equal to

$$T = T_{\mathrm{h}} + \frac{T_{\mathrm{m}} + T_{\mathrm{a}}}{2}\left(\frac{N}{B}\right) \tag{F.5}$$

where the time delays encountered account for the following operations:

T_{h} = Hashing function
T_{m} = Memory access
T_{a} = Arithmetic operations (e.g., comparison)

We must have $N > B$ and (F.5) becomes

$$T \approx \frac{T_{\mathrm{m}} + T_{\mathrm{a}}}{2}\left(\frac{N}{B}\right) \tag{F.6}$$

We see that the delays T_{m} and T_{a} dominate our table lookup delay. It is worthwhile investigating how to speed up the memory and the arithmetic operations required than investigating fast hashing functions.

F.3 Trees

The search time is not deterministic in hashing or in binary tree types of data organization. Balanced trees (B-tree) guarantee that the worst-case search time will not occur.

F.3.1 Binary Trees

Assume that we have to store numbers between 0 and 15. Later, we will want to check if a number has been stored or not. For that, we sort our numbers using binary sorting and store our data in a binary tree structure. Each node in the tree contains a *key* and the left branch is used if the number is smaller than the key. The right branch is used if the number is larger than the key.

Figure F.2 shows the binary search tree structure after several insertion and deletion operations. Each node stores data which doubles also as keys while performing a search.

It is apparent that the tree is unbalanced and the search time for a data item could be very short or very long. In the worst case, the depth of the tree could be equal to the number of data stored in it unless some ordering strategy is enforced. These techniques are called *balancing* and will be discussed in the next sections.

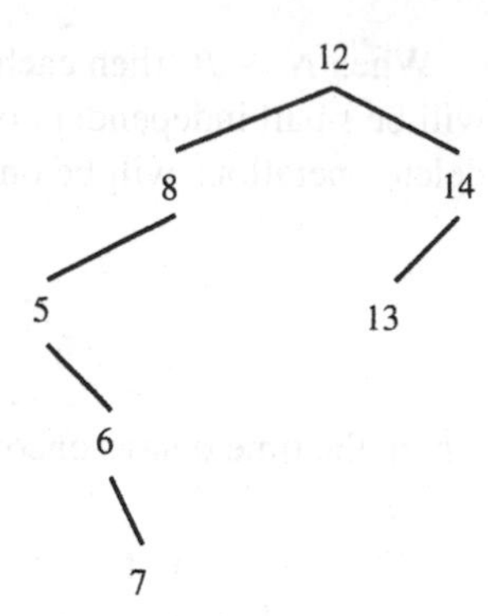

Fig. F.2 Binary search tree structure after several insertion and deletion operations

In addition to the worst-case time delay problem, we should mention that searching a tree implies comparing the search key to the contents of the nodes. This is really a *subtraction operation* and its delay depends on the speed of the adder in the switch processor. Typically, the number of bits used to represent the keys is 48 and this could require a significant adder delay.

F.3.2 Multiway Trees

Multiway trees offer a short search time for large databases where the worst-case time is guaranteed not to occur. In a multiway tree, each node stores several keys, not just one as in the binary tree such that each node can have up to m children. Figure F.3 shows a node in a multiway tree containing two keys k_0 and k_1.

Assuming $m-1$ keys and m branches per node, the keys are ordered in each node and the branches out of the node are for keys that have the following properties:

Branch	Property
0	$k < k_0$
1	$k_0 < k < k_1$
⋮	⋮
$m-2$	$k_{m-2} < k < k_{m-1}$
$m-1$	$k_{m-1} < k$

Multiway trees suffer from the same problem as binary trees when adding and removing data results in an unbalanced tree with potentially long search times. This

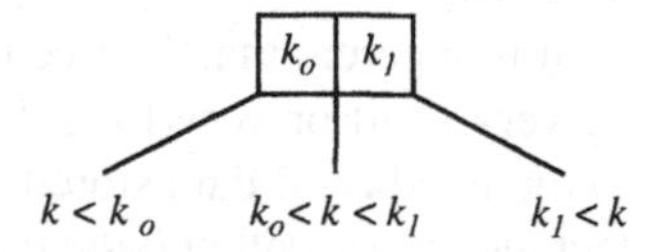

Fig. F.3 A node in a multiway tree where each node contains two keys

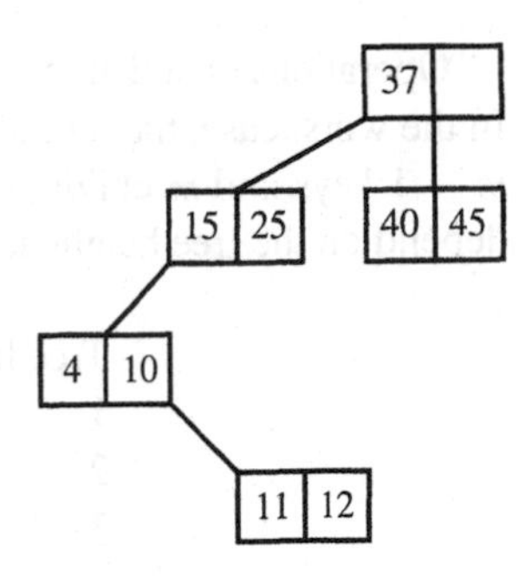

Fig. F.4 State of a multiway 2-tree after several insertion and deletion operations. Like the binary tree, it is also unbalanced

is illustrated in Fig. F.4 which shows the state of a 2-tree after some insertions and deletions. Multiway trees form the basis for the B-trees as explained in the next section.

F.3.3 B-Trees

There are many ways to define a B-tree. We choose a simple definition here. A B-tree of order m is a multiway tree with certain restrictions or properties that insure that the tree will always remain balanced. The properties of the B-tree are as follows:

1. The root node contains 1 to $2m$ keys.
2. The root node contains up to m children (pointers to other nodes).
3. Any other node that is not the root node contains $\lfloor m/2 \rfloor$ to $m - 1$ keys. Thus each node is at least half-full, except for the root node.
4. Any other node that is not the root node contains $\lceil m/2 \rceil$ to m children.

These properties ensure that no more than 50% of the memory space is wasted. In the worst case, all nodes are half-full and the parent node has only one key and two children.

Figure F.5 shows an example of a B-tree of order 5 after several insertion and deletion operations. Since $m = 5$, the number of keys in any node, except the root, is at most $m - 1 = 4$. The number of children per node must be between $m/2 = 3$ (after rounding up) and $m = 5$ as shown.

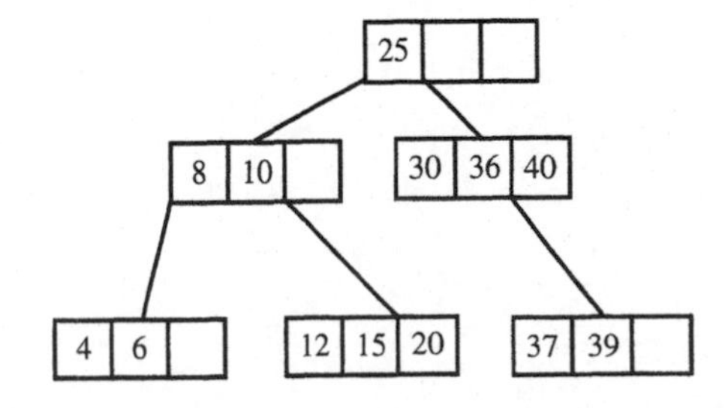

Fig. F.5 The structure of a B-tree of order 5 after some insertion and deletion operations

Operations on a B-tree such as search, insertion, and deletion can be found in [4]. In the worst case, the root node will have only one key and all other nodes will have $m-1$ keys and m children. In that case, the number of entries in the table N will depend on the tree height according to the following table [5]:

Tree height	Number of keys
1	1
2	$1+m$
3	$1+m+mc$
4	$1+m+mc+mc^2$
$\vdots$	$\vdots$
h	$1+m\sum_{i=0}^{h-2} c^i = c^{h-1}-1$

Reference

1. A.V. Aho, J.E. Hopcroft, and J.D. Ullman, *Data Structures and Algorithms*, Addison-Wesley, Reading, Massachusetts, 1983.
2. R. Sedgewick, *Algorithms*, Addison-Wesley, Reading, Massachusetts, 1984.
3. S. Baase, *Computer Algorithms*, Addison-Wesley, Reading, Massachusetts, 1983.
4. http://www.semaphorecorp.com/btp/algo.html.
5. N. Alechina, "B-Trees", http://www.cs.nott.ac.uk/ nza/G5BADS/l14.html

Index

J

K

L

N